Student's Solutions Manual

for use with

Algebra for College Students

Third Edition

by
Mark Dugopolski
Southeastern Louisiana University

 Higher Education

Boston   Burr Ridge, IL   Dubuque, IA   Madison, WI   New York   San Francisco   St. Louis
Bangkok   Bogotá   Caracas   Kuala Lumpur   Lisbon   London   Madrid   Mexico City
Milan   Montreal   New Delhi   Santiago   Seoul   Singapore   Sydney   Taipei   Toronto

The **McGraw·Hill** Companies

Student's Solutions Manual for use with
ALGEBRA FOR COLLEGE STUDENTS, THIRD EDITION
MARK DUGOPOLSKI

Published by McGraw-Hill Higher Education, an imprint of The McGraw-Hill Companies, Inc.,
1221 Avenue of the Americas, New York, NY 10020. Copyright © The McGraw-Hill Companies,
Inc., 2000, 2004. All rights reserved.

1 2 3 4 5 6 7 8 9 0 BKM BKM 0 9 8 7 6 5 4 3

ISBN 0-07-254653-0

www.mhhe.com

Preface

The ***Student's Solutions Manual for use with Algebra for College Students, Third Edition*** provides complete worked-out solutions to problems in the Warm-Ups, the Chapter Tests, and the Making Connections and to the odd-numbered problems in the end-of-section Exercises and the Chapter Reviews.

The ***Online Learning Center (OLC) to accompany Algebra for College Students, Third Edition*** at www.mhhe.com/dugopolski provides valuable student resources that include audio-visual tutorials, interactivities, algorithmically generated "bookmarkable" practice exercises (including hints), section-and chapter-level testing, and NetTutor™.

For student access to materials found on the OLC, students should purchase a new copy of the textbook. Passcodes can be found at the front of new texts.

The ***Video Series on CD-ROM for use with Algebra for College Students, Third Edition*** is based on problems taken directly from the exercises. A mathematics instructor presents selected problems and works through them, following the solution methodology employed in the text.

Student's Solutions Manual
Algebra for College Students

Third Edition

Table of Contents

1.1 WARM-UPS

1. False, since 5 is a counting number and $5 \notin A$.

2. False, since B has only 3 elements.

3. False, because the set has a specific number of members.

4. False, since $1 \notin B$.

5. True, since $3 \in A$.

6. True, because 3 and 4 are the only numbers that belong to both A and B.

7. True, because every member of C is also a member of B.

8. False, since $1 \in A$ but $1 \notin B$.

9. True, since $\emptyset$ is a subset of every set.

10. True, since $1 \in A$ and $1 \notin C$.

1.1 EXERCISES

1. A set is a collection of objects.

3. A Venn diagram is used to illustrate relationships between sets.

5. A is a subset of B if every member of set A is also a member of set B.

7. False, 6 is not odd.

9. True, because $1 \in A$ but $1 \notin B$.

11. True, because $C = \{1, 2, 3, 4, 5\}$.

13. False, because $5 \in A$.

15. False, because 0 is not a natural number.

17. False, because N is infinite and C is finite.

19. Note that $A = \{1, 3, 5, 7, 9\}$ and $B = \{2, 4, 6, 8\}$. Since A and B have no numbers in common, $A \cap B = \emptyset$.

21. Note that $A = \{1, 3, 5, 7, 9\}$ and $C = \{1, 2, 3, 4, 5\}$. Since 1, 3, and 5 are in both A and C, $A \cap C = \{1, 3, 5\}$.

23. The elements of B together with those of C give us $B \cup C = \{1, 2, 3, 4, 5, 6, 8\}$.

25. Since the empty set has no members, forming the union with another set does not add any new members. So $A \cup \emptyset = A$.

27. Since $\emptyset$ has no members in common with A, $A \cap \emptyset = \emptyset$.

29. Since every member of A is also a member of N, $A \cap N = A = \{1, 3, 5, 7, 9\}$.

31. Since the members of A are odd and the members of B are even, they have no members in common. So $A \cap B = \emptyset$.

33. $A \cup B = \{1, 2, 3, 4, 5, 6, 7, 8, 9\}$ from Exercise 20.

35. Take the elements that B has in common with C to get $B \cap C = \{2, 4\}$.

37. Since 3 is not in B, $3 \notin A \cap B$.

39. Since 4 is in both sets, $4 \in B \cap C$.

41. True, since each member of A is a counting number.

43. True, since both 2 and 3 are members of C.

45. True, since $6 \in B$ but $6 \notin C$.

47. True, since $\emptyset$ is a subset of every set.

49. False, since $1 \in A$ but $1 \notin \emptyset$.

51. True, since $A \cap B = \emptyset$ and $\emptyset$ is a subset of any set.

53. Using all numbers that belong to D or to E yields $D \cup E = \{2, 3, 4, 5, 6, 7, 8\}$.

55. Using only numbers that belong to both D and F gives $D \cap F = \{3, 5\}$.

57. Using all numbers that belong to E or to F gives $E \cup F = \{1, 2, 3, 4, 5, 6, 8\}$.

59. From Exercise 53, $D \cup E = \{2, 3, 4, 5, 6, 7, 8\}$ and $F = \{1, 2, 3, 4, 5\}$. Only 2, 3, 4, and 5 are in both sets. So $(D \cup E) \cap F = \{2, 3, 4, 5\}$.

61. Form the union of $E \cap F = \{2, 4\}$ with $D = \{3, 5, 7\}$ to get $D \cup (E \cap F) = \{2, 3, 4, 5, 7\}$.

63. Form the union of $D \cap F = \{3, 5\}$ with $E \cap F = \{2, 4\}$ to get $(D \cap F) \cup (E \cap F) = \{2, 3, 4, 5\}$.

65. Intersect $D \cup E = \{2, 3, 4, 5, 6, 7, 8\}$ with $D \cup F = \{1, 2, 3, 4, 5, 7\}$ to get $(D \cup E) \cap (D \cup F) = \{2, 3, 4, 5, 7\}$.

67. Use $\subseteq$, since every element of D is an odd natural number.

69. Use $\in$, since 3 is a member of D.

71. Use $\cap$, since D and E have no elements in common.

73. Use $\subseteq$, since every member of both D and F is a member of F.

75. Use $\cap$, since $8 \in E$ but not in $E \cap F$.

77. Use $\cup$, since $D \cup F$ and $F \cup D$ have exactly the same members.

79. The set of even natural numbers less than 20 is $\{2, 4, 6, \ldots, 18\}$.

81. The set of odd natural numbers greater than 11 is $\{13, 15, 17, \ldots\}$.

83. The set of even natural numbers between 4 and 79 is $\{6, 8, 10, \ldots, 78\}$. Note that 4 is not in the set because 4 is not between 4 and 79.

85. Answers may vary. Two possible answers are $\{x \mid x$ is a natural number between 2 and 7$\}$ or $\{x \mid x$ is a natural number between 2.5 and 6.3$\}$.

87. Answers may vary. One possibility is $\{x \mid x$ is an odd natural number greater than 4$\}$.

89. Answers may vary. One possibility is $\{x \mid x$ is an even natural number between 5 and 83$\}$

91. If A has n elements and B has m elements, then the union of the two sets cannot have more members than $n + m$. So the union is also finite.

93. a) $3 \in \{1, 2, 3\}$ **b)** $\{3\} \subseteq \{1, 2, 3\}$
c) This is tricky. Since $\{\emptyset\}$ is a set containing the empty set, $\{\emptyset\}$ is not empty. So $\emptyset \neq \{\emptyset\}$

1.2 WARM-UPS

1. False, because π is irrational.
2. True, because the set of real numbers is the union of the rationals and the irrationals.
3. False, 0 is rational and not irrational.
4. False, because every rational number is a real number that is not in the set of irrational numbers.
5. True, since it is a repeating decimal.
6. False, it is not repeating.
7. True, because the points on the number line correspond to the real numbers.
8. False, because 6 is not in the interval $(2, 6)$. The interval $(2, 6)$ consists of real numbers between 2 and 6, not include the endpoints.
9. True, because 3 is not in the interval $(1, 3)$ but it is in the interval $[3, 4)$. So every number in $(1, 4)$ is in either $(1, 3)$ or $[3, 4)$.
10. False, because 2 is in the interval $(1, 5)$ and $[2, 8]$ but not in the interval $(2, 8)$.

1.2 EXERCISES

1. The integers consist of the positive and negative counting numbers and zero.
3. The repeating or terminating decimal numbers are rational numbers.
5. The set of real numbers is the union of the rational and irrational numbers.
7. True, since -6 is rational.
9. False, since 0 is rational.
11. True, since a repeating decimal is rational.
13. True, since every natural number is rational.
15. The set of whole numbers smaller than 6 is the set $\{0, 1, 2, 3, 4, 5\}$. Note that 0 is a whole number.

17. The set of integers greater than -5 consists of every integer to the right of -5 on the number line:
$\{-4, -3, -2, -1, 0, 1, \ldots\}$.

19. The set of natural numbers between 0 and 5 does not include either 0 or 5:
$\{1, 2, 3, 4\}$.

21. The set of integers between -3 and 5 does not include either -3 or 5:
$\{-2, -1, 0, 1, 2, 3, 4\}$.

23. All of the numbers are real numbers.
25. The whole numbers include 0. $\{0, 8/2\}$
27. All of the numbers except $-\sqrt{10}$ and $\sqrt{2}$ are rational.
$\{-3, -5/2, -0.025, 0, 3\frac{1}{2}, 8/2\}$
29. True, since every rational number is also a real number.
31. False, since 0 is not irrational.
33. True, because all real numbers are either rational or irrational.
35. False, since nonrepeating decimals are irrational.
37. False, since repeating decimals are rational.
39. False, since repeating decimals are rational.

41. True, since π is irrational.

43. $N \subseteq W$, since every natural number is a whole number.

45. $J \nsubseteq N$, since $-9 \in J$ but $-9 \notin N$.

47. $Q \subseteq R$, since every rational number is a real number.

49. $\emptyset \subseteq I$, since $\emptyset$ is a subset of every set.

51. $N \subseteq R$, since every natural number is a real number.

53. $5 \in J$, since 5 is an integer.

55. $7 \in Q$, since 7 is a rational number.

57. $\sqrt{2} \in R$, since $\sqrt{2}$ is a real number.

59. $0 \notin I$, since 0 is rational.

61. $\{2, 3\} \subseteq Q$, since both 2 and 3 are rational.

63. $\left\{3, \sqrt{2}\right\} \subseteq R$, since both numbers are real.

65. The set of real numbers greater than 1 is written in interval notation as $(1, \infty)$ and graphed as follows.

67. The set of real numbers less than -1 is written in interval notation as $(-\infty, -1)$ and graphed as follows.

69. The set of real numbers between 3 and 4 is written in interval notation as $(3, 4)$ and graphed as follows.

71. The set of real numbers between 0 and 2 inclusive is written in interval notation as $[0, 2]$ and graphed as follows.

73. The set of real numbers greater than or equal to 1 and less than 3 is written in interval notation as $[1, 3)$ and graphed as follows.

75. The graph shows the real numbers between 5 and 7 inclusive, which is written in interval notation as $[5, 7]$.

77. The graph shows the real numbers greater than -3 and less than or equal to 0, which is written in interval notation as $(-3, 0]$.

79. The graph shows the real numbers greater than or equal to 60, which is written in interval notation as $[60, \infty)$.

81. The graph shows the real numbers less than -5, which is written in interval notation as $(-\infty, -5)$.

83. Graph the intervals $(1, 5)$ and $(4, 9)$ on the same number line to see that the union of $(1, 5)$ and $(4, 9)$ is $(1, 9)$.

85. Graph $(0, 3)$ and $(2, 8)$ on the same number line to see that the numbers they have in common are in the interval $(2, 3)$.

87. Graph $(-2, 4)$ and $(0, \infty)$ on the same number line to see that the union of these two intervals is $(-2, \infty)$.

89. Graph $(-\infty, 2)$ and $(0, 6)$ on the same number line to see that the numbers they have in common are in the interval $(0, 2)$.

91. Graph $[2, 5)$ and $(4, 9]$ on the same number line to see that the union of these intervals is $[2, 9]$.

93. Graph $[2, 6)$ and $[2, 8)$ on the same number line to see that the numbers they have in common are in the interval $[2, 6)$.

95. A rational number is a repeating or terminating decimal and an irrational number never repeats or terminates as a decimal number. Since $\sqrt{9} = 3$ it is rational. Since $\sqrt{3}$ is a nonterminating nonrepeating decimal number, it is irrational.

97. a) Find the decimal representations with a calculator. Note that each decimal is repeating.

b) The number of digits that repeats is the same as the number of nines in the denominator.

1.3 WARM-UPS

1. True, since $-6 + 6 = 0$.

2. True, -5 and 5 are opposites of each other.

3. False, $|6| = 6$.

4. True, since $b - a = b + (-a)$.

5. True, because the product of two numbers with opposite signs is negative.

6. False, since $6 + (-4) = 2$.

7. False, since $-3 - (-6) = -3 + 6 = 3$.

8. False, since $6 \div (-1/2) = 6(-2) = -12$.

9. False, since division by zero is undefined.
10. True, because 0 divided by any nonzero number is 0.

1.3 EXERCISES

1. The absolute value of a number is the number's distance from 0 on the number line.
3. Subtract their absolute values and use the sign of the number with the larger absolute value.
5. Multiply their absolute values, then affix a positive sign if the original numbers have the same sign and a negative sign if the original numbers have opposite signs.
7. Because -34 is a negative number,
$|-34| = -(-34) = 34.$
9. The only number with absolute value 0 is 0. So $|0| = 0.$
11. Since $|-6| = 6$ we have
$|-6| - |-6| = 6 - 6 = 0.$
13. Since $|-9| = 9$, we have $-|-9| = -9.$
15. $-(-9) = 9$
17. $-(-(-3)) = -(3) = -3$
19. $(-5) + 9 = 9 - 5 = 4$
21. $(-4) + (-3) = -(4 + 3) = -7$
23. $-6 + 4 = -(6 - 4) = -2$
25. $7 + (-17) = -(17 - 7) = -10$
27. $(-11) + (-15) = -(11 + 15) = -26$
29. $18 + (-20) = -(20 - 18) = -2$
31. $-14 + 9 = -(14 - 9) = -5$
33. $-4 + 4 = 0$
35. $-\frac{1}{10} + \frac{1}{5} = \frac{1}{5} - \frac{1}{10} = \frac{2}{10} - \frac{1}{10} = \frac{1}{10}$

37. $\frac{1}{2} + \left(-\frac{2}{3}\right) = \frac{3}{6} + \left(-\frac{4}{6}\right) = -\frac{1}{6}$

39. $-15 + 0.02 = -(15.00 - 0.02)$
$= -14.98$
41. $-2.7 + (-0.01) = -(2.70 + 0.01)$
$= -2.71$
43. $47.39 + (-44.587) = 2.803$
45. $0.2351 + (-0.5) = -0.2649$
47. $7 - 10 = -(10 - 7) = -3$
49. $-4 - 7 = -4 + (-7) = -11$
51. $7 - (-6) = 7 + 6 = 13$
53. $-1 - 5 = -1 + (-5) = -6$
55. $-12 - (-3) = -12 + 3 = -9$

57. $20 - (-3) = 20 + 3 = 23$
59. $\frac{9}{10} - \left(-\frac{1}{10}\right) = \frac{9}{10} + \frac{1}{10} = 1$

61. $1 - \frac{3}{2} = \frac{2}{2} - \frac{3}{2} = -\frac{1}{2}$

63. $2.00 - 0.03 = 1.97$
65. $5.3 - (-2) = 5.3 + 2 = 7.3$
67. $-2.44 - 48.29 = -50.73$
69. $-3.89 - (-5.16) = 1.27$
71. $(25)(-3) = -(25 \cdot 3) = -75$
73. $\left(-\frac{1}{3}\right)\left(-\frac{1}{2}\right) = \frac{1}{6}$
75. $(0.3)(-0.3) = -0.09$
77. $(-0.02)(-10) = 0.02(10) = 0.2$
79. The reciprocal of 20 is $\frac{1}{20}$ or 0.05.
81. The reciprocal of $-\frac{6}{5}$ is $-\frac{5}{6}$.
83. The reciprocal of -0.3 is $-\frac{1}{0.3}$ or $-\frac{10}{3}$.
85. $-6 \div 3 = -2$
87. $30 \div (-0.8) = -37.5$
89. $(-0.8)(0.1) = -0.08$
91. $(-0.1) \div (-0.4) = 0.25$

93. $9 \div \left(-\frac{3}{4}\right) = 9\left(-\frac{4}{3}\right) = -\frac{36}{3} = -12$

95. $-\frac{2}{3}\left(-\frac{9}{10}\right) = \frac{2}{3} \cdot \frac{9}{10} = \frac{18}{30} = \frac{3}{5}$

97. $(0.25)(-365) = -91.25$
99. $(-51) \div (-0.003) = 17,000$
101. $-62 + 13 = -(62 - 13) = -49$
103. $-32 - (-25) = -32 + 25 = -7$
105. $|-15| = -(-15) = 15$
107. $\frac{1}{2}(-684) = -342$
109. $\frac{1}{2} - \left(-\frac{1}{4}\right) = \frac{2}{4} + \frac{1}{4} = \frac{3}{4}$
111. $-57 \div 19 = -3$
113. $|-17| + |-3| = 17 + 3 = 20$
115. $0 \div (-0.15) = 0$
117. $27 \div (-0.15) = -180$
119. $-\frac{1}{3} + \frac{1}{6} = -\frac{2}{6} + \frac{1}{6} = -\frac{1}{6}$
121. $-63 + |8| = -63 + 8 = -55$

123. $-\frac{1}{2} + \left(-\frac{1}{2}\right) = -\frac{2}{2} = -1$

125. $-\frac{1}{2} - 19 = -\frac{1}{2} - \frac{38}{2} = -\frac{39}{2}$

4

127. $28 - 0.01 = 27.99$
129. $-29 - 0.3 = -29.3$
131. $(-2)(0.35) = -0.7$
133. $(-10)(-0.2) = 2$
135. $85{,}000 + (-45{,}000) + (-2300)$
$+ (-1500) + 1200 + 2(3500) = 44{,}400$
Net worth is \$44,400.
137. $14° - (-6°) = 20°F$
139. $-282 - (-1296) = 1014$ ft
141. We learn addition first, because subtraction is defined in terms of addition.

1.4 WARM-UPS

1. False, $2^3 = 8$.
2. True, because $2^2 = 4$ and $-1 \cdot 2^2 = -4$.
3. True, since $-2^2 = -(2^2) = -4$.
4. False, since $6 + 3 \cdot 2 = 6 + 6 = 12$.
5. False, since $(6 + 3) \cdot 2 = 9 \cdot 2 = 18$.
6. False, since $(6 + 3)^2 = 9^2 = 81$.
7. True, since $6 + 3^2 = 6 + 9 = 15$.
8. True, since $(-3)^3 = -27$ and $-3^3 = -27$.
9. False, since $|-3 - (-2)| = |-1| = 1$.
10. False, since $|7 - 8| = 1$ and
$|7| - |8| = 7 - 8 = -1$.

1.4 EXERCISES

1. An arithmetic expression is the result of writing numbers in a meaningful combination with the ordinary operations of arithmetic.
3. Grouping symbols are used to indicate the order in which operations are to be performed.
5. The order of operations tells us the order in which to perform operations when grouping symbols are omitted.
7. $(-3 \cdot 4) - (2 \cdot 5) = -12 - 10 = -22$
9. $4[5 - |3 - (2 \cdot 5)|] = 4[5 - 7] = -8$
11. $(6 - 8)(|2 - 3| + 6) = (-2)(7) = -14$
13. $2^5 = 2 \cdot 2 \cdot 2 \cdot 2 \cdot 2 = 32$
15. $(-1)^4 = (-1)(-1)(-1)(-1) = 1$
17. $\left(-\frac{1}{3}\right)^2 = \left(-\frac{1}{3}\right)\left(-\frac{1}{3}\right) = \frac{1}{9}$
19. $\sqrt{49} = 7$
21. $\sqrt{36 + 64} = \sqrt{100} = 10$
23. $\sqrt{4(7 + 9)} = \sqrt{64} = 8$
25. $4 - 6 \cdot 2 = 4 - 12 = -8$

27. $5 - 6(3 - 5) = 5 - 6(-2)$
$= 5 + 12 = 17$
29. $\left(\frac{1}{3} - \frac{1}{2}\right)\left(\frac{1}{4} - \frac{1}{2}\right) = \left(-\frac{1}{6}\right)\left(-\frac{1}{4}\right)$
$= \frac{1}{24}$
31. $-3^2 + (-8)^2 + 3 = -9 + 64 + 3 = 58$
33. $-(2 - 7)^2 = -(-5)^2 = -25$
35. $-5^2 \cdot 2^3 = -25 \cdot 8 = -200$
37. $(-5)(-2)^3 = -5(-8) = 40$
39. $-(3^2 - 4)^2 = -(9 - 4)^2 = -5^2 = -25$
41. $-60 \div 10 \cdot 3 \div 2 \cdot 5 \div 6$
$= -6 \cdot 3 \div 2 \cdot 5 \div 6$
$= -18 \div 2 \cdot 5 \div 6$
$= -9 \cdot 5 \div 6$
$= -45 \div 6 = -7.5$
43. $5.5 - 2.3^4 = -22.4841$
45. $(1.3 - 0.31)(2.9 - 4.88) = -1.9602$
47. $-388.8 \div (13.5)(9.6) = -276.48$
49. $\frac{2 - 6}{9 - 7} = \frac{-4}{2} = -2$
51. $\frac{-3 - 5}{6 - (-2)} = \frac{-8}{8} = -1$
53. $\frac{4 + 2 \cdot 7}{3 \cdot 2 - 9} = \frac{18}{-3} = -6$
55. $\frac{-3^2 - (-9)}{2 - 3^2} = \frac{-9 + 9}{2 - 9} = \frac{0}{-7} = 0$
57. $3^2 - 4(-1)(-4) = 9 - 16 = -7$
59. $\frac{-1 - 3}{-1 - (-4)} = \frac{-4}{-1 + 4} = -\frac{4}{3}$
61. $(-1 - 3)(-1 + 3) = (-4)(2) = -8$
63. $\sqrt{(-4)^2 - 2(-4) + 1} = \sqrt{16 + 8 + 1}$
$= 5$
65. $\frac{2}{-1} + \frac{3}{-4} - \frac{1}{-4} = -2 - \frac{3}{4} + \frac{1}{4}$

$= -\frac{8}{4} - \frac{3}{4} + \frac{1}{4} = -\frac{10}{4} = -\frac{5}{2}$

67. $|-1 - 3| = |-4| = 4$
69. $\frac{-6 - 4}{-7 - 2} = \frac{-10}{-9} = \frac{10}{9}$
71. $\frac{2 - (-1)}{1 - (-3)} = \frac{3}{4}$
73. $\frac{5.6 - 2.4}{4.7 - 5.9} = -\frac{8}{3} \approx -2.67$
75. $-2^2 + 5(3)^2 = -4 + 45 = 41$
77. $(-2 + 5)3^2 = 3 \cdot 9 = 27$
79. $\sqrt{5^2 - 4(1)(6)} = \sqrt{25 - 24} = 1$

81. $[13 + 2(-5)]^2 = 3^2 = 9$

83. $\dfrac{4 - (-1)}{-3 - 2} = \dfrac{5}{-5} = -1$

85. $3(-2)^2 - 5(-2) + 4 = 12 + 10 + 4 = 26$

87. $-4\left(\dfrac{1}{2}\right)^2 + 3\left(\dfrac{1}{2}\right) - 2$

$\qquad = -4\left(\dfrac{1}{4}\right) + \dfrac{3}{2} - 2$

$\qquad = -\dfrac{2}{2} + \dfrac{3}{2} - \dfrac{4}{2} = -\dfrac{3}{2}$

89. $-\dfrac{1}{2}\,|\,6 - 2\,| = -\dfrac{1}{2}(4) = -2$

91. $\dfrac{1}{2} - \dfrac{1}{3}\left|\dfrac{1}{4} - \dfrac{1}{2}\right| = \dfrac{1}{2} - \dfrac{1}{3} \cdot \dfrac{1}{4}$

$\qquad = \dfrac{1}{2} - \dfrac{1}{12} = \dfrac{5}{12}$

93. $|\,6 - 3 \cdot 7\,| + |\,7 - 5\,|$

$\qquad = |-15| + |\,2\,|$

$\qquad = 17$

95. $3 - 7[4 - (2 - 5)] = 3 - 7[4 + 3]$

$\qquad\qquad = 3 - 49 = -46$

97. $3 - 4(2 - |\,4 - 6\,|) = 3 - 4(2 - 2) = 3$

99. $4[2 - (5 - |-3|)^2] = 4[2 - (2^2)]$

$\qquad\qquad = 4[-2] = -8$

101. Find the target heart rates for both women and subtract them:

$0.65(220 - 25) - 0.65(220 - 65) = 26$

The target heart rate for a 25-yr old woman is 26 beats per minute larger than the target heart rate for a 65 yr-old woman. From the graph, a woman's target heart rate is 115 at about 43 years of age.

103. The perimeter is $2(34) + 2(18)$ or 104 feet.

105. a) Using the graph, the investment amounts to approximately $60,000.

b) The investment amounts to $10,000(1 + 0.062)^{30}$ or $60,776.47.

107. a) According to the graph she owes approximately $5,500.

b) The actual amount owed when the payments start is $4000(1 + 0.08)^4$ or $5441.96.

109. $45,600(1.08)^{10} - 74,278 \approx 24,169$

Using the formula in this exercise, a one-year stay costs $24,169 more.

1.5 WARM-UPS

1. True, because of the commutative properties.

2. False, since $8 \div (4 \div 2) = 4$ and $(8 \div 4) \div 2 = 1$.

3. False, since $10 \div 2 = 5$ and $2 \div 10 = 0.2$.

4. False, since $5 - 3 = 2$ and $3 - 5 = -2$.

5. False, since $10 - (7 - 3) = 6$ and $(10 - 7) - 3 = 0$.

6. False, since $4(6 \div 2) = 12$ and $(4 \cdot 6) \div (4 \cdot 2) = 3$.

7. True, since $(0.02)(50) = 1$.

8. True, because of Warm-up number 2.

9. False, because if $x = 0$ we get $3 = 0$.

10. True, because $\dfrac{1 \text{ car}}{0.04 \text{ hours}} = \dfrac{25 \text{ cars}}{1 \text{ hour}}$.

1.5 EXERCISES

1. The commutative property of addition says that $a + b = b + a$ and the commutative property of multiplication says that $a \cdot b = b \cdot a$.

3. The commutative property of addition says that you get the same result when you add two numbers in either order. The associative property of addition has to do with which two numbers are added first when adding three numbers.

5. Zero is the additive identity because adding zero to a number does not change the number.

7. $9 - 4 + 6 - 10 = 15 - 14 = 1$

9. $6 - 10 + 5 - 8 - 7 = 11 - 25 = -14$

11. $-4 - 11 + 6 - 8 + 13 - 20 = -43 + 19$

$\qquad\qquad\qquad\qquad\qquad = -24$

13. $-3.2 + 1.4 - 2.8 + 4.5 - 1.6$

$\qquad = -7.6 + 5.9 = -1.7$

15. $3.27 - 11.41 + 5.7 - 12.36 - 5$

$\qquad = 8.97 - 28.77 = -19.8$

17. $4(x - 6) = 4 \cdot x - 4 \cdot 6 = 4x - 24$

19. $2m + 10 = 2 \cdot m + 2 \cdot 5 = 2(m + 5)$

21. $a(3 + t) = 3a + at$

23. $-2(w - 5) = -2w - (-10) = -2w + 10$

25. $-2(3 - y) = -6 - (-2y) = -6 + 2y$

27. $5x - 5 = 5 \cdot x - 5 \cdot 1 = 5(x - 1)$

29. $-1(-2x - y) = (-1)(-2x) - (-1)y$

$\qquad\qquad\qquad = 2x + y$

31. $-3(-2w - 3y) = -3(-2w) - (-3)3y$
$= 6w + 9y$

33. $3y - 15 = 3 \cdot y - 3 \cdot 5 = 3(y - 5)$

35. $3a + 9 = 3 \cdot a + 3 \cdot 3 = 3(a + 3)$

37. $\frac{1}{2}(4x + 8) = \frac{4}{2}x + \frac{8}{2} = 2x + 4$

39. $-\frac{1}{2}(2x - 4) = -\frac{2}{2}x + \frac{4}{2} = -x + 2$

41. The reciprocal of $\frac{1}{2}$ is 2 because
$2 \cdot \frac{1}{2} = 1$.

43. The reciprocal of 1 is 1 because $1 \cdot 1 = 1$.

45. The reciprocal of 6 is $\frac{1}{6}$, because
$6 \cdot \frac{1}{6} = 1$.

47. Since $0.25 = \frac{1}{4}$, its reciprocal is 4.

49. Since $-0.7 = -\frac{7}{10}$, its reciprocal is $-\frac{10}{7}$.

51. Since $-1.8 = -\frac{18}{10}$ or $-\frac{9}{5}$, the reciprocal
of -1.8 is $-\frac{5}{9}$.

53. $\frac{1}{2.3} + \frac{1}{5.4} \approx 0.6200$

55. $\dfrac{\frac{1}{4.3}}{\frac{1}{5.6} + \frac{1}{7.2}} \approx 0.7326$

57. $\dfrac{1 \text{ mile}}{0.0006897 \text{ hours}} \approx 1450$ mph

59. $\frac{1}{0.01}$ bt/hr $+ \frac{1}{0.02}$ bt/hr $+ \frac{1}{0.015}$ bt/hr
≈ 217 buttons per hour

61. Because $3 + x = x + 3$ has the same form as $a + b = b + a$, the commutative property of addition is illustrated here.

63. Because $5(x - 7) = 5x - 35$ has the same form as $a(b + c) = ab + ac$, the distributive property is illustrated.

65. Because $3(xy) = (3x)y$ has the same form as $a(bc) = (ab)c$, the associative property of multiplication is illustrated.

67. Because 4 and 0.25 have a product of 1, the multiplicative inverse property is illustrated.

69. Because $y^3x = xy^3$ has the same form as $ab = ba$, the commutative property of multiplication is illustrated.

71. Because multiplying x by 1 does not change x, the multiplicative identity property is illustrated.

73. Because $2x + 3x = (2 + 3)x$ has the same form as $a(b + c) = ab + ac$, the distributive property is illustrated.

75. Because 7 and -7 have a sum of 0, the additive inverse property is illustrated.

77. Because the product of 0 and any number is 0, the multiplication property of zero is illustrated.

79. Because $xy + x = x(y + 1)$ has the same form as $a(b + c) = ab + ac$, the distributive property is illustrated.

81. Because $a + b = b + a$ for any real numbers, $5 + w = w + 5$.

83. Because $a(bc) = (ab)c$ for any real numbers, $5(xy) = (5x)y$.

85. Because $ab + ac = a(b + c)$ for any real numbers, $\frac{1}{2}x - \frac{1}{2} = \frac{1}{2}x - \frac{1}{2} \cdot 1 = \frac{1}{2}(x - 1)$.

87. Because $ab + ac = a(b + c)$ for any real numbers, $6x + 9 = 3 \cdot 2x + 3 \cdot 3 = 3(2x + 3)$.

89. Since 8 and 0.125 are reciprocals, $8(0.125) = 1$.

91. $0 = 5(0)$

93. $0.25(4) = 1$

95. Fortunately, the order in which the groceries are placed does not affect the total bill, because of the commutative and associative properties of addition.

1.6 WARM-UPS

1. True by the distributive property.

2. False, because $-4x + 8 = -4(x - 2)$.

3. True, because multiplying by -1 is equivalent to finding the opposite.

4. True, by the distributive property.

5. False, $(2x)(5x) = 10x^2$

6. True because of the distributive property.

7. False, $a + a = 2a$.

8. False, $b \cdot b = b^2$.

9. False, because 1 and $7x$ are not like terms.

10. True, because the like terms are combined correctly.

1.6 EXERCISES

1. A term is a single number or a product of a number and one or more variables.

3. The coefficient of a term is the number preceding the variables.

5. You can multiply and divide unlike terms.

7. $(45 \cdot 2) \cdot 100 = 90 \cdot 100 = 9000$

9. $\frac{4}{3}(0.75) = \frac{4}{3} \cdot \frac{3}{4} = 1$

11. $427 + (68 + 32) = 427 + 100 = 527$

13. $47 \cdot 4 + 47 \cdot 6 = 47(4 + 6) = 470$

15. $19 \cdot 2 \cdot 5 \cdot \frac{1}{5} = 19 \cdot 2 \cdot 1 = 19 \cdot 2 = 38$

17. $120 \div 4 \cdot 100 = 480 \cdot 100 = 48{,}000$

19. $13 \cdot 377 \cdot 0 = 0$

21. $348 + (5 + 45) = 348 + 50 = 398$

23. $\frac{2}{3} \cdot 1.5 = \frac{2}{3} \cdot \frac{3}{2} = 1$

25. $17 \cdot 101 - 17 \cdot 1 = 17(101 - 1) = 1700$

27. $354 + (7 + 3) + (8 + 2)$
$$= 354 + 10 + 10 = 374$$

29. $(567 + 874)(0) = 0$

31. $-4n + 6n = (-4 + 6)n = 2n$

33. $3w - (-4w) = 3w + 4w = 7w$

35. $4mw^2 - 15mw^2 = (4 - 15)mw^2$
$$= -11mw^2$$

37. $-5x - (-2x) = -5x + 2x = -3x$

39. Not like terms: $-4 - 7z$

41. $4t^2 + 5t^2 = (4 + 5)t^2 = 9t^2$

43. Not like terms: $-4ab + 3a^2b$

45. $9mn - mn = (9 - 1)mn = 8mn$

47. $x^3y - 3x^3y = 1x^3y - 3x^3y = -2x^3y$

49. $-kz^6 - kz^6 = -1kz^6 + (-1kz^6)$
$$= -2kz^6$$

51. $4(7t) = (4 \cdot 7)t = 28t$

53. $(-2x)(-5x) = (-2)(-5)x \cdot x = 10x^2$

55. $(-h)(-h) = (-1)(-1)h \cdot h = h^2$

57. $7w(-4) = -4 \cdot 7w = -28w$

59. $-x(1 - x) = -x \cdot 1 - (-x)(x)$
$$= -x + x^2$$

61. $5k \cdot 5k = 5 \cdot 5 \cdot k \cdot k = 25k^2$

63. $3 \cdot \frac{y}{3} = 3 \cdot \frac{1}{3} \cdot y = 1 \cdot y = y$

65. $9 \cdot \frac{2y}{9} = 9 \cdot \frac{1}{9} \cdot 2y = 1 \cdot 2y = 2y$

67. $\frac{6x^3}{2} = \frac{6}{2}x^3 = 3x^3$

69. $\frac{3x^2y + 15x}{3} = \frac{3x^2y}{3} + \frac{15x}{3} = x^2y + 5x$

71. $\frac{2x - 4}{-2} = \frac{2x}{-2} - \frac{4}{-2} = -x + 2$

73. $\frac{-xt + 10}{-2} = \frac{-xt}{-2} + \frac{10}{-2} = \frac{1}{2}xt - 5$

75. $a - (4a - 1) = a - 4a + 1 = -3a + 1$

77. $6 - (x - 4) = 6 - x + 4 = 10 - x$

79. $4m + 6 - m - 5 = 3m + 1$

81. $-5b + at - 7b = -12b + at$

83. $t^2 - 5w + 2w + t^2 = 2t^2 - 3w$

85. $x^2 - x^2 + y^2 + z = y^2 + z$

87. $(2x + 7x) + (3 + 5) = 9x + 8$

89. $(-3x + 4) + (5x - 6) = 2x - 2$

91. $4a^2 - 5c - 6a^2 + 7c = -2a^2 + 2c$

93. $5t^2 - 15w + 6w + 2t^2 = 7t^2 - 9w$

95. $-7m + 3m - 12 + 5m = m - 12$

97. $8 - 7k^3 - 21 - 4 = -7k^3 - 17$

99. $x - 0.04x - 0.04(50) = 0.96x - 2$

101. $0.10x + 0.5 - 0.04x - 2 = 0.06x - 1.5$

103. $3k + 5 - 6k + 8 - k + 3 = -4k + 16$

105. $5.7 - 4.5x + (4.5)(3.9) - 5.42$
$$= -4.5x + 17.83$$

107. $3 - 3xy - 2xy + 10 - 35 + xy$
$$= -4xy - 22$$

109. $3w^2 - 30w^2 - 2w^2 = -29w^2$

111. $3a^2w^2 - 5a^2w^2 - 4a^2w^2 = -6a^2w^2$

113. $\frac{1}{6} - \frac{1}{3}\left(-6x^2y - \frac{1}{2}\right) = \frac{1}{6} + 2x^2y + \frac{1}{6}$
$$= 2x^2y + \frac{1}{3}$$

115. $-\frac{1}{2}m\left(-\frac{1}{2}m\right) - \frac{1}{2}m - \frac{1}{2}m$
$$= \frac{1}{4}m^2 - m$$

117. $\frac{-8t^3}{-2} - \frac{6t^2}{-2} + \frac{2}{-2} = 4t^3 + 3t^2 - 1$

119. $\frac{-6xyz}{-3} - \frac{3xy}{-3} + \frac{9z}{-3}$
$$= 2xyz + xy - 3z$$

121. Perimeter is $s + s + 2 + s + 4$ or $3s + 6$ ft.

123. Perimeter is $2x + 2\left(x + \frac{1}{6}x\right)$ or $\frac{13}{3}x$ meters. Area is $x\left(x + \frac{1}{6}x\right) = \frac{7}{6}x^2$ m^2.

Enriching Your Mathematical Word Power
1. a   **2.** c   **3.** a   **4.** d   **5.** a
6. c   **7.** a   **8.** d   **9.** b   **10.** a
11. b   **12.** a   **13.** b   **14.** c

CHAPTER 1 REVIEW

1. True, because 3 is the only number common to A and B.

3. False, because $A \cup B = \{1, 2, 3, 4, 5\}$.

5. True, because $B \subseteq C$.

7. False, $A \cap \emptyset = \emptyset$.

9. True, because $A \cap B = \{3\}$.

11. True, because every member of B is also a member of C.

13. False, because $1 \in A$ and $1 \notin B$.

15. True, because 3 is a member of D.

17. False, because $0 \notin E$.

19. True, because $\emptyset$ is a subset of every set.

21. The set of whole numbers in the given set is $\{0, 1, 31\}$

23. The set of integers contained in the given set is $\{-1, 0, 1, 31\}$

25. The set of irrational numbers in the given set is $\left\{-\sqrt{2}, \sqrt{3}, \pi\right\}$.

27. The set of real numbers greater than 0 is written in interval notation as $(0, \infty)$ and graphed as follows:

29. The set of real numbers between 5 and 6 is written in interval notation as $(5, 6)$ and graphed as follows:

31. The set of real numbers greater than or equal to -1 and less than 2 is written in interval notation as $[-1, 2)$ and graphed as follows:

33. Graph the intervals $(0, 2)$ and $(1, 5)$ on the same number line to see that their union is $(0, 5)$.

35. Graph the intervals $(2, 4)$ and $(3, \infty)$ on the same number line to see that the numbers they have in common are in the interval $(3, 4)$.

37. Graph the intervals $[2, 6)$ and $(4, 8)$ on the same number line to see that their union is the interval $[2, 8)$.

39. $-4 + 9 = 9 - 4 = 5$

41. $25 - 37 = -12$

43. $(-4)(6) = -24$

45. $(-8) \div (-4) = 2$

47. $-\frac{1}{4} + \frac{1}{12} = -\frac{3}{12} + \frac{1}{12} = -\frac{2}{12} = -\frac{1}{6}$

49. $\frac{-20}{-2} = 10$

51. $-0.04 + 10 = 10.00 - 0.04 = 9.96$

53. $-6 - (-2) = -6 + 2 = -4$

55. $-0.5 + 0.5 = 0$

57. $3.2 \div (-0.8) = -4$

59. $0 \div (-0.3545) = 0$

61. $4 + 7 \cdot 5 = 4 + 35 = 39$

63. $(4 + 7)^2 = 11^2 = 121$

65. $5 + 3 \cdot |6 - 4 \cdot 3| = 5 + 3 \cdot |6 - 12|$
$\qquad = 5 + 3 \cdot 6 = 5 + 18$
$\qquad = 23$

67. $(6 - 8) - (5 - 9) = (-2) - (-4)$
$\qquad = -2 + 4 = 2$

69. $3 - 5(6 - 2 \cdot 5) = 3 - 5(6 - 10)$
$\qquad = 3 - 5(-4) = 3 + 20$
$\qquad = 23$

71. $5^2 - (6 + 5)^2 = 25 - (11)^2$
$\qquad = 25 - 121 = -96$

73. $\sqrt{3^2 + 4^2} = \sqrt{9 + 16} = \sqrt{25} = 5$

75. $\frac{-4 - 5}{7 - (-2)} = \frac{-9}{7 + 2} = \frac{-9}{9} = -1$

77. $1 - (0.8)(0.3) = 1.00 - 0.24 = 0.76$

79. $(-3)^2 - (4)(-1)(-2) = 9 - 8 = 1$

81. $3 - 2 |4 - 7| = 3 - 2 |-3|$
$\qquad = 3 - 2 \cdot 3 = 3 - 6 = -3$

83. $\frac{-5 - 3}{-4 + 2} = \frac{-8}{-2} = 4$

85. $\sqrt{3^2 - 4(-2)(-1)} = \sqrt{9 - 8}$
$\qquad = \sqrt{1} = 1$

87. $(-1 - 3)(-1 + 3) = -4 \cdot 2 = -8$

89. $(-2)^2 + 2(-2)(3) + 3^2 = 4 - 12 + 9$
$\qquad = 1$

91. $(-2)^3 - 3^3 = -8 - 27 = -35$

93. $\dfrac{3+(-1)}{-2+3} = \dfrac{2}{1} = 2$

95. $|-2-3| = |-5| = 5$

97. $(-2+3)(-1) = (1)(-1) = -1$

99. Commutative property of addition

101. Distributive property

103. Associative property of multiplication

105. Multiplicative identity property

107. Multiplicative inverse property

109. Multiplication property of zero

111. Additive identity property

113. Additive inverse property

115. $3x - 3a = 3(x - a)$

117. $3(w + 1) = 3 \cdot w + 3 \cdot 1 = 3w + 3$

119. $7x + 7 = 7x + 7 \cdot 1 = 7(x + 1)$

121. $5(x - 5) = 5x - 5 \cdot 5 = 5x - 25$

123. $-3(2x - 5) = -3 \cdot 2x - (-3)(5)$
$= -6x + 15$

125. $p - pt = p \cdot 1 - pt = p(1 - t)$

127. $3a + 7 + 4a - 5 = 7a + 2$

129. $5(t - 4) - 3(2t - 6)$
$= 5t - 20 - 6t + 18 = -t - 2$

131. $-(a - 2) + 2 - a = -a + 2 + 2 - a$
$= -2a + 4$

133. $5 - 3(x - 2) + 7(x + 4)$
$= 5 - 3x + 6 + 7x + 28 - 6$
$= 4x + 33$

135. $0.2(x + 0.1) - (x + 0.5)$
$= 0.2x + 0.02 - x - 0.50$
$= -0.8x - 0.48$

137. $0.05(x + 3) - 0.1(x + 20)$
$= 0.05x + 0.15 - 0.10x - 2$
$= -0.05x - 1.85$

139. $\frac{1}{2}(x + 4) - \frac{1}{4}(x - 8)$
$= \frac{1}{2}x + 2 - \frac{1}{4}x + 2 = \frac{1}{4}x + 4$

141. $\dfrac{-9x^2}{3} - \dfrac{6x}{3} + \dfrac{3}{3} = -3x^2 - 2x + 1$

143. $32(4)(-6 + 6) = 32(4)(0) = 0$,
additive inverse property, multiplication property of 0

145. $768(4) + 768(6) = 768(10) = 7680$,
distributive property

147. $(12 \cdot 4 + (-6)) + 6$
$= 12 \cdot 4 + (-6 + 6) = 48 + 0 = 48$,

associative property of addition, additive inverse property

149. $752(-6) + 752(6) = 752(-6 + 6)$
$= 752(0) = 0$, distributive property, additive inverse property

151. $(47 \cdot 6)\frac{4}{24} = 47\left(6 \cdot \frac{1}{6}\right) = 47 \cdot 1 = 47$,
associative property of multiplication, multiplicative inverse property, multiplicative identity property

153. $(-6 \cdot 24)\frac{1}{6} = (24 \cdot -6)\frac{1}{6}$
$= 24(-6 \cdot \frac{1}{6}) = 24(-1) = -24$,
commutative property of multiplication, associative property of multiplication, multiplicative inverse property

155. $5(-6 + 6)(4 + 24) = 5 \cdot 0 \cdot 28 = 0$,
additive inverse property, multiplication property of 0

157. $\dfrac{1}{0.2} + \dfrac{1}{0.5} = 7$ shingles per minute

159. a) $27{,}759(1.05)^8 \approx 41{,}013$
The car will cost \$41,013 in the year 2010.
b) From the graph, the first year in which the car is over \$50,000 will be 2015. Check using the formula: $27{,}759(1.05)^{12} \approx 49{,}851$ and $27{,}759(1.05)^{13} \approx 52{,}344$.

CHAPTER 1 TEST

1. List the elements in A or in B:
$A \cup B = \{2, 3, 4, 5, 6, 7, 8, 10\}$

2. Only 6 and 7 are in both B and C. So $B \cap C = \{6, 7\}$.

3. The intersection of $A = \{2, 4, 6, 8, 10\}$ with $B \cup C = \{3, 4, 5, 6, 7, 8, 9, 10\}$ is $\{4, 6, 8, 10\}$.

4. $\{0, 8\}$

5. $\{-4, 0, 8\}$

6. $\{-4, -\frac{1}{2}, 0, 1.65, 8\}$

7. $\left\{-\sqrt{3}, \sqrt{5}, \pi\right\}$

8. The integers between -3 and 5 are $-2, -1,$ $0, 1, 2, 3,$ and 4. The graph follows.

9. The interval of real numbers $(-3, 5]$ consists of all points on the number line greater than -3 and less than or equal to 5. The graph follows.

10. Graph $(-\infty, 2)$ and $(1, 4)$ on the same number line to see that their union is $(-\infty, 4)$.

11. Graph $(2, 8)$ and $[4, 9)$ on the same number line to see that the numbers they have in common are in the interval $[4, 8)$.

12. $6 + 3(-5) = 6 + (-15) = -9$

13. $\sqrt{(-2)^2 - 4(3)(-5)} = \sqrt{4 - (-60)}$
$$= \sqrt{64} = 8$$

14. $-5 + 6 - 12 = -17 + 6 = -11$

15. $0.02 - 2 = -1.98$

16. $\dfrac{-3 - (-7)}{3 - 5} = \dfrac{-3 + 7}{-2} = \dfrac{4}{-2} = -2$

17. $\dfrac{-6 - 2}{4 - 2} = \dfrac{-8}{2} = -4$

18. $\left(\dfrac{2}{3} - 1\right)\left(\dfrac{1}{3} - \dfrac{1}{2}\right) = \left(-\dfrac{1}{3}\right)\left(-\dfrac{1}{6}\right)$
$$= \dfrac{1}{18}$$

19. $-\dfrac{4}{7} - \dfrac{1}{2}\left(24 - \dfrac{8}{7}\right) = -\dfrac{4}{7} - 12 + \dfrac{4}{7}$
$$= -12$$

20. $|\, 3 - 5 \cdot 2 \,| = |\, 3 - 10 \,| = |\, -7 \,| = 7$

21. $5 - 2 \,|\, 6 - 10 \,| = 5 - 2 \,|\, -4 \,|$
$= 5 - 2 \cdot 4 = 5 - 8 = -3$

22. $(452 + 695)[-8 + 8] = (452 + 695)[0]$
$$= 0$$

23. $478(8 + 2) = 478(10) = 4780$

24. $-8 \cdot 3 - 4(6 - 9 \cdot 2^3)$
$= -24 - 4(6 - 9 \cdot 8)$
$= -24 - 4(-66) = -24 + 264$
$= 240$

25. $(-4)^2 - 4(-3)(2) = 16 + 24 = 40$

26. $\dfrac{(-3)^2 - (-4)^2}{-4 - (-3)} = \dfrac{9 - 16}{-1} = 7$

27. $\dfrac{(-3)(-4) - 6(2)}{(-4)^2 - 2^2} = \dfrac{12 - 12}{16 - 4} = \dfrac{0}{12} = 0$

28. Distributive property
29. Commutative property of multiplication
30. Associative property of addition
31. Additive inverse property
32. Commutative property of multiplication

33. $3(m - 5) - 4(-2m - 3)$
$= 3m - 15 + 8m + 12 = 11m - 3$

34. $x + 3 - 0.05x - 0.1 = 0.95x + 2.9$

35. $\dfrac{1}{2}x - \dfrac{4}{2} + \dfrac{1}{4}x + \dfrac{3}{4} = \dfrac{2}{4}x + \dfrac{1}{4}x - \dfrac{8}{4} + \dfrac{3}{4}$
$$= \dfrac{3}{4}x - \dfrac{5}{4}$$

36. $-3x^2 + 6y - 6y + 8x^2 = 5x^2$

37. $\dfrac{-6x^2}{-2} - \dfrac{4x}{-2} + \dfrac{2}{-2} = 3x^2 + 2x - 1$

38. $5x - 40 = 5x - 5 \cdot 8 = 5(x - 8)$

39. $7t - 7 = 7t - 7 \cdot 1 = 7(t - 1)$

40. $\dfrac{1 \text{ tree}}{0.0625 \text{ hours}} = 16 \text{ trees/hour}$

41. The perimeter is $2x + 2(x - 4)$ or $4x - 8$ feet. If $x = 9$, the perimeter is $4 \cdot 9 - 8$ or 28 feet. The area is $x^2 - 4x$. If $x = 9$, then the area is $9^2 - 4 \cdot 9$, or 45 ft^2.

42. $6(1.03)^{25} \approx 12.6$ billion

11

2.1 WARM-UPS

1. False, it is equivalent to $-2x = 5$.
2. True, because
$x - (x - 3) = x - x + 3 = 3$.
3. False, multiply each side by $\frac{4}{3}$.
4. True, because we can multiply each side by -1.
5. True, because all of the denominators divide into the LCD.
6. True, subtract 5 from each side and then divide each side by 3.
7. True, because it simplifies to $8 = 12$.
8. True, it is equivalent to $x = -3$.
9. True, it is equivalent to $0.8x = 0.8x$.
10. True, since it is equivalent to $3x - 12 = 0$.

2.1 EXERCISES

1. An equation is a sentence that expresses the equality of two algebraic expressions.
3. Equivalent equation are equations that have the same solution set.
5. If the equation involves fractions then multiply each side by the LCD.
7. A conditional equation is an equation that has at least one solution but is not an identity.
9. Yes, because $3 \cdot (-4) + 7 = -5$ is correct.
11. Yes, because $\frac{1}{2} \cdot 12 - 4 = \frac{1}{3} \cdot 12 - 2$.

13. No, because $0.2(200 - 50) = 30$ and $20 - 0.05 \cdot 200 = 10$.
15. No because $0.1 \cdot 80 - 30 = -22$ and $16 - 0.06 \cdot 80 = 11.2$.
17.
$$-72 - x = 15$$
$$-72 - x + 72 = 15 + 72$$
$$-x = 87$$
$$(-1)(-x) = -1 \cdot 87$$
$$x = -87$$
The solution set is $\{-87\}$.
19.
$$-3x - 19 = 5 - 2x$$
$$-3x = 24 - 2x$$
$$-3x + 2x = 24$$
$$-x = 24$$
$$x = -24$$
The solution set is $\{-24\}$.
21.
$$2x - 3 = 0$$
$$2x - 3 + 3 = 0 + 3$$

$$2x = 3$$
$$\frac{2x}{2} = \frac{3}{2}$$
$$x = \frac{3}{2}$$

The solution set is $\left\{ \frac{3}{2} \right\}$.

23.
$$-2x + 5 = 7$$
$$-2x + 5 - 5 = 7 - 5$$
$$-2x = 2$$
$$\frac{-2x}{-2} = \frac{2}{-2}$$
$$x = -1$$
The solution set is $\{-1\}$.
25.
$$-12x - 15 = 21$$
$$-12x - 15 + 15 = 21 + 15$$
$$-12x = 36$$
$$\frac{-12x}{-12} = \frac{36}{-12}$$
$$x = -3$$
The solution set is $\{-3\}$.
27.
$$26 = 4x + 16$$
$$26 - 16 = 4x + 16 - 16$$
$$10 = 4x$$
$$\frac{10}{4} = \frac{4x}{4}$$
$$\frac{5}{2} = x$$
The solution set is $\left\{ \frac{5}{2} \right\}$.

29.
$$-3(x - 16) = 12 - x$$
$$-3x + 48 = 12 - x$$
$$-3x = -36 - x$$
$$-2x = -36$$
$$x = 18$$
The solution set is $\{18\}$.
31.
$$2(x + 9) - x = 36$$
$$2x + 18 - x = 36$$
$$x + 18 = 36$$
$$x + 18 - 18 = 36 - 18$$
$$x = 18$$
The solution set is $\{18\}$.
33.
$$2 + 3(x - 1) = x - 1$$
$$2 + 3x - 3 = x - 1$$
$$3x - 1 = x - 1$$
$$3x - x = -1 + 1$$
$$2x = 0$$
$$x = 0$$
The solution set is $\{0\}$.

35. $-\frac{3}{7}x = 4$

$-\frac{7}{3}\left(-\frac{3}{7}x\right) = -\frac{7}{3}(4)$

$x = -\frac{28}{3}$

The solution set is $\left\{-\frac{28}{3}\right\}$.

37. $-\frac{5}{7}x - 1 = 3$

$-\frac{5}{7}x = 4$

$-\frac{7}{5}\left(-\frac{5}{7}x\right) = -\frac{7}{5}(4)$

$x = -\frac{28}{5}$

The solution set is $\left\{-\frac{28}{5}\right\}$.

39. $\frac{x}{3} + \frac{1}{2} = \frac{7}{6}$

$6\left(\frac{x}{3} + \frac{1}{2}\right) = 6 \cdot \frac{7}{6}$

$2x + 3 = 7$

$2x = 4$

$x = 2$

The solution set is $\{2\}$.

41. $\frac{2}{3}x + 5 = -\frac{1}{3}x + 17$

$3\left(\frac{2}{3}x + 5\right) = 3\left(-\frac{1}{3}x + 17\right)$

$2x + 15 = -x + 51$

$3x + 15 = 51$

$3x = 36$

$x = 12$

The solution set is $\{12\}$.

43. $\frac{1}{2}x + \frac{1}{4} = \frac{1}{4}\left(x - 6\right)$

$4\left(\frac{1}{2}x + \frac{1}{4}\right) = 4 \cdot \frac{1}{4}\left(x - 6\right)$

$4\left(\frac{1}{2}x + \frac{1}{4}\right) = 4 \cdot \frac{1}{4}\left(x - 6\right)$

$2x + 1 = x - 6$

$x + 1 = -6$

$x = -7$

The solution set is $\{-7\}$.

45. $8 - \frac{x-2}{2} = \frac{x}{4}$

$4\left(8 - \frac{x-2}{2}\right) = 4\left(\frac{x}{4}\right)$

$32 - 2(x - 2) = x$

$32 - 2x + 4 = x$

$-3x = -36$

$x = 12$

The solution set is $\{12\}$.

47. $\frac{y-3}{3} - \frac{y-2}{2} = -1$

$6\left(\frac{y-3}{3}\right) - 6\left(\frac{y-2}{2}\right) = 6(-1)$

$2y - 6 - 3y + 6 = -6$

$-y = -6$

$y = 6$

The solution set is $\{6\}$.

49. $x - 0.2x = 72$

$10(x - 0.2x) = 10(72)$

$10x - 2x = 720$

$8x = 720$

$x = 90$

The solution set is $\{90\}$.

51. $0.03(x + 200) + 0.05x = 86$

$0.03x + 0.03(200) + 0.05x = 86$

$0.08x + 6 = 86$

$0.08x = 80$

$x = 1000$

The solution set is $\{1000\}$.

53. $0.1x + 0.05(x - 300) = 105$

$0.1x + 0.05x - 0.05(300) = 105$

$0.15x - 15 = 105$

$0.15x = 120$

$x = 800$

The solution set is $\{800\}$.

55. $2(x + 1) = 2(x + 3)$

$2x + 2 = 2x + 6$

$2 = 6$

The solution set is $\emptyset$. The equation is inconsistent.

57. $x + x = 2x$

$2x = 2x$

The solution set is R. The equation is an identity.

59. $x + x = 2$

$2x = 2$

$x = 1$

The solution set is $\{1\}$ and the equation is conditional.

61. $\frac{4x}{4} = x$

$x = x$

The solution set is R and the equation is an identity.

63. $x \cdot x = x^2$
$$x^2 = x^2$$
The solution set is R and the equation is an identity.

65. $2(x+3) - 7 = 5(5-x) + 7(x+1)$
$$2x + 6 - 7 = 25 - 5x + 7x + 7$$
$$2x - 1 = 32 + 2x$$
$$-1 = 32$$
The solution set is $\emptyset$. The equation is inconsistent.

67. $2\left(\frac{1}{2}x + \frac{3}{2}\right) - \frac{7}{2} = \frac{3}{2}(x+1) - \left(\frac{1}{2}x + 2\right)$

$$x + 3 - \frac{7}{2} = \frac{3}{2}x + \frac{3}{2} - \frac{1}{2}x - 2$$
$$2x + 6 - 7 = 3x + 3 - x - 4$$
$$2x - 1 = 2x - 1$$
The solution set to this identity is R.

69. $2(0.5x + 1.5) - 3.5 = 3(0.5x + 0.5)$
$$x + 3 - 3.5 = 1.5x + 1.5$$
$$x - 0.5 = 1.5x + 1.5$$
$$10x - 5 = 15x + 15$$
$$-20 = 5x$$
$$-4 = x$$
The solution set is to this conditional equation is $\{-4\}$.

71. $4 - 6(2x - 3) + 1 = 3 + 2(5 - x)$
$$4 - 12x + 18 + 1 = 3 + 10 - 2x$$
$$-12x + 23 = 13 - 2x$$
$$-10x = -10$$
$$x = 1$$
The solution set is $\{1\}$.

73. $\frac{1}{2}\left(y - \frac{1}{6}\right) + \frac{2}{3} = \frac{5}{6} + \frac{1}{3}\left(\frac{1}{2} - 3y\right)$
$$\frac{1}{2}y - \frac{1}{12} + \frac{2}{3} = \frac{5}{6} + \frac{1}{6} - y$$
$$6y - 1 + 8 = 10 + 2 - 12y$$
$$6y + 7 = 12 - 12y$$
$$18y = 5$$
$$y = \frac{5}{18}$$
The solution set is $\left\{\frac{5}{18}\right\}$.

75. $8 - \frac{2}{3}(60x - 900) = \frac{1}{2}(400x + 6)$
$$8 - 40x + 600 = 200x + 3$$
$$-40x + 608 = 200x + 3$$
$$-240x = -605$$
$$x = \frac{605}{240} = \frac{121}{48}$$
The solution set is $\left\{\frac{121}{48}\right\}$.

77. $\frac{a-3}{4} - \frac{2a-5}{2} = \frac{a+1}{3} - \frac{1}{6}$

$$12 \cdot \frac{a-3}{4} - 12 \cdot \frac{2a-5}{2}$$
$$= 12 \cdot \frac{a+1}{3} - 12 \cdot \frac{1}{6}$$
$$3a - 9 - 12a + 30 = 4a + 4 - 2$$
$$-9a + 21 = 4a + 2$$
$$-13a = -19$$
$$a = \frac{19}{13}$$
The solution set is $\left\{\frac{19}{13}\right\}$.

79. $1.3 - 0.2(6 - 3x) = 0.1(0.2x + 3)$
$$1.3 - 1.2 + 0.6x = 0.02x + 0.3$$
$$130 - 120 + 60x = 2x + 30$$
$$10 + 60x = 2x + 30$$
$$58x = 20$$
$$x = \frac{20}{58} = \frac{10}{29}$$
The solution set is $\left\{\frac{10}{29}\right\}$.

81. $3x - 9 = 0$
$$3x = 9$$
$$x = 3$$
The solution set is $\{3\}$.

83. $7 - z = -9$
$$16 = z$$
The solution set is $\{16\}$.

85. $\frac{2}{3}x = \frac{1}{2}$
$$x = \frac{3}{2} \cdot \frac{1}{2} = \frac{3}{4}$$
The solution set is $\left\{\frac{3}{4}\right\}$.

87. $-\frac{3}{5}y = 9$
$$y = -\frac{5}{3} \cdot 9 = -15$$
The solution set is $\{-15\}$.

89. $3y + 5 = 4y - 1$
$$6 = y$$
The solution set is $\{6\}$.

91. $5x + 10(x + 2) = 110$
$$5x + 10x + 20 = 110$$
$$15x = 90$$
$$x = 6$$
The solution set is $\{6\}$.

93. $\dfrac{P+7}{3} - \dfrac{P-2}{5} = \dfrac{7}{3} - \dfrac{P}{15}$

$$15\left(\frac{P+7}{3}\right) - 15\left(\frac{P-2}{5}\right)$$
$$= 15\left(\frac{7}{3}\right) - 15\left(\frac{P}{15}\right)$$
$$5P + 35 - 3P + 6 = 35 - P$$
$$2P + 41 = 35 - P$$
$$3P = -6$$
$$P = -2$$

The solution set is $\{-2\}$.

95. $x - 0.06x = 50{,}000$
$$0.94x = 50{,}000$$
$$x = \frac{50{,}000}{0.94} \approx 53{,}191.49$$

The solution set is $\{53{,}191.49\}$.

97. $2.365x + 3.694 = 14.8095$
$$2.365x = 14.8095 - 3.694$$
$$2.365x = 11.1155$$
$$x = \frac{11.1155}{2.365} = 4.7$$

The solution set is $\{4.7\}$.

99. a) For 1992, $x = 7$.
$$0.45(7) + 39.05 = 42.2$$
Public school enrollment in 1992 was 42.2 million.

b) $0.45x + 39.05 = 50$
$$0.45x = 10.95$$
$$x \approx 24.3$$

So in 24.3 years (or during the 25th year) public school enrollment will reach 50 million. $1985 + 25 = 2010$.

c) Judging from the graph, enrollment is increasing.

101. To eliminate decimals multiply each side by an appropriate power of 10.

2.2 WARM-UPS

1. False, P is on both sides of the equation.
2. False, because we use the distributive property to factor out P on the right side.
3. False, because we divide each side by Pr to get $t = \dfrac{I}{Pr}$.
4. True, because $\dfrac{5 \cdot 6}{2} = 15$.
5. False, because $P = 2L + 2W$.
6. True, because $V = LWH$.
7. False, because $A = \frac{1}{2}h(b_1 + b_2)$.

8. True, $x - y = 5$ is equivalent to $x = 5 + y$, or $y = x - 5$.
9. True, because $-2(-3) - 4 = 6 - 4 = 2$.
10. False, perimeter is the distance around the outside edge.

2.2 EXERCISES

1. A formula is an equation involving two or more variables.
3. Solving for a variable means to rewrite the formula with the indicated variable isolated.
5. To find the value a variable, solve for that variable, then replace all other variables with the given numbers.

7. $I = Prt$
$$\frac{I}{Pr} = \frac{Prt}{Pr}$$
$$\frac{I}{Pr} = t$$
$$t = \frac{I}{Pr}$$

9. $F = \frac{9}{5}C + 32$
$$F - 32 = \frac{9}{5}C$$
$$\frac{5}{9}(F - 32) = \frac{5}{9} \cdot \frac{9}{5}C$$
$$C = \frac{5}{9}(F - 32)$$

11. $A = LW$
$$\frac{A}{L} = \frac{LW}{L}$$
$$\frac{A}{L} = W$$
$$W = \frac{A}{L}$$

13. $A = \frac{1}{2}(b_1 + b_2)$
$$2A = 2 \cdot \frac{1}{2}(b_1 + b_2)$$
$$2A = b_1 + b_2$$
$$2A - b_2 = b_1$$
$$b_1 = 2A - b_2$$

15. $P = 2L + 2W$
$$P - 2W = 2L$$
$$\frac{P - 2W}{2} = L$$
$$L = \frac{P - 2W}{2} \text{ or } L = \frac{1}{2}P - W$$

17. $V = \pi r^2 h$
$$\frac{V}{\pi r^2} = \frac{\pi r^2 h}{\pi r^2}$$
$$h = \frac{V}{\pi r^2}$$

19. $2x + 3y = 9$
$$3y = -2x + 9$$
$$\frac{3y}{3} = \frac{-2x}{3} + \frac{9}{3}$$
$$y = -\frac{2}{3}x + 3$$

21. $x - y = 4$
$$-y = -x + 4$$
$$\frac{-y}{-1} = \frac{-x}{-1} + \frac{4}{-1}$$
$$y = x - 4$$

23. $\frac{1}{2}x - \frac{1}{3}y = 2$
$$-\frac{1}{3}y = -\frac{1}{2}x + 2$$
$$-3\left(-\frac{1}{3}y\right) = -3\left(-\frac{1}{2}x + 2\right)$$
$$y = \frac{3}{2}x - 6$$

25. $y - 2 = \frac{1}{2}(x - 3)$
$$y - 2 = \frac{1}{2}x - \frac{3}{2}$$
$$y = \frac{1}{2}x - \frac{3}{2} + 2$$
$$y = \frac{1}{2}x + \frac{1}{2}$$

27. $A = P + Prt$
$$A - P = Prt$$
$$\frac{A - P}{Pr} = \frac{Prt}{Pr}$$
$$\frac{A - P}{Pr} = t$$
$$t = \frac{A - P}{Pr}$$

29. $ab + a = 1$
$$a(b + 1) = 1$$
$$\frac{a(b + 1)}{b + 1} = \frac{1}{b + 1}$$
$$a = \frac{1}{b + 1}$$

31. $xy + 5 = y - 7$
$$xy - y = -5 - 7$$
$$y(x - 1) = -12$$
$$y = \frac{-12}{x - 1}$$
$$y = \frac{-12(-1)}{(x - 1)(-1)}$$
$$y = \frac{12}{1 - x}$$

33. $xy^2 + xz^2 = xw^2 - 6$
$$xy^2 + xz^2 - xw^2 = -6$$
$$x(y^2 + z^2 - w^2) = -6$$
$$x = \frac{-6}{y^2 + z^2 - w^2}$$
$$x = \frac{6}{w^2 - y^2 - z^2}$$

35.
$$\frac{1}{R} = \frac{1}{R_1} + \frac{1}{R_2}$$
$$RR_1R_2\left(\frac{1}{R}\right) = RR_1R_2\left(\frac{1}{R_1}\right) + RR_1R_2\left(\frac{1}{R_2}\right)$$
$$R_1R_2 = RR_2 + RR_1$$
$$R_1R_2 - RR_1 = RR_2$$
$$R_1(R_2 - R) = RR_2$$
$$R_1 = \frac{RR_2}{R_2 - R}$$

37. $3.35x - 54.6 = 44.3 - 4.58x$
$$3.35x + 4.58x = 44.3 + 54.6$$
$$x(3.35 + 4.58) = 44.3 + 54.6$$
$$x = \frac{44.3 + 54.6}{3.35 + 4.58}$$
$$x \approx 12.472$$

39. $4.59x - 66.7 = 3.2(x - 5.67)$
$$4.59x - 66.7 = 3.2x - 3.2(5.67)$$
$$x(4.59 - 3.2) = 66.7 - 3.2(5.67)$$
$$x = \frac{66.7 - 3.2(5.67)}{4.59 - 3.2}$$
$$x \approx 34.932$$

41.
$$\frac{x}{19} - \frac{3}{23} = \frac{4}{31} + \frac{3x}{7}$$
$$\frac{x}{19} + \frac{3x}{7} = \frac{4}{31} + \frac{3}{23}$$
$$x\left(\frac{1}{19} + \frac{3}{7}\right) = \frac{4}{31} + \frac{3}{23}$$
$$x = \frac{\frac{4}{31} + \frac{3}{23}}{\frac{1}{19} + \frac{3}{7}} \approx 0.539$$

43. $2x - 3y = 5$
$$2(3) - 3y = 5$$
$$-3y = -1$$
$$y = \frac{1}{3}$$

45. $-4x + 2y = 1$
$$-4(3) + 2y = 1$$
$$2y = 13$$
$$y = \frac{13}{2}$$

47. $y = -2x + 5$
$$y = -2(3) + 5$$
$$y = -6 + 5$$
$$y = -1$$

49. $-x + 2y = 5$
$$-3 + 2y = 5$$
$$2y = 8$$
$$y = 4$$

51. $y - 1.046 = 2.63(x - 5.09)$
$$y - 1.046 = 2.63(3 - 5.09)$$
$$y - 1.046 = -5.4967$$
$$y = -4.4507$$

16

53.
$$wxy = 5$$
$$4x(2) = 5$$
$$8x = 5$$
$$x = \frac{5}{8}$$

55.
$$x + xz = 7$$
$$x + x(-3) = 7$$
$$-2x = 7$$
$$x = -\frac{7}{2}$$

57.
$$w(x - z) = y(x - 4)$$
$$4[x - (-3)] = 2(x - 4)$$
$$4x + 12 = 2x - 8$$
$$2x = -20$$
$$x = -10$$

59.
$$w = \frac{1}{2}xz$$
$$4 = \frac{1}{2}x(-3)$$
$$4 = -\frac{3}{2}x$$
$$-\frac{2}{3}(4) = -\frac{2}{3}\left(-\frac{3}{2}x\right)$$
$$-\frac{8}{3} = x$$

61.
$$\frac{1}{w} + \frac{1}{x} = \frac{1}{y}$$
$$\frac{1}{4} + \frac{1}{x} = \frac{1}{2}$$
$$4x \cdot \frac{1}{4} + 4x \cdot \frac{1}{x} = 4x \cdot \frac{1}{2}$$
$$x + 4 = 2x$$
$$4 = x$$

63. The formula for the area of a circle is $A = \pi r^2$.

65. Since $C = 2\pi r$, we can divide each side by 2π to get $r = \frac{C}{2\pi}$.

67. Because $P = 2L + 2W$, we can subtract $2L$ from each side to get $2W = P - 2L$, and then divide each side by 2 to get $W = \frac{P - 2L}{2}$.

69.
$$I = Prt$$
$$300 = 1000 \cdot 2 \cdot r$$
$$r = \frac{300}{2000} = 15\%$$

71.
$$A = LW$$
$$23 = L(4)$$
$$L = \frac{23}{4} = 5.75 \text{ yards}$$

73.
$$V = LWH$$
$$36 = 2(2.5)H$$
$$36 = 5H$$
$$H = 7.2 \text{ feet}$$

75.
$$V = 900 \text{ gal.} \left(\frac{1 \text{ ft}^3}{7.5 \text{ gal}}\right) = 120 \text{ ft}^3$$
$$V = LWH$$
$$120 = 4 \cdot 6 \cdot H$$
$$H = \frac{120}{24} = 5 \text{ feet}$$

77.
$$A = \frac{1}{2}bh$$
$$30 = \frac{1}{2}4h$$
$$30 = 2h$$
$$h = 15 \text{ feet}$$

79.
$$A = \frac{1}{2}h(b_1 + b_2)$$
$$300 = \frac{1}{2}(20)(16 + b_2)$$
$$300 = 10(16 + b_2)$$
$$30 = 16 + b_2$$
$$b_2 = 14 \text{ inches}$$

81.
$$P = 2L + 2W$$
$$600 = 2L + 2(132)$$
$$336 = 2L$$
$$L = 168 \text{ feet}$$

83.
$$C = 2\pi r$$
$$3\pi = 2\pi r$$
$$r = \frac{3\pi}{2\pi}$$
$$r = 1.5 \text{ meters}$$

85.
$$C = 2\pi r$$
$$r = \frac{C}{2\pi}$$
$$r = \frac{25{,}000}{2\pi}$$
$$r \approx 3979 \text{ miles}$$

87. If the diameter is 3 in., then the radius is 1.5 in.
$$V = \pi r^2 h$$
$$30 = \pi(1.5)^2 h$$
$$30 = 2.25\pi h$$
$$h = \frac{30}{2.25\pi} \approx 4.24 \text{ inches}$$

89. Let $W = 62 \text{ lb/ft}^3$, $D = 32$ ft, and $A = 48 \text{ ft}^2$ in $F = WDA$:
$$F = 62 \text{ lb/ft}^3 \cdot 32 \text{ ft} \cdot 48 \text{ ft}^2 = 95{,}232 \text{ lb}$$

91. a) $N = B + S - 1$
$$N = 2003 + 455 - 1 = 2457$$
b) $1452 = 1033 + S - 1$
$$420 = S$$

93. $\$1000 \div \$2 = 500$ feet
$$b_1 + b_2 = 500$$
$$A = \frac{1}{2}h(b_1 + b_2)$$
$$50{,}000 = \frac{1}{2}h \cdot 500$$
$$h = 200 \text{ feet}$$

95. $A = bh$
$60,000 = b \cdot 200$
$b = 300$ feet
There is 300 ft on each street. So 600 ft at \$2 each is a \$1200 assessment.

2.3 WARM-UPS

1. False, first identify what the variable stands for.
2. True, we must know what the letters represent.
3. False, you may have the wrong equation.
4. False, odd integers differ by 2.
5. True, because $x + 6 - x = 6$.
6. True, because $x + 7 - x = 7$.
7. False, $5x - 2 = 3(x + 20)$ since $5x$ is larger than $3(x + 20)$.
8. True, because 8% of x is $0.08x$.
9. False, because 10% of \$88,000 is \$8,800 and \$88,000 − \$8,800 is not \$80,000.
10. False, because the acid in the mixture must be between 10% and 14%.

2.3 EXERCISES

1. Three unknown consecutive integers are represented by x, $x + 1$, and $x + 2$.
3. The formula $P = 2L + 2W$ expresses the perimeter in terms of the length and width.
5. The commission is a percentage of the selling price.
7. Since two consecutive even integers differ by 2, we can use x and $x + 2$ to represent them.
9. The expressions x and $10 - x$ have a sum of 10, since $x + 10 - x = 10$.
11. If x is the selling price, then eighty-five percent of the selling price is $0.85x$.
13. Since $D = RT$, the distance is $3x$ miles.
15. Since the perimeter is twice the length $(x + 5)$ plus twice the width (x), we can represent the perimeter by $2(x + 5) + 2(x)$ or $4x + 10$.

17. Let $x =$ the first integer, $x + 1 =$ the second integer, and $x + 2 =$ the third integer. Their sum is 84:

$$x + x + 1 + x + 2 = 84$$
$$3x + 3 = 84$$
$$3x = 81$$
$$x = 27$$

If $x = 27$, then $x + 1 = 28$, and $x + 2 = 29$. The integers are 27, 28, and 29.

19. Let $x =$ the first even integer, $x + 2 =$ the second, and $x + 4 =$ the third. Their sum is 252:

$$x + x + 2 + x + 4 = 252$$
$$3x + 6 = 252$$
$$3x = 246$$
$$x = 82$$

If $x = 82$, then $x + 2 = 84$ and $x + 4 = 86$. The integers are 82, 84, and 86.

21. Let $x =$ the first odd integer and $x + 2 =$ the second. Their sum is 128:

$$x + x + 2 = 128$$
$$2x = 126$$
$$x = 63$$

If $x = 63$, then $x + 2 = 65$. The integers are 63 and 65.

23. Let $x =$ the width, and $2x + 1 =$ the length. Since $P = 2L + 2W$ we can write the following equation.

$$2(x) + 2(2x + 1) = 278$$
$$6x + 2 = 278$$
$$6x = 276$$
$$x = 46$$

If $x = 46$, then $2x + 1 = 93$. The width is 46 meters and the length is 93 meters.

25. Let $x =$ the length of the first side, $2x - 10 =$ the length of the second side, and $x + 50 =$ the length of the third side. Since the perimeter is 684, we have the following equation.

$$x + 2x - 10 + x + 50 = 684$$
$$4x + 40 = 684$$
$$4x = 644$$
$$x = 161$$

If $x = 161$ feet, then $2x - 10 = 312$ feet, and $x + 50 = 211$ feet.

27. Let x = the width, and $2x + 5$ = the length. To fence the 3 sides, we use 2 widths and 1 length:
$$2(x) + (2x + 5) = 50$$
$$4x = 45$$
$$x = 11.25$$
$$2x + 5 = 27.5$$
Width is 11.25 feet and the length is 27.5 feet.

29. Let x = the amount invested at 6% and $x + 1000$ = the amount invested at 10%. Since the interest on the investments is $0.06x$ and $0.10(x + 1000)$ we can write the equation
$$0.06x + 0.10(x + 1000) = 340$$
$$0.16x + 100 = 340$$
$$0.16x = 240$$
$$x = 1500$$
$$x + 1000 = 2500$$
He invested $1500 at 6% and $2500 at 10%.

31. Let x = the amount of his inheritance. He invests $\frac{1}{2}x$ at 10% and $\frac{1}{4}x$ at 12%. His total income of $6400 can be expressed as
$$0.10\left(\tfrac{1}{2}x\right) + 0.12\left(\tfrac{1}{4}x\right) = 6400$$
$$0.05x + 0.03x = 6400$$
$$0.08x = 6400$$
$$x = 80,000$$
His inheritance was $80,000.

33. Let x = the number of gallons of 5% solution. In the 5% solution there are $0.05x$ gallons of acid, and in the 20 gallons of 10% solution there are $0.10(20)$ gallons of acid. The final mixture consists of $x + 20$ gallons of which 8% is acid. The total acid in the mixture is the sum of the acid from each solution mixed together:
$$0.05x + 0.10(20) = 0.08(x + 20)$$
$$0.05x + 2 = 0.08x + 1.6$$
$$0.4 = 0.03x$$
$$x = \frac{0.4}{0.03} = \frac{40}{3}$$
Use $\frac{40}{3}$ gallons of 5% solution.

35. Let x = the number of gallons of pure acid. After mixing we will have $1 + x$ gallons of 6% solution. The original gallon has $0.05(1)$ gallons of acid in it. The equation totals up the acid:
$$0.05(1) + x = 0.06(1 + x)$$
$$0.05 + x = 0.06 + 0.06x$$

$$0.94x = 0.01$$
$$x = 0.010638 \text{ gallons}$$
Use 1 gallon = 128 ounces, to get approximately 1.36 ounces of pure acid.

37. Let x = his speed in the fog and $x + 30$ = his increased speed. Since $D = RT$, his distance in the fog was $3x$ and his distance later was $6(x + 30)$. The equation gives the total distance:
$$3x + 6(x + 30) = 540$$
$$9x + 180 = 540$$
$$9x = 360$$
$$x = 40$$
His speed in the fog was 40 mph.

39. Let x = the speed of the commuter bus and $x + 25$ = the speed of the express bus. Use $D = RT$ to get the distance traveled by each as $2x$ and $\frac{3}{4}(x + 25)$. The equation expresses the fact that they travel the same distance:
$$2x = \tfrac{3}{4}(x + 25)$$
$$4 \cdot 2x = 4 \cdot \tfrac{3}{4}(x + 25)$$
$$8x = 3x + 75$$
$$5x = 75$$
$$x = 15$$
The speed of the commuter bus was 15 mph.

41. If x = the selling price, then the commission is $0.08x$. The owner gets the selling price minus the commission:
$$x - 0.08x = 80,000$$
$$0.92x = 80,000$$
$$x \approx \$86,957$$

43. If x = the selling price, then $0.07x$ = the amount of sales tax. The selling price plus the sales tax is the total amount paid:
$$x + 0.07x = 9041.50$$
$$1.07x = 9041.50$$
$$x = \$8450$$

45. Let x = the distance from the base line to the service line and $x - 3$ = the distance from the service line to the net. The total is 39 feet:
$$x + x - 3 = 39$$
$$2x = 42$$
$$x = 21$$
The distance from service line to the net is 18 ft.

47. Let $x =$ the number of points scored by the Packers and $x - 25 =$ the number of points scored by the Chiefs.
$$x + x - 25 = 45$$
$$2x = 70$$
$$x = 35$$
The score was Packers 35, Chiefs 10.

49. Let $x =$ the price per pound for the blended coffee. The price of 0.75 lb of Brazilian coffee at $10 per lb is $7.50, and the price of 1.5 lb of Colombian coffee at $8 per lb is $12. The total price for 2.25 lb of blended coffee at x dollars per kg is $2.25x$ dollars. We write an equation expressing the total cost:
$$7.50 + 12 = 2.25x$$
$$19.50 = 2.25x$$
$$x \approx 8.67$$
The blended coffee should sell for $8.67 per lb to the nearest cent.

51. Let $x =$ the number of pounds of apricots. The total cost of 10 pounds of bananas at $3.20 per pound is $32, the total cost of x pounds of apricots at $4 per pound is $4x$, and the total cost $x + 10$ pounds of mix at $3.80 per pound is $3.80(x + 10)$. Write an equation expressing the total cost:
$$32 + 4x = 3.80(x + 10)$$
$$32 + 4x = 3.80x + 38$$
$$0.20x = 6$$
$$x = 30$$
The mix should contain 30 pounds of apricots.

53. Let $x =$ the number of quarts to be drained out. In the original 20 qt radiator there are $0.30(20)$ qts of antifreeze. In the x qts that are drained there are $0.30x$ qts of antifreeze, but in the x qts put back in there are x qts of antifreeze. The equation accounts for all of the antifreeze:
$$0.30(20) - 0.30x + x = 0.50(20)$$
$$6 + 0.70x = 10$$
$$0.7x = 4$$
$$x = \frac{4}{0.7} = \frac{40}{7}$$

The amount drained should be $\frac{40}{7}$ qts.

55. If $x =$ the profit in 2000. Since the profit in 2001 was 18.8% greater and the profit in

2001 was $12,735 million, we have the following.
$$x + 0.188x = 12,735$$
$$1.188x = 12,735$$
$$x = \frac{12,735}{1.188} \approx 10,720$$
The total profit in 2000 was $10,720 million.

57. Let $x =$ Brian's inheritance, $\frac{1}{2}x =$ Daniel's inheritance, and $\frac{1}{3}x - 1000 =$ Raymond's inheritance. The sum of the three amounts is $25,400:
$$x + \frac{1}{2}x + \frac{1}{3}x - 1000 = 25400$$
$$\frac{11}{6}x = 26400$$
$$x = \frac{6}{11} \cdot 26400$$
$$= 14,400$$
$$\frac{1}{2}x = 7200$$
$$\frac{1}{3}x - 1000 = 3800$$
Brian gets $14,400, Daniel $7200, and Raymond $3800.

59. Let $x =$ the first integer and $x + 1 =$ the second. Subtract the larger from twice the smaller to get 21:
$$2x - (x + 1) = 21$$
$$x - 1 = 21$$
$$x = 22$$
$$x + 1 = 23$$
The integers are 22 and 23.

61. Let $x =$ Berenice's time and $x - 2 =$ Jarrett's time. Berenice's distance is $50x$ and Jarrett's distance is $56(x - 2)$.
$$50x + 56(x - 2) = 683$$
$$106x - 112 = 683$$
$$106x = 795$$
$$x = 7.5$$
Berenice drove for 7.5 hours.

63. Let $x =$ the length of a side of the square. She will use x meters of fencing for each of 3 sides, but only $\frac{1}{2}x$ meters for the side with the opening.
$$3x + \frac{1}{2}x = 70$$
$$\frac{7}{2}x = 70$$
$$x = \frac{2}{7} \cdot 70 = 20$$
The square will by 20 meters by 20 meters.

65. Let x = the amount invested at 8% and $3000 - x$ = the amount invested at 10%. Income on the first investment is $0.08x$ and income on the second is $0.10(3000 - x)$. The total income is $290:

$$0.08x + 0.10(3000 - x) = 290$$
$$0.08x + 300 - 0.10x = 290$$
$$-0.02x = -10$$
$$x = \frac{-10}{-0.02} = 500$$
$$3000 - x = 2500$$

She invested $500 at 8% and $2,500 at 10%.

67. Let x = the number of gallons of 5% alcohol and $5 - x$ = the number of gallons of 10% alcohol. The alcohol in the final 5 gallons is the sum of the alcohol in the two separate quantities that are mixed together:

$$0.05x + 0.10(5 - x) = 0.08(5)$$
$$0.05x + 0.5 - 0.10x = 0.4$$
$$-0.05x = -0.1$$
$$x = \frac{-0.1}{-0.05} = 2$$
$$5 - x = 3$$

Use 2 gallons of 5% solution and 3 gallons of 10% solution.

69. Let x = Darla's age now and $78 - x$ = Todd's age now. In 6 years Todd will be $78 - x + 6$ or $84 - x$, and 6 years ago Darla was $x - 6$. Todd's age in 6 years is twice what Darla's age was 6 years ago:

$$84 - x = 2(x - 6)$$
$$84 - x = 2x - 12$$
$$-3x = -96$$
$$x = 32$$
$$78 - x = 46$$

Todd is 46 now and Darla is 32 now.

2.4 WARM-UPS

1. False, $0 = 0$.
2. False, $-300 < -2$.
3. True, because $-60 = -60$.
4. False, since $6 < x$ is equivalent to $x > 6$.
5. False, $-2x < 10$ is equivalent to $x > -5$.
6. False, $3x \geq -12$ is equivalent to $x \geq -4$.
7. True, multiply each side by -1.
8. True, because of the trichotomy property.
9. True, because of the trichotomy property.
10. True, because $3 - 4(-2) \leq 11$ is correct.

2.4 EXERCISES

1. An inequality is a sentence that expresses inequality between two algebraic expressions.
3. If a is less than b, then a lies to the left of b on the number line.
5. When you multiply or divide by a negative number, the inequality symbol is reversed.
7. False, because $-3 > -9$.
9. True, because $0 < 8$.
11. True, because $-60 > -120$.
13. True, because $9 - (-3) = 12$.
15. Yes, because $2(-3) - 4 < 8$ simplifies to $-10 < 8$.
17. No, because $2(5) - 3 \leq 3(5) - 9$ simplifies to $7 \leq 6$.
19. No, because $5 - (-1) < 4 - 2(-1)$ simplifies to $6 < 6$.
21. The solution set is the interval $(-\infty, -1]$. To graph it, shade the numbers to the left of -1 and including -1.

23. The solution set is the interval $(20, \infty)$. To graph it, shade numbers to the right of 20.

25. The solution set is the interval $[3, \infty)$ because $3 \leq x$ is equivalent to $x \geq 3$. We shade the numbers to the right of 3, including 3.

27. The solution set is the interval $(-\infty, 2.3)$. Shade to the left of 2.3.

29. $x + 5 > 12$ is equivalent to $x > 7$.
31. $-x < 6$ is equivalent to $x > -6$.
33. $-2x \geq 8$ is equivalent to $x \leq -4$.
35. $4 < x$ is equivalent to $x > 4$.

37. $7x > -14$
$\quad x > -2$
Solution set is $(-2, \infty)$.

39. $-3x \leq 12$
$x \geq -4$
Solution set is $[-4, \infty)$.

41. $2x - 3 > 7$
$2x > 10$
$x > 5$
Solution set is $(5, \infty)$.

43. $18 \geq 3 - 5x$
$15 \geq -5x$
$-3 \leq x$
$x \geq -3$
Solution set is $[-3, \infty)$.

45. $\frac{x-3}{-5} < -2$
$x - 3 > 10$
$x > 13$
Solution set is $(13, \infty)$.

47. $2 \geq \frac{5 - 3x}{4}$
$8 \geq 5 - 3x$
$3 \geq -3x$
$-1 \leq x$
$x \geq -1$
Solution set is $[-1, \infty)$.

49. $3 - \frac{1}{4}x \geq 2$
$4\left(3 - \frac{1}{4}x\right) \geq 4(2)$
$12 - x \geq 8$
$-x \geq -4$
$x \leq 4$
Solution set is $(-\infty, 4]$.

51. $\frac{1}{4}x - \frac{1}{2} < \frac{1}{2}x - \frac{2}{3}$
$12\left(\frac{1}{4}x - \frac{1}{2}\right) < 12\left(\frac{1}{2}x - \frac{2}{3}\right)$
$3x - 6 < 6x - 8$
$-3x < -2$
$x > \frac{2}{3}$
Solution set is $(2/3, \infty)$.

53. $\frac{y-3}{2} > \frac{1}{2} - \frac{y-5}{4}$
$4 \cdot \frac{y-3}{2} > 4 \cdot \frac{1}{2} - 4 \cdot \frac{y-5}{4}$
$2y - 6 > 2 - y + 5$
$3y > 13$
$y > \frac{13}{3}$
Solution set is $(13/3, \infty)$.

55. $3(x + 2) \leq 9 + 3x$
$3x + 6 \leq 9 + 3x$
$6 \leq 9$
Solution set is $(-\infty, \infty)$.

57. $-2(5x - 1) \leq -5(5 + 2x)$
$-10x + 2 \leq -25 - 10x$
$2 \leq -25$
Solution set is $\emptyset$.

59. $3x - (4 - 2x) < 5 - (2 - 5x)$
$3x - 4 + 2x < 5 - 2 + 5x$
$5x - 4 < 3 + 5x$
$-4 < 3$
Solution set is $(-\infty, \infty)$.

61. $\frac{1}{2}x + \frac{1}{4}x < \frac{1}{8}(6x - 4)$
$\frac{3}{4}x < \frac{3}{4}x - \frac{1}{2}$
$0 < -\frac{1}{2}$
Solution set is $\emptyset$.

63. If $x = $ Tony's height in feet, then $x > 6$.

65. If $s = $ Wilma's salary in dollars, then $s < 80,000$.

67. If $v = $ speed of the Concorde in mph, then $v \leq 1450$.

69. If $a = $ amount in dollars that Julie can afford, then $a \leq 400$.

71. If $b = $ Burt's height in feet, then $b \leq 5$.

73. If $t = $ Tina's hourly wage in dollars, then $t \leq 8.20$.

75. Let x = the price of the car in dollars and $0.08x$ = the amount of tax. To spend less than \$10,000 we must satisfy the inequality

$$x + 0.08x + 172 < 10,000$$
$$1.08x < 9828$$
$$x < 9100$$

The price range for the car is $x < 9100$.

77. Let x = the price of the truck in dollars and $0.09x$ = the amount of sales tax. The total cost of at least \$10,000 is expressed as

$$x + 0.09x + 80 \geq 10,000$$
$$1.09x \geq 9920$$
$$x \geq 9100.9174$$

The price range for the truck is $x \geq 9100.92$ to the nearest cent.

79. a) Decreasing

b)
$$-0.52n + 71.1 < 55$$
$$-0.52n < -16.1$$
$$n > 30.96$$

Round to 31 years and add 31 to 1980 to get 2011 as the first year in which the number of births will be less than 55 per 1000 women.

81. Let x = the final exam score. One-third of the midterm plus two-thirds of the final must be at least 70:

$$\tfrac{1}{3}(56) + \tfrac{2}{3}x \geq 70$$
$$3\left(\tfrac{1}{3}(56) + \tfrac{2}{3}x\right) \geq 3(70)$$
$$56 + 2x \geq 210$$
$$2x \geq 154$$
$$x \geq 77$$

The final exam score must satisfy $x \geq 77$.

83. Let x = the price of a pair of A-Mart jeans and $x + 50$ = the price of a pair of designer jeans. Four pairs of A-Mart jeans cost less than one pair of designer jeans is written as follows.

$$4x < x + 50$$
$$3x < 50$$
$$x < 16.6666$$

The price range for A-Mart jeans is $x < \$16.67$.

85. a) Write the inequality $x \geq 2$ and subtract -6 from each side to get

$$x - (-6) \geq 2 - (-6), \text{ or } x + 6 \geq 8. \text{ So the}$$

results are all greater than or equal to 8, which is the interval $[8, \infty)$.

b) Write the inequality $x < -3$ and multiply each side by 2 to get $2x < -6$. So the results are all less than -6, which is the interval $(-\infty, -6)$.

c) If every number in $(8, \infty)$ is divided by 4 the result is the interval $(2, \infty)$.

d) If every number in $(6, \infty)$ is multiplied by -2, the result is the interval $(-\infty, -12)$

e) If every number in $(-\infty, -10)$ is divided by -5, the result is the interval $(2, \infty)$.

2.5 WARM-UPS

1. True, because both inequalities are true.

2. True, because both inequalities are correct.

3. False, because $3 > 5$ is incorrect.

4. True, because $3 \leq 10$ is correct.

5. True, because both inequalities are correct.

6. True, because both are correct.

7. False, because $0 < -2$ is incorrect.

8. True, because only numbers larger than 8 are larger than 3 and larger than 8.

9. False, because $(3, \infty) \cup [8, \infty) = (3, \infty)$.

10. True, because the numbers greater than -2 and less than 9 are between -2 and 9.

2.5 EXERCISES

1. A compound inequality consists of two inequalities joined with the words "and" or "or."

3. A compound inequality using or is true when either one or the other or both inequalities is true.

5. The inequality $a < b < c$ means that $a < b$ and $b < c$.

7. No, because $-6 > -3$ is incorrect.

9. Yes, because both inequalities are correct.

11. No, because both inequalities are incorrect.

13. No, because $-4 > -3$ is incorrect.

15. Yes, because $-4 - 3 \geq -7$ is correct.

17. Yes, because $2(-4) - 1 < -7$ is correct.

19. The solution set is the set of numbers between -1 and 4:

23

21. Numbers that satisfy $x \geq 2$ or $x \geq 5$ are greater than or equal to 2. Graph the union of the two solution sets:

23. The union of $(-\infty, 6]$ with $(-2, \infty)$ is the interval $(-\infty, \infty)$. The union of the two solution sets consists of all real numbers:

25. The solution set is $\emptyset$, because no number is greater than 9 and less than or equal to 6. There is no graph.

27. The union of the two solution sets is graphed as follows:

29. The solution set is $\emptyset$, because there is no intersection to the two solution sets. There is no graph.

31. $x - 3 > 7$ or $3 - x > 2$
$\qquad x > 10$ or $\quad -x > -1$
$\qquad x > 10$ or $\qquad x < 1$
$(-\infty, 1) \cup (10, \infty)$

33. $3 < x$ and $1 + x > 10$
$\qquad x > 3$ and $\qquad x > 9$
$(9, \infty)$

35. $\frac{1}{2}x > 5$ or $-\frac{1}{3}x < 2$
$\qquad x > 10$ or $\quad x > -6$
$(-6, \infty)$

37. $2x - 3 \leq 5$ and $x - 1 > 0$
$\qquad 2x \leq 8$ and $\qquad x > 1$
$\qquad x \leq 4$ and $\qquad x > 1$
$(1, 4]$

39. $\frac{1}{2}x - \frac{1}{3} \geq -\frac{1}{6}$ or $\frac{2}{7}x \leq \frac{1}{10}$
$\qquad 3x - 2 \geq -1$ or $\quad x \leq \frac{7}{2} \cdot \frac{1}{10}$

$x \geq \frac{1}{3}$ or $x \leq \frac{7}{20}$
$(-\infty, \infty)$

41. $0.5x < 2$ and $-0.6x < -3$
$\qquad x < 4$ and $\qquad x > 5$
The solution set is $\emptyset$, because there are no numbers that are less than 4 and greater than 5. There is no graph.

43. $5 < 2x - 3 < 11$
$\qquad 5 + 3 < 2x - 3 + 3 < 11 + 3$
$\qquad 8 < 2x < 14$
$\qquad 4 < x < 7$
$(4, 7)$

45. $-1 < 5 - 3x \leq 14$
$\qquad -6 < -3x \leq 9$
$\qquad \frac{-6}{-3} > \frac{-3x}{-3} \geq \frac{9}{-3}$
$\qquad 2 > x \geq -3$
$\qquad -3 \leq x < 2$
$[-3, 2)$

47. $-3 < \frac{3m + 1}{2} \leq 5$
$\qquad 2(-3) < 2 \cdot \frac{3m + 1}{2} \leq 2 \cdot 5$
$\qquad -6 < 3m + 1 \leq 10$
$\qquad -7 < 3m \leq 9$
$\qquad -\frac{7}{3} < m \leq 3$
$(-7/3, 3]$

49. $-2 < \frac{1 - 3x}{-2} < 7$
$\qquad -2(-2) > -2 \cdot \frac{1 - 3x}{-2} > -2(7)$
$\qquad 4 > 1 - 3x > -14$
$\qquad 3 > -3x > -15$
$\qquad -1 < x < 5$
$(-1, 5)$

24

51.
$$3 \le 3 - 5(x - 3) \le 8$$
$$3 \le 3 - 5x + 15 \le 8$$
$$3 \le 18 - 5x \le 8$$
$$-15 \le -5x \le -10$$
$$3 \ge x \ge 2$$
$$2 \le x \le 3$$
$[2, 3]$

53. $(2, \infty) \cup (4, \infty) = (2, \infty)$
55. $(-\infty, 5) \cap (-\infty, 9) = (-\infty, 5)$
57. $(-\infty, 4] \cap [2, \infty) = [2, 4]$
59. $(-\infty, 5) \cup [-3, \infty) = (-\infty, \infty)$
61. $(3, \infty) \cap (-\infty, 3] = \emptyset$
63. $(3, 5) \cap [4, 8) = [4, 5)$
65. $[1, 4) \cup (2, 6] = [1, 6]$
67. The graph shows real numbers to the right of 2: $x > 2$
69. The graph shows the real numbers to the left of 3: $x < 3$
71. This graph is the union of the numbers greater than 2 with the numbers less than or equal to -1: $x > 2$ or $x \le -1$
73. This graph shows real numbers between -2 and 3, including -2: $-2 \le x < 3$
75. The graph shows real numbers greater than or equal to -3: $x \ge -3$
77. Let $x =$ the final exam score. We write an inequality expressing the fact that 1/3 of the midterm plus 2/3 of the final must be between 70 and 79 inclusive.
$$70 \le \tfrac{1}{3} \cdot 64 + \tfrac{2}{3}x \le 79$$
$$210 \le 64 + 2x \le 237$$
$$146 \le 2x \le 173$$
$$73 \le x \le 86.5$$
79. Let $x =$ the price of the truck. The total spent will be $x + 0.08x + 84$.
$$12,000 \le x + 0.08x + 84 \le 15,000$$
$$12,000 \le 1.08x + 84 \le 15,000$$
$$11,916 \le 1.08x \le 14,916$$
$$\$11,033 \le x \le \$13,811$$
81. Let $x =$ the number of cigarettes smoked on the run, giving the equivalent of $\tfrac{1}{2}x$ cigarettes smoked. Thus, she smokes $3 + \tfrac{1}{2}x$ whole cigarettes per day and this number is between 5 and 12 inclusive:

$$5 \le 3 + \tfrac{1}{2}x \le 12$$
$$10 \le 6 + x \le 24$$
$$4 \le x \le 18$$
She smokes from 4 to 18 cigarettes on the run.
83. a) In 2000, we have $n = 10$.
$$16.45(10) + 1062.45 = 1226.95$$
In 2000, there were 1,226,950 bachelors degrees awarded.
b) $\quad 16.45n + 1062.45 = 1,400$
$$16.45n = 337.55$$
$$n \approx 21$$
$$1990 + 21 = 2011$$
c) $16.45n + 1062.45 > 1,400$
$$16.45n > 337.55$$
$$n > 20.52$$

$$7.79n + 326.82 > 550$$
$$7.79n > 223.18$$
$$n > 28.64$$
Both happen in the year $1990 + 29$, or 2019.
d) Either happens in the year $1990 + 21$, or 2011.
85. If $a < b$ and $a < -x < b$, we can multiply each part of this inequality by -1 to get $-a > x > -b$ or $-b < x < -a$. In words, x is between $-b$ and $-a$.
87. a) If $3 < x < 8$, then $12 < 4x < 32$. $(12, 32)$
b) If $-2 \le x < 4$, then
$$(-5)(-2) \ge -5x > (-5)(4)$$
$$10 \ge -5x > -20$$
$$-20 < -5x \le 10.$$
$(-20, 10]$
c) If $-3 < x < 6$, then $0 < x + 3 < 9$. $(0, 9)$
d) If $3 \le x \le 9$, then
$$\frac{3}{-3} \ge \frac{x}{-3} \ge \frac{9}{-3}$$
$$-1 \ge \frac{x}{-3} \ge -3$$
$$-3 \le \frac{x}{-3} \le -1.$$
$[-3, -1]$

2.6 WARM-UPS

1. True, because both 2 and -2 have absolute value 2.
2. False, because $|x| = 0$ has only one solution and $|x| = -1$ has no solutions.

3. False, because it is equivalent to $2x - 3 = 7$ or $2x - 3 = -7$.

4. True, because $|x| > 5$ means that x is more than 5 units away from 0 and that is true to the right of 5 or to the left of -5.

5. False, because this equation has no solution.

6. True, because only 3 satisfies the equation.

7. False, because only inequalities that express x between two numbers are written this way.

8. False, because $|x| < 7$ is equivalent to $-7 < x < 7$.

9. True, because 2 is subtracted from each side.

10. False, because if $x = 0$, then $|x| = 0$ and 0 is not positive.

2.6 EXERCISES

1. Absolute value of a number is the number's distance from 0 on the number line.

3. Since both 4 and -4 are four units from 0, $|x| = 4$ has two solutions.

5. Since the distance from 0 for every number on the number line is greater than or equal to 0, $|x| \geq 0$.

7. $|a| = 5$
$a = 5$ or $a = -5$
Solution set: $\{-5, 5\}$

9. $|x - 3| = 1$
$\quad x - 3 = 1 \quad$ or $\quad x - 3 = -1$
$\qquad x = 4 \quad$ or $\qquad x = 2$
Solution set: $\{2, 4\}$

11. $|3 - x| = 6$
$\quad 3 - x = 6 \quad$ or $\quad 3 - x = -6$
$\qquad -x = 3 \quad$ or $\qquad -x = -9$
$\qquad x = -3 \quad$ or $\qquad x = 9$
Solution set: $\{-3, 9\}$

13. $|3x - 4| = 12$
$\quad 3x - 4 = 12$ or $3x - 4 = -12$
$\qquad 3x = 16 \qquad$ or $\qquad 3x = -8$
$\qquad x = \frac{16}{3} \qquad$ or $\qquad x = -\frac{8}{3}$
Solution set: $\left\{-\frac{8}{3}, \frac{16}{3}\right\}$

15. $\left|\frac{2}{3}x - 8\right| = 0$
$\qquad \frac{2}{3}x - 8 = 0$
$\qquad \frac{2}{3}x = 8$
$\qquad x = 12$

Solution set: $\{12\}$

17. $6 - 0.2x = 10 \qquad$ or $\quad 6 - 0.2x = -10$
$\quad -0.2x = 4 \qquad$ or $\qquad -0.2x = -16$
$\qquad x = -20 \quad$ or $\qquad\qquad x = 80$
Solution set: $\{-20, 80\}$

19. Since absolute value is nonnegative, the solution set is $\emptyset$.

21. $\qquad |2(x - 4) + 3| = 5$
$2(x - 4) + 3 = 5 \qquad$ or $2(x - 4) + 3 = -5$
$\quad 2x - 5 = 5 \qquad$ or $\quad 2x - 5 = -5$
$\qquad 2x = 10 \qquad$ or $\qquad 2x = 0$
$\qquad x = 5 \qquad$ or $\qquad x = 0$
Solution set: $\{0, 5\}$

23. $|7.3x - 5.26| = 4.215$
$\qquad 7.3x - 5.26 = 4.215$
$\qquad\qquad 7.3x = 9.475$
$\qquad\qquad\qquad x \approx 1.298$
or $\quad 7.3x - 5.26 = -4.215$
$\qquad\qquad 7.3x = 1.045$
$\qquad\qquad\qquad x \approx 0.143$
Solution set: $\{0.143, 1.298\}$

25. $3 + |x| = 5$
$\qquad |x| = 2$
$\quad x = 2$ or $x = -2$
Solution set: $\{-2, 2\}$

27. $2 - |x + 3| = -6$
$\qquad -|x + 3| = -8$
$\qquad |x + 3| = 8$
$\quad x + 3 = 8 \quad$ or $x + 3 = -8$
$\qquad x = 5 \quad$ or $\qquad x = -11$
Solution set: $\{-11, 5\}$

29. $5 - \dfrac{|3 - 2x|}{3} = 4$
$\quad 15 - |3 - 2x| = 12$
$\qquad -|3 - 2x| = -3$
$\qquad |3 - 2x| = 3$
$\quad 3 - 2x = 3 \quad$ or $\quad 3 - 2x = -3$
$\quad -2x = 0 \qquad$ or $\qquad -2x = -6$
$\qquad x = 0 \qquad$ or $\qquad x = 3$
The solution set is $\{0, 3\}$.

31. $x - 5 = 2x + 1 \qquad$ or $\quad x - 5 = -(2x + 1)$
$\qquad -6 = x \qquad\qquad$ or $\quad x - 5 = -2x - 1$
$\qquad\qquad\qquad\qquad\qquad\qquad 3x = 4$
$\qquad\qquad\qquad\qquad\qquad\qquad x = 4/3$
Solution set: $\left\{-6, \frac{4}{3}\right\}$

33. $\left|\frac{5}{2} - x\right| = \left|2 - \frac{x}{2}\right|$

$\frac{5}{2} - x = 2 - \frac{x}{2}$ or $\frac{5}{2} - x = -\left(2 - \frac{x}{2}\right)$

$5 - 2x = 4 - x$ or $\frac{5}{2} - x = -2 + \frac{x}{2}$

$1 = x$ or $5 - 2x = -4 + x$

$-3x = -9$

$x = 3$

Solution set: $\{1, 3\}$

35. $|x - 3| = |3 - x|$

$x - 3 = 3 - x$ or $x - 3 = -(3 - x)$

$2x = 6$ or $x - 3 = -3 + x$

$x = 3$ or $x - 3 = x - 3$

The second equation is an identity. So the solution set is the set of real numbers, which is written as R or $(-\infty, \infty)$.

37. The graph shows the real numbers between -2 and 2: $|x| < 2$

39. The graph shows numbers greater than 3 or less than -3: $|x| > 3$

41. The graph shows numbers between -1 and 1 inclusive: $|x| \leq 1$

43. The graph shows numbers two or more units away from 0: $|x| \geq 2$

45. No, because $|x| < 3$ is equivalent to $-3 < x < 3$.

47. Yes.

49. No, because $|x - 3| \geq 1$ is equivalent to $x - 3 \geq 1$ or $x - 3 \leq -1$.

51. Yes, because the following compound inequalities are equivalent.

$|4 - x| < 1$

$4 - x < 1$ and $4 - x > -1$

$4 - x < 1$ and $-(4 - x) < 1$

53. $|x| > 6$

$x > 6$ or $x < -6$

$(-\infty, -6) \cup (6, \infty)$

$\begin{array}{c}\leftarrow\!\!+\!)\!\!+\!+\!+\!+\!+\!+\!+\!+\!+\!(\!+\!\!\rightarrow \\ {\scriptstyle -8\,-6\,-4\,-2\ 0\ 2\ 4\ 6\ 8}\end{array}$

55. $|2a| < 6$

$-6 < 2a < 6$

$-3 < a < 3$

$(-3, 3)$

$\begin{array}{c}\leftarrow\!(\!+\!+\!+\!+\!+\!)\!\!\rightarrow \\ {\scriptstyle -3\,-2\,-1\ 0\ 1\ 2\ 3}\end{array}$

57. $x - 2 \geq 3$ or $x - 2 \leq -3$

$x \geq 5$ or $x \leq -1$

$(-\infty, -1] \cup [5, \infty)$

59. $\frac{1}{5}|2x - 4| < 1$

$|2x - 4| < 5$

$-5 < 2x - 4 < 5$

$-1 < 2x < 9$

$-\frac{1}{2} < x < \frac{9}{2}$

$\left(-\frac{1}{2}, \frac{9}{2}\right)$

61. $-2|5 - x| \geq -14$

$|5 - x| \leq 7$

$-7 \leq 5 - x \leq 7$

$-12 \leq -x \leq 2$

$12 \geq x \geq -2$

$[-2, 12]$

63. $2|3 - 2x| - 6 \geq 18$

$2|3 - 2x| \geq 24$

$|3 - 2x| \geq 12$

$3 - 2x \geq 12$ or $3 - 2x \leq -12$

$-2x \geq 9$ or $-2x \leq -15$

$x \leq -\frac{9}{2}$ or $x \geq \frac{15}{2}$

$(-\infty, -\frac{9}{2}] \cup [\frac{15}{2}, \infty)$

$\begin{array}{c}\leftarrow\!]\!+\!+\!+\!+\!+\!+\!+\!+\!+\!+\![\!\!\rightarrow \\ {\scriptstyle -4\,-2\ 0\ 2\ 4\ 6\ 8}\end{array}$

65. $|x - 2| > 0$, except when $x = 2$.

$(-\infty, 2) \cup (2, \infty)$

$\begin{array}{c}\leftarrow\!+\!+\!+\!×\!+\!+\!+\!\rightarrow \\ {\scriptstyle -1\ 0\ 1\ 2\ 3\ 4\ 5}\end{array}$

67. Since the absolute value of any quantity is greater than or equal to zero, the solution set is R, or $(-\infty, \infty)$.

69. $-2|3x - 7| > 6$ is equivalent to $|3x - 7| < -3$.

Absolute value of an expression cannot be less than a negative number. Solution set is $\emptyset$.

71. $|2x + 3| + 6 > 0$

$|2x + 3| > -6$

Since the absolute value of any expression is greater than or equal to zero, it is greater than any negative number. The solution set is R or $(-\infty, \infty)$.

73.
$$1 < |x+2|$$
$$|x+2| > 1$$
$$x+2 > 1 \quad \text{or} \quad x+2 < -1$$
$$x > -1 \quad \text{or} \quad x < -3$$
$$(-\infty, -3) \cup (-1, \infty)$$

75.
$$5 > |x+1|$$
$$|x| + 1 < 5$$
$$|x| < 4$$
$$-4 < x < 4$$
$$(-4, 4)$$

77.
$$3 - 5|x| > -2$$
$$-5|x| > -5$$
$$|x| < 1$$
$$-1 < x < 1$$
$$(-1, 1)$$

79.
$$|5.67x - 3.124| < 1.68$$
$$-1.68 < 5.67x - 3.124 < 1.68$$
$$1.444 < 5.67x < 4.804$$
$$0.255 < x < 0.847$$
$$(0.255, 0.847)$$

81. Let $x =$ the year of the battle of Orleans. Since the difference between x and 1415 is 14 years, we have $|x - 1415| = 14$.
$$x - 1415 = 14 \quad \text{or} \quad x - 1415 = -14$$
$$x = 1429 \quad \text{or} \quad x = 1401$$
The battle Agincourt was either in 1401 or 1429.

83. Let $x =$ the weight of Kathy. The difference between their weights is less than 6 pounds is expressed by the absolute value inequality
$$|x - 127| < 6$$
$$-6 < x - 127 < 6$$
$$121 < x < 133$$
Kathy weighs between 121 and 133 pounds.

85. a) From the graph it appears that the balls are at the same height when $t = 1$ second.
b) Height of first ball is $S = -16t^2 + 50t$ and height of second ball is
$S = -16t^2 + 40t + 10$.
When the balls are at the same height we have
$$-16t^2 + 50t = -16t^2 + 40t + 10$$
$$50t = 40t + 10$$
$$10t = 10$$
$$t = 1$$
The balls are at the same height when $t = 1$ sec.

c) The difference between the heights is less than 5 feet when
$$|-16t^2 + 50t - (-16t^2 + 40t + 10)| < 5$$
$$|10t - 10| < 5$$
$$-5 < 10t - 10 < 5$$
$$5 < 10t < 15$$
$$0.5 < t < 1.5$$

87. a) The equation $|m - n| = |n - m|$ is satisfied for all real numbers, because $m - n$ and $n - m$ are opposites of each other and opposites always have the same absolute value. So both m and n can be in the interval $(-\infty, \infty)$.
b) $|mn| = |m| \cdot |n|$ is satisfied for all real numbers, because of the rules for multiplying real numbers. So both m and n can be in the interval $(-\infty, \infty)$.
c) Since you cannot have 0 in a denominator, the equation is satisfied by all real numbers except if $n = 0$.

Enriching Your Mathematical Word Power
1. c **2.** b **3.** d **4.** c
5. c **6.** d **7.** d **8.** a
9. d **10.** d **11.** d **12.** c
13. d

CHAPTER 2 REVIEW

1.
$$2x - 7 = 9$$
$$2x = 16$$
$$x = 8$$
Solution set: $\{8\}$

3.
$$5 - 4x = 11$$
$$-4x = 6$$
$$x = \frac{6}{-4} = -\frac{3}{2}$$
Solution set: $\left\{-\frac{3}{2}\right\}$

5. $x - 6 - (x - 6) = 0$
$$x - 6 - x + 6 = 0$$
$$0 = 0$$
Solution set: R or $(-\infty, \infty)$

7.
$$2(x - 3) - 5 = 5 - (3 - 2x)$$
$$2x - 6 - 5 = 5 - 3 + 2x$$
$$2x - 11 = 2 + 2x$$
$$-11 = 2$$
Solution set: $\emptyset$

9. $\frac{3}{17}x = 0$

$\quad\quad x = 0$

Solution set: $\{0\}$

11. $\frac{1}{4}x - \frac{1}{5} = \frac{1}{5}x + \frac{4}{5}$

$20\left(\frac{1}{4}x - \frac{1}{5}\right) = 20\left(\frac{1}{5}x + \frac{4}{5}\right)$

$\quad\quad 5x - 4 = 4x + 16$

$\quad\quad\quad\quad x = 20$

Solution set: $\{20\}$

13.

$\quad\quad \frac{t}{2} - \frac{t-2}{3} = \frac{3}{2}$

$6\left(\frac{t}{2}\right) - 6\left(\frac{t-2}{3}\right) = 6\left(\frac{3}{2}\right)$

$\quad\quad 3t - 2t + 4 = 9$

$\quad\quad\quad\quad t + 4 = 9$

$\quad\quad\quad\quad\quad\quad t = 5$

Solution set: $\{5\}$

15. $1 - 0.4(x - 4) + 0.6(x - 7) = -0.6$

$1 - 0.4x + 1.6 + 0.6x - 4.2 = -0.6$

$\quad\quad\quad\quad\quad 0.2x - 1.6 = -0.6$

$\quad\quad\quad\quad\quad\quad\quad 0.2x = 1$

$\quad\quad\quad\quad\quad\quad\quad\quad x = 5$

The solution set is $\{5\}$.

17. $\quad ax + b = 0$

$\quad\quad\quad ax = -b$

$\quad\quad\quad\quad x = \frac{-b}{a}$

$\quad\quad\quad\quad x = -\frac{b}{a}$

19. $\quad ax + 2 = cx$

$\quad ax - cx = -2$

$\quad x(a - c) = -2$

$\quad \frac{x(a-c)}{a-c} = \frac{-2}{a-c}$

$\quad\quad\quad\quad x = \frac{2}{c-a}$

21. $\quad mwx = P$

$\quad \frac{mwx}{mw} = \frac{P}{mw}$

$\quad\quad\quad x = \frac{P}{mw}$

23. $\quad\quad \frac{1}{x} + \frac{1}{2} = w$

$2x\left(\frac{1}{x}\right) + 2x\left(\frac{1}{2}\right) = 2x(w)$

$\quad\quad\quad\quad 2 + x = 2xw$

$\quad\quad\quad x - 2xw = -2$

$\quad\quad x(1 - 2w) = -2$

$\quad\quad\quad\quad x = \frac{-2}{1 - 2w}$

$\quad\quad\quad\quad x = \frac{2}{2w - 1}$

25. $\quad 3x - 2y = -6$

$\quad\quad -2y = -3x - 6$

$\quad\quad\quad y = \frac{-3x - 6}{-2}$

$\quad\quad\quad y = \frac{3}{2}x + 3$

27. $\quad y - 2 = -\frac{1}{3}x + 2$

$\quad\quad\quad y = -\frac{1}{3}x + 4$

29. $\frac{1}{2}x - \frac{1}{4}y = 5$

$4\left(\frac{1}{2}x - \frac{1}{4}y\right) = 4(5)$

$\quad\quad 2x - y = 20$

$\quad\quad\quad -y = -2x + 20$

$\quad\quad\quad\quad y = 2x - 20$

31. Let $W =$ the width and $W + 5.5 =$ the length. Since $2L + 2W = P$, we have

$\quad 2(W + 5.5) + 2W = 45$

$\quad\quad 2W + 11 + 2W = 45$

$\quad\quad\quad\quad\quad\quad 4W = 34$

$\quad\quad\quad\quad\quad\quad\quad W = 8.5$

Length is 14 inches and width is 8.5 inches.

33. Let $x =$ the wife's income and $x + 8000 =$ Roy's income. Roy saves $0.10(x + 8000)$ and his wife saves $0.08x$. Since the total saved is \$5660, we can write

$\quad 0.10(x + 8000) + 0.08x = 5660$

$\quad\quad 0.10x + 800 + 0.08x = 5660$

$\quad\quad\quad\quad\quad\quad\quad 0.18x = 4860$

$\quad\quad\quad\quad\quad\quad\quad\quad x = 27,000$

Roy's wife earns \$27,000 and Roy earns \$35,000.

35. Let $x =$ the list price and $0.20x =$ the discount. The list price minus the discount is equal to \$7,600.

$\quad\quad x - 0.20x = 7600$

$\quad\quad\quad\quad 0.80x = 7600$

$\quad\quad\quad\quad\quad\quad x = 9500$

The list price was \$9500.

37. Let $x =$ the number of nickels and $15 - x =$ the number of dimes. The value of x nickels is $5x$ cents and the value of $15 - x$ dimes is $10(15 - x)$ cents. Since she has a total of 95 cents, we can write the equation

$$5x + 10(15 - x) = 95$$
$$5x + 150 - 10x = 95$$
$$-5x = -55$$
$$x = 11$$
$$15 - x = 4$$

She has 11 nickels and 4 dimes.

39. Let $x =$ her walking speed and $x + 9 =$ her riding speed. Since she walked for 3 hours, $3x$ is the number of miles that she walked. Since she rode for 5 hours, $5(x + 9)$ is the number of miles that she rode. Her total distance was 85 miles.

$$3x + 5(x + 9) = 85$$
$$3x + 5x + 45 = 85$$
$$8x = 40$$
$$x = 5$$

She walked 5 miles per hour for 3 hours, so she walked 15 miles.

41. $\quad 3 - 4x < 15$
$$-4x < 12$$
$$x > -3$$
$(-3, \infty)$

43. $\quad 2(x - 3) > -6$
$$2x - 6 > -6$$
$$2x > 0$$
$$x > 0$$
$(0, \infty)$

45. $\quad -\frac{3}{4}x \geq 6$
$$-3x \geq 24$$
$$x \leq -8$$
$(-\infty, -8]$

47. $\quad 3(x + 2) > 5(x - 1)$
$$3x + 6 > 5x - 5$$
$$-2x > -11$$

$$x < \frac{11}{2}$$
$(-\infty, 11/2)$

49. $\quad \frac{1}{2}x + 7 \leq \frac{3}{4}x - 5$
$$4\left(\frac{1}{2}x + 7\right) \leq 4\left(\frac{3}{4}x - 5\right)$$
$$2x + 28 \leq 3x - 20$$
$$48 \leq x$$
$[48, \infty)$

51. $\quad x + 2 > 3 \qquad$ or $\quad x - 6 < -10$
$$x > 1 \qquad \text{or} \qquad x < -4$$
$(-\infty, -4) \cup (1, \infty)$

53. $\quad x > 0$ and $x - 6 < 3$
$$x > 0 \text{ and } x < 9$$
$(0, 9)$

55. $\quad 6 - x < 3 \qquad$ or $\quad -x < 0$
$$-x < -3 \qquad \text{or} \qquad -x < 0$$
$$x > 3 \qquad \text{or} \qquad x > 0$$
$(0, \infty)$

57. $\quad 2x < 8$ and $2(x - 3) < 6$
$$2x < 8 \text{ and } 2x - 6 < 6$$
$$x < 4 \quad \text{and} \qquad 2x < 12$$
$$x < 4 \quad \text{and} \qquad x < 6$$
$(-\infty, 4)$

59. $\quad x - 6 > 2$ and $6 - x > 0$
$$x > 8 \qquad \text{and} \qquad -x > -6$$
$$x > 8 \qquad \text{and} \qquad x < 6$$
No number is greater than 8 and less than 6.
$\emptyset$

61. $\quad 0.5x > 10$ or $0.1x < 3$
$$x > 20 \qquad \text{or} \qquad x < 30$$
Every number is either greater than 20 or less than 30. Solution set is R or $(-\infty, \infty)$

63. $-2 \leq \dfrac{2x-3}{10} \leq 1$

$10(-2) \leq 10 \cdot \dfrac{2x-3}{10} \leq 10(1)$

$-20 \leq 2x - 3 \leq 10$

$-17 \leq 2x \leq 13$

$-\dfrac{17}{2} \leq x \leq \dfrac{13}{2}$

$\left[-\dfrac{17}{2}, \dfrac{13}{2}\right]$

65. $[1, 4) \cup (2, \infty) = [1, \infty)$

67. $(3, 6) \cap [2, 8] = (3, 6)$

69. $(-\infty, 5) \cup [5, \infty) = (-\infty, \infty)$

71. $(-3, -1] \cap [-2, 5] = [-2, -1]$

73. $|x| + 2 = 16$

$|x| = 14$

$x = 14$ or $x = -14$

$\{-14, 14\}$

75. $|4x - 12| = 0$

$4x - 12 = 0$

$4x = 12$

$x = 3$

$\{3\}$

77. Since $|x| \geq 0$ for any real number, the solution set is $\emptyset$.

79. $|2x - 1| - 3 = 0$

$|2x - 1| = 3$

$2x - 1 = 3$ or $2x - 1 = -3$

$2x = 4$ or $2x = -2$

$x = 2$ or $x = -1$

$\{-1, 2\}$

81. $|2x| \geq 8$

$2x \geq 8$ or $2x \leq -8$

$x \geq 4$ or $x \leq -4$

$(-\infty, -4] \cup [4, \infty)$

83. $\left|1 - \dfrac{x}{5}\right| > \dfrac{9}{5}$

$1 - \dfrac{x}{5} > \dfrac{9}{5}$ or $1 - \dfrac{x}{5} < -\dfrac{9}{5}$

$5 - x > 9$ or $5 - x < -9$

$-x > 4$ or $-x < -14$

$x < -4$ or $x > 14$

$(-\infty, -4) \cup (14, \infty)$

85. Since $|x - 3| \geq 0$ for any value of x, the solution set is $\emptyset$.

87. Since $|x + 4| \geq 0$ for any value of x, $|x + 4| \geq -1$ for any x. Solution set: R

89. $1 - \dfrac{3}{2}|x - 2| < -\dfrac{1}{2}$

$2 - 3|x - 2| < -1$

$-3|x - 2| < -3$

$|x - 2| > 1$

$x - 2 > 1$ or $x - 2 < -1$

$x > 3$ or $x < 1$

$(-\infty, 1) \cup (3, \infty)$

91. Let $x =$ the rental price, $0.45x =$ the overhead per tape, and $x - 0.45x$ or $0.55x =$ the profit per tape. The rental price must be less than or equal to \$5 and satisfy the inequality

$0.55x \geq 1.65$

$x \geq 3$

The range of the rental price is \$3 $\leq x \leq$ \$5.

93. Since $150 < h < 180$, we have

$150 < 60.089 + 2.238F < 180$

$89.911 < 2.238F < 119.911$

$40.2 < F < 53.6$

The length of the femur is in $(40.2, 53.6)$.

95. Let $x =$ the number on the mile marker where Dane was picked up. We can write the absolute value equation

$|x - 86| = 5$

$x - 86 = 5$ or $x - 86 = -5$

$x = 91$ or $x = 81$

He was either at 81 or 91.

97. $b = 0.20(300{,}000 - b)$

$b = 60{,}000 - 0.20b$

$1.2b = 60{,}000$

$b = 50{,}000$

Bonus is \$50,000 according to the accountant, and \$60,000 according the employees.

99. Let $x =$ the number of cows in Washington County and $3600 - x =$ the number of cows in Cade County. We have the equation

$$0.30x + 0.60(3600 - x) = 0.50(3600)$$
$$-0.30x + 2160 = 1800$$
$$-0.30x = -360$$
$$x = 1200$$
$$3600 - x = 2400$$

There are 1200 cows in Washington County and 2400 in Cade County.

101. The numbers to the right of 1 are described by the inequality $x > 1$.

103. The number 2 satisfies the equation $|x - 2| = 0$.

105. The numbers 3 and -3 both satisfy the equation $|x| = 3$.

107. The numbers to the left of and including -1 satisfy $x \leq -1$.

109. The numbers between -2 and 2 including the endpoints satisfy $|x| \leq 2$.

111. $x \leq 2$ or $x \geq 7$

113. The numbers greater than 3 or less than -3 satisfy $|x| > 3$.

115. $5 < x < 7$

117. Every number except 0 has a positive absolute value and satisfies $|x| > 0$.

CHAPTER 2   TEST

1. $-10x - 5 + 4x = -4x + 3$
$$-6x - 5 = -4x + 3$$
$$-2x = 8$$
$$x = -4$$
Solution set: $\{-4\}$

2. $\dfrac{y}{2} - \dfrac{y - 3}{3} = \dfrac{y + 6}{6}$
$$6\left(\dfrac{y}{2}\right) - 6\left(\dfrac{y - 3}{3}\right) = 6\left(\dfrac{y + 6}{6}\right)$$
$$3y - 2y + 6 = y + 6$$
$$y + 6 = y + 6$$
The equation is an identity.   Solution set: R

3. $|w| + 3 = 9$
$$|w| = 6$$
$$w = 6 \text{ or } w = -6$$
Solution set: $\{-6, 6\}$

4. $|3 - 2(5 - x)| = 3$
$$|-7 + 2x| = 3$$
$$-7 + 2x = 3 \text{ or } -7 + 2x = -3$$
$$2x = 10 \qquad \text{or} \qquad 2x = 4$$
$$x = 5 \qquad \text{or} \qquad x = 2$$
Solution set: $\{2, 5\}$

5. $2x - 5y = 20$
$$-5y = -2x + 20$$
$$y = \tfrac{2}{5}x - 4$$

6. $y = 3xy + 5$
$$y - 3xy = 5$$
$$y(1 - 3x) = 5$$
$$y = \dfrac{5}{1 - 3x}$$

7. $|m - 6| \leq 2$
$$-2 \leq m - 6 \leq 2$$
$$4 \leq m \leq 8$$
$[4, 8]$

8. $2|x - 3| - 5 > 15$
$$2|x - 3| > 20$$
$$|x - 3| > 10$$
$$x - 3 > 10 \qquad \text{or } x - 3 < -10$$
$$x > 13 \qquad \text{or} \qquad x < -7$$
$(-\infty, -7) \cup (13, \infty)$

9. $2 - 3(w - 1) < -2w$
$$2 - 3w + 3 < -2w$$
$$-w < -5$$
$$w > 5$$
$(5, \infty)$

10. $2 < \dfrac{5 - 2x}{3} < 7$
$$3(2) < 3\left(\dfrac{5 - 2x}{3}\right) < 3(7)$$
$$6 < 5 - 2x < 21$$
$$1 < -2x < 16$$
$$-\tfrac{1}{2} > x > -8$$
$\left(-8, -\tfrac{1}{2}\right)$

11. $3x - 2 < 7 \qquad$ and $-3x \leq 15$
$$3x < 9 \qquad \text{and} \qquad x \geq -5$$
$$x < 3 \qquad \text{and} \qquad x \geq -5$$
$[-5, 3)$

32

12. $\frac{2}{3}y < 4$ or $y - 3 < 12$

$\frac{3}{2}\left(\frac{2}{3}y\right) < \frac{3}{2}(4)$ or $y - 3 < 12$

$y < 6$ or $y < 15$

$(-\infty, 15)$

11 12 13 14 15 16 17

13. The equation $\mid 2x - 7 \mid \; = -3$ has no solution because the absolute value of any real number is greater than or equal to zero. The solution set is $\emptyset$.

14. $x - 4 > 1$ or $x < 12$

      $x > 5$ or $x < 12$

$(-\infty, \infty)$

15. $3x < 0$ and $x - 5 > 2$

      $x < 0$ and $x > 7$

No real number is both less than 0 and greater than 7. Solution set: $\emptyset$

16. Since no real number satisfies $|2x - 5| < 0$, we need only solve $|2x - 5| = 0$, which is equivalent to $2x - 5 = 0$, or $x = 2.5$.
Solution set: $\{\, 2.5\}$

17. Since no real number satisfies $|x - 3| < 0$, the solution set is $\emptyset$.

18. Since $x + 3x = 4x$ is equivalent to $4x = 4x$, the solution set is R or $(-\infty, \infty)$.

19. $2(x + 7) = 2x + 9$

      $2x + 14 = 2x + 9$

          $14 = 9$

Solution set is $\emptyset$.

20. Since $|x - 6| \geq 0$ for any real number x, $\mid x - 6 \mid \; \geq -6$ for any real number x. The solution set is R.

21. $x - 0.04(x - 10) = 96.4$

      $x - 0.04x + 0.4 = 96.4$

         $0.96x = 96$

            $x = 100$

$\{100\}$

22. Let $W =$ the width and $W + 16 =$ the length. Use $2W + 2L = P$ to write the equation

$$2W + 2(W + 16) = 84$$
$$4W + 32 = 84$$
$$4W = 52$$
$$W = 13$$

The width is 13 meters.

23. Let $h =$ the height.   Use the formula $A = \frac{1}{2}bh$ to write the following equation.

$$\frac{1}{2} \cdot 3h = 21$$
$$3h = 42$$
$$h = 14$$

The height is 14 inches.

24. Let $x =$ the original price and $0.30x =$ the amount of discount. The original price minus the discount is equal to the price she paid.

$$x - 0.30x = 210$$
$$0.70x = 210$$
$$x = 300$$

The original price was $300.

25. Let $x =$ the number of liters of 11% alcohol. The mixture will be $x + 60$ liters of 7% alcohol. There are $0.11x$ liters of alcohol in the 11% solution, 0.05(60) liters of alcohol in the 5% solution, and $0.07(x + 60)$ liters of alcohol in the 7% solution.

$$0.11x + 0.05(60) = 0.07(x + 60)$$
$$0.11x + 3 = 0.07x + 4.2$$
$$0.04x = 1.2$$
$$x = 30.$$

Use 30 liters of 11% alcohol solution.

26. Let $b =$ Brenda's salary.

$$\mid b - 28{,}000 \mid \; > 3000$$

$b - 28{,}000 > 3000$ or $b - 28{,}000 < -3000$

      $b > 31{,}000$ or $b < 25{,}000$

Brenda's salary is either greater than $31,000 or less than $25,000.

Making Connections

Chapters 1-2

1. $5x + 6x = 11x$
2. $5x \cdot 6x = 30x^2$
3. $\frac{6x + 2}{2} = \frac{6x}{2} + \frac{2}{2} = 3x + 1$
4. $5 - 4(2 - x) = 5 - 8 + 4x = 4x - 3$
5. $(30 - 1)(30 + 1) = 29 \cdot 31 = 899$
6. $(30 + 1)^2 = 31^2 = 961$
7. $(30 - 1)^2 = 29^2 = 841$
8. $(2 + 3)^2 = 5^2 = 25$
9. $2^2 + 3^2 = 4 + 9 = 13$
10. $(8 - 3)(3 - 8) = 5(-5) = -25$
11. $(-1)(3 - 8) = -1(-5) = 5$
12. $-2^2 = -(2^2) = -4$
13. $3x + 8 - 5(x - 1) = 3x + 8 - 5x + 5$
    $= -2x + 13$
14. $(-6)^2 - 4(-3)2 = 36 + 24 = 60$
15. $3^2 \cdot 2^3 = 9 \cdot 8 = 72$
16. $4(-6) - (-5)(3) = -24 + 15 = -9$
17. $-3x \cdot x \cdot x = -3x^3$
18. $(-1)^6 = 1$

19. $\quad 5x + 6x = 8x$
    $\quad\quad\quad 11x = 8x$
    $\quad\quad\quad\; 3x = 0$
    $\quad\quad\quad\;\; x = 0$
    Solution set: $\{0\}$

20. $\quad 5x + 6x = 11x$
    $\quad\quad\quad 11x = 11x$
    This equation is an identity. Solution set: R

21. $\quad 5x + 6x = 0$
    $\quad\quad\quad 11x = 0$
    $\quad\quad\quad\;\; x = 0$
    Solution set: $\{0\}$

22. $\quad 5x + 6 = 11x$
    $\quad\quad\quad -6x = -6$
    $\quad\quad\quad\;\; x = 1$
    Solution set: $\{1\}$

23. $\quad 3x + 1 = 0$
    $\quad\quad\quad 3x = -1$
    $\quad\quad\quad\; x = -\frac{1}{3}$
    Solution set: $\left\{ -\frac{1}{3} \right\}$

24. $\quad 5 - 4(2 - x) = 1$
    $\quad\quad 5 - 8 + 4x = 1$
    $\quad\quad\quad\quad\; 4x = 4$
    $\quad\quad\quad\quad\;\; x = 1$
    Solution set: $\{1\}$

25. $\quad 3x + 6 = 3(x + 2)$
    $\quad 3x + 6 = 3x + 6$
    This equation is an identity. Solution set: R

26. $\quad x - 0.01x = 990$
    $\quad\quad\; 0.99x = 990$
    $\quad\quad\quad\quad x = 1000$
    Solution set: $\{1000\}$

27. $\quad |\, 5x + 6 \,| = 11$
    $\quad 5x + 6 = 11 \quad\quad$ or $\;\; 5x + 6 = -11$
    $\quad\quad\; 5x = 5 \quad\quad$ or $\quad\quad 5x = -17$
    $\quad\quad\quad x = 1 \quad\quad$ or $\quad\quad\; x = -17/5$
    Solution set: $\{-17/5, \, 1\}$

28. a) From the graph it appears that the cost of renting and buying are equal at around 87,500 copies.

b) If $x =$ the number of copies made in 5 years, then the cost for renting is
$R = 60(75) + 0.06x$ dollars
or $R = 4500 + 0.06x$ dollars.
The cost if the copier is purchased is
$P = 8000 + 0.02x$ dollars.

c) $\quad 60(75) + 0.06x = 8000 + 0.02x$
$\quad\quad\quad\quad 0.04x = 3500$
$\quad\quad\quad\quad\quad\;\; x = 87,500$
Five-year cost is same for 87,500 copies.

d) If 120,000 copies are made, then renting cost is \$11,700 and buying cost is \$10,400. So buying is \$1300 cheaper.

e) $|60(75) + 0.06x - (8000 + 0.02x)| < 500$
$\quad\quad\quad |-3500 + 0.04x| < 500$
$\quad\quad\; -500 < -3500 + 0.04x < 500$
$\quad\quad\quad 3000 < 0.04x < 4000$
$\quad\quad\; 75,000 < x < 100,000$
If the number of copies is between 75,000 and 100,000, then the plans differ by less than \$500.

3.1 WARM-UPS

1. False, because $3(5) - 2(2) = -4$ is not correct.

2. False, the vertical axis is called the y-axis.

3. False, because the point $(0, 0)$ is not in any quadrant.

4. True, because its first coordinate is 0.

5. True, because the graph of $x = k$ for any real number k is a vertical line.

6. True, because $8 - y = 0$ is equivalent to $y = 8$ and the graph of $y = k$ for any real number k is a horizontal line.

7. True, because if $x = 0$, then $y = -3$.

8. True, because if $n = 2$, then $C = 3(2) + 4 = 10$.

9. False, because if $P = 12$, then $3x = 12$ and $x = 4$.

10. True, because we usually put the dependent variable on the vertical axis.

3.1 EXERCISES

1. The origin is the point where the x-axis and y-axis intersect.

3. Intercepts are points where a graph crosses the axes.

5. The graph of an equation of the type $x = k$ where k is a fixed number is a vertical line.

7. To plot $(2, 5)$, start at the origin and go 2 units to the right and up 5. The point is in quadrant I.

9. To plot $(-3, -1/2)$, start at the origin and go 3 units to the left and down 1/2 unit. The point is in quadrant III.

11. To plot $(0, 4)$, start at the origin and go 4 units upward. The point is on the y-axis.

13. To plot $(\pi, -1)$, start at the origin and go approximately 3.14 units to the right and 1 unit downward. The point is in quadrant IV.

15. To plot $(-4, 3)$, start at the origin and go 4 units to the left and 3 units upward. The point is in quadrant II.

17. To plot $(3/2, 0)$, start at the origin and go approximately 3/2 units to the right. The point is on the x-axis.

19. To plot $(0, -7/3)$, start at the origin and go 7/3 units downward. The point is on the y-axis.

21. Use the midpoint formula:
$$\left(\frac{2+5}{2}, \frac{5+1}{2}\right) = \left(\frac{7}{2}, \frac{6}{2}\right) = \left(\frac{7}{2}, 3\right)$$
Use the distance formula to find the length:
$$\sqrt{(5-2)^2 + (1-5)^2} = \sqrt{9+16} = \sqrt{25}$$
$$= 5$$

23. Use the midpoint formula:
$$\left(\frac{-2+6}{2}, \frac{4+(-2)}{2}\right) = \left(\frac{4}{2}, \frac{2}{2}\right) = (2, 1)$$
Use the distance formula to find the length:
$$\sqrt{(6-(-2))^2 + (-2-4)^2} = \sqrt{64+36}$$
$$= \sqrt{100} = 10$$

25. Use the midpoint formula:
$$\left(\frac{3+(-9)}{2}, \frac{1+(-4)}{2}\right) = \left(\frac{-6}{2}, \frac{-3}{2}\right)$$
$$= \left(-3, -\frac{3}{2}\right)$$
Use the distance formula to find the length:
$$\sqrt{(-9-3)^2 + (-4-1)^2} = \sqrt{144+25}$$
$$= \sqrt{169} = 13$$

27. Use the midpoint formula:
$$\left(\frac{-1+1}{2}, \frac{4+1}{2}\right) = \left(\frac{0}{2}, \frac{5}{2}\right) = \left(0, \frac{5}{2}\right)$$
Use the distance formula to find the length:
$$\sqrt{(1-(-1))^2 + (1-4)^2} = \sqrt{4+9}$$
$$= \sqrt{13}$$

29. $y = x + 1$
If $x = 0$, $y = 0 + 1 = 1$.
If $x = 1$, $y = 1 + 1 = 2$.
If $x = 2$, $y = 2 + 1 = 3$.
If $x = -1$, $y = -1 + 1 = 0$.
Plot $(0, 1)$, $(1, 2)$, $(2, 3)$, $(-1, 0)$, and draw a line through the points.

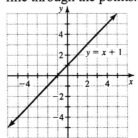

31. $y = -2x + 3$
If $x = 0$, $y = -2(0) + 3 = 3$.
If $x = 1$, $y = -2(1) + 3 = 1$.

If $x = 2$, $y = -2(2) + 3 = -1$.
If $x = -1$, $y = -2(-1) + 3 = 5$.
Plot $(0, 3)$, $(1, 1)$, $(2, -1)$, and $(-1, 5)$ and draw a line through the points.

33. Since the x and y-coordinates are equal in $y = x$, plot $(0, 0)$, $(1, 1)$, $(2, 2)$, $(-1, -1)$ and draw a line through the points.

35. In $y = 3$, the x-coordinate can be any number as long as the y-coordinate is 3. Plot $(0, 3)$, $(1, 3)$, $(2, 3)$, $(-1, 3)$, and draw a horizontal line through the points.

37. For the equation $y = 1 - x$, plot the points $(0, 1)$, $(1, 0)$, $(2, -1)$, and $(-2, 3)$. Draw a line through the points.

39. For $x = 2$, the y-coordinate can be any number, but the x-coordinate is 2. Plot $(2, 0)$, $(2, 1)$, $(2, 3)$, $(2, -1)$, and draw a vertical line through the points.

41. For the equation $y = \frac{1}{2}x - 1$, plot the points $(0, -1)$, $(2, 0)$, $(4, 1)$, and $(-4, -3)$. Draw a line through the points.

43.
$$y + 3 = 0$$
$$y = -3$$
Plot $(-2, -3)$, $(0, -3)$, $(1, -3)$, and $(3, -3)$ and draw a line through the points.

45. $x - 4 = 0$
$$x = 4$$
Plot $(4, 5)$, $(4, 2)$, $(4, -1)$, $(4, 1)$, and draw a line through the points.

47. For $y = \frac{1}{2}x$ plot $(0,0), (2,1), (4,2), (6,3)$ and draw a line through the points.

49. $3x + y = 5$

$y = -3x + 5$

Plot $(0,5), (1,2), (2,-1), (-1,8)$, and draw a line through the points.

51. $6x + 3y = 0$

$3y = -6x$

$y = -2x$

Plot the points $(0,0), (1,-2), (2,-4), (-2,4)$ and draw a line through the points.

53. For $y = 2x - 20$ plot the points $(0,-20), (5,-10), (10,0), (-5,-30)$ and draw a line through them.

55. If $x = 0$ in $4x - 3y = 12$, then $4(0) - 3y = 12, -3y = 12,$ or $y = -4$.

If $y = 0$, in $4x - 3y = 12$, then $4x - 3(0) = 12, 4x = 12,$ or $x = 3$. The intercepts are $(0, -4)$ and $(3, 0)$. Draw a line through these points.

57. If $x = 0$, in $x - y + 5 = 0$, then $0 - y + 5 = 0$ or $5 = y$. If $y = 0$, then $x - 0 + 5 = 0$ or $x = -5$. The intercepts are $(0, 5)$ and $(-5, 0)$. Draw a line through these points.

59. If $x = 0$, in $2x + 3y = 5$, then $2(0) + 3y = 5$ or $y = 5/3$. If $y = 0$, $2x + 3(0) = 5$ or $x = 5/2$. Plot the intercepts $(0, 5/3)$ and $(5/2, 0)$, and draw a line through the points.

61. If $x = 0$, in $2x - 3y = 60$, then $2(0) - 3y = 60$, or $y = -20$. If $y = 0$, then $2x = 60$ or $x = 30$. Plot $(0, -20)$ and $(30, 0)$, and draw a line through the points.

63. If $x = 0$ in $y = 2x - 4$, then $y = -4$. If $y = 0$, then $2x - 4 = 0$, or $x = 2$. Draw a line through the intercepts $(0, -4)$ and $(2, 0)$.

65. If $x = 0$ in $y = -\frac{1}{2}x - 20$, then $y = -20$. If $y = 0$, then $-\frac{1}{2}x - 20 = 0$, or $x = -40$. Draw a line through the intercepts $(0, -20)$ and $(-40, 0)$.

67. If $x = 0$ in $3x - 5y = 15$, then $-5y = 15$ or $y = -3$. If $y = 0$, then $3x = 15$ or $x = 5$. The intercepts are $(0, -3)$ and $(5, 0)$.

69. If $x = 0$ in $y = 5x$, then $y = 5(0) = 0$. If $y = 0$, then $5x = 0$ or $x = 0$. The intercept is $(0, 0)$.

71. The equation $6x + 3 = 0$ is equivalent to $x = -\frac{1}{2}$, which is a vertical line. It has no y-intercept. The x-intercept is $\left(-\frac{1}{2}, 0\right)$.

73. The equation $12 + 18y = 0$ is equivalent to $y = -\frac{2}{3}$, which is a horizontal line. It has no x-intercept. The y-intercept is $\left(0, -\frac{2}{3}\right)$.

75. If $x = 0$ in $2 - 4y = 8x$ then $2 - 4y = 0$ or $y = \frac{1}{2}$. If $y = 0$ then $2 = 8x$ or $x = \frac{1}{4}$. The intercepts are $\left(0, \frac{1}{2}\right)$ and $\left(\frac{1}{4}, 0\right)$.

77. If $x = 2$, then $y = -3(2) + 6 = 0$. If $y = -3$, then $-3 = -3x + 6$, or $x = 3$. The points are $(2, 0)$ and $(3, -3)$.

79. If $x = -4$, then $\frac{1}{2}(-4) - \frac{1}{3}y = 9$, or $y = -33$. If $y = 6$, then $\frac{1}{2}x - \frac{1}{3} \cdot 6 = 9$, or $x = 22$. The points are $(-4, -33)$ and $(22, 6)$.

81. a) Use $n = 8$ in $P = 320n + 22{,}765$ to get $P = 320(8) + 22{,}765 = 25{,}325$ In 2008 the car will cost \$25,325.

b) If $n = 9$, the price is \$25,645. Every year the price goes up by \$320.

c)

83. a) If $m = 400$ in $C = 0.26m + 42$, then $C = 0.26(400) + 42 = 146$. The charge for a car driven 400 miles is \$146.

b) If $m = 0$, $C = 0.26(0) + 42 = 42$. If $m = 1000$, $C = 0.26(1000) + 42 = 302$. Plot $(0, 42)$ and $(1000, 302)$, and draw a line segment with those endpoints.

85. a) $C = 0.50(5) + 8.95 = \$11.45$.

b)
$$0.50t + 8.95 = 14.45$$
$$0.50t = 5.50$$
$$t = 11$$
A pizza that sells for \$14.45 has 11 toppings.

87. Let $n =$ then number of note pads and $b =$ the number of binders. Since the total cost is \$100, we have $1n + 2b = 100$ or $n + 2b = 100$. If he orders 30 note pads, then $30 + 2b = 100$ or $b = 35$. The graph follows.

38

Note that either n or b could be placed on the horizontal axis.

89. a) If $x = 850$, then
$C = 0.55(850) + 50 = \$517.50$
$R = 1.50(850) = \$1275$
$P = 0.95(850) - 50 = \$757.50$. We have found her cost, revenue, and profit if she sells 850 roses.

b) If $P = 995$, then
$$0.95x - 50 = 995$$
$$0.95x = 1045$$
$$x = 1100$$
Here profit was $995 when she sold 1100 roses.

c) If $x = 1100$, then $R = \$1650$ and $C = \$655$. So $R - C = \$995$. The difference between revenue and cost is $995, which is her profit and this agrees with part (b).

91.

93.

95.

1. True, because the closer the line is to vertical, the larger its slope.
2. False, slope is rise divided by run.
3. False, because it is horizontal line with 0 slope.
4. True, because it is a vertical line.
5. False, slope of a line can be any real number.
6. False, the slope is $2/5$.
7. False, it has slope 1.
8. True, because it is rising as you go from left to right.
9. False, lines with slope $2/3$ and $-3/2$ are perpendicular.
10. False, because two vertical parallel lines do not have equal slopes.

3.2 EXERCISES

1. Slope measures the steepness of a line.
3. A horizontal line has zero slope because it has no rise.
5. If m_1 and m_2 are the slopes of perpendicular lines, then $m_1 = -1/m_2$.
7. In going from $(-3, 0)$ to $(0, 2)$ we rise 2 and run 3. The slope is $2/3$.
9. The slope for this line is undefined because it is a vertical line.
11. The slope of any horizontal line is 0.
13. In going from $(-2, 3)$ to $(3, -3)$, we rise -5 and run 5. The slope is $-5/5$ or -1.
15. In going from $(2, 0)$ to $(4, 3)$ we rise 3 and run 2. The slope is $3/2$.
17. In going from $(-2, 4)$ to $(4, -2)$ we rise -6 and run 6. The slope is $-6/6$ or -1.
19. $m = \dfrac{6 - 1}{2 - 5} = -\dfrac{5}{3}$
21. $m = \dfrac{-1 - 3}{-3 - 4} = \dfrac{-4}{-7} = \dfrac{4}{7}$
23. $m = \dfrac{2 - 7}{-2 - (-1)} = \dfrac{-5}{-1} = 5$
25. $m = \dfrac{-5 - 0}{3 - 0} = \dfrac{-5}{3} = -\dfrac{5}{3}$
27. $m = \dfrac{3 - 0}{0 - 5} = -\dfrac{3}{5}$

29. $m = \dfrac{-1 - (-\frac{1}{2})}{\frac{3}{4} - (-\frac{1}{2})} = \dfrac{-\frac{1}{2}}{\frac{5}{4}}$

$\qquad = -\dfrac{1}{2} \cdot \dfrac{4}{5} = -\dfrac{2}{5}$

31. $m = \dfrac{212 - 209}{6 - 7} = \dfrac{3}{-1} = -3$

33. $m = \dfrac{7 - 7}{4 - (-12)} = \dfrac{0}{16} = 0$

35. $\dfrac{6 - (-6)}{2 - 2} = \dfrac{12}{0}$ The slope is undefined because 0 is in the denominator.

37. $m = \dfrac{11.9 - 8.4}{24.3 - 3.57} = \dfrac{3.5}{20.73} \approx 0.169$

39. $m = \dfrac{1 - 0}{\frac{\pi}{4} - \frac{\pi}{2}} \approx \dfrac{1}{-0.7854} \approx -1.273$

41. The line through $(-5, 1)$ and $(3, -2)$ has slope

$$m = \dfrac{1 - (-2)}{-5 - 3} = \dfrac{3}{-8} = -\dfrac{3}{8}.$$

Any line perpendicular to this line has slope 8/3 because 8/3 is the opposite of the reciprocal of $-3/8$. Line l has slope 8/3.

43. The slope of the line through $(-3, -2)$ and $(4, 1)$ is

$$m = \dfrac{-2 - 1}{-3 - 4} = \dfrac{3}{7}.$$

Any line parallel to this line also has slope 3/7. So line l has slope 3/7.

45. The opposite of the reciprocal of $4/5$ is $-5/4$. Line l has slope $-5/4$.

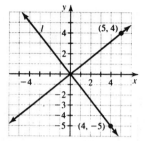

47. Draw a quadrilateral with the four given points as vertices. Find the slope of each side.

$m_1 = \dfrac{3 - 1}{0 - (-6)} = \dfrac{1}{3}$ $m_2 = \dfrac{3 - 1}{0 - 4} = -\dfrac{1}{2}$

$m_3 = \dfrac{1 - (-1)}{4 - (-2)} = \dfrac{1}{3}$

$m_4 = \dfrac{-1 - 1}{-2 - (-6)} = -\dfrac{1}{2}$

Since opposite sides of the quadrilateral have equal slopes, the quadrilateral is a parallelogram.

49. Draw a quadrilateral with the four given points as vertices. Find the slope of each side.

$m_1 = \dfrac{6 - 2}{3 - (-3)} = \dfrac{2}{3}$

$m_2 = \dfrac{6 - 4}{3 - 6} = -\dfrac{2}{3}$

$m_3 = \dfrac{4 - (-1)}{6 - (-1)} = \dfrac{5}{7}$

$m_4 = \dfrac{-1 - 2}{-1 - (-3)} = -\dfrac{3}{2}$

Since there are no parallel sides, this quadrilateral is not a trapezoid.

51. Draw a triangle with the three given points as vertices. Find the slope of each side.

$m_1 = \dfrac{6 - 3}{-1 - (-3)} = \dfrac{3}{2}$ $m_2 = \dfrac{6 - 0}{-1 - 0} = -6$

$m_3 = \dfrac{3 - 0}{-3 - 0} = -1$

From the slopes we see that none of the line segments are perpendicular. The triangle is not a right triangle.

53. a) $m = \dfrac{23,635 - 21,135}{2002 - 1998}$

$\qquad = 625$ dollars per year

b) Using the graph we might guess that the price in 2010 will be $29,000.

c) The slope is the average yearly increase in price. So from 2002 to 2010 there will be 8 increases of $625 each time, for an increase of $5000. So the price in 2010 should be $23,635 + $5000, or $28,635.

55. Use graph paper to solve this problem. The slope 4 means that we can rise 4 and run 1 to get additional points on this line. From $(2, 1)$ rise 4 and run 1 to get to $(3, 5)$. From $(2, 1)$ we can rise -4 and run -1 to get to $(1, -3)$. From $(1, -3)$ we can rise -4 and run -1 to get to $(0, -7)$.

57. Find the slope:
$$\frac{k - (-5)}{2 - (-3)} = \frac{1}{2}$$
$$\frac{k + 5}{5} = \frac{1}{2}$$
$$2k + 10 = 5$$
$$k = -\frac{5}{2}$$

59. The slope of the line perpendicular to the line with slope 0.247 is the opposite of the reciprocal of 0.247:
$$-\frac{1}{0.247} \approx -4.049$$

61. A horizontal line has zero slope and a vertical line has undefined slope.

63. Draw a quadrilateral with the four given vertices. The slope of the diagonal joining $(-3, -1)$ and $(5, 3)$ is
$$m_1 = \frac{3 - (-1)}{5 - (-3)} = \frac{1}{2}.$$
The slope of the diagonal joining $(0, 3)$ and $(2, -1)$ is
$$m_2 = \frac{3 - (-1)}{0 - 2} = -2.$$
Since the opposite of the reciprocal of $1/2$ is -2, the diagonals are perpendicular.

65. Increasing m makes the graph increase faster. The slopes of these lines are 1, 2, 3, and 4.

3.3 WARM-UPS

1. True, because there is a one-to-one correspondence between the slopes of lines through a fixed point and the real numbers.

2. False, the line through (a, b) with slope m has equation $y - b = m(x - a)$.

3. False, $y = mx + b$ goes through $(0, b)$ and has slope m.

4. True, because the y-intercept has x-coordinate 0.

5. True, because the x-intercept has y-coordinate 0.

6. False, because vertical lines do not have equations in slope intercept form.

7. True, because if we solve the equation for y, we get $y = (-3/2)x + (7/2)$.

8. False, because a line perpendicular to a line with slope 3 has slope $-1/3$.

9. False, because if $x = 0$ in this equation, we get $y = 5/2$.

10. True, because even the vertical lines can be expressed in standard form.

3.3 EXERCISES

1. Point-slope form is $y - y_1 = m(x - x_1)$ where m is the slope and (x_1, y_1) is a point on the line.

3. To write an equation of a line we need the slope and a point on the line. We might have to find the slope from two points or from knowing the line is parallel or perpendicular to another line.

5. To find the slope from standard form solve the equation for y to get the form $y = mx + b$ where m is the slope.

7. Use the given point and the given slope in the point-slope formula:
$$y - (-3) = 2(x - 2)$$
$$y + 3 = 2x - 4$$
$$y = 2x - 7$$

9. Use the point $(-2, 3)$ and the slope $-1/2$ in the point-slope formula:
$$y - 3 = -\frac{1}{2}(x - (-2))$$
$$y - 3 = -\frac{1}{2}x - 1$$
$$y = -\frac{1}{2}x + 2$$

11. The slope of the line through $(-2, -1)$ and $(3, -6)$ is
$$m = \frac{-6 - (-1)}{3 - (-2)} = -1.$$
Use $(3, -6)$ and slope -1 in the point-slope formula:
$$y - (-6) = -1(x - 3)$$
$$y + 6 = -x + 3$$
$$y = -x - 3$$

13. The slope of the line through $(2, 2)$ and $(-1, 1)$ is
$$m = \frac{1 - 2}{-1 - 2} = \frac{1}{3}.$$

Use $(2, 2)$ and slope $1/3$ in the point-slope formula:

$$y - 2 = \tfrac{1}{3}(x - 2)$$
$$y - 2 = \tfrac{1}{3}x - \tfrac{2}{3}w$$
$$y = \tfrac{1}{3}x - \tfrac{2}{3} + 2$$
$$y = \tfrac{1}{3}x + \tfrac{4}{3}$$

15. The slope of the line through $(-3, 1)$ and $(5, -1)$ is

$$m = \frac{1 - (-1)}{-3 - 5} = -\tfrac{1}{4}.$$

So the line we want has slope 4 and goes through $(-1, -12)$:

$$y - (-12) = 4(x - (-1))$$
$$y + 12 = 4x + 4$$
$$y = 4x - 8$$

17. The line through $(9, -3)$ and $(-3, 6)$ has slope

$$m = \frac{6 - (-3)}{-3 - 9} = -\tfrac{3}{4}$$

Use the same slope $-3/4$ and $(0, 0)$ in the point-slope formula:

$$y - 0 = -\tfrac{3}{4}(x - 0)$$
$$y = -\tfrac{3}{4}x$$

19. The line goes through $(0, 2)$ and $(-4, 0)$. The slope is $\tfrac{1}{2}$ and the y-intercept is $(0, 2)$. Using slope-intercept form, we can write $y = \tfrac{1}{2}x + 2$.

21. This line has 0 slope and y-intercept $(0, -2)$. So its equation in slope-intercept form is $y = 0 \cdot x - 2$, or $y = -2$.

23. This is a vertical line with an x-intercept of $(1, 0)$ and so its equation is $x = 1$.

25. This line has a y-intercept of $(0, 0)$ and also goes through $(3, -3)$. Its slope is -1. Its equation in slope-intercept form is $y = -x$.

27. This line goes through $(2, 0)$ and $(0, -3)$. Its slope is $3/2$ and its equation is $y = \tfrac{3}{2}x - 3$.

29. This line goes through $(0, 2)$ and $(2, 0)$. Since its slope is -1, its equation is $y = -x + 2$.

31.
$$y = \tfrac{1}{3}x - 2$$
$$3(y) = 3(\tfrac{1}{3}x - 2)$$
$$3y = x - 6$$
$$-x + 3y = -6$$

33.
$$x - 3y = 6$$
$$y - 5 = \tfrac{1}{2}(x + 3)$$
$$2(y - 5) = 2 \cdot \tfrac{1}{2}(x + 3)$$
$$2y - 10 = x + 3$$
$$-x + 2y = 13$$
$$x - 2y = -13$$

35.
$$y + \tfrac{1}{2} = \tfrac{1}{3}(x - 4)$$
$$6\left(y + \tfrac{1}{2}\right) = 6 \cdot \tfrac{1}{3}(x - 4)$$
$$6y + 3 = 2(x - 4)$$
$$-2x + 6y = -11$$
$$2x - 6y = 11$$

37.
$$0.05x + 0.06y - 8.9 = 0$$
$$100(0.05x + 0.06y - 8.9) = 100(0)$$
$$5x + 6y - 890 = 0$$
$$5x + 6y = 890$$

39.
$$2x + 5y = 1$$
$$5y = -2x + 1$$
$$y = -\tfrac{2}{5}x + \tfrac{1}{5}$$

The slope is $-\tfrac{2}{5}$ and the y-intercept is $(0, 1/5)$.

41.
$$3x - y - 2 = 0$$
$$3x - 2 = y$$
$$y = 3x - 2$$

The slope is 3 and the y-intercept is $(0, -2)$.

43.
$$y + 3 = 5$$
$$y = 2$$

The slope is 0 and the y-intercept is $(0, 2)$.

45.
$$y - 2 = 3(x - 1)$$
$$y - 2 = 3x - 3$$
$$y = 3x - 1$$

The slope is 3 and the y-intercept is $(0, -1)$.

47.
$$\frac{y - 5}{x + 4} = \frac{3}{2}$$
$$2(y - 5) = 3(x + 4)$$
$$2y - 10 = 3x + 12$$
$$2y = 3x + 22$$
$$y = \tfrac{3}{2}x + 11$$

The slope is $\tfrac{3}{2}$ and the y-intercept is $(0, 11)$.

49.
$$y - \tfrac{1}{2} = \tfrac{1}{3}\left(x + \tfrac{1}{4}\right)$$
$$y - \tfrac{1}{2} = \tfrac{1}{3}x + \tfrac{1}{12}$$
$$y = \tfrac{1}{3}x + \tfrac{1}{12} + \tfrac{1}{2}$$
$$y = \tfrac{1}{3}x + \tfrac{7}{12}$$

The slope is $\tfrac{1}{3}$ and the y-intercept is $\left(0, \tfrac{7}{12}\right)$.

51.
$$y - 6000 = 0.01(x + 5700)$$
$$y - 6000 = 0.01x + 57$$
$$y = 0.01x + 6057$$
The slope is 0.01 and the y-intercept is $(0, 6057)$.

53. Since the y-intercept is $(0, 5)$ and the slope is 1/2, we can use slope-intercept form to write the equation.
$$y = \tfrac{1}{2}x + 5$$
To get integral coefficients multiply by 2.
$$2y = x + 10$$
$$-x + 2y = 10$$
$$x - 2y = -10 \quad \text{(Standard form)}$$

55. The slope of the line through $(2, 0)$ and $(0, 4)$ is
$$m = \frac{4 - 0}{0 - 2} = -2.$$
Use slope-intercept form to write the equation of the line with slope -2 and y-intercept $(0, 4)$:
$$y = -2x + 4$$
$$2x + y = 4 \quad \text{(Standard form)}$$

57. Any line parallel to $y = 2x + 6$ has slope 2. Use point-slope formula to find the equation of a line with slope 2 and going through $(-3, -1)$:
$$y - (-1) = 2(x - (-3))$$
$$y + 1 = 2x + 6$$
$$-2x + y = 5$$
$$2x - y = -5 \quad \text{(Standard form)}$$

59. Write $2x + 4y = 1$ in slope-intercept form to identify the slope:
$$4y = -2x + 1$$
$$y = -\tfrac{1}{2}x + \tfrac{1}{4}$$
Any line parallel to $2x + 4y = 1$ has slope $-1/2$. Use the point-slope formula to find the equation of the line with slope $-1/2$ through $(-3, 5)$:
$$y - 5 = -\tfrac{1}{2}(x - (-3))$$
$$y - 5 = -\tfrac{1}{2}x - \tfrac{3}{2}$$
$$2y - 10 = -x - 3$$
$$x + 2y = 7 \quad \text{(Standard form)}$$

61. A line perpendicular to $y = (1/2)x - 3$ has slope -2. Use the point-slope formula to find the equation of a line with slope -2 going through $(1, 1)$:

$$y - 1 = -2(x - 1)$$
$$y - 1 = -2x + 2$$
$$2x + y = 3 \quad \text{(Standard form)}$$

63. Write $x + 3y = 4$ in slope-intercept form to determine its slope:
$$3y = -x + 4$$
$$y = -\tfrac{1}{3}x + \tfrac{4}{3}$$
The slope of $x + 3y = 4$ is $-1/3$. Any line perpendicular to $x + 3y = 4$ has slope 3. Use the point-slope formula to find the equation of the line with slope 3 through $(-4, -3)$.
$$y - (-3) = 3(x - (-4))$$
$$y + 3 = 3x + 12$$
$$-3x + y = 9$$
$$3x - y = -9 \quad \text{(Standard form)}$$

65. Any line parallel to the x-axis has slope 0. If it goes through $(2, 5)$, it has y-intercept $(0, 5)$. Using slope-intercept form we get
$$y = 0 \cdot x + 5$$
$$y = 5$$

67. The slope of $y = (1/2)x$ is 1/2 and the y-intercept is $(0, 0)$. To graph the line start at $(0, 0)$ and go up 1 unit and 2 units to the right to find another point on the line, $(2, 1)$. Draw the line through $(0, 0)$ and $(2, 1)$.

69. The slope of $y = 2x - 3$ is 2 and the y-intercept is $(0, -3)$. The slope of 2 is obtained from $2/1$. To graph the line start at $(0, -3)$ and rise 2 units and go 1 unit to the right to get to the point $(1, -1)$. Draw a line through the points $(0, -3)$ and $(1, -1)$.

71. The slope of $y = (-2/3)x + 2$ is $-2/3$ and the y-intercept is $(0, 2)$. To graph the line start at $(0, 2)$ and go down 2 units and then 3 units to the right to locate the point $(3, 0)$. Draw a line through $(0, 2)$ and $(3, 0)$.

73. The equation $3y + x = 0$ can be written as $y = (-1/3)x$. Its slope is $-1/3$ and its y-intercept is $(0, 0)$. To graph the line start at $(0, 0)$ and go 1 unit down and 3 units to the right to locate the point $(3, -1)$. Draw a line through $(0, 0)$ and $(3, -1)$.

75. The graph of $y = x + 3$ has y-intercept $(0, 3)$ and slope 1. The graph of $y = x + 2$ has y-intercept $(0, 2)$ and slope 1. The lines are parallel.

77. The graph of $y = 3x + 1$ has y-intercept $(0, 1)$ and slope 3. The graph of $y = -\frac{1}{3}x + 1$ has y-intercept $(0, 1)$ and slope $-\frac{1}{3}$. The lines are perpendicular.

79. The graph of $y = \frac{2}{3}x - 1$ has y-intercept $(0, -1)$ and slope $\frac{2}{3}$. The graph of $y = \frac{2}{3}x + 1$ has y-intercept $(0, 1)$ and slope $\frac{2}{3}$. The lines are parallel.

81. The graph of $y = \frac{3}{4}x + 3$ has y-intercept $(0, 3)$ and slope $\frac{3}{4}$. The graph of $y = -\frac{4}{3}x + 1$ has y-intercept $(0, 1)$ and slope $-\frac{4}{3}$. The lines are perpendicular.

83. Solve $x + 3y = 7$ for y to get $y = -\frac{1}{3}x + \frac{7}{3}$. The lines are perpendicular because there slopes are 3 and $-1/3$.

85. Solve $2x - 4y = 9$ for y to get $y = \frac{1}{2}x - \frac{9}{4}$. Solve $\frac{1}{3}x = \frac{2}{3}y - 8$ for y to get $y = \frac{1}{2}x + 12$. The lines are parallel because there slopes are equal.

87. Solve $2y = x + 6$ for y to get $y = \frac{1}{2}x + 3$. Solve $y - 2x = 4$ for y to get $y = 2x + 4$. The lines are neither parallel nor perpendicular.

89. Solve $x - 6 = 9$ for x to get $x = 15$. Solve $y - 4 = 12$ for y to get $y = 16$. The lines are perpendicular, because one is horizontal and the other is vertical.

91. The slope of the line through $(0, 60)$ and $(120, 200)$ is
$$m = \frac{200 - 60}{120 - 0} = \frac{140}{120} = \frac{7}{6}.$$

Using slope-intercept form with a t-intercept of $(0, 60)$ we get $t = \frac{7}{6}s + 60$.
After 30 seconds the temperature will be

$$t = \frac{7}{6}(30) + 60 = 35 + 60 = 95°\text{F}.$$

93. a) $m = \dfrac{26 - 14}{2000 - 1970} = \dfrac{12}{30} = \dfrac{2}{5} = 0.4$

$$y - 14 = 0.4(x - 1970)$$
$$y - 14 = 0.4x - 788$$
$$y = 0.4x - 774$$

b) If $x = 2010$, then
$$y = 0.4(2010) - 774 = 30.$$
So in 2010 worldwide emission of CO_2 will be 30 billion tons

95. a) The slope of the line through $(9.14, 1230)$ and $(7.84, 826)$ is

$$m = \frac{1230 - 826}{9.14 - 7.84} \approx 310.77$$

Use the point-slope formula for the equation of a line with slope 310.77 and the point $(7.84, 826)$.
$$w - 826 = 310.77(d - 7.84)$$
$$w - 826 = 310.77d - 2436.4368$$
$$w = 310.77d - 1610.44$$
b) Use $d = 8.25$ in the formula to find w.

$$w = 310.77(8.25) - 1610.44 \approx 953$$
When depth is 8.25 ft, the flow is 953 ft^3/sec.
c) As the depth increases, we can see from the graph that the flow increases.

97. a) If $y = 0$, then $x/4 = 1$ or $x = 4$. If $x = 0$, then $y/6 = 1$ or $y = 6$. So the intercepts are $(4, 0)$ and $(0, 6)$.
b) If $y = 0$, then $x/a = 1$ or $x = a$. If $x = 0$, then $y/b = 1$ or $y = b$. So the intercepts are $(a, 0)$ and $(0, b)$.
c) Since the intercepts are $(0, 3)$ and $(-5, 0)$ the equation is $\frac{x}{-5} + \frac{y}{3} = 1$.
d) To be written in intercept form a line must have two intercepts. So horizontal, vertical, and lines through $(0, 0)$ cannot be written in intercept form.

99. The x-intercept is $(3000, 0)$ and the y-intercept is $(0, -3000)$. So a viewing window such as x in $[-4000, 4000]$ and y in $[-4000, 4000]$ would show both intercepts. The are many other possibilities.

101. The lines intersect at $(50, 97)$. This intersection can be seen if x is in $[0, 100]$ and y is in $[0, 200]$.

3.4 WARM-UPS
1. True, because $-3 > -3(2) + 2$.
2. False, because $3x - y > 2$ is equivalent to $y < 3x - 2$, which is below the line $3x - y = 2$.
3. True, because $3x + y < 5$ is equivalent to $y < -3x + 5$.
4. False, the region $x < -3$ is to the left of the vertical line $x = -3$.
5. True, because the word "and" is used.
6. True, because the word "or" is used.
7. False, because $(2, -5)$ does not satisfy the inequality $y > -3x + 5$.
Note that $-5 > -3(2) + 5$ is incorrect.
8. True, because $(-3, 2)$ satisfies $y \leq x + 5$.
9. False, it is equivalent to the compound inequality $2x - y \leq 4$ and $2x - y \geq -4$.
10. True, because in general $|x| > k$ (for a positive k) is equivalent to $x > k$ or $x < -k$.

3.4 EXERCISES

1. A linear inequality is an inequality of the form $Ax + By \leq C$ (or using $<$, $>$, or $\geq$) where A, B, and C are real numbers and A and B are not both zero.

3. If the inequality includes equality then the line should be solid.

5. The test point method is used to determine which side of the boundary line to shade.

7. Graph the equation $y = x + 2$. Use its slope of 1 and y-intercept of $(0, 2)$.

The graph of $y < x + 2$ is the region below the line $y = x + 2$. The line is dashed because of the inequality symbol. Points on the line satisfy the equation but not the inequality.

9. First graph the line $y = -2x + 1$. Start at its y-intercept $(0, 1)$ and use a slope of -2. Go down 2 units and 1 to the right to locate a second point on the line.

The graph of the inequality $y \leq -2x + 1$ is the region below the line and including the line. For this reason the line is drawn solid.

11. The inequality $x + y > 3$ is equivalent to $y > -x + 3$. First use slope of -1 and a y-intercept of $(0, 3)$ to graph $y = -x + 3$.

The graph of $y > -x + 3$ is the region above the line. The line is drawn dashed because of the inequality symbol.

13. The inequality $2x + 3y < 9$ is equivalent to $y < -\frac{2}{3}x + 3$. Use a slope of $-2/3$ and a y-intercept of 3 to graph $y = -\frac{2}{3}x + 3$.

The graph of the inequality is the region below the line. The line is not included in the graph and so it is drawn dashed.

15. Solve the inequality for y:
$$3x - 4y \leq 8$$
$$-4y \leq -3x + 8$$
$$y \geq \frac{3}{4}x - 2$$

To graph the equation $y = \frac{3}{4}x - 2$, draw a solid line with y-intercept $(0, -2)$ and slope $3/4$. Since the inequality symbol is $\geq$, shade the region above the line.

17. If we solve $x - y > 0$ for y we get $y < x$. To graph $y = x$, use a y-intercept of $(0, 0)$ and a slope of 1. Draw a dashed line and shade below the line for the graph of $y < x$.

19. The graph of $x \geq 1$ consists of the vertical line $x = 1$ together with the region to the right of the line. Draw a solid vertical line through $(1, 0)$ and shade the region to the right.

21. The graph of $y < 3$ is the region below the horizontal line $y = 3$. Draw a dashed horizontal line through $(0, 3)$ and shade the region below.

23. To graph $2x - 3y < 5$ using a test point, we can use the x and y-intercepts to determine the graph of the line $2x - 3y = 5$.
If $x = 0$, $2(0) - 3y = 5$ or $y = -5/3$.
If $y = 0$, $2x - 3(0) = 5$ or $x = 5/2$.
Locate the points $(0, -5/3)$ and $(5/2, 0)$ and draw a dashed line through them. Test the point $(0, 0)$ in the inequality:
$$2(0) - 3(0) < 5$$
Since $(0, 0)$ satisfies the inequality, we shade the side of the line that includes $(0, 0)$.

25. Graph $x + y + 3 = 0$ by using the x and y-intercepts, $(0, -3)$ and $(-3, 0)$. Draw a solid line through these two points. Test $(0, 0)$ in the inequality: $0 + 0 + 3 \geq 0$
Since $(0, 0)$ satisfies the inequality, we shade the side of the line that includes $(0, 0)$.

27. First graph the line $y - 2x = 0$ or $y = 2x$, which has slope 2 and y-intercept $(0, 0)$. Now test a point, say $(3, 0)$ in $y - 2x \leq 0$. Since $0 - 2(3) \leq 0$ is correct, we shade the region containing $(3, 0)$.

29. First graph the dashed line $3x - 2y = 0$ or $y = \frac{3}{2}x$, which has slope 3/2 and y-intercept $(0, 0)$. Now test a point, say $(4, 0)$ in $3x - 2y > 0$. Since $3(4) - 2(0) > 0$ is correct, we shade the region containing $(4, 0)$.

47

31. Find x and y-intercepts for $\frac{1}{2}x + \frac{1}{3}y = 1$:
If $x = 0$, we get $\frac{1}{3}y = 1$ or $y = 3$.
If $y = 0$, we get $\frac{1}{2}x = 1$ or $x = 2$.
Draw a dashed line through the intercepts $(0,3)$ and $(2,0)$. Test the point $(0,0)$ in the inequality:
$$\frac{1}{2}(0) + \frac{1}{3}(0) < 1$$
Since $(0,0)$ satisfies the inequality, we shade the region that includes $(0,0)$.

33. To graph $y > x$ and $y > -2x + 3$, we first draw dashed lines for $y = x$ and for $y = -2x + 3$. Test one point in each of the four regions to see if it satisfies the compound inequality. Test $(5,0)$, $(0,5)$, $(-5,0)$ and $(0,-5)$. Only $(0,5)$ satisfies both inequalities. So shade the region containing $(0,5)$.

$y > x$ and $y > -2x + 3$

35. First graph the equations $y = x + 3$ and $y = -x + 2$. Test the points $(0,5)$, $(5,0)$, $(0,-5)$, and $(-5,0)$ in the compound inequality. Only $(-5,0)$ fails to satisfy the compound inequality. So shade all regions except the one containing $(-5,0)$.

$y < x + 3$ or
$y > -x + 2$

37. First graph the lines $x - 4y = 0$ and $3x + 2y = 6$. Use a dashed line for the first and a solid line for the second. Select a point in each region as a test point. Use $(0,5)$, $(0,1)$, $(0,-5)$, and $(6,0)$. Only $(0,5)$ satisfies the compound sentence $x - 4y < 0$ and $3x + 2y \geq 6$. So shade the region that contains $(0,5)$.

$x - 4y < 0$ and
$3x + 2y \geq 6$

39. First graph the equations $x + y = 5$ and $x - y = 3$. Test one point in each of the 4 regions. Only points in the region containing $(0,0)$ satisfy both inequalities. Shade that region including the boundary lines.

$x + y \leq 5$ and
$x - y \leq 3$

41. Graph the equations $x - 2y = 4$ and $2x - 3y = 6$. Only the region containing the point $(0,-5)$ fails to satisfy the compound inequality. Shade the other three regions including the boundary lines.

$x - 2y \leq 4$ or
$2x - 3y \leq 6$

43. Graph the horizontal line $y = 2$ and the vertical line $x = 3$. Only points in the region containing $(0,5)$ satisfy both inequalities. Shade that region with dashed boundary lines.

$y > 2$ and $x < 3$

45. Graph $y = x$ and $x = 2$. Only points in the region containing $(0, 5)$ satisfy both inequalities. Shade that region and include the boundary lines.

$y \geq x$ and $x \leq 2$

47. Graph $2x = y + 3$ and $y = 2 - x$. Only points in the region containing $(0, -5)$ fail to satisfy the compound inequality. Shade all regions except the one containing $(0, -5)$. Use dashed lines for the boundaries because of the inequality symbols.

$2x < y + 3$ or $y > 2 - x$

49. Graph the lines $y = x - 1$ and $y = x + 3$. Only points in the region containing $(0, 0)$ satisfy the compound inequality. Shade that region and use dashed boundary lines.

$x - 1 < y < x + 3$

51. Graph the lines $y = 0$, $y = x$, and $x = 1$. Only points inside the triangular region

bounded by the lines satisfy the compound inequality. Shade that region and use solid boundary lines.

$0 \leq y \leq x$ and $x \leq 1$

53. Graph $x = 1$, $x = 3$, $y = 2$, and $y = 5$.

$1 \leq x \leq 3$ and $2 \leq y \leq 5$

Only points inside the rectangular region satisfy the compound inequality. Shade that region and include the boundary lines.

55. Graph the equations $x + y = 2$ and $x + y = -2$. Only points between these two parallel lines satisfy the absolute value inequality.

$|x + y| < 2$

57. Graph the parallel lines $2x + y = 1$ and $2x + y = -1$. Points between the lines do not satisfy the absolute value inequality. So shade the other two regions.

$|2x + y| \geq 1$

49

59. The inequality $|y - x| > 2$ is equivalent to
$$y - x > 2 \qquad \text{or} \qquad y - x < -2$$
$$y > x + 2 \qquad \text{or} \qquad y < x - 2$$
Graph the lines $y = x + 2$ and $y = x - 2$. Points above $y = x + 2$ together with points below $y = x - 2$ satisfy the absolute value inequality.

61. Graph the parallel lines $x - 2y = 4$ and $x - 2y = -4$. Points in the region between the lines satisfy the inequality.

63. Graph the vertical lines $x = 2$ and $x = -2$. Points in the region between the lines do not satisfy the inequality but points in the other two regions do.

65. Graph the horizontal lines $y = 1$ and $y = -1$. Only points in the region between the lines satisfy $|y| < 1$.

$|y| < 1$

67. Graph the lines $x = 2$, $x = -2$, $y = 3$, and $y = -3$. Only points inside the rectangular region bounded by the lines satisfy the compound inequality $|x| < 2$ and $|y| < 3$.

$|x| < 2$ and $|y| < 3$

69. The inequality $|x - 3| < 1$ is equivalent to $-1 < x - 3 < 1$, or $2 < x < 4$. The inequality $|y - 2| < 1$ is equivalent to $-1 < y - 2 < 1$, or $1 < y < 3$. Graph the lines $x = 2$, $x = 4$, $y = 1$, and $y = 3$. Only points inside the square region bounded by these lines satisfy the inequalities $|x - 3| < 1$ and $|y - 2| < 1$.

$|x - 3| < 1$ and $|y - 2| < 1$

71. Let $x =$ the number of compact cars and $y =$ the number of full-size cars. We have
$$15,000x + 20,000y \leq 120,000$$
$$3x + 4y \leq 24$$
We also have $x \geq 0$ and $y \geq 0$ because they cannot purchase a negative number of cars. Graph $3x + 4y \leq 24$, $x \geq 0$, and $y \geq 0$.

$x \geq 0, \ y \geq 0, \ 3x + 4y \leq 24$

73. Graph $3x + 4y \leq 24$, $x \geq 0$, $y \geq 0$, and $y > x$:

$x \geq 0, \ y \geq x, \ 3x + 4y \leq 24$

75. Graph $h \leq 187 - 0.85a$ and $h \geq 154 - 0.70a$ for $20 \leq a \leq 75$.

$h \leq 187 - 0.85a$,
$h \geq 154 - 0.70a$,
$a \geq 20, \ a \leq 75$

77. Let $d =$ the number of days of the *Daily Chronicle* advertising and $t =$ the number times an ad is aired on TV.

$$300d + 1000t \leq 9000$$
$$3d + 10t \leq 90$$

Graph $3d + 10t \leq 90$, $d \geq 0$, and $t \geq 0$:

$d \geq 0, \ t \geq 0$,
$3d + 10t \leq 90$

79. The solution set to a compound inequality using "and" is the intersection of the two solution sets. The solution set to a compound inequality using "or" is the union of the two solution sets.

3.5 WARM-UPS

1. False, $\{(1, 2), (1, 3)\}$ is not a function.
2. True, because $C = \pi D$.

3. True, because no two ordered pairs have the same first coordinate and different second coordinates.
4. False, because $(1, 5)$ and $(1, 7)$ have the same first coordinate and different second coordinates.
5. True, because every x-coordinate in $y = x^2$ corresponds to a unique y-coordinate.
6. False, because $\{(1, 5), (3, 6), (1, 7)\}$ is a relation that is not a function.
7. True, because of the definition of domain.
8. False, because the domain of a relation or function is the set of first coordinates.
9. True, because $\sqrt{x}$ is a real number only if $x \geq 0$.
10. True, because if $h(x) = x^2 - 3$, then $h(-2) = (-2)^2 - 3 = 4 - 3 = 1$.

3.5 EXERCISES

1. To say that b is a function of a means that b is uniquely determined by a.
3. A relation is any set of ordered pairs.
5. The range of relation is the set of all possible second coordinates.
7. The number of gallons that you get for $10 is uniquely determined by the price per gallon. So b is a function of a.
9. A student's test score is not determined by the number of hours spent studying, because two students can study for the same amount of time and score differently on the test. So b is not a function of a.
11. The Fahrenheit temperature is uniquely determined by the Celsius temperature by a formula. So b is a function of a.
13. Two items with the same price can have different universal product codes. So the product code cannot be determined from the cost and b is not a function of a.
15. Since a vertical line can be drawn to cross this graph more than once, this graph is not the graph of a function.
17. Since no vertical line can be drawn to cross this graph more than once, this graph is the graph of a function.

19. Since a vertical line can be drawn to cross this graph more than once, this graph is not the graph of a function.

21. A set of ordered pairs is a function unless there are two pairs with the same first coordinate and different second coordinates. So this set is a function.

23. This set of ordered pairs is not a function because $(-2, 4)$ and $(-2, 6)$ have the same x-coordinate and different y-coordinates.

25. This set of ordered pairs is not a function because $(\pi, -1)$ and $(\pi, 1)$ have the same x-coordinate and different y-coordinates.

27. This set is a function because no two ordered pairs have the same first coordinate and different second coordinates.

29. Note that the value of y^2 is the same for a number and its opposite. So if $y = \pm 1$, $x = 2(\pm 1)^2 = 2$. So $(2, 1)$ and $(2, -1)$ both satisfy the equation.

31. Note that the value of $|2y|$ is the same for a number and its opposite. So if $y = \pm 4$, $x^2 = |2(\pm 4)| = 8$. So $(8, 4)$ and $(8, -4)$ both satisfy the equation.

33. Note that the value of x^2 or y^2 is the same for a number and its opposite. So if $y = \pm 1$, $x^2 + (\pm 1)^2 = 1$, or $x^2 = 0$. Now 0 satisfies $x^2 = 0$. So $(0, 1)$ and $(0, -1)$ both satisfy the equation.

35. Note that the value of y^4 is the same for a number and its opposite. So if $y = \pm 2$, $x = (\pm 2)^4 = 16$. So $(16, 2)$ and $(16, -2)$ both satisfy the equation.

37. Note that the value of $|y|$ is the same for a number and its opposite. So if $y = \pm 1$, $x - 2 = |\pm 1| = 1$, or $x = 3$. So $(3, 1)$ and $(3, -1)$ both satisfy the equation.

39. This relation is a function because for each value of x, $y = x^2$ determines only one value of y.

41. This relation is not a function because $(2, 1)$ and $(2, -1)$ both satisfy $x = |y| + 1$.

43. This relation is a function because for each value of x, $y = x$ determines only one value of y.

45. This relation is not a function because $(2, 1)$ and $(2, -1)$ both satisfy $x = y^4 + 1$.

47. This relation is a function because for each value of x, $y = \sqrt{x}$ determines only one value of y.

49. This relation is not a function because $(2, 1)$ and $(2, -1)$ both satisfy $|x| = |2y|$.

51. This relation is not a function because $(0, 3)$ and $(0, -3)$ both satisfy $x^2 + y^2 = 9$.

53. This relation is a function because for each value of x, $x = 2\sqrt{y}$ determines only one value of y.

55. This relation is a function because $(0, 5)$ and $(0, -5)$ both satisfy $x + 5 = |y|$.

57. The domain is the set of first coordinates, $\{4, 7\}$, and the range is the set of second coordinates, $\{1\}$.

59. The domain is the set of first coordinates, $\{2\}$, and the range is the set of second coordinates, $\{3, 5, 7\}$.

61. Since any number can be use in place of x in $y = x + 1$, the domain is the set of all real numbers, $(-\infty, \infty)$. Since any number can be used in place of y the range is also $(-\infty, \infty)$.

63. Since any number can be use in place of x in $y = 5 - x$, the domain is the set of all real numbers, $(-\infty, \infty)$. Since any number can be used in place of y the range is also $(-\infty, \infty)$.

65. Since $y = \sqrt{x - 2}$, we must have $x - 2 \geq 0$ or $x \geq 2$. So the domain is $[2, \infty)$. The values of y must be nonnegative. The range is $[0, \infty)$.

67. Since $y = \sqrt{2x}$, we must have $2x \geq 0$ or $x \geq 0$. So the domain is $[0, \infty)$. The values of y must be nonnegative. The range is $[0, \infty)$.

69. $f(0) = 3(0) - 2 = -2$

71. $f(4) = 3(4) - 2 = 12 - 2 = 10$

73. $g(-2) = -(-2)^2 + 3(-2) - 2$
$$= -4 - 6 - 2 = -12$$

75. $h(-3) = |-3 + 2| = |-1| = 1$

77. $h(-4.236) = |-4.236 + 2|$
$$= |-2.236| = 2.236$$

79. $f(2) = 3(2) - 2 = 4$
$g(3) = -3^2 + 3(3) - 2 = -2$
$f(2) + g(3) = 4 + (-2) = 2$

81. $g(2) = -2^2 + 3(2) - 2 = 0$
$h(-3) = |-3 + 2| = 1$
$\dfrac{g(2)}{h(-3)} = \dfrac{0}{1} = 0$

83. $f(-1) = 3(-1) - 2 = -5$
$h(-4) = |-4 + 2| = 2$
$f(-1) \cdot h(-4) = -5 \cdot 2 = -10$

85. The graph of the function $f(x) = 2x - 3$ is a line with slope 2 and y-intercepts $(0, -3)$.

87. The graph of the function $f(x) = 5 - 2x$ is a line with slope -2 and y-intercept $(0, 5)$.

89. The graph of the function $f(x) = 3$ is a line with slope 0 and y-intercept $(0, 3)$.

91. a) $h(2) = 256 - 16(2)^2 = 192$ ft
b) $h(4) = 256 - 16(4)^2 = 0$ ft
93. The area of a square is found by squaring the length of a side: $A = s^2$ or $A(s) = s^2$.
95. The cost of the purchase, C, is the price per yard times the number of yards purchased: $C(x) = 3.98x$, $C(3) = 3.98(3) = \$11.94$.
97. The total cost, C, is found by multiplying the number of toppings, n, by the cost for each topping, \$0.50, and adding on the \$14.95 base charge: $\quad C(n) = 0.50n + 14.95$
$C(6) = 0.50(6) + 14.95 = \$17.95$
99. The relation $y = x + 2$ is a function because there is only one y for any value of x.

The relation $y > x + 2$ is not a function because $(2, 5)$ and $(2, 6)$ both satisfy $y > x + 2$.

Enriching Your Mathematical Word Power
1. d   **2.** a   **3.** a   **4.** b   **5.** c
6. a   **7.** b   **8.** b   **9.** a   **10.** c
11. b   **12.** c   **13.** b   **14.** a   **15.** d
16. b

CHAPTER 3 REVIEW

1. Since both coordinates are negative, the point $(-3, -2)$ lies in quadrant III.
3. The point $(\pi, 0)$ lies on the x-axis. The quadrants do not include points on any axis.
5. The point $(0, -1)$ is on the y-axis.
7. Since the first coordinate is positive and the second negative, the point $\left(\sqrt{2}, \ -3 \right)$ is in quadrant IV.
9. Use the midpoint formula:
$\left(\frac{8+2}{2}, \frac{-2+6}{2} \right) = \left(\frac{10}{2}, \frac{4}{2} \right) = (5, 2)$
Use the distance formula to find the length:
$\sqrt{(8-2)^2 + (-2-6)^2} = \sqrt{36 + 64}$
$\qquad\qquad = \sqrt{100} = 10$
11. Use the midpoint formula:
$\left(\frac{2+3}{2}, \frac{-2+1}{2} \right) = \left(\frac{5}{2}, -\frac{1}{2} \right)$
Use the distance formula to find the length:
$\sqrt{(2-3)^2 + (-2-1)^2} = \sqrt{1 + 9}$
$\qquad\qquad = \sqrt{10}$
13. If $x = 0$, $y = -3(0) + 2 = 2$.
The ordered pair is $(0, 2)$.
If $y = 0$, then $0 = -3x + 2$.
$\qquad 3x = 2$
$\qquad x = \frac{2}{3}$
The ordered pair is $\left(\frac{2}{3}, 0 \right)$.
If $x = 4$, $y = -3(4) + 2 = -10$.
The ordered pair is $(4, -10)$.
If $y = -3$, then $-3 = -3x + 2$.
$\qquad 3x = 5$
$\qquad x = \frac{5}{3}$
The ordered pair is $\left(\frac{5}{3}, -3 \right)$.
15. $m = \dfrac{9 - 6}{-2 - (-5)} = \dfrac{3}{3} = 1$

17. $m = \frac{-2-1}{-3-4} = \frac{-3}{-7} = \frac{3}{7}$

19. $m = \frac{-1-(-4)}{5-(-3)} = \frac{3}{8}$

The slope of any line parallel to one with slope 3/8 is also 3/8.

21. $m = \frac{-6-5}{4-(-3)} = \frac{-11}{7} = -\frac{11}{7}$

The slope of any line perpendicular to one with slope $-11/7$ is $7/11$.

23. The slope for $y = -3x + 4$ is -3, and the y-intercept is $(0, 4)$.

25. Write the equation in slope-intercept form.
$$y - 3 = \tfrac{2}{3}(x - 1)$$
$$y - 3 = \tfrac{2}{3}x - \tfrac{2}{3}$$
$$y = \tfrac{2}{3}x + \tfrac{7}{3}$$
The slope is $\frac{2}{3}$ and the y-intercept is $\left(0, \frac{7}{3}\right)$.

27.
$$y = \tfrac{2}{3}x - 4$$
$$3(y) = 3(\tfrac{2}{3}x - 4)$$
$$3y = 2x - 12$$
$$-2x + 3y = -12$$
$$2x - 3y = 12 \quad \text{(Standard form)}$$

29.
$$y - 1 = \tfrac{1}{2}(x + 3)$$
$$2(y - 1) = 2 \cdot \tfrac{1}{2}(x + 3)$$
$$2y - 2 = x + 3$$
$$-x + 2y = 5$$
$$x - 2y = -5 \quad \text{(Standard form)}$$

31. Start with point-slope form:
$$y - (-3) = \tfrac{1}{2}(x - 1)$$
$$y + 3 = \tfrac{1}{2}x - \tfrac{1}{2}$$
$$2y + 6 = x - 1$$
$$-x + 2y = -7$$
$$x - 2y = 7$$

33. Start with point-slope form:
$$y - 6 = -\tfrac{3}{4}(x - (-2))$$
$$y - 6 = -\tfrac{3}{4}x - \tfrac{3}{2}$$
$$4(y - 6) = 4(-\tfrac{3}{4}x - \tfrac{3}{2})$$
$$4y - 24 = -3x - 6$$
$$3x + 4y = 18$$

35. Any line with slope 0 is a horizontal line. Since it contains $(3, 5)$ its equation is $y = 5$.

37. To graph $y = 2x - 3$ note that the y-intercept is $(0, -3)$ the slope is $2 = 2/1$. Start at $(0, -3)$ and go up 2 units and 1 unit to the right to locate the second point $(1, -1)$. Draw a line through the two points.

39. If $x = 0$, $3(0) - 2y = -6$, or $y = 3$
If $y = 0$, then $3x - 2(0) = -6$, $x = -2$.
Draw a line through the intercepts $(0, 3)$ and $(-2, 0)$.

41. The equation $y - 3 = 10$ is equivalent to $y = 13$. Its graph is a horizontal line with a y-intercept of $(0, 13)$.

43. If $x = 0$, then $5(0) - 3y = 7$, or $y = -7/3$. If $y = 0$, then $5x - 3(0) = 7$, or $x = 7/5$. Draw a line through the intercepts $(0, -7/3)$ and $(7/5, 0)$.

45. For $5x + 4y = 100$, the intercepts are $(0, 25)$ and $(20, 0)$.

47. For $x - 80y = 400$ the intercepts are $(0, -5)$ and $(400, 0)$.

49. Use the slope $3 = 3/1$ and y-intercept $(0, -2)$ to graph the equation $y = 3x - 2$. Start at $(0, -2)$ and go up 3 units and 1 to the right. Draw a dashed line through the two points $(0, -2)$ and $(1, 1)$. Shade the region above the line to indicate the graph of $y > 3x - 2$.

51. First graph $x - y = 5$ using the intercepts $(0, -5)$ and $(5, 0)$. The line should be solid because of the inequality symbol $\leq$. Test the point $(0, 0)$ in the inequality $x - y \leq 5$:
$$0 - 0 \leq 5$$
Since $(0, 0)$ satisfies the inequality, shade the region containing $(0, 0)$.

53. The inequality $3x > 2$ is equivalent to $x > \frac{2}{3}$. First graph the vertical line $x = \frac{2}{3}$ as a dashed line through $\left(\frac{2}{3}, 0\right)$. The graph of $x > \frac{2}{3}$ is the region to the right of this line.

55. The inequality $4y \leq 0$ is equivalent to $y \leq 0$. Graph the line $y = 0$. The line $y = 0$ coincides with the x-axis. Draw the line solid. Points on the line and in the region below the line satisfy the inequality $y \leq 0$.

57. Solve $4x - 2y \geq 6$ for y:
$$-2y \geq -4x + 6$$
$$y \leq 2x - 3$$
All points on the line $y = 2x - 3$ together with the points below the line satisfy the inequality. Use y-intercept $(0, -3)$ and slope 2 to graph the solid line.

59. The x and y-intercepts for the line $5x - 2y = 9$ are $\left(0, -\frac{9}{2}\right)$ and $\left(\frac{9}{5}, 0\right)$. Draw a dashed line through the intercepts. Check $(0, 0)$ in the inequality $5x - 2y < 9$:
$$5(0) - 2(0) < 9 \text{ is correct.}$$
Shade the region containing $(0, 0)$.

$5x - 2y < 9$

61. First graph the lines $y = 3$ and $y - x = 5$.

$y > 3$ and $y - x < 5$

Test the points $(0, 0)$, $(0, 4)$, $(0, 6)$, and $(-6, 0)$. Only $(0, 4)$ satisfies $y > 3$ and $y - x < 5$. Shade the region containing $(0, 4)$.

63. First graph the lines $3x + 2y = 8$ and $3x - 2y = 6$. Test the points $(0, 0)$, $(0, 5)$, $(0, -6)$, and $(5, 0)$. The points $(0, 0)$, $(0, 5)$, and $(5, 0)$ satisfy the compound inequality $3x + 2y \geq 8$ or $3x - 2y \leq 6$. Shade the regions containing those points.

65. First graph the equations $x + 2y = 10$ and $x + 2y = -10$. Test the points $(0, 0)$, $(0, 8)$, and $(0, -8)$ in the inequality $|x + 2y| < 10$. Only $(0, 0)$ satisfies the inequality. So shade the region containing $(0, 0)$.

$|x + 2y| < 10$

67. First graph the two vertical lines $x = 5$ and $x = -5$.

$|x| \leq 5$

Test the points $(0, 0)$, $(8, 0)$, $(-8, 0)$ in the inequality $|x| \leq 5$. Only $(0, 0)$ satisfies the inequality. So shade the region containing $(0, 0)$, the region between the parallel lines.

69. First graph the lines $y - x = 2$ and $y - x = -2$. Test the points $(0, 0)$, $(0, 4)$, and $(0, -4)$ in the inequality $|y - x| > 2$. The points $(0, 4)$ and $(0, -4)$ satisfy the inequality. Shade the regions containing those points.

71. The relation is not a function because two ordered pairs $(5, 7)$ and $(5, 10)$ have the same first coordinate and different second coordinates.

73. The relation is a function because no two ordered pairs have the same x-coordinate and different y-coordinates.

75. The relation is a function because the condition that $y = x^2$ guarantees that no two ordered pairs will have the same x-coordinate and different y-coordinates.

77. This relation is not a function because the ordered pairs $(16, 2)$ and $(16, -2)$ are both in the relation.

79. The domain is the set of x-coordinates $\{3, 4, 5\}$, and the range is the set of y-coordinates $\{1, 5, 9\}$.

81. Since we can use any number for x in the equation $y = x + 1$, the domain is $(-\infty, \infty)$. Since y can be any real number in $y = x + 1$, the range is $(-\infty, \infty)$.

83. In the equation $y = \sqrt{x + 5}$, the value of $x + 5$ must be nonnegative. So $x + 5 \geq 0$ or $x \geq -5$. The domain is $[-5, \infty)$. Since y is equal to a square root, y must be nonnegative. The range is $[0, \infty)$.

85. $f(0) = 2(0) - 5 = -5$

87. $g(0) = 0^2 + 0 - 6 = -6$

89. $g\left(\frac{1}{2}\right) = \left(\frac{1}{2}\right)^2 + \frac{1}{2} - 6$
$$= \frac{1}{4} + \frac{2}{4} - \frac{24}{4} = -\frac{21}{4}$$

91. This line contains the points $(2, 0)$ and $(0, -6)$. The line has slope
$$m = \frac{-6 - 0}{0 - 2} = 3.$$
In slope-intercept form its equation is
$$y = 3x - 6,$$
and in standard form $3x - y = 6$.

93. Use the point-slope formula.
$$y - 4 = -\frac{1}{2}(x - (-1))$$

$$y - 4 = -\frac{1}{2}x - \frac{1}{2}$$
$$2y - 8 = -x - 1$$
$$x + 2y = 7$$

95. The line through $(2, -6)$ and $(2, 5)$ is a vertical line. All vertical lines are of the form $x = k$. So the equation is $x = 2$.

97. Since $x = 5$ is a vertical line, a line perpendicular to it is horizontal. The equation of a horizontal line through $(0, 0)$ is $y = 0$.

99. Any line parallel to $y = 2x + 1$ has slope 2. Use the point-slope formula with the point $(-1, 4)$ and slope 2:
$$y - 4 = 2(x - (-1))$$
$$y - 4 = 2x + 2$$
$$-2x + y = 6$$
$$2x - y = -6$$

101. The line with y-intercept $(0, 6)$ and slope 3 has equation $y = 3x + 6$. Written in standard form it is $3x - y = -6$.

103. A horizontal line has slope 0. If it goes through $(2, 5)$, its y-intercept is $(0, 5)$. Its equation is $y = 0 \cdot x + 5$ or simply $y = 5$.

105. Draw a quadrilateral with the 4 points as vertices. Find the slope of each side:
$$m_1 = \frac{-5 - (-1)}{-5 - (-3)} = \frac{-4}{-2} = 2$$

$$m_2 = \frac{-1 - 2}{-3 - 6} = \frac{-3}{-9} = \frac{1}{3}$$

$$m_3 = \frac{2 - (-2)}{6 - 4} = \frac{4}{2} = 2$$

$$m_4 = \frac{-2 - (-5)}{4 - (-5)} = \frac{3}{9} = \frac{1}{3}$$

Since the slopes of the opposite sides are equal, it is a parallelogram.

107. Plot the four points and find the slope of each side of the quadrilateral.
$$m_1 = \frac{2 - 6}{-2 - 2} = 1 \qquad m_2 = \frac{6 - 4}{2 - 4} = -1$$
$$m_3 = \frac{4 - 0}{4 - 0} = 1 \qquad m_4 = \frac{2 - 0}{-2 - 0} = -1$$
Since opposite sides have the same slope, the quadrilateral is a parallelogram. Since the slopes of the adjacent sides are 1 and -1, the adjacent sides are perpendicular. Therefore, the quadrilateral is a rectangle.

109. a) Find the slope of the line through $(20, 200)$ and $(70, 150)$.
$$m = \frac{200 - 150}{20 - 70} = \frac{50}{-50} = -1$$
Use the point-slope formula with $m = -1$ and the point $(20, 200)$:
$$h - h_1 = m(a - a_1)$$
$$h - 200 = -1(a - 20)$$
$$h - 200 = -a + 20$$
$$h = 220 - a$$
b) If $a = 40$, then $h = 220 - 40 = 180$.
c) The maximum heart rate decreases as you get older.

111. To determine the number of days it would take for the rental charge to equal $1080, we solve the equation

$$1080 = 26 + 17d.$$
$$1054 = 17d$$
$$62 = d$$

It will take 62 days for the rental charge to equal the cost of the air hammer.
If $d = 1$, then $C = 26 + 17(1) = 43$.
If $d = 30$, then $C = 26 + 17(30) = 536$.
The graph of this function for d ranging from 1 to 30 is a straight line segment joining the two points $(1, 43)$ and $(30, 536)$.

CHAPTER 3 TEST

1. If $x = 0$, then $2(0) + y = 5$.
$$y = 5$$
If $y = 0$, then $2x + 0 = 5$.
$$x = 5/2$$
If $x = 4$, then $2(4) + y = 5$.
$$y = -3$$
If $y = -8$, then $2x + (-8) = 5$.
$$2x = 13$$
$$x = 13/2$$
The pairs are $(0, 5)$, $\left(\frac{5}{2}, 0\right)$, and

$\left(\frac{13}{2}, -8\right)$.

2. $m = \frac{7 - 1}{-3 - 2} = \frac{6}{-5} = -\frac{6}{5}$

3. Use the midpoint formula:
$$\left(\frac{2 + (-3)}{2}, \frac{0 + (-1)}{2}\right) = \left(-\frac{1}{2}, -\frac{1}{2}\right)$$
Use the distance formula to find the length:
$$\sqrt{(2 - (-3))^2 + (0 - (-1))^2} = \sqrt{25 + 1}$$
$$= \sqrt{26}$$

4. Solve the equation $8x - 5y = -10$ for y:
$$-5y = -8x - 10$$
$$y = \frac{8}{5}x + 2$$
The slope is $\frac{8}{5}$ and the y-intercept is $(0, 2)$.

5. Plot the four points and find the slopes of the sides of the quadrilateral.
$$m_1 = \frac{2 - 0}{6 - 0} = \frac{1}{3} \qquad m_2 = \frac{2 - 0}{6 - 5} = 2$$

$$m_3 = \frac{-2 - 0}{-1 - 5} = \frac{1}{3} \qquad m_4 = \frac{-2 - 0}{-1 - 0} = 2$$

Since the opposite sides have equal slopes, the quadrilateral is a parallelogram.

6. The first coordinate is a and the second coordinate is V. We want the equation of the line containing the two points $(0, 22{,}000)$ and $(3, 16{,}000)$. The slope of the line is
$$m = \frac{22{,}000 - 16{,}000}{0 - 3} = -2000.$$
Using the slope and y-intercept, we can write the equation $V = -2000a + 22{,}000$.

7. The equation of the line with y-intercept $(0, 3)$ and slope $-1/2$ is
$$y = -\frac{1}{2}x + 3$$
$$2y = -x + 6$$
$$x + 2y = 6$$

8. Use the point-slope form to write the equation of the line through $(-3, 5)$ with slope -4:
$$y - 5 = -4(x - (-3))$$
$$y - 5 = -4(x + 3)$$
$$y - 5 = -4x - 12$$
$$4x + y = -7$$

9. Solve $3x - 5y = 7$ for y:
$$-5y = -3x + 7$$
$$y = \frac{3}{5}x - \frac{7}{5}$$
Since this line has slope $3/5$, any line perpendicular to it has slope $-5/3$. The line

58

through $(2, 3)$ with slope $-5/3$ has equation
$$y - 3 = -\frac{5}{3}(x - 2)$$
$$y - 3 = -\frac{5}{3}x + \frac{10}{3}$$
$$3y - 9 = -5x + 10$$
$$5x + 3y = 19$$

10. The line shown goes through the points $(0, 2)$ and $(-4, 0)$. We can see from the graph that its slope is $2/4$ or $1/2$ and its y-intercept is $(0, 2)$. In slope-intercept form its equation is
$$y = \tfrac{1}{2}x + 2.$$
$$2y = x + 4$$
$$-x + 2y = 4$$
$$x - 2y = -4$$

11. The graph of $y = 4$ is a horizontal line through $(0, 4)$.

12. The graph of $x = 3$ is a vertical line through $(3, 0)$.

13. The graph of $3x + 4y = 12$ has x-intercept $(4, 0)$ and y-intercept $(0, 3)$.

14. The graph of $y = \frac{2}{3}x - 2$ has y-intercept $(0, -2)$ and slope $2/3$.

15. First graph the equation $y = (-1/2)x + 3$, using y-intercept of $(0, 3)$, a slope of $-1/2$, and a dashed line. The graph of $y > (-1/2)x + 3$ is the region above this line.

16. Graph the vertical line $x = 2$ using a dashed line. Graph $x + y = 0$ using a dashed line. Test the points $(-2, 0)$, $(1, 0)$, $(5, 0)$, and $(3, -5)$. Only $(5, 0)$ satisfies $x > 2$ and $x + y > 0$. So shade the region containing $(5, 0)$.

17. First graph the parallel lines $2x + y = 3$ and $2x + y = -3$, using solid lines. Test the points $(-5, 0)$, $(0, 0)$ and $(5, 0)$ in $|2x + y| \geq 3$. Both $(-5, 0)$ and $(5, 0)$ satisfy the inequality, so shade the regions containing those points.

18. This set of ordered pairs is a function because no two of the ordered pairs have the same first coordinate and different second coordinates.

19. If $f(x) = -2x + 5$, then $f(-3) = -2(-3) + 5 = 11$.

20. Since $x - 7$ must be nonnegative, we have $x - 7 \geq 0$, or $x \geq 7$. So the domain is $[7, \infty)$. Since y is equal to a square root, y must be nonnegative also. So the range is $[0, \infty)$.

21. The shipping and handling fee S is a linear function of the weight of the order n, $S = 0.50n + 3$.

22. $A(2) = -16(2)^2 + 32(2) + 6 = 6$ ft

Making Connections

Chapters 1 - 3

1. $2^3 \cdot 4^2 = 8 \cdot 16 = 128$

2. $2^7 - 2^6 = 128 - 64 = 64$.

3. $3^2 - 4(5)(-2) = 9 + 40 = 49$

4. $3 - 2|5 - 7 \cdot 3| = 3 - 2 \cdot 16 = -29$

5. $\dfrac{2 - (-3)}{5 - 6} = \dfrac{5}{-1} = -5$

6. $\dfrac{-3 - 7}{-1 - (-3)} = \dfrac{-10}{2} = -5$

7. $3t \cdot 4t = 12t^2$

8. $3t + 4t = 7t$

9. $\dfrac{4x + 8}{4} = x + 2$

10. $\dfrac{-8y}{-4} - \dfrac{10y}{-2} = 2y - (-5y) = 7y$

11. $3(x - 4) - 4(5 - x)$
$\quad = 3x - 12 - 20 + 4x = 7x - 32$

12. $-2(3x^2 - x) + 3(2x - 5x^2)$
$\quad = -6x^2 + 2x + 6x - 15x^2$
$\quad = -21x^2 + 8x$

13. $\quad 15(b - 27) = 0$
$\quad\quad 15b - 405 = 0$
$\quad\quad\quad\quad 15b = 405$
$\quad\quad\quad\quad\quad b = 27$
The solution set is $\{27\}$.

14. $0.05a - 0.04(a - 50) = 4$
$\quad\quad\quad 0.01a + 2 = 4$
$\quad\quad\quad\quad 0.01a = 2$
$\quad\quad\quad\quad\quad a = 200$
The solution set is $\{200\}$.

15. $\quad |\,3v - 7\,| = 0$
$\quad\quad 3v - 7 = 0$
$\quad\quad\quad 3v = 7$
$\quad\quad\quad\quad v = \dfrac{7}{3}$
The solution set is $\left\{\dfrac{7}{3}\right\}$.

16. $\quad |\,3u - 7\,| = 3$
$\quad 3u - 7 = 3$ or $3u - 7 = -3$
$\quad\quad 3u = 10$ or $\quad 3u = 4$
$\quad\quad u = \dfrac{10}{3}$ or $\quad u = \dfrac{4}{3}$
The solution set is $\{4/3, 10/3\}$.

17. $\quad |\,3x - 7\,| = -77$
The absolute value of any quantity is nonnegative. So the equation has no solution. The solution set is $\emptyset$.

18. $\quad |\,3x - 7\,| + 1 = 8$
$\quad\quad |\,3x - 7\,| = 7$
$\quad 3x - 7 = 7$ or $3x - 7 = -7$
$\quad\quad 3x = 14$ or $\quad 3x = 0$
$\quad\quad x = \dfrac{14}{3}$ or $\quad x = 0$
The solution set is $\{0, 14/3\}$.

19. $\quad 2x - 1 > 7$
$\quad\quad 2x > 8$
$\quad\quad\quad x > 4$

20. $\quad 5 - 3x \leq -1$
$\quad\quad -3x \leq -6$
$\quad\quad\quad x \geq 2$

21. $\quad x - 5 \leq 4$ and $3x - 1 < 8$
$\quad\quad x \leq 9$ and $\quad 3x < 9$
$\quad\quad x \leq 9$ and $\quad x < 3$

The last compound inequality is equivalent to the simple inequality $x < 3$.

22. $2x \le -6$ or $5 - 2x < -7$

$\quad\quad x \le -3 \quad\quad$ or $\quad -2x < -12$

$\quad\quad x \le -3 \quad\quad$ or $\quad\quad x > 6$

23. $|x - 3| < 2$

$\quad\quad -2 < x - 3 < 2$

$\quad\quad\quad 1 < x < 5$

24. $|1 - 2x| \ge 7$

$\quad 1 - 2x \ge 7 \quad$ or $1 - 2x \le -7$

$\quad\quad -2x \ge 6 \quad$ or $\quad -2x \le -8$

$\quad\quad\quad x \le -3 \quad$ or $\quad\quad x \ge 4$

25.   First graph the dashed line $y = 2x - 1$ through $(0, -1)$ with slope 2. Now shade below the line for $y < 2x - 1$.

26.   First graph the solid line $3x - y = 2$ or $y = 3x - 2$ through $(0, -2)$ with slope 3. Now test $(0, 0)$ in $3x - y \le 2$. Since $(0, 0)$ satisfies $3x - y \le 2$, we shade the side of the line containing $(0, 0)$.

27.   Graph the dashed line $y = x$ through $(0, 0)$ and $(1, 1)$. Next graph the dashed lin $y = -3x + 5$ through $(0, 5)$ with slope -3. Now test a point in each of the four regions. Only points that lie above $y = x$ and below $y = -3x + 5$ satisfy both inequalities. So shade that region.

28. The graph of $y < 2$ consists of the points below the horizontal line $y = 2$. The graph of $x > -3$ consists of the points to the right of the vertical line $x = -3$. The region that lies below $y = 2$ and to the right of $x = -3$ is shaded on the graph. The boundary lines are drawn solid because of the $\le$ and $\ge$ symbols.

29.   **a)** For ages 62 through 64
$b = 500a - 24,000$.
For ages 65 through 67
$b = 667a - 34,689$.
For ages 68 through 70
$b = 800a - 43,600$.
b) At age 64 $b = 500(64) - 24,000 = \$8000$
c) If a person get \$11,600 then
$$11,600 = 800a - 43,600$$
$$55,200 = 800a$$
$$69 = a$$
d) The slopes 500, 667, and 800 indicate the additional amount per year received beyond the basic amount in each category.

4.1 WARM-UPS

1. True, because if $x = 1$ and $y = 2$, then $2(1) + 2 = 4$ is correct.

2. False, because if $x = 1$ and $y = 2$, then $3(1) - 2 = 6$ is incorrect. Both equations must be satisfied for the compound statement using "and" to be true.

3. False, because $(2, 3)$ does not satisfy $4x - y = -5$.

4. True.

5. True, because when we substitute, one of the variables is eliminated.

6. True, because each of these lines has slope 3 and they are parallel.

7. True, because the lines are not parallel and they have different y-intercepts.

8. True, because the lines are parallel and have different y-intercepts.

9. True, because dependent equations have the same solution sets.

10. True, because a system of independent equations has one point in its solution set.

4.1 EXERCISES

1. The intersection point of the graphs is the solution to an independent system.

3. The graphing method can be very inaccurate.

5. If the equation you get after substituting turns out to be incorrect, such as $0 = 9$, then the system has no solution.

7. The graph of $y = 2x$ is a straight line with y-intercept $(0, 0)$ and slope 2. The graph of $y = -x + 3$ is a straight line with y-intercept $(0, 3)$ and slope -1. The graphs appear to intersect at $(1, 2)$. Check that $(1, 2)$ satisfies both equations. The solution set is $\{(1, 2)\}$.

9. The graph of $y = 2x - 1$ is a straight line with y-intercept $(0, -1)$ and slope 2. The graph of $2y = x - 2$ or $y = \frac{1}{2}x - 1$ is a straight line with y-intercept $(0, -1)$ and slope $\frac{1}{2}$. The graphs intersect at $(0, -1)$. The solution set to the system is $\{(0, -1)\}$.

11. The graph of $y = x - 3$ is a line with y-intercept $(0, -3)$ and slope 1. The graph of

$x - 2y = 4$ or $y = \frac{1}{2}x - 2$ is a line with y-intercept $(0, -2)$ and slope $\frac{1}{2}$. The lines appear to intersect at $(2, -1)$. To be sure, check that $(2, -1)$ satisfies both of the original equations. The solution set is $\{(2, -1)\}$.

13. If we rewrite these equations in slope-intercept form, we get $y = x + 1$ and $y = x + 3$. The graphs are parallel lines with slopes of 1 and y-intercepts of $(0, 1)$ and $(0, 3)$. Since the lines do not intersect, the solution set is the empty set, $\emptyset$.

15. Solving $x + 2y = 8$ for y yields $y = -\frac{1}{2}x + 4$, which is the same as the first equation. So the equations have the same graph. All points on the line satisfy both equations. The solution set is $\{(x, y) \mid x + 2y = 8\}$.

17. The graph of $y = 2x + 4$ is a line with slope 2 and y-intercept $(0, 4)$. The graph of $3x + y = -1$ or $y = -3x - 1$ is a line with slope -3 and y-intercept $(0, -1)$. The lines appear to intersect at $(-1, 2)$. Check that $(-1, 2)$ satisfies both equations. The solution set is $\{(-1, 2)\}$

19. The graph of $y = -\frac{1}{4}x$ is a line with slope $-1/4$ and y-intercept $(0, 0)$. The graph of $x + 4y = 8$ or $y = -\frac{1}{4}x + 2$ is a line with slope $-1/4$ and y-intercept $(0, 2)$. The lines have the same slope and do not intersect. The solution set is the empty set, $\emptyset$.

21. The lines in graph (c) intersect at $(3, -2)$ and $(3, -2)$ satisfies both equations of the given system.

23. The lines in graph (b) intersect at $(-3, -2)$ and $(-3, -2)$ satisfies both equations.

25. Substitute $y = x - 5$ into $2x - 5y = 1$.
$$2x - 5(x - 5) = 1$$
$$2x - 5x + 25 = 1$$
$$-3x = -24$$
$$x = 8$$
Use $x = 8$ in $y = x - 5$ to find y.
$$y = 8 - 5$$
$$y = 3$$

The solution set is $\{(8, 3)\}$. The equations are independent.

27. Substitute $x = 2y - 7$ into $3x + 2y = -5$.
$$3(2y - 7) + 2y = -5$$
$$6y - 21 + 2y = -5$$
$$8y = 16$$
$$y = 2$$
Use $y = 2$ in $x = 2y - 7$ to find x.
$$x = 2(2) - 7 = -3$$
The solution set is $\{(-3, 2)\}$. The equations are independent.

29. Substitute $y = 2x - 30$ into the other equation.
$$\tfrac{1}{5}x - \tfrac{1}{2}(2x - 30) = -1$$
$$\tfrac{1}{5}x - x + 15 = -1$$
$$5\left(\tfrac{1}{5}x - x + 15\right) = 5(-1)$$
$$x - 5x + 75 = -5$$
$$-4x = -80$$
$$x = 20$$
If $x = 20$, then
$$y = 2x - 30 = 2(20) - 30 = 10.$$
The solution set is $\{(20, 10)\}$ and the equations are independent.

31. Write $x - y = 5$ as $x = y + 5$ and substitute into $2x = 2y + 14$.
$$2(y + 5) = 2y + 14$$
$$2y + 10 = 2y + 14$$
$$10 = 14$$
Since this last equation is incorrect no matter what values are used for x and y, the equations are inconsistent and the solution set is $\emptyset$.

33. Substitute $y = 2x - 5$ into
$y + 1 = 2(x - 2)$.
$$2x - 5 + 1 = 2(x - 2)$$
$$2x - 4 = 2x - 4$$
Since the last equation is an identity, the equations are dependent. Any ordered pair that satisfies one equation will satisfy the other. The solution set is
$\{(x, y) \mid y = 2x - 5\}$.

35. Write $2x + y = 9$ as $y = -2x + 9$ and substitute into $2x - 5y = 15$.
$$2x - 5(-2x + 9) = 15$$
$$2x + 10x - 45 = 15$$
$$12x - 45 = 15$$
$$12x = 60$$
$$x = 5$$
Use $x = 5$ in $y = -2x + 9$ to find y.
$$y = -2(5) + 9 = -1$$

The solution set is $\{(5, -1)\}$. The equations are independent.

37. Write $x - y = 0$ as $y = x$ and substitute into $2x + 3y = 35$.
$$2x + 3x = 35$$
$$5x = 35$$
$$x = 7$$
Since $y = x$, $y = 7$ also. The solution set is $\{(7, 7)\}$ and the equations are independent.

39. Write $x + y = 40$ as $x = 40 - y$ and substitute into $0.2x + 0.8y = 23$.
$$0.2(40 - y) + 0.8y = 23$$
$$8 - 0.2y + 0.8y = 23$$
$$0.6y = 15$$
$$y = 25$$
Since $x = 40 - y$, we get $x = 40 - 25 = 15$. The solution set is $\{(15, 25)\}$ and the equations are independent.

41. Write $2x = 3y - 8$ as $y = \tfrac{2}{3}x + \tfrac{8}{3}$.
Substitute $y = \tfrac{2}{3}x + \tfrac{8}{3}$ into $5y = 10x - 5$.
$$5\left(\tfrac{2}{3}x + \tfrac{8}{3}\right) = 10x - 5$$
$$\tfrac{10}{3}x + \tfrac{40}{3} = 10x - 5$$
$$10x + 40 = 30x - 15$$
$$-20x = -55$$
$$x = \tfrac{11}{4}$$
$$y = \tfrac{2}{3}x + \tfrac{8}{3}$$
$$y = \tfrac{2}{3}\left(\tfrac{11}{4}\right) + \tfrac{8}{3} = \tfrac{11}{6} + \tfrac{16}{6} = \tfrac{9}{2}$$

The solution set is $\left\{\left(\tfrac{11}{4}, \tfrac{9}{2}\right)\right\}$. The system is independent.

43. Substitute $y = \tfrac{5}{7}x$ into $x = -\tfrac{2}{3}y$:
$$x = -\tfrac{2}{3}\left(\tfrac{5}{7}x\right)$$
$$x = -\tfrac{10}{21}x$$
$$21x = -10x$$
$$31x = 0$$
$$x = 0$$
$$y = \tfrac{5}{7}x$$
$$y = \tfrac{5}{7} \cdot 0 = 0$$

The solution set is $\{(0, 0)\}$. The system is independent.

63

45. Write $x - y = 5$ as $x = y + 5$ and substitute into $x + y = 4$.
$$y + 5 + y = 4$$
$$2y = -1$$
$$y = -1/2$$
If $y = -1/2$, then
$$x = y + 5 = -1/2 + 5 = 9/2.$$
The solution set is $\left\{ \left(\frac{9}{2}, -\frac{1}{2} \right) \right\}$ and the equations are independent.

47. First simplify.
$$3(y - 1) = 2(x - 3)$$
$$3y - 3 = 2x - 6$$
$$3y - 2x = -3$$
By simplifying the first equation we see that it is the same as the second equation. There is no need to substitute now. The equations are dependent and the solution set is
$\{(x, y) \mid 3y - 2x = -3\}$.

49. Substitute $y = 3x$ into $y = 3x + 1$.
$$3x = 3x + 1$$
$$0 = 1$$
Since the last equation is false, the equations are inconsistent. The solution set to the system is the empty set, $\emptyset$.

51. Write $x - y = 0.1$ as $x = y + 0.1$ and substitute into $2x - 3y = -0.5$.
$$2(y + 0.1) - 3y = -0.5$$
$$2y + 0.2 - 3y = -0.5$$
$$-y = -0.7$$
$$y = 0.7$$
If $y = 0.7$, then $x = 0.7 + 0.1 = 0.8$. The solution set is $\{(0.8, 0.7)\}$ and the equations are independent.

53. Let x represent the width and y represent the length, then $2x + 2y = 82$ and $y = x + 15$.
$$2x + 2(x + 15) = 82$$
$$4x + 30 = 82$$
$$4x = 52$$
$$x = 13$$
$$y = 13 + 15 = 28$$
The length is 28 feet and the width is 13 feet.

55. Let $x = $ the amount invested at 5% and $y = $ the amount invested at 10%. The interest earned on the investments is $0.05x$ and $0.10y$ respectively. We can write two equations, one expressing the total amount invested, and the other expressing the total of the interest earned.
$$x + y = 30,000$$
$$0.05x + 0.10y = 2,300$$
Write the first equation as $y = 30,000 - x$ and substitute into the second.
$$0.05x + 0.10(30,000 - x) = 2,300$$
$$0.05x + 3,000 - 0.10x = 2,300$$
$$-0.05x = -700$$
$$x = 14,000$$
If $x = 14,000$, then
$y = 30,000 - 14,000 = 16,000$. She invested $14,000 at 5% and $16,000 at 10%.

57. Let x and y be the numbers. The fact that their sum is 2 and their difference is 26 gives us a system of equations:
$$x + y = 2$$
$$x - y = 26$$
Write $x + y = 2$ as $x = 2 - y$ and substitute.
$$2 - y - y = 26$$
$$-2y = 24$$
$$y = -12$$
If $y = -12$, then
$x = 2 - y = 2 - (-12) = 14$. The numbers are -12 and 14.

59. Let $x = $ the number of toasters and $y = $ the number of vacation coupons. We write one equation expressing the fact that the total number of prizes is 100 and the other expressing the total bill of $708.
$$x + y = 100$$
$$6x + 24y = 708$$
Write $x + y = 100$ as $x = 100 - y$ and substitute.
$$6(100 - y) + 24y = 708$$
$$600 - 6y + 24y = 708$$
$$18y = 108$$
$$y = 6$$
If $y = 6$, then $x = 100 - y = 100 - 6 = 94$. He gave away 94 toasters and 6 vacation coupons.

61. $s = 0.05(100,000 - f)$
$f = 0.30(100,000 - s)$

Simplify:
$$s = 5000 - 0.05f$$
$$f = 30,000 - 0.30s$$

Substitute:
$$f = 30,000 - 0.30(5000 - 0.05f)$$
$$f = 30,000 - 1500 + 0.015f$$
$$0.985f = 28,500$$
$$f \approx 28,934$$
$$s = 5,000 - 0.05(28,934)$$
$$\approx 3553$$
Rounding to the nearest dollar, the state tax is $3,553 and federal tax is $28,934.

63.
$$B = 0.20N$$
$$N = 120,000 - B$$
Substitute:
$$N = 120,000 - 0.20N$$
$$1.2N = 120,000$$
$$N = 100,000$$
$$B = 0.20 \cdot 100,000 = 20,000$$
The bonus is $20,000.

65. a) Use $x = 10,000$ in $C = 10x + 400,000$ to get the cost of 10,000 textbooks. The cost is $500,000.
b) Use $x = 10.000$ in $F = 30x$ to get the revenue for 10,000 textbooks. The revenue for 10,000 textbooks is $300,000.
c) The cost and revenue are equal when
$$30x = 10x + 400,000$$
$$20x = 400,000$$
$$x = 20,000.$$
The cost is equal to the revenue for 20,000 textbooks.
d) Let $x = 0$ in $C = 10x + 400,000$ to get $C = 400,000$. The fixed cost is $400,000.

67. Solve each equation for y to get $y = \frac{2}{3}x - 2$ in every case except (a), which is $y = \frac{2}{3}x + 2$.

69. a) Graph the equations and use the intersect feature of the calculator to find the point of intersection: $(2.8, 2.6)$.
b) Solve the equations for y and then graph them on a calculator. Use the intersect feature to find the point of intersection: $(1.0, -0.2)$.

4.2 WARM-UPS

1. True, because when we add the equations, the y-terms add up to 0 and one variable is eliminated.

2. False, multiply the first by -2 and the second by 3 to eliminate x.
3. True, multiplying the second by 2 and adding will eliminate both x and y.
4. True, because both ordered pairs satisfy both equations.
5. False, solution set is $\{(x, y) \mid 4x - 2y = 20\}$.
6. True, because the equations are inconsistent.
7. True.
8. False, as long as we add the left sides and the right sides, the form does not matter.
9. True.
10. True.

4.2 EXERCISES

1. In this section we learned the addition method.
3. In some cases we multiply one or both of the equations on each side to change the coefficients of the variable that we are trying to eliminate.
5. If an identity (such as $0 = 0$) results form addition of the equations, then the equations are dependent.

7.
$$\begin{aligned} x + y &= 7 \\ \underline{x - y} &= \underline{9} \\ 2x &= 16 \\ x &= 8 \end{aligned}$$
Use $x = 8$ in $x + y = 7$ to find y.
$$8 + y = 7$$
$$y = -1$$
The solution set is $\{(8, -1)\}$.

9.
$$\begin{aligned} x - y &= 12 \\ \underline{2x + y} &= \underline{3} \\ 3x &= 15 \\ x &= 5 \end{aligned}$$
Use $x = 5$ in $x - y = 12$.
$$5 - y = 12$$
$$-7 = y$$
The solution set is $\{(5, -7)\}$.

11.
$$\begin{aligned} 3x - y &= 5 \\ \underline{5x + y} &= \underline{-2} \\ 8x &= 3 \\ x &= \frac{3}{8} \end{aligned}$$

Use $x = \frac{3}{8}$ in $5x + y = -2$.
$$5 \cdot \frac{3}{8} + y = -2$$
$$y = -2 - \frac{15}{8} = -\frac{31}{8}$$
The solution set is $\left\{ \left(\frac{3}{8}, -\frac{31}{8} \right) \right\}$.

13. If we multiply $2x - y = -5$ by 2, we get $4x - 2y = -10$. Add this equation to the second equation.
$$4x - 2y = -10$$
$$3x + 2y = 3$$
$$\overline{7x \quad\;\; = -7}$$
$$x = -1$$
Use $x = -1$ in $2x - y = -5$.
$$2(-1) - y = -5$$
$$-2 - y = -5$$
$$3 = y$$
The solution set is $\{(-1, 3)\}$.

15. If we multiply the first equation $-3x + 5y = 1$ by 3, we get $-9x + 15y = 3$. Add this equation to the second equation.
$$-9x + 15y = 3$$
$$9x - 3y = 5$$
$$\overline{\quad\;\; 12y = 8}$$
$$y = \frac{2}{3}$$
Use $y = 2/3$ in $9x - 3y = 5$:
$$9x - 3 \cdot \frac{2}{3} = 5$$
$$9x - 2 = 5$$
$$9x = 7$$
$$x = \frac{7}{9}$$
The solution set is $\left\{ \left(\frac{7}{9}, \frac{2}{3} \right) \right\}$.

17. Multiply the first equation by 4 and the second by 5.
$$4(2x - 5y) = 4(13)$$
$$5(3x + 4y) = 5(-15)$$

$$8x - 20y = 52$$
$$15x + 20y = -75$$
$$\overline{23x \quad\quad\; = -23}$$
$$x = -1$$
Use $x = -1$ in $2x - 5y = 13$.
$$2(-1) - 5y = 13$$
$$-2 - 5y = 13$$
$$-5y = 15$$
$$y = -3$$
The solution set is $\{(-1, -3)\}$.

19. Rewrite the first equation to match the form of the second.
$$2x - 3y = 11$$
$$7x - 4y = 6$$
Multiply the first equation by -7 and the second by 2.
$$-7(2x - 3y) = -7(11)$$
$$2(7x - 4y) = 2(6)$$

$$-14x + 21y = -77$$
$$14x - 8y = 12$$
$$\overline{\quad\;\; 13y = -65}$$
$$y = -5$$
Use $y = -5$ in $2x = 3y + 11$.
$$2x = 3(-5) + 11$$
$$2x = -4$$
$$x = -2$$
The solution set is $\{(-2, -5)\}$.

21.
$$-12(x + y) = -12(48)$$
$$12x + 14y = 628$$

$$-12x - 12y = -576$$
$$12x + 14y = 628$$
$$\overline{\quad\;\; 2y = 52}$$
$$y = 26$$
$$x + 26 = 48$$
$$x = 22$$
The solution set is $\{(22, 26)\}$.

23.
$$3x - 4y = 9$$
$$-3x + 4y = 12$$
$$\overline{\quad\;\; 0 = 21}$$
Since we obtained an incorrect equation by addition, the equations are inconsistent and the solution set is $\emptyset$.

25. Multiply $5x - y = 1$ by -2, to get $-10x + 2y = -2$. Add this equation to the second equation.
$$-10x + 2y = -2$$
$$10x - 2y = 2$$
$$\overline{\quad\;\; 0 = 0}$$
Since we obtained an identity by adding the equations, the equations are dependent and the solution set is $\{(x, y) \mid 5x - y = 1\}$.

27.
$$2x - y = 5$$
$$2x + y = 5$$
$$\overline{4x \quad\;\; = 10}$$
$$x = 10/4 = 5/2$$

Use $x = 5/2$ in $2x + y = 5$ to find y.
$$2(5/2) + y = 5$$
$$5 + y = 5$$
$$y = 0$$
The solution set is $\{(5/2, 0)\}$ and the equations are independent.

29. Multiplying the first equation by 12 to eliminate the fractions gives us the following system.
$$3x + 4y = 60$$
$$x - y = 6$$
Multiply the second equation by 4 and then add.

$$3x + 4y = 60$$
$$\underline{4x - 4y = 24}$$
$$7x \qquad = 84$$
$$x = 12$$
Use $x = 12$ in $x - y = 6$ to find y.
$$12 - y = 6$$
$$6 = y$$
The solution set is $\{(12, 6)\}$.

31. If we multiply the first equation by 12 and the second by 24 to eliminate fractions, we get the following system.
$$3x - 4y = -48$$
$$\underline{3x + 4y = 0}$$
$$6x \qquad = -48$$
$$x = -8$$
Use $x = -8$ in $3x + 4y = 0$ to find y.
$$3(-8) + 4y = 0$$
$$4y = 24$$
$$y = 6$$
The solution set is $\{(-8, 6)\}$.

33. Multiply the first equation by 8 and the second by -16 to get the following system.
$$x + 2y = 40$$
$$\underline{-x - 8y = -112}$$
$$-6y = -72$$
$$y = 12$$
$$x + 2(12) = 40$$
$$x = 16$$
The solution set is $\{(16, 12)\}$.

35. Multiply the first equation by 6 and the second by 12 to get the following system.
$$2x + 3y = 2$$
$$10x - 9y = 2$$

Now multiply the first equation by -5 to get the following system:
$$-10x - 15y = -10$$
$$\underline{10x - 9y = 2}$$
$$-24y = -8$$
$$y = \frac{1}{3}$$
Use $y = 1/3$ in $\frac{1}{3}x + \frac{1}{2}y = \frac{1}{3}$
$$\frac{1}{3}x + \frac{1}{2} \cdot \frac{1}{3} = \frac{1}{3}$$
$$2x + 1 = 2$$
$$2x = 1$$
$$x = \frac{1}{2}$$
The solution set is $\left\{\left(\frac{1}{2}, \frac{1}{3}\right)\right\}$.

37. Multiply the first equation by 100 and the second by -10 to get the following system.
$$5x + 10y = 130$$
$$\underline{-10x - 10y = -190}$$
$$-5x \qquad = -60$$
$$x = 12$$
Use $x = 12$ in $x + y = 19$ to get $y = 7$. The solution set is $\{(12, 7)\}$.

39. Multiply the first equation by -9 and the second by 100 to get the following system.
$$-9x - 9y = -10{,}800$$
$$\underline{12x + 9y = 12{,}000}$$
$$3x \qquad = 1200$$
$$x = 400$$
Use $x = 400$ in $x + y = 1200$ to get $y = 800$. The solution set is $\{(400, 800)\}$.

41. Multiply the first equation by -2 to get the following system.
$$-3x + 4y = 0.5$$
$$\underline{3x + 1.5y = 6.375}$$
$$5.5y = 6.875$$
$$y = 1.25$$
Use $y = 1.25$ in $3x + 1.5y = 6.375$ to find x.
$$3x + 1.5(1.25) = 6.375$$
$$3x + 1.875 = 6.375$$
$$3x = 4.5$$
$$x = 1.5$$
The solution set is $\{(1.5, 1.25)\}$.

43. Multiply both equations by 100 to get the following system.
$$24x + 60y = 58$$
$$80x - 12y = 52$$

Multiply the second equation by 5 and add:
$$24x + 60y = 58$$
$$\underline{400x - 60y = 260}$$
$$424x = 318$$
$$x = \frac{318}{424} = \frac{3}{4}$$
Use $x = 3/4$ in $24x + 60y = 58$ to find y.
$$24 \cdot \frac{3}{4} + 60y = 58$$
$$60y = 40$$
$$y = \frac{40}{60} = \frac{2}{3}$$
The solution set is $\left\{ \left(\frac{3}{4}, \frac{2}{3} \right) \right\}$.

45. Substitute $y = x + 1$ into $2x - 5y = -20$:
$$2x - 5(x + 1) = -20$$
$$-3x - 5 = -20$$
$$-3x = -15$$
$$x = 5$$
If $x = 5$, then $y = 5 + 1 = 6$. The solution set is $\{(5, 6)\}$.

47.
$$x - y = 19$$
$$\underline{2x + y = -13}$$
$$3x = 6$$
$$x = 2$$
Use $x = 2$ in $x - y = 19$.
$$2 - y = 19$$
$$-17 = y$$
The solution set is $\{(2, -17)\}$.

49. Substitute $x = y - 1$ into $2y = x + 2$:
$$2y = y - 1 + 2$$
$$y = 1$$
If $y = 1$, then $x = 1 - 1 = 0$. The solution set is $\{(0, 1)\}$.

51.
$$2y - 3x = -1$$
$$\underline{5y + 3x = 29}$$
$$7y = 28$$
$$\cdot y = 4$$
Use $y = 4$ in $5y + 3x = 29$.
$$5(4) + 3x = 29$$
$$3x = 9$$
$$x = 3$$
The solution set is $\{(3, 4)\}$.

53. Substitute $y = \frac{2}{3}x$ into $6x + 3y = 4$:
$$6x + 3\left(\frac{2}{3}x \right) = 4$$
$$6x + 2x = 4$$
$$8x = 4$$
$$x = \frac{1}{2}$$

$$y = \frac{2}{3}x$$
$$y = \frac{2}{3} \cdot \frac{1}{2} = \frac{1}{3}$$
The solution set is $\left\{ \left(\frac{1}{2}, \frac{1}{3} \right) \right\}$.

55. Substitute $y = 3x + 1$ into $x = \frac{1}{3}y + 5$:
$$x = \frac{1}{3}(3x + 1) + 5$$
$$x = x + \frac{1}{3} + 5$$
$$0 = \frac{16}{3}$$
The solution set is the empty set, $\emptyset$.

57.
$$x - y = 0$$
$$\underline{x + y = 2x}$$
$$2x = 2x$$
The system is dependent and the solution set is $\{(x, y) | \, y = x\}$.

59. Let $x =$ the price of one doughnut and $y =$ the price of one cup of coffee. His bills for Monday and Tuesday give us the following system of equations.

$$3x + 2y = 3.40$$
$$2x + 3y = 3.60$$

Multiply the first equation by -2 and the second by 3.
$$-6x - 4y = -6.80$$
$$\underline{6x + 9y = 10.80}$$
$$5y = 4.00$$
$$y = 0.80$$
Use $y = 0.80$ in $3x + 2y = 3.40$ to find x.
$$3x + 2(0.80) = 3.40$$
$$3x + 1.60 = 3.40$$
$$3x = 1.80$$
$$x = 0.60$$

Doughnuts are $0.60 each and a cup of coffee is $0.80. So his total bill on Wednesday was $1.40.

61. Let $x =$ the number of boys and $y =$ the number of girls at Freemont High. We get a system of equations from the number of boys and girls who attended the dance and game.

$$\frac{1}{2}x + \frac{1}{3}y = 570$$
$$\frac{1}{3}x + \frac{1}{2}y = 580$$

Multiply the first equation by 12 and the second equation by -18.

$$6x + 4y = 6840$$
$$\underline{-6x - 9y = -10440}$$
$$-5y = -3600$$
$$y = 720$$

Use $y = 720$ in $6x + 4y = 6840$ to find x.

$$6x + 4(720) = 6840$$
$$6x + 2880 = 6840$$
$$6x = 3960$$
$$x = 660$$

There are 660 boys and 720 girls at Freemont High, or 1380 students.

63. Let $x =$ the number of dimes and $y =$ the number of nickels. We write one equation about the total number of the coins and the other about the value of the coins (in cents).

$$x + y = 35$$
$$10x + 5y = 330$$

Multiply the first equation by -5.

$$-5x - 5y = -175$$
$$\underline{10x + 5y = 330}$$
$$5x = 155$$
$$x = 31$$

Use $x = 31$ in $x + y = 35$ to find that $y = 4$. He has 31 dimes and 4 nickels.

65. a) From the graph it appears that there should be about 20 pounds of Chocolate fudge and 30 pounds of Peanut Butter fudge.
b) Let $x =$ the number of pounds of Chocolate fudge and $y =$ the number of pounds of Peanut Butter fudge.

$$x + y = 50$$
$$0.35x + 0.25y = 0.29(50)$$

$$y = 50 - x$$
$$0.35x + 0.25(50 - x) = 14.5$$
$$0.10x = 2$$
$$x = 20$$
$$20 + y = 50$$
$$y = 30$$

Use 20 pounds of Chocolate fudge and 30 pounds of Peanut Butter fudge.

67. Let $a =$ the time from Allentown to Harrisburg and $h =$ the time from Harrisburg to Pittsburgh.

$$a + h = 6$$
$$42a + 51h = 288$$

$$-51a - 51h = -306$$
$$\underline{42a + 51h = 288}$$
$$-9a = -18$$
$$a = 2$$
$$h = 6 - 2 = 4$$

It took 4 hours to go from Harrisburg to Pittsburgh.

69. Let $p =$ the probability of rain and $q =$ the probability that it doesn't rain:

$$p + q = 1$$
$$p = 4q$$

$$4q + q = 1$$
$$5q = 1$$
$$q = 0.20 = 20\%$$
$$p = 80\%$$

The probability of rain is 80%.

71. Let $W =$ the width and $L =$ the length.

$$W = 0.75L$$
$$2W + 2L = 700$$

$$2(0.75L) + 2L = 700$$
$$1.5L + 2L = 700$$
$$3.5L = 700$$
$$L = 200$$
$$W = 0.75(200) = 150$$

The width is 150 m and the length is 200 m.

73. If one equation is already solved for x or y then substitution is usually easier. If both equations are in standard form then addition is usually easier.

75. a) Find the equation of the line through the two points. A system satisfied by both $(-1, 2)$ and $(4, 5)$ is $y = \frac{3}{5}x + \frac{13}{5}$ and $5y = 3x + 13$.
b) The equations are dependent.
c) There is not independent system satisfied by both lines, because two points determines a unique line.

4.3 WARM-UPS

1. False, because $1 + (-2) - 3 = 4$ is incorrect.

2. False, because there are infinitely many ordered triples that satisfy $x + y - z = 4$.

3. True, because if $x = 1$, $y = -1$, and $z = 2$ then each of the equations is satisfied.

4. False, we can use substitution and addition to eliminate variables.

5. False, because two planes never intersect in a single point.

6. True, because if we multiply the second one by -1 and add, then we get $0 = 2$.

7. True, because -2 times the first equation is the same as the second equation.

8. False, because the graph is a plane.

9. False, because the value of x nickels, y dimes, and z quarters is $5x + 10y + 25z$ cents.

10. False, because $x = -2$, $z = 3$, and $-2 + y + 3 = 6$ implies that $y = 5$.

4.3 EXERCISES

1. A linear equation in three variables is an equation of the form $Ax + By + Cz = D$ where A, B, and C cannot all be zero.

3. A solution to a system of linear equations in three variable is an ordered triple that satisfies all of the equations in the system.

5. The graph of a linear equation in three variables is a plane in a three-dimensional coordinate system.

7. Add the first two equations.

$$\begin{array}{l} x + y + z = 2 \\ x + 2y - z = 6 \\ \hline 2x + 3y \phantom{{}+z} = 8 \quad \text{A} \end{array}$$

Add the first and third equations.

$$\begin{array}{l} x + y + z = 2 \\ 2x + y - z = 5 \\ \hline 3x + 2y \phantom{{}+z} = 7 \quad \text{B} \end{array}$$

Multiply equation A by 3 and equation B by -2.

$$\begin{array}{ll} 6x + 9y = 24 & 3 \times \text{A} \\ -6x - 4y = -14 & -2 \times \text{B} \\ \hline 5y = 10 & \\ y = 2 & \end{array}$$

Use $y = 2$ in the equation $3x + 2y = 7$.

$$\begin{array}{l} 3x + 2(2) = 7 \\ 3x = 3 \\ x = 1 \end{array}$$

Use $x = 1$ and $y = 2$ in $x + y + z = 2$.

$$\begin{array}{l} 1 + 2 + z = 2 \\ z = -1 \end{array}$$

The solution set is $\{(1, 2, -1)\}$.

9. Multiply the first equation by -1 to get $-x + 2y - 4z = -3$. Now add this equation and the second, and this equation and the third.

$$\begin{array}{ll}
\begin{array}{l} -x + 2y - 4z = -3 \\ x + 3y - 2z = 6 \\ \hline 5y - 6z = 3 \end{array}
&
\begin{array}{l} -x + 2y - 4z = -3 \\ x - 4y + 3z = -5 \\ \hline -2y - z = -8 \end{array}
\end{array}$$

Multiply $-2y - z = -8$ by -6 and add the result to $5y - 6z = 3$.

$$\begin{array}{l} 12y + 6z = 48 \\ 5y - 6z = 3 \\ \hline 17y \phantom{{}+ 6z} = 51 \\ y = 3 \end{array}$$

Use $y = 3$ in $5y - 6z = 3$.

$$\begin{array}{l} 5(3) - 6z = 3 \\ -6z = -12 \\ z = 2 \end{array}$$

Use $y = 3$ and $z = 2$ in $x + 3y - 2z = 6$.

$$\begin{array}{l} x + 3(3) - 2(2) = 6 \\ x = 1 \end{array}$$

The solution set is $\{(1, 3, 2)\}$.

11. Multiply $2x - y + z = 10$ by 2 to get $4x - 2y + 2z = 20$. Add this equation and the second and this equation and the third.

$$\begin{array}{ll}
\begin{array}{l} 4x - 2y + 2z = 20 \\ 3x - 2y - 2z = 7 \\ \hline 7x - 4y \phantom{{}+ 2z} = 27 \end{array}
&
\begin{array}{l} 4x - 2y + 2z = 20 \\ x - 3y - 2z = 10 \\ \hline 5x - 5y \phantom{{}+ 2z} = 30 \\ x - y \phantom{{}+ 2z} = 6 \end{array}
\end{array}$$

Multiply $x - y = 6$ by -4 to get $-4x + 4y = -24$. Add this to $7x - 4y = 27$.

$$\begin{array}{l} -4x + 4y = -24 \\ 7x - 4y = 27 \\ \hline 3x \phantom{{}+ 4y} = 3 \\ x = 1 \end{array}$$

Use $x = 1$ in $7x - 4y = 27$ to find y.

$$\begin{array}{l} 7(1) - 4y = 27 \\ -4y = 20 \\ y = -5 \end{array}$$

Use $x = 1$ and $y = -5$ in $2x - y + z = 10$.

$$\begin{array}{l} 2(1) - (-5) + z = 10 \\ z = 3 \end{array}$$

The solution set is $\{(1, -5, 3)\}$.

13. Multiply $x - y + 2z = -5$ by 2 to get $2x - 2y + 4z = -10$. Add this equation and the second equation, and add the first and second equations.

$$2x - 2y + 4z = -10 \qquad 2x - 3y + z = -9$$
$$\underline{-2x + y - 3z = 7} \qquad \underline{-2x + y - 3z = 7}$$
$$\quad -y + z = -3 \qquad\qquad -2y - 2z = -2$$
$$\qquad\qquad\qquad\qquad\qquad\quad y + z = 1$$

Add the last two equations to eliminate y.

$$-y + z = -3$$
$$\underline{y + z = 1}$$
$$\quad\; 2z = -2$$
$$\quad\;\; z = -1$$

Use $z = -1$ in $y + z = 1$ to get $y = 2$. Now use $z = -1$ and $y = 2$ in $x - y + 2z = -5$.
$$x - 2 + 2(-1) = -5$$
$$x = -1$$
The solution set is $\{(-1, 2, -1)\}$.

15. Multiply the first equation by -2 and add the result to the last equation.
$$-4x + 10y - 4z = -32$$
$$\underline{4x - 3y + 4z = 18}$$
$$\quad 7y \qquad\quad = -14$$
$$\qquad y \quad\; = -2$$
Multiply the first equation by 3 and the second by 2 and add the results.
$$6x - 15y + 6z = 48$$
$$\underline{6x + 4y - 6z = -38}$$
$$12x - 11y \qquad = 10$$
Use $y = -2$ in the last equation to find x.
$$12x - 11(-2) = 10$$
$$12x = -12$$
$$x = -1$$
Use $x = -1$ and $y = -2$ in $2x - 5y + 2z = 16$.
$$2(-1) - 5(-2) + 2z = 16$$
$$2z = 8$$
$$z = 4$$
The solution set is $\{(-1, -2, 4)\}$.

17. If we add the last two equations we get $x + 2y = 7$. Multiply the first equation by -1

to get $-x - y = -4$. Add these two equations.
$$x + 2y = 7$$
$$\underline{-x - y = -4}$$
$$\quad\; y = 3$$
Use $y = 3$ in $x + y = 4$ to get $x = 1$. Use $y = 3$ in $y - z = -2$.
$$3 - z = -2$$
$$5 = z$$
The solution set is $\{(1, 3, 5)\}$.

19. Multiply the first equation by -1 and add the result to the second equation.
$$-x - y \qquad = -7$$
$$\underline{\quad\;\; y - z = -1}$$
$$-x \qquad - z = -8$$
Add this result to the last equation.
$$-x - z = -8$$
$$\underline{x + 3z = 18}$$
$$\quad\;\; 2z = 10$$
$$\quad\;\; z = 5$$
Use $z = 5$ in $x + 3z = 18$ to get $x + 15 = 18$ or $x = 3$. Use $x = 3$ in $x + y = 7$ to get $3 + y = 7$ or $y = 4$. The solution set is $\{(3, 4, 5)\}$.

21. Add the first two equations.
$$x - y + 2z = 3$$
$$\underline{2x + y - z = 5}$$
$$3x \qquad + z = 8 \qquad \text{A}$$
Multiply the second equation by 3 and add the result to the last equation.
$$6x + 3y - 3z = 15$$
$$\underline{3x - 3y + 6z = 4}$$
$$9x \qquad + 3z = 19 \qquad \text{B}$$
Multiply A by -3 and add the result to B.
$$-9x - 3z = -24$$
$$\underline{9x + 3z = 19}$$
$$0 = -5$$
Since the last equation is false no matter what values the variables have, the solution set is $\emptyset$.

23. The second equation is the same as 3 times the first equation, and the third equation is the same as -4 times the first equation. So the system is a dependent system. All three equations are equivalent. The solution set is $\{(x, y, z) \mid 3x - y + z = 5\}$.

25. Add the first and second equation to get $x + z = 11$. Multiply this equation by -2 and then add the result to the last equation.

$$\begin{array}{r} -2x - 2z = -22 \\ 2x + 2z = 7 \\ \hline 0 = -15 \end{array}$$

Since the last result is false, the solution set is $\emptyset$.

27. Multiply the first equation by 300 to get $30x + 24y - 12z = 900$. Multiply the second equation by 6 to get $30x + 24y - 12z = 900$. Multiply the third equation by 100 to get $30x + 24y - 12z = 900$. Since all three of these equations are different forms of the same equation, the system is dependent. The solution set is
$\{(x, y, z) \mid 5x + 4y - 2z = 150\}$.

29. Multiply the second equation by 10 and add the result to the first equation.

$$\begin{array}{r} 37x - 2y + 0.5z = 4.1 \\ 3x + 2y - 0.4z = 0.1 \\ \hline 40x \quad\quad + 0.1z = 4.2 \quad \text{A} \end{array}$$

Multiply the second equation by 19 and add the result to the last equation.

$$\begin{array}{r} 70.3x - 3.8y + 0.95z = 7.79 \\ -2x + 3.8y - 2.1z = -3.26 \\ \hline 68.3x \quad\quad - 1.15z = 4.53 \quad \text{B} \end{array}$$

Multiply equation A by 11.5 and add the result to equation B.

$$\begin{array}{r} 460x + 1.15z = 48.3 \\ 68.3x - 1.15z = 4.53 \\ \hline 528.3x \quad\quad = 52.83 \\ x = 0.1 \end{array}$$

Use $x = 0.1$ in $40x + 0.1z = 4.2$.

$$40(0.1) + 0.1z = 4.2$$
$$0.1z = 0.2$$
$$z = 2$$

Use $x = 0.1$ and $z = 2$ in
$3x + 2y - 0.4z = 0.1$.

$$3(0.1) + 2y - 0.4(2) = 0.1$$
$$0.3 + 2y - 0.8 = 0.1$$
$$2y = 0.6$$
$$y = 0.3$$

The solution set is $\{(0.1, 0.3, 2)\}$.

31. Let $x =$ the price of the Chevy, $y =$ the price of the Ford, and $z =$ the price of the Toyota. Write the following system.

$$x + y + z = 66,000$$
$$y = x + 2000$$
$$z = y + 2000$$

By substitution $z = x + 4000$ and

$$x + (x + 2000) + (x + 4000) = 66,000$$
$$3x = 60,000$$
$$x = 20,000$$
$$y = x + 2,000 = 22,000$$
$$z = y + 2,000 = 24,000$$

The Chevrolet was $20,000, the Ford was $22,000, and the Toyota was $24,000.

33. Let $x =$ the number of hours driven on the first day, $y =$ the number of hours for the second day, and $z =$ the number of hours for the third day. His distance each day was $64x$, $62y$, and $58z$.

$$x + y + z = 36$$
$$64x + 62y + 58z = 2196$$
$$z = x + 4$$

Substitute:

$$x + y + (x + 4) = 36$$
$$64x + 62y + 58(x + 4) = 2196$$

Simplify:

$$2x + y = 32$$
$$122x + 62y = 1964$$

Multiply the first equation by -62:

$$\begin{array}{r} -124x - 62y = -1984 \\ 122x + 62y = 1964 \\ \hline -2x \quad\quad = -20 \\ x = 10 \end{array}$$

$$z = x + 4 = 10 + 4 = 14$$
$$x + y + z = 36$$
$$10 + y + 14 = 36$$
$$y = 12$$

So he drove 10 hours on the first day, 12 hours on the second day, and 14 hours on the third day.

35. Let $x =$ her investment in stocks, $y =$ her investment in bonds, and $z =$ her investment in a mutual funds. We can write 3 equations concerning x, y, and z.

$$x + y + z = 12,000$$
$$0.10x + 0.08y + 0.12z = 1230$$
$$x + y = z$$

Substitute $z = x + y$ into the first equation.

$$x + y + x + y = 12000$$
$$2x + 2y = 12000$$
$$x + y = 6000 \quad \text{A}$$

Substitute $z = x + y$ into the second equation.
$$0.10x + 0.08y + 0.12(x + y) = 1230$$
$$0.22x + 0.20y = 1230$$
$$-5(0.22x + 0.20y) = -5(1230)$$
$$-1.1x - y = -6150 \quad \text{B}$$
Add equations A and B.
$$x + y = 6000$$
$$\underline{-1.1x - y = -6150}$$
$$-0.1x = -150$$
$$x = 1500$$
Use $x = 1500$ in $x + y = 6000$ to get
$y = 4500$. Since $z = x + y$, we must have
$z = 6000$. So Ann invested \$1500 in stocks,
\$4500 in bonds, and \$6000 in a mutual fund.

37. Let $x = $ Anna's weight, $y = $ Bob's weight,
and $z = $ Chris's weight. We are given three
equations.
$$x + y = 226$$
$$y + z = 210$$
$$x + z = 200$$
From the first equation we get $y = 226 - x$.
Substitute this equation into the second
equation to eliminate y.
$$226 - x + z = 210$$
$$-x + z = -16$$
Add this last equation to $x + z = 200$ to
eliminate x.
$$-x + z = -16$$
$$\underline{x + z = 200}$$
$$2z = 184$$
$$z = 92$$
If $z = 92$ and $x + z = 200$, we get $x = 108$. If
$z = 92$ and $y + z = 210$, we get $y = 118$. So
Anna weighs 108 pounds, Bob weighs 118
pounds, and Chris weighs 92 pounds.

39. Let $x = $ the number of nickels, $y = $ the
number of dimes, and $z = $ the number of
quarters. Since he has 13 coins altogether,
$x + y + z = 13$. Since the value of the coins
in cents is 175, we have
$5x + 10y + 25z = 175$ or $x + 2y + 5z = 35$.
Since the number of dimes is twice the number
of nickels, $y = 2x$.
$$x + y + z = 13$$
$$x + 2y + 5z = 35$$
$$y = 2x$$
Substitute $y = 2x$ into the other two equations
to get the following system.

$$3x + z = 13$$
$$5x + 5z = 35$$
Divide the second equation by -5 and add:
$$3x + z = 13$$
$$\underline{-x - z = -7}$$
$$2x = 6$$
$$x = 3$$
If $x = 3$, then $y = 6$. If $x = 3$ and $y = 6$, then
$3 + 6 + z = 13$, or $z = 4$. So he used 3
nickels, 6 dimes, and 4 quarters.

41. Let $x = $ her income from teaching,
$y = $ her income from house painting, and
$z = $ her royalties.
$$x + y + z = 48,000$$
$$x - y = 6000$$
$$z = \tfrac{1}{7}(x + y)$$
The last equation can be written as
$x + y = 7z$. Replacing $x + y$ in the first
equation by $7z$ gives us $7z + z = 48,000$, or
$z = 6000$. If $z = 6000$, then $x + y = 42,000$.
Add $x + y = 42000$ and the second equation.
$$x + y = 42,000$$
$$\underline{x - y = 6000}$$
$$2x = 48,000$$
$$x = 24,000$$
Use $x = 24,000$ in $x - y = 6000$ to get
$y = 18,000$. So she made \$24,000 teaching,
\$18,000 house painting, and \$6,000 from
royalties.

43. Let $x = $ Edwin's age, $y = $ his father's age
and $z = $ his grandfather's age. Since their
average age is 53, we can write
$$\frac{x + y + z}{3} = 53$$
or $\qquad x + y + z = 159$.
We can also write
$$\tfrac{1}{2}z + \tfrac{1}{3}y + \tfrac{1}{4}x = 65$$
or $\quad 12\left(\tfrac{1}{2}z + \tfrac{1}{3}y + \tfrac{1}{4}x\right) = 12(65)$
or $\quad 6z + 4y + 3x = 780$.
Four years ago Edwin was $x - 4$ years old and
his grandfather was $z - 4$ years old. Since 4
years ago the grandfather was 4 times as old as
Edwin, we can write
$$z - 4 = 4(x - 4)$$
$$z - 4 = 4x - 16$$
$$z = 4x - 12.$$

Substitute $z = 4x - 12$ into the two equations above.

$$x + y + 4x - 12 = 159$$
$$6(4x - 12) + 4y + 3x = 780$$

$$5x + y = 171$$
$$27x + 4y = 852$$

Substitute $y = 171 - 5x$ into $27x + 4y = 852$.

$$27x + 4(171 - 5x) = 852$$
$$7x + 684 = 852$$
$$7x = 168$$
$$x = 24$$

Since $z = 4x - 12$ we have $z = 84$. Since $x + y + z = 159$, we have $24 + y + 84 = 159$, or $y = 51$. So Edwin is 24, his father is 51, and his grandfather is 84 years old.

4.4 WARM-UPS

1. True
2. True.
3. True. Replace R_2 of (a) by $R_1 + R_2$ to get matrix (b).
4. False, because matrix (c) corresponds to an inconsistent system and (d) corresponds to a dependent system.
5. True, because the last row represents the equation $0 = 7$.
6. False. Replace R_2 by $2R_1 + R_2$ to get $0 = -3$ which is inconsistent.
7. False, because the system corresponding to (d) consists of two equations equivalent to $x + 3y = 5$.
8. False, because the augmented matrix is a 2×3 matrix.
9. True.
10. False, because it means to interchange R_1 and R_2.

4.4 EXERCISES

1. A matrix is a rectangular array of numbers.
3. The order of a matrix is the number of rows and columns.
5. An augmented matrix is a matrix where the entries in the first column are the coefficients of x, the entries in the second column are the coefficients of y, and the entries in the third column are the constants from a system of two linear equations in two unknowns.
7. The matrix is a 2×2 matrix because it has 2 rows and 2 columns.
9. The matrix is a 3×2 matrix because it has 3 rows and 2 columns.
11. The matrix is a 3×1 matrix because it has 3 rows and 1 column.
13. Use the coefficients 2 and -3, and the constant 9 as the first row. Use the coefficients -3 and 1, and the constant -1 as the second row.

$$\begin{bmatrix} 2 & -3 & 9 \\ -3 & 1 & -1 \end{bmatrix}$$

15. Use the coefficients 1, -1, and 1, and the constant 1 as the first row. Use the coefficients 1, 1, and -2, and the constant 3 as the second row. Use the coefficients 0, 1, and -3, and the constant 4 as the third row.

$$\begin{bmatrix} 1 & -1 & 1 & 1 \\ 1 & 1 & -2 & 3 \\ 0 & 1 & -3 & 4 \end{bmatrix}$$

17. The entries in the first row $(5, 1, -1)$ represent the equation $5x + y = -1$. The entries in the second row $(2, -3, 0)$ represent the equation $2x - 3y = 0$. So the matrix represents the following system.

$$5x + y = -1$$
$$2x - 3y = 0$$

19. The entries in the first row $(1, 0, 0, 6)$ represent the equation $x = 6$. The entries in the second row $(-1, 0, 1, -3)$ represent the equation $-x + z = -3$. The entries in the third row $(1, 1, 0, 1)$ represent the equation $x + y = 1$. So the matrix represents the following system.

$$x = 6$$
$$-x + z = -3$$
$$x + y = 1$$

21. Row 1 and row 2 are interchanged. This operation is written in symbols as $R_1 \leftrightarrow R_2$.

23. The second row is multiplied by $\frac{1}{5}$. In symbols $\frac{1}{5}R_2 \rightarrow R_2$

25. $\begin{bmatrix} 1 & 1 & 3 \\ -3 & 1 & -1 \end{bmatrix}$

$\begin{bmatrix} 1 & 1 & 3 \\ 0 & 4 & 8 \end{bmatrix}$ $\quad 3R_1 + R_2 \rightarrow R_2$

$\begin{bmatrix} 1 & 1 & 3 \\ 0 & 1 & 2 \end{bmatrix}$ $\quad \frac{1}{4}R_2 \rightarrow R_2$

$\begin{bmatrix} 1 & 0 & 1 \\ 0 & 1 & 2 \end{bmatrix}$ $\quad -R_2 + R_1 \rightarrow R_1$

The solution set is $\{(1, 2)\}$.

27. $\begin{bmatrix} 2 & -1 & 3 \\ 1 & 1 & 9 \end{bmatrix}$

$\begin{bmatrix} 1 & 1 & 9 \\ 2 & -1 & 3 \end{bmatrix}$ $\quad R_1 \leftrightarrow R_2$

$\begin{bmatrix} 1 & 1 & 9 \\ 0 & -3 & -15 \end{bmatrix}$ $\quad -2R_1 + R_2 \rightarrow R_2$

$\begin{bmatrix} 1 & 1 & 9 \\ 0 & 1 & 5 \end{bmatrix}$ $\quad R_2 \div (-3) \rightarrow R_2$

$\begin{bmatrix} 1 & 0 & 4 \\ 0 & 1 & 5 \end{bmatrix}$ $\quad -R_2 + R_1 \rightarrow R_1$

The solution set is $\{(4, 5)\}$.

29. $\begin{bmatrix} 3 & -1 & 4 \\ 2 & 1 & 1 \end{bmatrix}$

$\begin{bmatrix} 1 & -2 & 3 \\ 2 & 1 & 1 \end{bmatrix}$ $\quad -R_2 + R_1 \rightarrow R_1$

$\begin{bmatrix} 1 & -2 & 3 \\ 0 & 5 & -5 \end{bmatrix}$ $\quad -2R_1 + R_2 \rightarrow R_2$

$\begin{bmatrix} 1 & -2 & 3 \\ 0 & 1 & -1 \end{bmatrix}$ $\quad R_2 \div 5 \rightarrow R_2$

$\begin{bmatrix} 1 & 0 & 1 \\ 0 & 1 & -1 \end{bmatrix}$ $\quad 2R_2 + R_1 \rightarrow R_1$

The solution set is $\{(1, -1)\}$.

31. $\begin{bmatrix} 6 & -7 & 0 \\ 2 & 1 & 20 \end{bmatrix}$

$\begin{bmatrix} 0 & -10 & -60 \\ 2 & 1 & 20 \end{bmatrix}$ $\quad -3R_2 + R_1 \rightarrow R_1$

$\begin{bmatrix} 0 & 1 & 6 \\ 2 & 1 & 20 \end{bmatrix}$ $\quad R_1 \div (-10) \rightarrow R_1$

$\begin{bmatrix} 2 & 1 & 20 \\ 0 & 1 & 6 \end{bmatrix}$ $\quad R_1 \leftrightarrow R_2$

$\begin{bmatrix} 2 & 0 & 14 \\ 0 & 1 & 6 \end{bmatrix}$ $\quad -R_2 + R_1 \rightarrow R_1$

$\begin{bmatrix} 1 & 0 & 7 \\ 0 & 1 & 6 \end{bmatrix}$ $\quad R_1 \div 2 \rightarrow R_1$

The solution set is $\{(7, 6)\}$.

33. $\begin{bmatrix} 2 & -3 & 4 \\ -2 & 3 & 5 \end{bmatrix}$

$\begin{bmatrix} 2 & -3 & 4 \\ 0 & 0 & 9 \end{bmatrix}$ $\quad R_1 + R_2 \rightarrow R_2$

Since the second row represents the equation $0 = 9$, the solution set is the empty set, $\emptyset$.

35. $\begin{bmatrix} 1 & 2 & 1 \\ 3 & 6 & 3 \end{bmatrix}$

$\begin{bmatrix} 1 & 2 & 1 \\ 0 & 0 & 0 \end{bmatrix}$ $\quad -3R_1 + R_2 \rightarrow R_2$

Since the system is equivalent to the single equation $x + 2y = 1$, the equations are dependent and the solution set is $\{(x, y) \mid x + 2y = 1\}$.

37. $\begin{bmatrix} 1 & 1 & 1 & 6 \\ 1 & -1 & 1 & 2 \\ 0 & 2 & -1 & 1 \end{bmatrix}$

$\begin{bmatrix} 1 & 1 & 1 & 6 \\ 0 & -2 & 0 & -4 \\ 0 & 2 & -1 & 1 \end{bmatrix}$ $\quad -R_1 + R_2 \rightarrow R_2$

$$\begin{bmatrix} 1 & 1 & 1 & 6 \\ 0 & 1 & 0 & 2 \\ 0 & 2 & -1 & 1 \end{bmatrix}$$ $R_2 \div (-2) \to R_2$

$-R_2 + R_1 \to R_1$

$$\begin{bmatrix} 1 & 0 & 1 & 4 \\ 0 & 1 & 0 & 2 \\ 0 & 0 & -1 & -3 \end{bmatrix}$$

$-2R_2 + R_3 \to R_3$

$$\begin{bmatrix} 1 & 0 & 1 & 4 \\ 0 & 1 & 0 & 2 \\ 0 & 0 & 1 & 3 \end{bmatrix}$$

$R_3 \div (-1) \to R_3$

$$\begin{bmatrix} 1 & 0 & 0 & 1 \\ 0 & 1 & 0 & 2 \\ 0 & 0 & 1 & 3 \end{bmatrix}$$ $-R_3 + R_1 \to R_1$

The solution set is $\{(1, 2, 3)\}$.

39. $$\begin{bmatrix} 2 & 1 & 1 & 4 \\ 1 & 1 & -1 & 1 \\ 1 & -1 & 2 & 2 \end{bmatrix}$$

$$\begin{bmatrix} 1 & 1 & -1 & 1 \\ 2 & 1 & 1 & 4 \\ 1 & -1 & 2 & 2 \end{bmatrix}$$ $R_1 \leftrightarrow R_2$

$$\begin{bmatrix} 1 & 1 & -1 & 1 \\ 0 & -1 & 3 & 2 \\ 0 & -2 & 3 & 1 \end{bmatrix}$$ $-2R_1 + R_2 \to R_2$

$-R_1 + R_3 \to R_3$

$$\begin{bmatrix} 1 & 1 & -1 & 1 \\ 0 & 1 & -3 & -2 \\ 0 & -2 & 3 & 1 \end{bmatrix}$$ $-1 \cdot R_2 \to R_2$

$$\begin{bmatrix} 1 & 0 & 2 & 3 \\ 0 & 1 & -3 & -2 \\ 0 & 0 & -3 & -3 \end{bmatrix}$$ $-1 \cdot R_2 + R_1 \to R_1$

$2R_2 + R_3 \to R_3$

$$\begin{bmatrix} 1 & 0 & 2 & 3 \\ 0 & 1 & -3 & -2 \\ 0 & 0 & 1 & 1 \end{bmatrix}$$ $-\frac{1}{3} \cdot R_3 \to R_3$

$$\begin{bmatrix} 1 & 0 & 0 & 1 \\ 0 & 1 & 0 & 1 \\ 0 & 0 & 1 & 1 \end{bmatrix}$$ $-2R_3 + R_1 \to R_1$

$3R_3 + R_2 \to R_2$

The solution set is $\{(1, 1, 1)\}$.

41. $$\begin{bmatrix} 1 & 1 & -3 & 3 \\ 2 & -1 & 1 & 0 \\ 1 & -1 & 1 & -1 \end{bmatrix}$$

$$\begin{bmatrix} 1 & 1 & -3 & 3 \\ 0 & -3 & 7 & -6 \\ 0 & -2 & 4 & -4 \end{bmatrix}$$ $-2R_1 + R_2 \to R_2$

$-R_1 + R_3 \to R_3$

$$\begin{bmatrix} 1 & 1 & -3 & 3 \\ 0 & 1 & -1 & 2 \\ 0 & -2 & 4 & -4 \end{bmatrix}$$ $-2R_3 + R_2 \to R_2$

$$\begin{bmatrix} 1 & 0 & -2 & 1 \\ 0 & 1 & -1 & 2 \\ 0 & 0 & 2 & 0 \end{bmatrix}$$ $-R_2 + R_1 \to R_1$

$2R_2 + R_3 \to R_3$

$$\begin{bmatrix} 1 & 0 & -2 & 1 \\ 0 & 1 & -1 & 2 \\ 0 & 0 & 1 & 0 \end{bmatrix}$$ $R_3 \div 2 \to R_3$

$2R_3 + R_1 \to R_1$

$$\begin{bmatrix} 1 & 0 & 0 & 1 \\ 0 & 1 & 0 & 2 \\ 0 & 0 & 1 & 0 \end{bmatrix}$$ $R_3 + R_2 \to R_2$

The solution set is $\{(1, 2, 0)\}$.

43. $$\begin{bmatrix} 1 & -1 & -4 & -3 \\ -1 & 3 & 1 & 0 \\ 1 & 1 & 2 & 3 \end{bmatrix}$$

$$\begin{bmatrix} 1 & -1 & -4 & -3 \\ 0 & 2 & -3 & -3 \\ 0 & 2 & 6 & 6 \end{bmatrix}$$ $R_1 + R_2 \to R_2$

$-R_1 + R_3 \to R_3$

$$\begin{bmatrix} 1 & -1 & -4 & -3 \\ 0 & 2 & -3 & -3 \\ 0 & 0 & 9 & 9 \end{bmatrix}$$ $-R_2 + R_3 \to R_3$

$$\begin{bmatrix} 1 & -1 & -4 & -3 \\ 0 & 2 & -3 & -3 \\ 0 & 0 & 1 & 1 \end{bmatrix} \quad R_3 \div 9 \to R_3$$

$$4R_3 + R_1 \to R_1$$
$$\begin{bmatrix} 1 & -1 & 0 & 1 \\ 0 & 2 & 0 & 0 \\ 0 & 0 & 1 & 1 \end{bmatrix} \quad 3R_3 + R_2 \to R_2$$

$$\begin{bmatrix} 1 & -1 & 0 & 1 \\ 0 & 1 & 0 & 0 \\ 0 & 0 & 1 & 1 \end{bmatrix} \quad R_2 \div 2 \to R_2$$

$$\begin{bmatrix} 1 & 0 & 0 & 1 \\ 0 & 1 & 0 & 0 \\ 0 & 0 & 1 & 1 \end{bmatrix} \quad R_2 + R_1 \to R_1$$

The solution set is $\{(1, 0, 1)\}$.

45. $\begin{bmatrix} 1 & -1 & 1 & 1 \\ 2 & -2 & 2 & 2 \\ -3 & 3 & -3 & -3 \end{bmatrix}$

$$\begin{bmatrix} 1 & -1 & 1 & 1 \\ 0 & 0 & 0 & 0 \\ 0 & 0 & 0 & 0 \end{bmatrix} \quad -2R_1 + R_2 \to R_2$$
$$3R_1 + R_3 \to R_3$$

Since the system of equations is equivalent to the first equation, the solution set is
$\{(x, y, z) \mid x - y + z = 1\}$.

47. $\begin{bmatrix} 1 & 1 & -1 & 2 \\ 2 & -1 & 1 & 1 \\ 3 & 3 & -3 & 8 \end{bmatrix}$

$$\begin{bmatrix} 1 & 1 & -1 & 2 \\ 0 & -3 & 3 & -3 \\ 0 & 0 & 0 & 2 \end{bmatrix} \quad -2R_1 + R_2 \to R_2$$
$$-3R_1 + R_3 \to R_3$$

Since the third equation is $0 = 2$, there is no solution to the system. The solution set is $\emptyset$.

4.5 WARM-UPS

1. True. because the determinant is
$(-1)(-5) - (2)(3) = -1$.
2. False, because the determinant is
$2 \cdot 8 - (-4)(4) = 32$.

3. False, because if $D = 0$, then Cramer's rule fails to give us the solution.
4. True, because the determinant is the value of $ad - bc$.
5. True, because this is the case where Cramer's rule fails to give the precise solution.
6. True.
7. True, because this is precisely when Cramer's rule works.
8. True, because of the definition of determinant of a 3×3 matrix.
9. True.
10. True.

4.5 EXERCISES
1. A determinant is a real number associated with a square matrix.
3. Cramer's rule works on systems that have exactly one solution.
5. A minor for an element is obtained by deleting the row and column of the element and then finding the determinant of the 2×2 matrix that remains.

7. $\begin{vmatrix} 2 & 5 \\ 3 & 7 \end{vmatrix} = 2 \cdot 7 - 3 \cdot 5 = -1$

9. $\begin{vmatrix} 0 & 3 \\ 1 & 5 \end{vmatrix} = 0 \cdot 5 - 1 \cdot 3 = -3$

11. $\begin{vmatrix} -3 & -2 \\ -4 & 2 \end{vmatrix} = -3 \cdot 2 - (-4)(-2) = -14$

13. $\begin{vmatrix} 0.05 & 0.06 \\ 10 & 20 \end{vmatrix} = 0.05(20) - 0.06(10)$
$$= 0.4$$

15. $D = \begin{vmatrix} 2 & -1 \\ 3 & 2 \end{vmatrix} = 7$

$$D_x = \begin{vmatrix} 5 & -1 \\ -3 & 2 \end{vmatrix} = 7$$

$$D_y = \begin{vmatrix} 2 & 5 \\ 3 & -3 \end{vmatrix} = -21$$

$$x = \frac{D_x}{D} = \frac{7}{7} = 1 \qquad y = \frac{D_y}{D} = \frac{-21}{7} = -3$$

The solution set is $\{(1, -3)\}$.

17. $D = \begin{vmatrix} 3 & -5 \\ 2 & 3 \end{vmatrix} = 19$

$D_x = \begin{vmatrix} -2 & -5 \\ 5 & 3 \end{vmatrix} = 19$

$D_y = \begin{vmatrix} 3 & -2 \\ 2 & 5 \end{vmatrix} = 19$

$x = \dfrac{D_x}{D} = \dfrac{19}{19} = 1 \qquad y = \dfrac{D_y}{D} = \dfrac{19}{19} = 1$

The solution set is $\{(1, 1)\}$.

19. $D = \begin{vmatrix} 4 & -3 \\ 2 & 5 \end{vmatrix} = 26$

$D_x = \begin{vmatrix} 5 & -3 \\ 7 & 5 \end{vmatrix} = 46 \qquad D_y = \begin{vmatrix} 4 & 5 \\ 2 & 7 \end{vmatrix} = 18$

$x = \dfrac{D_x}{D} = \dfrac{46}{26} = \dfrac{23}{13} \qquad y = \dfrac{D_y}{D} = \dfrac{18}{26} = \dfrac{9}{13}$

The solution set is $\left\{ \left(\dfrac{23}{13}, \dfrac{9}{13} \right) \right\}$.

21. $D = \begin{vmatrix} 0.5 & 0.2 \\ 0.4 & -0.6 \end{vmatrix} = -0.38$

$D_x = \begin{vmatrix} 8 & 0.2 \\ -5 & -0.6 \end{vmatrix} = -3.8$

$D_y = \begin{vmatrix} 0.5 & 8 \\ 0.4 & -5 \end{vmatrix} = -5.7$

$x = \dfrac{D_x}{D} = \dfrac{-3.8}{-0.38} = 10$

$y = \dfrac{D_y}{D} = \dfrac{-5.7}{-0.38} = 15$

The solution set is $\{(10, 15)\}$.

23. Multiply the first equation by 4 and the second by 6 to eliminate the fractions.

$2x + y = 20$
$2x - 3y = -6$

$D = \begin{vmatrix} 2 & 1 \\ 2 & -3 \end{vmatrix} = -8$

$D_x = \begin{vmatrix} 20 & 1 \\ -6 & -3 \end{vmatrix} = -54$

$D_y = \begin{vmatrix} 2 & 20 \\ 2 & -6 \end{vmatrix} = -52$

$x = \dfrac{D_x}{D} = \dfrac{-54}{-8} = \dfrac{27}{4} \quad y = \dfrac{D_y}{D} = \dfrac{-52}{-8} = \dfrac{13}{2}$

The solution set is $\left\{ \left(\dfrac{27}{4}, \dfrac{13}{2} \right) \right\}$.

25. The minor for 3 is the determinant of the matrix obtained by deleting the row and column containing 3, the first row and first column.

$\begin{vmatrix} -3 & 7 \\ 1 & -6 \end{vmatrix} = 11$

27. The minor for 5 is the determinant of the matrix obtained by deleting the first row and third column.

$\begin{vmatrix} 4 & -3 \\ 0 & 1 \end{vmatrix} = 4$

29. The minor for 7 is the determinant of the matrix obtained by deleting the second row and third column.

$\begin{vmatrix} 3 & -2 \\ 0 & 1 \end{vmatrix} = 3$

31. The minor for 1 is the determinant of the matrix obtained by deleting the third row and second column.

$\begin{vmatrix} 3 & 5 \\ 4 & 7 \end{vmatrix} = 1$

33. $1\begin{vmatrix} 3 & 1 \\ 1 & 5 \end{vmatrix} - 2\begin{vmatrix} 1 & 2 \\ 1 & 5 \end{vmatrix} + 3\begin{vmatrix} 1 & 2 \\ 3 & 1 \end{vmatrix}$

$= 1(14) - 2(3) + 3(-5) = -7$

35. $2\begin{vmatrix} 0 & 1 \\ 1 & 2 \end{vmatrix} - 1\begin{vmatrix} 1 & 0 \\ 1 & 2 \end{vmatrix} + 3\begin{vmatrix} 1 & 0 \\ 0 & 1 \end{vmatrix}$

$= 2(-1) - 1(2) + 3(1) = -1$

37. $-2\begin{vmatrix} 3 & 1 \\ 4 & 0 \end{vmatrix} + 3\begin{vmatrix} 1 & 2 \\ 4 & 0 \end{vmatrix} - 5\begin{vmatrix} 1 & 2 \\ 3 & 1 \end{vmatrix}$

$= -2(-4) + 3(-8) - 5(-5) = 9$

39. $1\begin{vmatrix} 3 & 2 \\ 2 & 3 \end{vmatrix} - 0\begin{vmatrix} 1 & 5 \\ 2 & 3 \end{vmatrix} + 0\begin{vmatrix} 1 & 5 \\ 3 & 2 \end{vmatrix}$

$= 1(5) - 0(-7) + 0(-13) = 5$

41. Expand by minors about the second column because it has two zeros in it.

$$-1\begin{vmatrix} 2 & 6 \\ 4 & 1 \end{vmatrix} + 0\begin{vmatrix} 3 & 5 \\ 4 & 1 \end{vmatrix} - 0\begin{vmatrix} 3 & 5 \\ 2 & 6 \end{vmatrix}$$
$$= -1(-22) = 22$$

43. Expand by minors about the first column because it has one zero in it.

$$-2\begin{vmatrix} 1 & -1 \\ -4 & -3 \end{vmatrix} - 0\begin{vmatrix} 1 & 3 \\ -4 & -3 \end{vmatrix} + 2\begin{vmatrix} 1 & 3 \\ 1 & -1 \end{vmatrix}$$
$$= -2(-7) + 2(-4) = 6$$

45. Expand by minors about the third column because it has two zeros in it.

$$0\begin{vmatrix} 4 & -1 \\ 0 & 3 \end{vmatrix} - 0\begin{vmatrix} -2 & -3 \\ 0 & 3 \end{vmatrix} + 5\begin{vmatrix} -2 & -3 \\ 4 & -1 \end{vmatrix}$$
$$= 5(14) = 70$$

47. Expand by minors about the second column.

$$-1\begin{vmatrix} 0 & 5 \\ 5 & 4 \end{vmatrix} + 0\begin{vmatrix} 2 & 1 \\ 5 & 4 \end{vmatrix} - 0\begin{vmatrix} 2 & 1 \\ 0 & 5 \end{vmatrix}$$
$$= -1(-25) = 25$$

49. $D = \begin{vmatrix} 1 & 1 & 1 \\ 1 & -1 & 1 \\ 2 & 1 & 1 \end{vmatrix} = 2$

$$D_x = \begin{vmatrix} 6 & 1 & 1 \\ 2 & -1 & 1 \\ 7 & 1 & 1 \end{vmatrix} = 2$$

$$D_y = \begin{vmatrix} 1 & 6 & 1 \\ 1 & 2 & 1 \\ 2 & 7 & 1 \end{vmatrix} = 4$$

$$D_z = \begin{vmatrix} 1 & 1 & 6 \\ 1 & -1 & 2 \\ 2 & 1 & 7 \end{vmatrix} = 6$$

$$x = \frac{D_x}{D} = \frac{2}{2} = 1,\ y = \frac{D_y}{D} = \frac{4}{2} = 2,$$

$$z = \frac{D_z}{D} = \frac{6}{2} = 3$$

The solution set is $\{(1, 2, 3)\}$.

51. $D = \begin{vmatrix} 1 & -3 & 2 \\ 1 & 1 & 1 \\ 1 & -1 & 1 \end{vmatrix} = -2$

$$D_x = \begin{vmatrix} 0 & -3 & 2 \\ 2 & 1 & 1 \\ 0 & -1 & 1 \end{vmatrix} = 2$$

$$D_y = \begin{vmatrix} 1 & 0 & 2 \\ 1 & 2 & 1 \\ 1 & 0 & 1 \end{vmatrix} = -2$$

$$D_z = \begin{vmatrix} 1 & -3 & 0 \\ 1 & 1 & 2 \\ 1 & -1 & 0 \end{vmatrix} = -4$$

$$x = \frac{D_x}{D} = \frac{2}{-2} = -1,\ y = \frac{D_y}{D} = \frac{-2}{-2} = 1,$$

$$z = \frac{D_z}{D} = \frac{-4}{-2} = 2$$

The solution set is $\{(-1, 1, 2)\}$.

53. $D = \begin{vmatrix} 1 & 1 & 0 \\ 0 & 2 & -1 \\ 1 & 1 & 1 \end{vmatrix} = 2$

$$D_x = \begin{vmatrix} -1 & 1 & 0 \\ 3 & 2 & -1 \\ 0 & 1 & 1 \end{vmatrix} = -6$$

$$D_y = \begin{vmatrix} 1 & -1 & 0 \\ 0 & 3 & -1 \\ 1 & 0 & 1 \end{vmatrix} = 4$$

$$D_z = \begin{vmatrix} 1 & 1 & -1 \\ 0 & 2 & 3 \\ 1 & 1 & 0 \end{vmatrix} = 2$$

$$x = \frac{D_x}{D} = \frac{-6}{2} = -3,\ y = \frac{D_y}{D} = \frac{4}{2} = 2,$$

$$z = \frac{D_z}{D} = \frac{2}{2} = 1$$

The solution set is $\{(-3, 2, 1)\}$.

55. $D = \begin{vmatrix} 1 & 1 & -1 \\ 2 & 2 & 1 \\ 1 & -3 & 0 \end{vmatrix} = 12$

$D_x = \begin{vmatrix} 0 & 1 & -1 \\ 6 & 2 & 1 \\ 0 & -3 & 0 \end{vmatrix} = 18$

$D_y = \begin{vmatrix} 1 & 0 & -1 \\ 2 & 6 & 1 \\ 1 & 0 & 0 \end{vmatrix} = 6$

$D_z = \begin{vmatrix} 1 & 1 & 0 \\ 2 & 2 & 6 \\ 1 & -3 & 0 \end{vmatrix} = 24$

$x = \dfrac{D_x}{D} = \dfrac{18}{12} = \dfrac{3}{2}, \quad y = \dfrac{D_y}{D} = \dfrac{6}{12} = \dfrac{1}{2},$

$z = \dfrac{D_z}{D} = \dfrac{24}{12} = 2$

The solution set is $\left\{ \left(\dfrac{3}{2}, \dfrac{1}{2}, 2 \right) \right\}$.

57. $D = \begin{vmatrix} 1 & 1 & 1 \\ 0 & 2 & 2 \\ 3 & -1 & 0 \end{vmatrix} = 2$

$D_x = \begin{vmatrix} 0 & 1 & 1 \\ 0 & 2 & 2 \\ -1 & -1 & 0 \end{vmatrix} = 0$

$D_y = \begin{vmatrix} 1 & 0 & 1 \\ 0 & 0 & 2 \\ 3 & -1 & 0 \end{vmatrix} = 2$

$D_z = \begin{vmatrix} 1 & 1 & 0 \\ 0 & 2 & 0 \\ 3 & -1 & -1 \end{vmatrix} = -2$

$x = \dfrac{D_x}{D} = \dfrac{0}{2} = 0, \quad y = \dfrac{D_y}{D} = \dfrac{2}{2} = 1,$

$z = \dfrac{D_z}{D} = \dfrac{-2}{2} = -1$

The solution set is $\{(0, 1, -1)\}$.

59. a) From the graph it appears that there should be approximately 9 servings of peas and 11 servings of beets.

b) Let $x =$ the number of servings of canned peas and $y =$ the number of servings of canned beets. In the first equation we find the total grams of protein and in the second the total grams of carbohydrates.

$$3x + y = 38$$
$$11x + 8y = 187$$

$D = \begin{vmatrix} 3 & 1 \\ 11 & 8 \end{vmatrix} = 13 \quad D_x = \begin{vmatrix} 38 & 1 \\ 187 & 8 \end{vmatrix} = 117$

$D_y = \begin{vmatrix} 3 & 38 \\ 11 & 187 \end{vmatrix} = 143$

$x = \dfrac{D_x}{D} = \dfrac{117}{13} = 9 \quad y = \dfrac{D_y}{D} = \dfrac{143}{13} = 11$

To get the required grams of protein and carbohydrates we need 9 servings of peas and 11 servings of beets.

61. Let $x =$ the price of a gallon of milk and $y =$ the price of the magazine. Note that she paid $0.30 tax. The first equation expresses the total price of the goods and the second expresses the total tax.

$$x + y = 4.65$$
$$0.05x + 0.08y = 0.30$$

$D = \begin{vmatrix} 1 & 1 \\ 0.05 & 0.08 \end{vmatrix} = 0.03$

$D_x = \begin{vmatrix} 4.65 & 1 \\ 0.3 & 0.08 \end{vmatrix} = 0.072$

$D_y = \begin{vmatrix} 1 & 4.65 \\ 0.05 & 0.3 \end{vmatrix} = 0.0675$

$x = \dfrac{D_x}{D} = \dfrac{0.072}{0.03} = 2.4$

$y = \dfrac{D_y}{D} = \dfrac{0.0675}{0.03} = 2.25$

The price of the milk was $2.40 and the price of the magazine was $2.25.

63. Let $x =$ the number of singles and $y =$ the number of doubles. Write one equation for the patties and one equation for the tomato slices.

$$x + 2y = 32$$
$$2x + y = 34$$

$$D = \begin{vmatrix} 1 & 2 \\ 2 & 1 \end{vmatrix} = -3 \quad D_x = \begin{vmatrix} 32 & 2 \\ 34 & 1 \end{vmatrix} = -36$$

$$D_y = \begin{vmatrix} 1 & 32 \\ 2 & 34 \end{vmatrix} = -30$$

$$x = \frac{D_x}{D} = \frac{-36}{-3} = 12 \quad y = \frac{D_y}{D} = \frac{-30}{-3} = 10$$

He must sell 12 singles and 10 doubles.

65. Let $x =$ Gary's age and $y =$ Harry's age. Since Gary is 5 years older than Harry, $x = y + 5$. Twenty-nine years ago Gary was $x - 29$ and Harry was $y - 29$. Gary was twice as old as Harry (29 years ago) is expressed as $x - 29 = 2(y - 29)$. These equations can be rewritten as follows.

$$x - y = 5$$
$$x - 2y = -29$$

$$D = \begin{vmatrix} 1 & -1 \\ 1 & -2 \end{vmatrix} = -1$$

$$D_x = \begin{vmatrix} 5 & -1 \\ -29 & -2 \end{vmatrix} = -39$$

$$D_y = \begin{vmatrix} 1 & 5 \\ 1 & -29 \end{vmatrix} = -34$$

$$x = \frac{D_x}{D} = \frac{-39}{-1} = 39 \quad y = \frac{D_y}{D} = \frac{-34}{-1} = 34$$

So Gary is 39 and Harry is 34.

67. Let $x =$ the length of a side of the square and $y =$ the length of a side of the equilateral triangle. Since the perimeters are to be equal, $4x = 3y$. Since the total of the two perimeters is 80, $4x + 3y = 80$. Rewrite the equations as follows.

$$4x - 3y = 0$$
$$4x + 3y = 80$$

$$D = \begin{vmatrix} 4 & -3 \\ 4 & 3 \end{vmatrix} = 24$$

$$D_x = \begin{vmatrix} 0 & -3 \\ 80 & 3 \end{vmatrix} = 240$$

$$D_y = \begin{vmatrix} 4 & 0 \\ 4 & 80 \end{vmatrix} = 320$$

$$x = \frac{D_x}{D} = \frac{240}{24} = 10 \quad y = \frac{D_y}{D} = \frac{320}{24} = \frac{40}{3}$$

The length of the side of the square should be 10 feet and the length of the side of the triangle should be $40/3$ feet.

69. Let $x =$ the number of gallons of 10% solution and $y =$ the number of gallons of 25% solution. Since the total mixture is to be 30 gallons, $x + y = 30$. The next equation comes from the fact that the chlorine in the two parts is equal to the chlorine in the 30 gallons, $0.10x + 0.25y = 0.20(30)$. Rewrite the equations as follows.

$$x + y = 30$$
$$0.10x + 0.25y = 6$$

$$D = \begin{vmatrix} 1 & 1 \\ 0.10 & 0.25 \end{vmatrix} = 0.15$$

$$D_x = \begin{vmatrix} 30 & 1 \\ 6 & 0.25 \end{vmatrix} = 1.5$$

$$D_y = \begin{vmatrix} 1 & 30 \\ 0.10 & 6 \end{vmatrix} = 3$$

$$x = \frac{D_x}{D} = \frac{1.5}{0.15} = 10$$

$$y = \frac{D_y}{D} = \frac{3}{0.15} = 20$$

Use 10 gallons of 10% solution and 20 gallons of 25% solution.

71. Let $x =$ Mimi's weight, $y =$ Mitzi's weight, and $z =$ Cassandra's weight. We can write 3 equations.

$$x + y + z = 175$$
$$x \quad\;\;\; + z = 143$$
$$y + z = 139$$

81

$$D = \begin{vmatrix} 1 & 1 & 1 \\ 1 & 0 & 1 \\ 0 & 1 & 1 \end{vmatrix} = -1$$

$$D_x = \begin{vmatrix} 175 & 1 & 1 \\ 143 & 0 & 1 \\ 139 & 1 & 1 \end{vmatrix} = -36$$

$$D_y = \begin{vmatrix} 1 & 175 & 1 \\ 1 & 143 & 1 \\ 0 & 139 & 1 \end{vmatrix} = -32$$

$$D_z = \begin{vmatrix} 1 & 1 & 175 \\ 1 & 0 & 143 \\ 0 & 1 & 139 \end{vmatrix} = -107$$

$$x = \frac{D_x}{D} = \frac{-36}{-1} = 36,$$

$$y = \frac{D_y}{D} = \frac{-32}{-1} = 32,$$

$$z = \frac{D_z}{D} = \frac{-107}{-1} = 107$$

So Mimi weights 36 pounds, Mitzi weighs 32 pounds, and Cassandra weighs 107 pounds.

73. Let $x =$ the number of degrees in the larger of the two acute angles, $y =$ the number of degrees in the smaller acute angle, and $z =$ the number of degrees in the right angle. We can write 3 equations about the sizes of the angles.

$$\begin{aligned} x + y + z &= 180 \\ x - y &= 12 \\ z &= 90 \end{aligned}$$

$$D = \begin{vmatrix} 1 & 1 & 1 \\ 1 & -1 & 0 \\ 0 & 0 & 1 \end{vmatrix} = -2$$

$$D_x = \begin{vmatrix} 180 & 1 & 1 \\ 12 & -1 & 0 \\ 90 & 0 & 1 \end{vmatrix} = -102$$

$$D_y = \begin{vmatrix} 1 & 180 & 1 \\ 1 & 12 & 0 \\ 0 & 90 & 1 \end{vmatrix} = -78$$

$$D_z = \begin{vmatrix} 1 & 1 & 180 \\ 1 & -1 & 12 \\ 0 & 0 & 90 \end{vmatrix} = -180$$

$$x = \frac{D_x}{D} = \frac{-102}{-2} = 51,$$

$$y = \frac{D_y}{D} = \frac{-78}{-2} = 39,$$

$$z = \frac{D_z}{D} = \frac{-180}{-2} = 90$$

The measures of the three angles of the triangle are 39°, 51°, and 90°.

75. If $D = 0$, then use one of the other methods.

77. No, because Cramer's rule only works on linear systems and the given system is not linear.

4.7 WARM-UPS

1. False, because $x \geq 0$ consists of the points on or to the right of the y-axis.

2. False, because $y \geq 0$ consists of the points on or above the x-axis.

3. False, because $x + y \leq 6$ consists of the points on or below the line $x + y = 6$.

4. False, because the x-intercept is $(15, 0)$ and the y-intercept is $(0, 10)$.

5. False, because the solution set to a system is the intersection of the solution sets.

6. True, because that is the definition of constraint.

7. False, because a linear function does not have an x^2 in it.

8. True, because $R(2, 4) = 3(2) + 5(4) = 26$.

9. False, because
$C(0, 5) = 12(0) + 10(5) = 50$.

10. True, because the maximum or minimum of a linear function occurs at a vertex.

4.7 EXERCISES

1. A constraint is an inequality that restricts the values of the variables.

3. Constraints may be limitations on the amount of available supplies, money, or other resources.

5. The maximum or minimum of a linear function subject to linear constraints occurs at a vertex of the region determined by the constraints.

7. The graph of $x \geq 0$ is the region on or to the right of the y-axis. The graph of $y \geq 0$ is the region on or above the x-axis. The graph of $x + y \leq 5$ is the region on or below the line $x + y = 5$. The intersection of these 3 regions is shaded here.

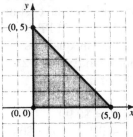

9. The graph of $x \geq 0$ is the region on or to the right of the y-axis. The graph of $y \geq 0$ is the region on or above the x-axis. The graph of $2x + y \leq 4$ is the region on or below the line $2x + y = 4$. The graph of $x + y \leq 3$ is the region on or below the line $x + y = 3$. The intersection of these regions is shaded below.

11. The graph of $x \geq 0$ is the region on or to the right of the y-axis. The graph of $y \geq 0$ is the region on or above the x-axis. The graph of $2x + y \geq 3$ is the region on or above the line $2x + y = 3$. The graph of $x + y \geq 2$ is the region on or above the line $x + y = 2$. The intersection of these regions is shaded below.

13. The graph of $x \geq 0$ is the region on or to the right of the y-axis. The graph of $y \geq 0$ is the region on or above the x-axis. The graph of $x + 3y \leq 15$ is the region on or below the line $x + 3y = 15$. The graph of $2x + y \leq 10$ is the region on or below the line $2x + y = 10$. The intersection of these regions is shaded below.

15. The graph of $x \geq 0$ is the region on or to the right of the y-axis. The graph of $y \geq 0$ is the region on or above the x-axis. The graph of $x + y \geq 4$ is the region on or above the line $x + y = 4$. The graph of $3x + y \geq 6$ is the region on or above the line $3x + y = 6$. The intersection of these regions is shaded below.

17. a) $A(0,0) = 0$
$A(0,80) = 9000(0) + 4000(80) = 320,000$
$A(50,0) = 9000(50) + 4000(0) = 450,000$
$A(30,60) = 9000(30) + 4000(60) = 510,000$
b) To maximize the audience they should use 30 TV ads and 60 radio ads.

19. Let x = the number of doubles and y = the number of triples. We can write four inequalities.

$x \geq 0, \quad y \geq 0$
$2x + 3y \leq 24$
$6x + 3y \leq 48$

Graph the system of inequalities. The vertices of the region are $(0, 8)$, $(6, 4)$, $(8, 0)$, and $(0, 0)$. Since Doubles are \$1.20 each and Triples are \$1.50 each, the revenue in dollars from x Doubles and y Triples is given by $R(x, y) = 1.2x + 1.5y$.

Find the revenue at each vertex.

$R(0, 0) = 0$
$R(0, 8) = 1.2(0) + 1.5(8) = \$12.00$
$R(6, 4) = 1.2(6) + 1.5(4) = \$13.20$
$R(8, 0) = 1.2(8) + 1.5(0) = \$9.60$

The maximum revenue occurs at $(6, 4)$. So they should make 6 Doubles and 4 Triples to maximize their revenue.

21. Let x = the number of Doubles and y = the number of Triples. The constraints and the graph are the same as in Exercise 19. Since the Doubles now sell for \$1.00 each and the Triples now sell for \$2.00 each, the revenue in dollars is given by the function is $R(x, y) = x + 2y$. Find the revenue at each vertex.

$R(0, 0) = 0$
$R(0, 8) = 0 + 2(8) = \$16.00$
$R(6, 4) = 6 + 2(4) = \$14.00$
$R(8, 0) = 8 + 2(0) = \$8.00$

The maximum revenue occurs at $(0, 8)$. So to maximize the revenue they should make no Doubles and 8 Triples.

23. Let x = the number of cups of Doggie Dinner, and y = the number of cups of Puppie Power. We can write four inequalities.

$x \geq 0, \quad y \geq 0$
$20x + 30y \geq 200$
$40x + 20y \geq 180$

Graph the region that satisfies the constraints. Since one cup of Doggie Dinner costs 16 cents and one cup of Puppie Power costs 20 cents, the total cost in dollars for x cups of DD and y cups of PP is $C(x, y) = 0.16x + 0.20y$.

Find the cost at each vertex.

$C(0, 9) = 0.16(0) + 0.20(9) = \$1.80$
$C(1.75, 5.5) = 0.16(1.75) + 0.20(5.5) = \$1.38$
$C(10, 0) = 0.16(10) + 0.20(0) = \$1.60$

The minimum cost occurs at $(1.75, 5.5)$. To minimize the cost and satisfy the constraints she should use 1.75 cups of DD and 5.5 cups of PP.

25. If the cost of one cup of DD is 4 cents and one cup of PP is 10 cents then the total cost in dollars of x cups of DD and y cups of PP is $C(x, y) = 0.04x + 0.10y$. The inequalities and the region are the same as in Exercise 23. Find the cost at each vertex of the region.

$C(0, 9) = 0.04(0) + 0.10(9) = \$0.90$
$C(1.75, 5.5) = 0.04(1.75) + 0.10(5.5) = \$0.62$
$C(10, 0) = 0.04(10) + 0.10(0) = \$0.40$

The minimum cost occurs at $(10, 0)$. She should use 10 cups of DD and 0 cups of PP to satisfy the constraints and minimize the cost.

27. Let x = the amount invested in the laundromat and y = the amount invested in the car wash. We can write four inequalities.

$x \geq 0, \quad y \geq 0$
$x + y \leq 24,000$
$2x \leq y \leq 3x$

Graph the region that satisfies the system of inequalities. Since she makes 18% on the money invested in the laundry and 12% on the money invested in the car wash, the income in dollars is given by the function $I(x, y) = 0.18x + 0.12y$.

Find the income at each vertex.

$I(0, 0) = 0$
$I(8000, 16000) = 0.18(8000) + 0.12(16000)$
$\qquad\qquad\qquad = \$3,360$
$I(6000, 18000) = 0.18(6000) + 0.12(18000)$
$\qquad\qquad\qquad = \$3,240$

The maximum income occurs at $(8000, 16000)$. So she should invest \$8000 in the laundromat and \$16,000 in the car wash.

84

Enriching Your Mathematical Word Power
1. c   **2.** a   **3.** a   **4.** d
5. b   **6.** c   **7.** a   **8.** c
9. d   **10.** b   **11.** a   **12.** d

CHAPTER 4 REVIEW

1. The graph of $y = 2x - 1$ has y-intercept $(0, -1)$ and slope 2. the graph of $y = -x + 2$ has y-intercept $(0, 2)$ and slope -1. The graphs appear to intersect at $(1, 1)$. Check that $(1, 1)$ satisfies both equations. The solution set is $\{(1, 1)\}$ and the system is independent.

3. Solve $x + 2y = 4$ for y to get $y = -\frac{1}{2}x + 2$, which is the same as the second equation. So the two equations have the same graph and all points on that graph satisfy the system. The solution set to this dependent system is $\{(x, y) | x + 2y = 4\}$.

5. The graph of $y = -x$ is a line with slope -1 and y-intercept $(0, 0)$. The graph of $y = -x + 3$ is a line with slope -1 and y-intercept $(0, 3)$. Since these lines are parallel, there is no solution to the system. The system is inconsistent and the solution set is $\emptyset$.

7. Substitute $y = 3x + 11$ into $2x + 3y = 0$.
$$2x + 3(3x + 11) = 0$$
$$11x + 33 = 0$$
$$11x = -33$$
$$x = -3$$
$$y = 3(-3) + 11 = 2$$
The solution set is $\{(-3, 2)\}$ and the system is independent.

9. Substitute $x = y + 5$ into $2x - 2y = 12$.
$$2(y + 5) - 2y = 12$$
$$2y + 10 - 2y = 12$$
$$10 = 12$$
The solution set is $\emptyset$ and the system is inconsistent.

11. Write $2x - y = 3$ as $y = 2x - 3$. Substitute into $6x - 9 = 3y$.
$$6x - 9 = 3(2x - 3)$$
$$6x - 9 = 6x - 9$$
Since the last equation is an identity, the solution set is $\{(x, y) \mid 2x - y = 3\}$. The system is dependent.

13. Substitute $y = \frac{1}{2}x - 3$ into $y = \frac{1}{3}x + 2$.

$$\tfrac{1}{3}x + 2 = \tfrac{1}{2}x - 3$$
$$2x + 12 = 3x - 18$$
$$30 = x$$
$$y = \tfrac{1}{3}x + 2$$
$$y = \tfrac{1}{3} \cdot 30 + 2 = 12$$
The solution set to this independent system is $\{(30, 12)\}$.

15. Substitute $x = 1 - 2y$ into $8x + 6y = 4$:
$$8(1 - 2y) + 6y = 4$$
$$8 - 16y + 6y = 4$$
$$-10y = -4$$
$$y = \tfrac{2}{5}$$
$$x = 1 - 2y$$
$$x = 1 - 2 \cdot \tfrac{2}{5} = \tfrac{1}{5}$$
The solution set to this independent system is

17. Multiply the first equation by 2 and the second by 3, and then add the resulting equations.
$$10x - 6y = -40$$
$$\underline{9x + 6y = 21}$$
$$19x \quad\quad = -19$$
$$x = -1$$
Use $x = -1$ in $3x + 2y = 7$.
$$3(-1) + 2y = 7$$
$$2y = 10$$
$$y = 5$$
The solution set is $\{(-1, 5)\}$ and the system is independent.

19. Rewrite the first equation.
$$2(y - 5) + 4 = 3(x - 6)$$
$$2y - 10 + 4 = 3x - 18$$
$$-3x + 2y = -12$$

Add this last equation to the original second equation.
$$-3x + 2y = -12$$
$$\underline{3x - 2y = 12}$$
$$0 = 0$$

The two equations are just different forms of an equation for the same straight line. The solution set is $\{(x, y) \mid 3x - 2y = 12\}$ and the system is dependent.

21. Rewrite the first equation in the same form as the last so that they can be added.
$$3x - 4(y - 5) = x + 2$$
$$3x - 4y + 20 = x + 2$$
$$2x - 4y = -18$$
$$x - 2y = -9$$
$$-2y + x = -9$$

Add this last equation with the second equation.
$$-2y + x = -9$$
$$\underline{2y - x = 7}$$
$$0 = -2$$

Since the result of the addition is a false statement, the system is inconsistent. The solution set is $\emptyset$.

23. Multiply the first equation by 8 and the second equation by 2 to clear the fractions:

$$2x + 3y = 3$$
$$5x - 12y = 14$$

Multiply the first equation by 4 and add:
$$8x + 12y = 12$$
$$\underline{5x - 12y = 14}$$
$$13x \qquad = 26$$
$$x = 2$$
Use $x = 2$ in $2x + 3y = 3$ to find y.
$$2(2) + 3y = 3$$
$$4 + 3y = 3$$
$$3y = -1$$
$$y = -\frac{1}{3}$$
The solution set is $\left\{\left(2, -\frac{1}{3}\right)\right\}$. The system is independent.

25. Multiply the first equation by 100 and the second equation by 100 to clear the decimals:
$$40x + 6y = 1160$$
$$80x - 5y = 1300$$
Multiply the first equation by -2 and add:
$$-80x - 12y = -2320$$
$$\underline{80x - 5y = 1300}$$
$$-17y = -1020$$
$$y = 60$$
Use $y = 60$ in $40x + 6y = 1160$ to find x.
$$40x + 6(60) = 1160$$
$$40x = 800$$
$$x = 20$$

The solution set is $\{(20, 60)\}$. The system is independent.

27. Add the first and second equations.
$$x - y + z = 4$$
$$\underline{-x + 2y - z = 0}$$
$$y \qquad = 4$$
Add the first and third equations:
$$x - y + z = 4$$
$$\underline{-x + y - 3z = -16}$$
$$-2z = -12$$
$$z = 6$$
Use $y = 4$ and $z = 6$ in $x - y + z = 4$.
$$x - 4 + 6 = 4$$
$$x = 2$$
The solution set is $\{(2, 4, 6)\}$.

29. Add the first and second equations.
$$2x - y - z = 3$$
$$\underline{3x + y + 2z = 4}$$
$$5x \qquad + z = 7 \qquad \text{A}$$
Multiply the first equation by 2 and add the result to the last equation.
$$4x - 2y - 2z = 6$$
$$\underline{4x + 2y - z = -4}$$
$$8x \qquad - 3z = 2 \qquad \text{B}$$
Multiply equation A by 3 and add the result to equation B.
$$15x + 3z = 21$$
$$\underline{8x - 3z = 2}$$
$$23x \qquad = 23$$
$$x = 1$$
Use $x = 1$ in $8x - 3z = 2$.
$$8(1) - 3z = 2$$
$$-3z = -6$$
$$z = 2$$
Use $x = 1$ and $z = 2$ in $3x + y + 2z = 4$.
$$3(1) + y + 2(2) = 4$$
$$y = -3$$
The solution set is $\{(1, -3, 2)\}$.

31. Add the first two equations.
$$x - 3y + z = 5$$
$$\underline{2x - 4y - z = 7}$$
$$3x - 7y \qquad = 12 \qquad \text{A}$$
Multiply the second equation by 2 and add the result to the last equation.
$$4x - 8y - 2z = 14$$
$$\underline{2x - 6y + 2z = 6}$$
$$6x - 14y \qquad = 20 \qquad \text{B}$$

Multiply equation A by -2 and add the result to equation B.

$$-6x + 14y = -24$$
$$\underline{6x - 14y = 20}$$
$$0 = -4$$

The solution set is $\emptyset$.

33. $\begin{bmatrix} 1 & 1 & 7 \\ -1 & 2 & 5 \end{bmatrix}$

$\begin{bmatrix} 1 & 1 & 7 \\ 0 & 3 & 12 \end{bmatrix}$ $R_1 + R_2 \to R_2$

$\begin{bmatrix} 1 & 1 & 7 \\ 0 & 1 & 4 \end{bmatrix}$ $R_2 \div 3 \to R_2$

$\begin{bmatrix} 1 & 0 & 3 \\ 0 & 1 & 4 \end{bmatrix}$ $-1R_2 + R_1 \to R_1$

The solution set is $\{(3, 4)\}$.

35. $\begin{bmatrix} 1 & -3 & 14 \\ 2 & 1 & 0 \end{bmatrix}$

$\begin{bmatrix} 1 & -3 & 14 \\ 0 & 7 & -28 \end{bmatrix}$ $-2R_1 + R_2 \to R_2$

$\begin{bmatrix} 1 & -3 & 14 \\ 0 & 1 & -4 \end{bmatrix}$ $R_2 \div 7 \to R_2$

$\begin{bmatrix} 1 & 0 & 2 \\ 0 & 1 & -4 \end{bmatrix}$ $3R_2 + R_1 \to R_1$

The solution set is $\{(2, -4)\}$.

37. $\begin{bmatrix} 1 & 1 & -1 & 0 \\ 1 & -1 & 2 & 4 \\ 2 & 1 & -1 & 1 \end{bmatrix}$

$\begin{bmatrix} 1 & 1 & -1 & 0 \\ 0 & -2 & 3 & 4 \\ 0 & -1 & 1 & 1 \end{bmatrix}$ $-R_1 + R_2 \to R_2$

$-2R_1 + R_3 \to R_3$

$\begin{bmatrix} 1 & 1 & -1 & 0 \\ 0 & -1 & 1 & 1 \\ 0 & -2 & 3 & 4 \end{bmatrix}$ $R_2 \leftrightarrow R_3$

$\begin{bmatrix} 1 & 1 & -1 & 0 \\ 0 & 1 & -1 & -1 \\ 0 & -2 & 3 & 4 \end{bmatrix}$ $-R_2 \to R_2$

$\begin{bmatrix} 1 & 0 & 0 & 1 \\ 0 & 1 & -1 & -1 \\ 0 & 0 & 1 & 2 \end{bmatrix}$ $-R_2 + R_1 \to R_1$

$2R_2 + R_3 \to R_3$

$\begin{bmatrix} 1 & 0 & 0 & 1 \\ 0 & 1 & 0 & 1 \\ 0 & 0 & 1 & 2 \end{bmatrix}$ $R_3 + R_2 \to R_2$

The solution set is $\{(1, 1, 2)\}$.

39. $\begin{vmatrix} 1 & 3 \\ 0 & 2 \end{vmatrix} = 1 \cdot 2 - 0 \cdot 3 = 2$

41. $\begin{vmatrix} 0.01 & 0.02 \\ 50 & 80 \end{vmatrix} = 0.01(80) - 0.02(50)$

$$= 0.8 - 1 = -0.2$$

43. $D = \begin{vmatrix} 2 & -1 \\ 3 & 1 \end{vmatrix} = 5$

$D_x = \begin{vmatrix} 0 & -1 \\ -5 & 1 \end{vmatrix} = -5$

$D_y = \begin{vmatrix} 2 & 0 \\ 3 & -5 \end{vmatrix} = -10$

$x = \frac{D_x}{D} = \frac{-5}{5} = -1$

$y = \frac{D_y}{D} = \frac{-10}{5} = -2$

The solution set is $\{(-1, -2)\}$.

45. Write the system in standard form.

$$-2x + y = -3$$
$$3x - 2y = 4$$

$D = \begin{vmatrix} -2 & 1 \\ 3 & -2 \end{vmatrix} = 1$

$D_x = \begin{vmatrix} -3 & 1 \\ 4 & -2 \end{vmatrix} = 2$

$D_y = \begin{vmatrix} -2 & -3 \\ 3 & 4 \end{vmatrix} = 1$

$x = \frac{D_x}{D} = \frac{2}{1} = 2 \qquad y = \frac{D_y}{D} = \frac{1}{1} = 1$

The solution set is $\{(2, 1)\}$.

47. Expand by minors using the first column.

$$2\begin{vmatrix} 2 & 4 \\ 1 & 1 \end{vmatrix} - (-1)\begin{vmatrix} 3 & 1 \\ 1 & 1 \end{vmatrix} + 6\begin{vmatrix} 3 & 1 \\ 2 & 4 \end{vmatrix}$$

$$= 2(-2) + 2 + 6(10) = 58$$

49. Expand by minors using the second column.

$$-3\begin{vmatrix} 2 & 4 \\ -1 & 3 \end{vmatrix} + 0\begin{vmatrix} 2 & -2 \\ -1 & 3 \end{vmatrix} - 0\begin{vmatrix} 2 & -2 \\ 2 & 4 \end{vmatrix}$$

$$= -3(10) = -30$$

51. $D = \begin{vmatrix} 1 & 1 & 0 \\ 1 & 1 & 1 \\ 1 & -1 & -1 \end{vmatrix} = 2$

$$D_x = \begin{vmatrix} 3 & 1 & 0 \\ 0 & 1 & 1 \\ 2 & -1 & -1 \end{vmatrix} = 2$$

$$D_y = \begin{vmatrix} 1 & 3 & 0 \\ 1 & 0 & 1 \\ 1 & 2 & -1 \end{vmatrix} = 4$$

$$D_z = \begin{vmatrix} 1 & 1 & 3 \\ 1 & 1 & 0 \\ 1 & -1 & 2 \end{vmatrix} = -6$$

$$x = \frac{D_x}{D} = \frac{2}{2} = 1, \quad y = \frac{D_y}{D} = \frac{4}{2} = 2,$$

$$z = \frac{D_z}{D} = \frac{-6}{2} = -3$$

The solution set is $\{(1, 2, -3)\}$.

53. The graph of $x \geq 0$ consists of points on and to the right of the y-axis. The graph of $y \geq 0$ consists of points on and above the x-axis. The graph of $x + 2y \leq 6$ consists of points on and below the line $x + 2y = 6$. The graph of $x + y \leq 5$ consists of points on and below the line $x + y = 5$. Points that satisfy all 4 inequalities are shown in the following graph.

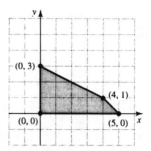

The vertices of the region are $(0, 3)$, $(4, 1)$, $(5, 0)$, and $(0, 0)$.

55. First graph the region that satisfies the four inequalities. The vertices of the region are $(0,0)$, $(0,3)$, $(2,2)$, and $(3,0)$.
Evaluate $R(x, y) = 6x + 9y$ at each vertex.
$R(0, 0) = 0$
$R(0, 3) = 6(0) + 9(3) = 27$
$R(2, 2) = 6(2) + 9(2) = 30$
$R(3, 0) = 6(3) + 9(0) = 18$
The maximum value of $R(x, y)$ is 30.

57. Let $x =$ the tens digit and $y =$ the ones digit. The value of the number is $10x + y$. If the digits are reversed the value will be $10y + x$.
We can write the following two equations.

$$x + y = 15$$
$$10x + y = 10y + x - 9$$

$$x + y = 15$$
$$9x - 9y = -9$$

$$\begin{aligned} x + y &= 15 \\ x - y &= -1 \\ \hline 2x \phantom{{}- y} &= 14 \\ x &= 7 \\ y &= 8 \quad \text{Since } x + y = 15 \end{aligned}$$

The number is 78.

59. Let $b =$ his boat's speed in still water and $c =$ the speed of the current. Let $t =$ his time to go the same distance in the lake.

	Distance	Rate	Time
Down	$\frac{1}{2}(b + c)$	$b + c$	$\frac{1}{2}$
Back	$\frac{3}{4}(b - c)$	$b - c$	$\frac{3}{4}$
Lake	bt	b	t

Since the distance down the stream is the same as the distance going back, we can write the following equation.

$$\frac{1}{2}(b+c) = \frac{3}{4}(b-c)$$
$$\frac{1}{2}b + \frac{1}{2}c = \frac{3}{4}b - \frac{3}{4}c$$
$$2b + 2c = 3b - 3c$$
$$5c = b$$

Since $bt = \frac{1}{2}(b+c)$, we can substitute $b = 5c$ into this equation.

$$(5c)t = \frac{1}{2}(5c + c)$$
$$5ct = 3c$$
$$t = \frac{3c}{5c} = \frac{3}{5}$$

So the time in the lake is $3/5$ of an hour or 36 minutes.

61. Let $x =$ the number of liters of solution A, $y =$ the number of liters of solution B, and $z =$ the number of liters of solution C. We can write 3 equations.

$$x + y + z = 20$$
$$0.30x + 0.20y + 0.60z = 0.38(20)$$
$$z = 2x$$

Substitute $z = 2x$ into the other two equations.

$$x + y + 2x = 20$$
$$0.3x + 0.2y + 0.6(2x) = 7.6$$

$$3x + y = 20$$
$$1.5x + 0.2y = 7.6$$

Substitute $y = 20 - 3x$ into the second equation.

$$1.5x + 0.2(20 - 3x) = 7.6$$
$$0.9x + 4 = 7.6$$
$$0.9x = 3.6$$
$$x = 4$$

If $x = 4$, then $y = 20 - 3(4) = 8$ and $z = 2(4) = 8$. She should use 4 liters of 30% solution (A), 8 liters of 20% solution (B), and 8 liters of 60% solution (C).

63. Let $x =$ the number of servings of beets and $y =$ the number of servings of beans. We can write two equations.

$$x + 6y = 21$$
$$6x + 20y = 78$$

Substitute $x = 21 - 6y$ into $6x + 20y = 78$.

$$6(21 - 6y) + 20y = 78$$
$$126 - 36y + 20y = 78$$
$$-16y = -48$$

$$y = 3$$

If $y = 3$, then $x = 21 - 6(3) = 3$. It would take 3 servings of each.

CHAPTER 4 TEST

1. The graph of $y = -x + 4$ is a straight line with y-intercept $(0, 4)$ and slope -1. The graph of $y = 2x + 1$ is a straight line with y-intercept $(0, 1)$ and slope 2. The graphs appear to intersect at $(1, 3)$. After checking that $(1, 3)$ satisfies both equations, we can be sure that the solution set is $\{(1, 3)\}$.

2. Substitute $y = 2x - 8$ into $4x + 3y = 1$.

$$4x + 3(2x - 8) = 1$$
$$10x = 25$$
$$x = \frac{5}{2}$$

If $x = 5/2$, then $y = 2(5/2) - 8 = -3$. The solution set is $\left\{ \left(\frac{5}{2}, -3 \right) \right\}$.

3. Substitute $y = x - 5$ into the second equation.

$$3x - 4(y - 2) = 28 - x$$
$$3x - 4(x - 5 - 2) = 28 - x$$
$$3x - 4(x - 7) = 28 - x$$
$$28 - x = 28 - x$$

Since the last equation is an identity, the solution set is $\{(x, y) \mid y = x - 5\}$.

4. Multiply the first equation by 3 and the second by 2 and add the results.

$$9x + 6y = 9$$
$$\underline{8x - 6y = -26}$$
$$17x \qquad = -17$$
$$x = -1$$

Use $x = -1$ in $3x + 2y = 3$ to find y.

$$3(-1) + 2y = 3$$
$$2y = 6$$
$$y = 3$$

The solution set is $\{(-1, 3)\}$.

5. Multiply the first equation by 2 and then add the result to the second equation.

$$6x - 2y = 10$$
$$\underline{-6x + 2y = 1}$$
$$0 = 11$$

The solution set to the system is $\emptyset$.

6. The lines both have slope 3 and different y-intercepts, so they are parallel. There is no solution to the system. The system is inconsistent.

7. If we multiply the second equation by 2, the result is the same as the first equation. These are two equations for the same straight line. The system is dependent.

8. The lines have different slopes and different y-intercepts. They will intersect in exactly one point. The system is independent.

9. Add the first two equations.
$$\begin{aligned} x + y - z &= 2 \\ 2x - y + 3z &= -5 \\ \hline 3x \qquad + 2z &= -3 \quad \text{A} \end{aligned}$$

Multiply the first equation by 3 and add the result to the last equation.

$$\begin{aligned} 3x + 3y - 3z &= 6 \\ x - 3y + z &= 4 \\ \hline 4x \qquad - 2z &= 10 \quad \text{B} \end{aligned}$$

Add equations A and B.
$$\begin{aligned} 3x + 2z &= -3 \\ 4x - 2z &= 10 \\ \hline 7x \qquad &= 7 \\ x &= 1 \end{aligned}$$
Use $x = 1$ in $3x + 2z = -3$.
$$\begin{aligned} 3(1) + 2z &= -3 \\ 2z &= -6 \\ z &= -3 \end{aligned}$$

Use $x = 1$ and $z = -3$ in $x + y - z = 2$.
$$\begin{aligned} 1 + y - (-3) &= 2 \\ y &= -2 \end{aligned}$$
The solution set is $\{(1, -2, -3)\}$.

10.
$$\begin{bmatrix} 1 & 2 & 12 \\ 3 & -1 & 1 \end{bmatrix}$$

$$\begin{bmatrix} 1 & 2 & 12 \\ 0 & -7 & -35 \end{bmatrix} \quad -3R_1 + R_2 \to R_2$$

$$\begin{bmatrix} 1 & 2 & 12 \\ 0 & 1 & 5 \end{bmatrix} \quad R_2 \div (-7) \to R_2$$

$$\begin{bmatrix} 1 & 0 & 2 \\ 0 & 1 & 5 \end{bmatrix} \quad -2R_2 + R_1 \to R_1$$
The solution set is $\{(2, 5)\}$.

11.
$$\begin{bmatrix} 1 & -1 & -1 & 1 \\ -1 & -1 & 2 & -2 \\ -1 & -3 & 1 & -5 \end{bmatrix}$$

$$\begin{bmatrix} 1 & -1 & -1 & 1 \\ 0 & -2 & 1 & -1 \\ 0 & -4 & 0 & -4 \end{bmatrix} \quad \begin{aligned} R_1 + R_2 \to R_2 \\ R_1 + R_3 \to R_3 \end{aligned}$$

$$\begin{bmatrix} 1 & -1 & -1 & 1 \\ 0 & -4 & 0 & -4 \\ 0 & -2 & 1 & -1 \end{bmatrix} \quad R_2 \leftrightarrow R_3$$

$$\begin{bmatrix} 1 & -1 & -1 & 1 \\ 0 & 1 & 0 & 1 \\ 0 & -2 & 1 & -1 \end{bmatrix} \quad R_2 \div (-4) \to R_2$$

$$\begin{bmatrix} 1 & 0 & -1 & 2 \\ 0 & 1 & 0 & 1 \\ 0 & 0 & 1 & 1 \end{bmatrix} \quad \begin{aligned} R_2 + R_1 \to R_1 \\ 2R_2 + R_3 \to R_3 \end{aligned}$$

$$\begin{bmatrix} 1 & 0 & 0 & 3 \\ 0 & 1 & 0 & 1 \\ 0 & 0 & 1 & 1 \end{bmatrix} \quad R_3 + R_1 \to R_1$$

The solution set is $\{(3, 1, 1)\}$.

12. $\begin{vmatrix} 2 & 3 \\ 4 & -3 \end{vmatrix} = 2(-3) - 4(3) = -18$

13. Expand by minors using the third column.

$$-1\begin{vmatrix} 2 & 3 \\ 1 & 1 \end{vmatrix} - 1\begin{vmatrix} 1 & -2 \\ 1 & 1 \end{vmatrix} + 0\begin{vmatrix} 1 & -2 \\ 2 & 3 \end{vmatrix}$$

$$= -1(-1) - 1(3) + 0(7) = -2$$

14. $D = \begin{vmatrix} 2 & -1 \\ 3 & 1 \end{vmatrix} = 5$

$D_x = \begin{vmatrix} -4 & -1 \\ -1 & 1 \end{vmatrix} = -5$

$D_y = \begin{vmatrix} 2 & -4 \\ 3 & -1 \end{vmatrix} = 10$

$x = \frac{D_x}{D} = \frac{-5}{5} = -1 \qquad y = \frac{D_y}{D} = \frac{10}{5} = 2$

The solution set is $\{(-1, 2)\}$.

15. $D = \begin{vmatrix} 1 & 1 & 0 \\ 1 & -1 & 2 \\ 2 & 1 & -1 \end{vmatrix} = 4$

$D_x = \begin{vmatrix} 0 & 1 & 0 \\ 6 & -1 & 2 \\ 1 & 1 & -1 \end{vmatrix} = 8$

$D_y = \begin{vmatrix} 1 & 0 & 0 \\ 1 & 6 & 2 \\ 2 & 1 & -1 \end{vmatrix} = -8$

$D_z = \begin{vmatrix} 1 & 1 & 0 \\ 1 & -1 & 6 \\ 2 & 1 & 1 \end{vmatrix} = 4$

$x = \frac{D_x}{D} = \frac{8}{4} = 2, \qquad y = \frac{D_y}{D} = \frac{-8}{4} = -2,$

$z = \frac{D_z}{D} = \frac{4}{4} = 1$ The solution set is $\{(2, -2, 1)\}$.

16. Let $x =$ the price of a single and $y =$ the price of a double. We can write an equation for each night.

$5x + 12y = 390$
$9x + 10y = 412$

Multiply the first equation by -9 and the second by 5, then add the results.

$$-45x - 108y = -3510$$
$$\underline{45x + 50y = 2060}$$
$$-58y = -1450$$
$$y = 25$$

Use $y = 25$ in $5x + 12y = 390$ to find x.
$$5x + 12(25) = 390$$
$$5x + 300 = 390$$
$$5x = 90$$
$$x = 18$$
So singles rent for \$18 per night and doubles rent for \$25 per night.

17. Let $x =$ Jill's study time, $y =$ Karen's study time, and $z =$ Betsy's study time. We can write three equations.

$$x + y + z = 93$$
$$x + y = \tfrac{1}{2}z$$
$$x = y + 3$$

Rewrite the equations as follows.
$$x + y + z = 93$$
$$2x + 2y - z = 0$$
$$x - y = 3$$
Adding the first two equations to eliminate z gives us $3x + 3y = 93$, or $x + y = 31$. Add this result to the third equation.
$$x + y = 31$$
$$\underline{x - y = 3}$$
$$2x = 34$$
$$x = 17$$

Since $x + y = 31$, we get $y = 14$. Use $x = 17$ and $y = 14$ in the first equation.
$$17 + 14 + z = 93$$
$$z = 62$$
So Jill studied 17 hours, Karen studied 14 hours, and Betsy studied 62 hours.

18. First graph the system of inequalities. The vertices of the region are $(0, 0)$, $(0, 4)$, $(3, 2)$, and $(5, 0)$. Evaluate the function $P(x, y) = 8x + 10y$ at each vertex.
$P(0, 0) = 0$
$P(0, 4) = 8(0) + 10(4) = 40$
$P(3, 2) = 8(3) + 10(2) = 44$
$P(5, 0) = 8(5) + 10(0) = 40$
The maximum value of the function $P(x, y)$ is 44.

Making Connections

1. $-3^4 = -(3^4) = -81$

2. $\frac{1}{3}(3) + 6 = 1 + 6 = 7$

3. $(-5)^2 - 4(-2)(6) = 25 - (-48) = 73$

4. $6 - (0.2)(0.3) = 6 - 0.06 = 5.94$

5. $5(t - 3) - 6(t - 2) = 5t - 15 - 6t + 12$
$$= -t - 3$$

6. $0.1(x - 1) - (x - 1) = 0.1x - 0.1 - x + 1$
$$= -0.9x + 0.9$$

7. $\dfrac{-9x^2 - 6x + 3}{-3} = 3x^2 + 2x - 1$

8. $\dfrac{4y - 6}{2} - \dfrac{3y - 9}{3} = 2y - 3 - (y - 3)$
$$= y$$

9. $3x - 5y = 7$
$$-5y = -3x + 7$$
$$y = \frac{3}{5}x - \frac{7}{5}$$

10. $Cx - Dy = W$
$$-Dy = -Cx + W$$
$$y = \frac{C}{D}x - \frac{W}{D}$$

11. $Cy = Wy - K$
$$K = Wy - Cy$$
$$K = y(W - C)$$
$$y = \frac{K}{W - C}$$

12. $A = \frac{1}{2}b(w - y)$
$$2A = bw - by$$
$$by = bw - 2A$$
$$y = \frac{bw - 2A}{b}$$

13. $2x + 3(x - 5) = 5$
$$5x - 15 = 5$$
$$5x = 20$$
$$x = 4$$
$$y = x - 5 = 4 - 5 = -1$$
The solution set is $\{(4, -1)\}$.

14. $y = 1200 - x$
$$0.05x + 0.06(1200 - x) = 67$$
$$-0.01x + 72 = 67$$
$$-0.01x = -5$$
$$x = 500$$
$$y = 1200 - x = 1200 - 500 = 700$$
The solution set is $\{(500, 700)\}$.

15. $x = 5y - 17$
$$3(5y - 17) - 15y = -51$$
$$15y - 51 - 15y = -51$$
$$-51 = -51$$

The equations are equivalent and the solution set is $\{(x, y) \mid x + 17 = 5y\}$.

16. Multiply the first equation by 100 to get $7a + 30b = 670$. This equation is inconsistent with the equation $7a + 30b = 67$. So the solution set is the empty set $\emptyset$.

17. The slope of the line is $55/99$ or $5/9$. Since the y-intercept is $(0, 55)$, the equation is $y = \frac{5}{9}x + 55$.

18. The slope of the line through $(2, -3)$ and $(-4, 8)$ is $11/-6$ or $-11/6$. Use the point-slope formula:
$$y - 8 = -\frac{11}{6}(x - (-4))$$
$$y - 8 = -\frac{11}{6}x - \frac{44}{6}$$
$$y = -\frac{11}{6}x - \frac{22}{3} + \frac{24}{3}$$
$$y = -\frac{11}{6}x + \frac{2}{3}$$

19. Any line parallel to $y = 5x$ has slope 5.
$$y - 6 = 5(x - (-4))$$
$$y - 6 = 5x + 20$$
$$y = 5x + 26$$

20. Any line perpendicular to $y = -2x + 1$ has slope $1/2$.
$$y - 7 = \frac{1}{2}(x - 4)$$
$$y - 7 = \frac{1}{2}x - 2$$
$$y = \frac{1}{2}x + 5$$

21. Any line parallel to the x-axis has slope 0 and is of the form $y = k$. Since it must go through $(3, 5)$ it is $y = 5$.

22. Any line perpendicular to the x-axis is of the form $x = k$. So the equation is $x = -7$.

23. a) From the y-intercept we see that the purchase price for A is \$4000 and for B is \$2000. So A has the larger purchase price.
b) Machine B makes 300,000 copies for \$12,000, or \$0.04 per copy. Machine A makes 300,000 copies for \$9,000 or \$0.03 per copy.
c) The slope of the line for machine B is 0.04 and the slope of the line for machine A is 0.03, which is the per copy cost for each machine.
d) Machine B: $y = 0.04x + 2000$
Machine A: $y = 0.03x + 4000$
e) $0.04x + 2000 = 0.03x + 4000$
$$0.01x = 2000$$
$$x = 200,000$$
The total costs are equal at 200,000 copies.

5.1 WARM-UPS

1. True, because $3^5 \cdot 3^4 = 3^{5+4} = 3^9$.
2. False, because $2x^{-4} = 2 \cdot \dfrac{1}{x^4} = \dfrac{2}{x^4}$.
3. False, because $10^{-3} = \dfrac{1}{1000} = 0.001$.
4. False, because $\dfrac{x^5}{x^{-2}} = x^{5-(-2)} = x^7$.
5. True, because $\dfrac{2^5}{2^{-2}} = 2^{5-(-2)} = 2^7$.
6. False, because $2^3 \cdot 5^2 = 8 \cdot 25 = 200$.
7. True, because $-2^{-2} = -\dfrac{1}{2^2} = -\dfrac{1}{4}$.
8. True, $46.7 \times 10^5 = 4.67 \times 10^1 \times 10^5$.
9. True, $0.512 \times 10^{-3} = 5.12 \times 10^{-1} \times 10^{-3}$.
10. False, because $\dfrac{8 \times 10^{30}}{2 \times 10^{-5}} = 4 \times 10^{35}$.

5.1 EXERCISES

1. An exponential expression is an expression of the form a^n.
3. The product rule says that $a^m a^n = a^{m+n}$.
5. To convert a number in scientific notation to standard notation move the decimal point n places to the left if the exponent on 10 is $-n$ or move the decimal point n places to the right if the exponent on 10 is n, assuming that n is a positive integer.
7. $2^2 = 4, -2^2 = -(2^2) = -4, (-2)^2 = 4,$
$2^{-2} = \dfrac{1}{2^2} = \dfrac{1}{4}, -2^{-2} = -\dfrac{1}{2^2} = -\dfrac{1}{4},$
$(-2)^{-2} = \dfrac{1}{(-2)^2} = \dfrac{1}{4}$
9. $2^3 = 8, -2^3 = -(2^3) = -8, (-2)^3 = -8,$
$2^{-3} = \dfrac{1}{2^3} = \dfrac{1}{8}, -2^{-3} = -\dfrac{1}{2^3} = -\dfrac{1}{8},$
$(-2)^{-3} = \dfrac{1}{(-2)^3} = -\dfrac{1}{8}$
11. $\dfrac{1}{5^2} = \dfrac{1}{25}, \dfrac{1}{-5^2} = -\dfrac{1}{25},$
$\dfrac{1}{(-5)^2} = \dfrac{1}{25}, \dfrac{1}{5^{-2}} = 5^2 = 25,$
$\dfrac{1}{-5^{-2}} = -5^2 = -25, \dfrac{1}{(-5)^{-2}} = (-5)^2 = 25$
13. $7^1 = 7, -7^1 = -7, (-7)^1 = -7,$
$7^{-1} = \dfrac{1}{7}, -7^{-1} = -\dfrac{1}{7^1} = -\dfrac{1}{7},$
$(-7)^{-1} = \dfrac{1}{(-7)^1} = -\dfrac{1}{7}$
15. $\left(\dfrac{1}{2}\right)^2 = \dfrac{1}{4}, \left(-\dfrac{1}{2}\right)^2 = \dfrac{1}{4},$

$-\left(\dfrac{1}{2}\right)^2 = -\dfrac{1}{4}, \left(\dfrac{1}{2}\right)^{-2} = \left(\dfrac{2}{1}\right)^2 = 4,$
$\left(-\dfrac{1}{2}\right)^{-2} = \left(-\dfrac{2}{1}\right)^2 = 4,$
$-\left(\dfrac{1}{2}\right)^{-2} = -\left(\dfrac{2}{1}\right)^2 = -4$

17. $\left(\dfrac{2}{3}\right)^3 = \dfrac{8}{27}, \left(-\dfrac{2}{3}\right)^3 = -\dfrac{8}{27},$
$-\left(\dfrac{2}{3}\right)^3 = -\dfrac{8}{27}, \left(\dfrac{2}{3}\right)^{-3} = \left(\dfrac{3}{2}\right)^3 = \dfrac{27}{8},$
$\left(-\dfrac{2}{3}\right)^{-3} = \left(-\dfrac{3}{2}\right)^3 = -\dfrac{27}{8},$
$-\left(\dfrac{2}{3}\right)^{-3} = -\left(\dfrac{3}{2}\right)^3 = -\dfrac{27}{8}$
19. $2^5 \cdot 2^{12} = 2^{5+12} = 2^{17}$
21. $-2x^{-7} \cdot 3x = -6x^{-7+1} = -6x^{-6} = -\dfrac{6}{x^6}$
23. $-7b^{-7}(-3b^{-3}) = 21b^{-10} = \dfrac{21}{b^{10}}$
25. $3^0 = 1, -3^0 = -(3^0) = -1, (-3)^0 = 1,$
$-(-3)^0 = -1$
27. $(2+3)^0 = 5^0 = 1,$
$2^0 + 3^0 = 1 + 1 = 2,$
$(2^0 + 3)^0 = (1+3)^0 = 1$
29. $3st^0 = 3s \cdot 1 = 3s,$
$3(st)^0 = 3 \cdot 1 = 3, (3st)^0 = 1$
31. $2w^{-3}(w^7 \cdot w^{-4}) = 2w^0 = 2$
33. $\dfrac{2}{4^{-2}} = 2 \cdot 4^2 = 2 \cdot 16 = 32$
35. $\dfrac{3^{-1}}{10^{-2}} = \dfrac{10^2}{3^1} = \dfrac{100}{3}$
37. $\dfrac{2x^{-3}(4x)}{5y^{-2}} = \dfrac{8x^{-2}}{5y^{-2}} = \dfrac{8y^2}{5x^2}$
39. $\dfrac{4^{-2}x^3 x^{-6}}{3x^{-3}x^2} = \dfrac{x^{-3}}{16 \cdot 3x^{-1}} = \dfrac{x^{-2}}{48} = \dfrac{1}{48x^2}$
41. $x^{5-3} = x^2$
43. $3^{6-(-2)} = 3^8$
45. $\dfrac{4a^{-5-(-2)}}{12} = \dfrac{a^{-3}}{3} = \dfrac{1}{3a^3}$
47. $\dfrac{-6w^{-5}}{2w^3} = -3w^{-5-3} = -3w^{-8} = \dfrac{-3}{w^8}$
49. $\dfrac{3^3 w^{-2} w^5}{3^{-5}w^{-3}} = \dfrac{3^3 w^3}{3^{-5}w^{-3}} = 3^8 w^6$
51. $\dfrac{3x^{-6} \cdot x^2 y^{-1}}{6x^{-5}y^{-2}} = \dfrac{3x^{-4}y}{6x^{-5}}$
$= \dfrac{x^{-4-(-5)}y}{2} = \dfrac{xy}{2}$
53. $3^{-1} \cdot \left(\dfrac{1}{3}\right)^{-3} = \dfrac{1}{3} \cdot 3^3 = 9$
55. $-2^4 + \left(\dfrac{1}{2}\right)^{-1} = -16 + 2 = -14$
57. $-(-2)^{-3} \cdot 2^{-1} = -\left(-\dfrac{1}{8}\right)\dfrac{1}{2} = \dfrac{1}{16}$
59. $7 \cdot 2^{-3} - 2 \cdot 4^{-1} = \dfrac{7}{8} - \dfrac{1}{2} = \dfrac{3}{4}$

61. $(1 + 2^{-1})^{-2} = \left(1 + \frac{1}{2}\right)^{-2}$
$$= \left(\frac{3}{2}\right)^{-2} = \left(\frac{2}{3}\right)^2 = \frac{4}{9}$$

63. $-5a^2 \cdot 3a^2 = -15a^4$

65. $5a \cdot a^2 + 3a^{-2} \cdot a^5 = 5a^3 + 3a^3 = 8a^3$

67. $\dfrac{-3a^5(-2a^{-1})}{6a^3} = \dfrac{6a^4}{6a^3} = a$

69. $\dfrac{(-3x^3y^2)(-2xy^{-3})}{-9x^2y^{-5}} = \dfrac{6x^4y^{-1}}{-9x^2y^{-5}}$
$$= \dfrac{-2x^2y^4}{3}$$

71. $\dfrac{2^{-1} + 3^{-1}}{2^{-1}} = \left(\frac{1}{2} + \frac{1}{3}\right) \cdot \frac{2}{1} = \frac{5}{6} \cdot \frac{2}{1} = \frac{5}{3}$

73. $8 = 2^3$

75. $\frac{1}{4} = 2^{-2}$

77. $16 = \left(\frac{1}{2}\right)^{-4}$

79. $10^{-3} = 0.001$

81. Since the exponent in 4.86×10^8 is positive, we move the decimal point 8 places to the right: 486,000,000

83. Since the exponent in 2.37×10^{-6} is negative, we move the decimal point 6 places to the left: 0.00000237

85. Since the exponent in 4×10^6 is positive, we move the decimal point 6 places to the right: 4,000,000

87. Since the exponent in 5×10^{-6} is negative, we move the decimal point 6 places to the left: 0.000005

89. The decimal point in 320,000 must be moved 5 places to the left for scientific notation. Since the original number is larger than 10 the exponent is positive: 3.2×10^5

91. The decimal point in 0.00000071 must be moved 7 places to the right to get scientific notation. Since the original number is smaller than 1, the exponent is negative: 7.1×10^{-7}

93. Move the decimal point 5 places to the right: 7.03×10^{-5}

95. $205 \times 10^5 = 2.05 \times 10^2 \times 10^5$
$$= 2.05 \times 10^7$$

97. $(4000)(5000)(0.0003)$
$= 4 \times 10^3 \cdot 5 \times 10^3 \cdot 3 \times 10^{-4} = 60 \times 10^2$
$$= 6 \times 10^1 \times 10^2 = 6 \times 10^3$$

99. $\dfrac{(5,000,000)(0.0003)}{2000}$
$$= \dfrac{5 \times 10^6 \cdot 3 \times 10^{-4}}{2 \times 10^3} = \dfrac{7.5 \times 10^2}{10^3}$$
$$= 7.5 \times 10^{-1}$$

101. By the quotient rule, we subtract exponents when we divide exponential expressions: $3 \times 10^{40-18} = 3 \times 10^{22}$

103. $\dfrac{(-4 \times 10^5)(6 \times 10^{-9})}{2 \times 10^{-16}}$
$= -12 \times 10^{5+(-9)-(-16)}$
$= -12 \times 10^{12}$
$= -1.2 \times 10^1 \times 10^{12} = -1.2 \times 10^{13}$

105. Enter the numbers into a calculator using scientific notation. The calculator will give answers to the computations in scientific notation. 1.578×10^5

107. Use a calculator to get 9.187×10^{-5}.

109. Use a calculator to get 3.828×10^{30}.

111. $9.3 \times 10^7 \text{miles} \times \dfrac{5280 \text{ feet}}{1 \text{ mile}}$
$$= 4.910 \times 10^{11} \text{ ft}$$

113. Since $T = \dfrac{D}{R}$, $T = \dfrac{4.6 \times 10^{12} \text{ km}}{1.2 \times 10^5 \text{ kps}}$
$$\approx 3.83 \times 10^7 \text{ seconds}$$

115. $\dfrac{8.71 \times 10^7 \text{ ton} \times \dfrac{2000 \text{ lb}}{1 \text{ ton}}}{1.80863 \times 10^8 \text{ people}} \div 365 \text{ day} \approx$

2.6 lb/person/day

117. a) The only integers for which $2^m \cdot 3^n = 6^{m+n}$ is correct are $m = 0$ and $n = 0$. There are other values that would work, but they are not integers and it would require logarithms to find them.
b) If m and n are nonzero integers, then $2^m \cdot 3^n \neq 6^{m+n}$.

119. Suppose that a is *positive*. The value of $-a^n$ is always negative because $-a^n = -(a^n)$. The value of $(-a)^n$ is negative and equal to $-a^n$ when n is odd. The value of $(-a)^n$ is positive and not equal to $-a^n$ when n is even. Similar statement can be made if a is negative.

5.2 WARM-UPS

1. False, because $(2^2)^3 = 2^6$.
2. True, because $(2^{-3})^{-1} = 2^3 = 8$.
3. True, because $(x^{-3})^3 = x^{-3 \cdot 3} = x^{-9}$.
4. False, because $2^3 \cdot 2^3 = 2^6$ and $(2^3)^3 = 2^9$.
5. False, because $(2x)^3 = 2^3 x^3 = 8x^3$.
6. False, because $(-3y^3)^2 = 9y^6$.

7. True, the reciprocal of $\frac{2}{3}$ is $\frac{3}{2}$.

8. True, because $\left(\frac{2}{3}\right)^3 = \frac{2^3}{3^3} = \frac{8}{27}$.

9. True, because $\left(\frac{x^2}{2}\right)^3 = \frac{x^6}{2^3} = \frac{x^6}{8}$.

10. True, because $\left(\frac{2}{x}\right)^{-2} = \left(\frac{x}{2}\right)^2 = \frac{x^2}{4}$.

5.2 EXERCISES

1. The power of a power rule says that $(a^m)^n = a^{mn}$.

3. The power of a quotient rule says that $(a/b)^m = a^m/b^m$.

5. To compute the amount A when interest is compounded annually use $A = P(1 + i)^n$ where P is the principal, i is the annual interest rate, and n is the number of years.

7. $(2^2)^3 = 2^6 = 64$

9. $(y^2)^5 = y^{2 \cdot 5} = y^{10}$

11. $(x^2)^{-4} = x^{(2)(-4)} = x^{-8} = \frac{1}{x^8}$

13. $(m^{-3})^{-6} = m^{(-3)(-6)} = m^{18}$

15. $(x^{-2})^3 (x^{-3})^{-2} = x^{-6}x^6 = x^0 = 1$

17. $\frac{(x^3)^{-4}}{(x^2)^{-5}} = \frac{x^{-12}}{x^{-10}} = x^{-2} = \frac{1}{x^2}$

19. $(-9y)^2 = (-9)^2 y^2 = 81y^2$

21. $(-5w^3)^2 = (-5)^2 w^6 = 25w^6$

23. $(x^3 y^{-2})^3 = x^9 y^{-6} = \frac{x^9}{y^6}$

25. $(3ab^{-1})^{-2} = 3^{-2} a^{-2} b^2 = \frac{b^2}{9a^2}$

27. $\frac{2xy^{-2}}{(3x^2 y)^{-1}} = \frac{2xy^{-2}}{3^{-1} x^{-2} y^{-1}}$

$\qquad = 2 \cdot 3 x^3 y^{-1} = \frac{6x^3}{y}$

29. $\frac{(2ab)^{-2}}{2ab^2} = \frac{2^{-2} a^{-2} b^{-2}}{2ab^2}$

$\qquad = 2^{-3} a^{-3} b^{-4} = \frac{1}{8a^3 b^4}$

31. $\left(\frac{w}{2}\right)^3 = \frac{w^3}{2^3} = \frac{w^3}{8}$

33. $\left(-\frac{3a}{4}\right)^3 = \frac{(-3)^3 a^3}{4^3} = -\frac{27a^3}{64}$

35. $\left(\frac{2x^{-1}}{y}\right)^{-2} = \frac{2^{-2} x^2}{y^{-2}} = \frac{x^2 y^2}{4}$

37. $\left(\frac{-3x^3}{y}\right)^{-2} = \frac{(-3)^{-2} (x^3)^{-2}}{y^{-2}} = \frac{x^{-6} y^2}{(-3)^2}$

$\qquad = \frac{y^2}{9x^6}$

39. $\left(\frac{2}{5}\right)^{-2} = \left(\frac{5}{2}\right)^2 = \frac{25}{4}$

41. $\left(-\frac{1}{2}\right)^{-2} = (-2)^2 = 4$

43. $\left(-\frac{2x}{3}\right)^{-3} = \left(-\frac{3}{2x}\right)^3$

$\qquad = -\frac{3^3}{2^3 x^3} = -\frac{27}{8x^3}$

45. $\left(\frac{2x^2}{3y}\right)^{-3} = \left(\frac{3y}{2x^2}\right)^3 = \frac{3^3 y^3}{2^3 (x^2)^3} = \frac{27y^3}{8x^6}$

47. $5^{2t} \cdot 5^{4t} = 5^{6t}$

49. $(2^{-3w})^{-2w} = 2^{6w^2}$

51. $\frac{7^{2m+6}}{7^{m+3}} = 7^{2m+6-m-3} = 7^{m+3}$

53. $8^{2a-1}(8^{a+4})^3 = 8^{2a-1} \cdot 8^{3a+12}$

$\qquad = 8^{5a+11}$

55. Add the exponents: $6x^9$

57. Multiply the exponents: $-8x^6$

59. Eliminate negative exponents: $\frac{3z}{x^2 y}$

61. The reciprocal of $-\frac{2}{3}$ is $-\frac{3}{2}$.

63. Multiply the exponents: $\frac{4x^6}{9}$

65. $(-2x^{-2})^{-1} = (-2)^{-1} x^2 = -\frac{x^2}{2}$

67. $\left(\frac{2x^2 y}{xy^2}\right)^{-3} = \left(\frac{xy^2}{2x^2 y}\right)^3 = \frac{x^3 y^6}{8x^6 y^3} = \frac{y^3}{8x^3}$

69. $\frac{(5a^{-1} b^2)^3}{(5ab^{-2})^4} = \frac{5^3 a^{-3} b^6}{5^4 a^4 b^{-8}}$

$\qquad = 5^{-1} a^{-7} b^{14} = \frac{b^{14}}{5a^7}$

71. $\frac{(2x^{-2} y)^{-3}}{(2xy^{-1})^2} \cdot 2x^2 y^{-7} = \frac{2^{-3} x^6 y^{-3}}{2^2 x^2 y^{-2}} \cdot 2x^2 y^{-7}$

$\qquad = 2^{-4} x^6 y^{-8} = \frac{x^6}{16y^8}$

73. $\left(\frac{6a^{-2} b^3}{2c^4}\right)^{-2} (3a^{-1} b^2)^3$

$= \frac{6^{-2} a^4 b^{-6}}{2^{-2} c^{-8}} \cdot 3^3 a^{-3} b^6 = 3a^1 b^0 c^8 = 3ac^8$

75. $32 \cdot 64 = 2^5 \cdot 2^6 = 2^{11}$

77. $81 \cdot 6^{-4} = 3^4 (2 \cdot 3)^{-4} = 3^4 2^{-4} 3^{-4} = 2^{-4}$

79. $4^{3n} = (2^2)^{3n} = 2^{6n}$

81. $\left(\frac{1}{(2^4)^{-m}}\right)^m = \left(\frac{1}{2^{-4m}}\right)^m = (2^{4m})^m$

$\qquad = 2^{4m^2}$

83. $\frac{1}{5^{-2}} = 25$

85. $2^{-1} + 2^{-2} = 0.75$

87. $(0.036)^{-2} + (4.29)^3 \approx 850.559$

89. $\dfrac{(5.73)^{-1} + (4.29)^{-1}}{(3.762)^{-1}} \approx 1.533$

91. $40{,}000(1 + 0.12)^3 = 40{,}000(1.12)^3$
$\qquad\qquad = \$56{,}197.12$

93. $10{,}000(1 + 0.07)^{-18} \approx \$2{,}958.64$

95. a) $L = 72.2(1.002)^{20} \approx 75.1$ yr

b) $L = 72.2(1.002)^{60} \approx 81.4$ yr

97. If both a and b are nonzero, then $(ab)^{-1} = a^{-1}b^{-1}$. If $a = b = 2$, then $2 + 2)^{-1} = \frac{1}{4}$ and $2^{-1} + 2^{-1} = 1$.

99. All expressions have value -1 except (d) which has value 1.

101. a)

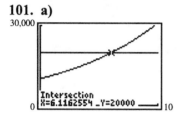

b) $(6.116,\ 20{,}000)$

c) 6.116 years

5.3 WARM-UPS

1. False, because it has a negative exponent.

2. False, because the coefficient of x is -5.

3. False, because the highest power of x is 3.

4. True, because $5^2 - 3 = 22$.

5. False, because $P(0) = 30(0) + 10 = 10$.

6. True, because the two trinomials are added correctly.

7. False, because
$$(x^2 - 5x) - (x^2 - 3x) = -2x$$
for any value of x.

8. True, because the monomial and binomial are correctly multiplied.

9. True, because $-(a - b) = b - a$ for any values of a and b.

10. False, because $-(y + 5) = -y - 5$ for any value of y.

5.3 EXERCISES

1. A term of a polynomial is a single number or the product of a number and one or more variables raised to whole number powers.

3. A constant is simply a number.

5. The degree of a polynomial in one variable is the highest power of the variable in the polynomial.

7. Yes, it is a monomial.

9. No, because polynomials do not have negative exponents.

11. Yes, it is a trinomial.

13. No, because $\frac{1}{x}$ is not allowed in a polynomial.

15. The degree is 4 because 4 is the highest power. The coefficient of x^3 is -8. It is a binomial because it has two terms.

17. The degree of -8 is 0 and the coefficient of x^3 is 0. It is a monomial.

19. The degree is 7 because 7 is the highest power. The coefficient of x^3 is 0, because x^3 is missing. It is a monomial because it has one term.

21. The degree is 6 because 6 is the highest power. The coefficient of x^3 is 1. It is a trinomial because it has three terms.

23. Replace x by 3 in $P(x) = x^4 - 1$.
$\qquad P(3) = 3^4 - 1 = 81 - 1 = 80$

25. Replace x by -2:
$\qquad M(-2) = -3(-2)^2 + 4(-2) - 9 = -29$

27. Replace x by 1:
$\qquad R(1) = 1^5 - 1^4 + 1^3 - 1^2 + 1 - 1 = 0$

29. $(2a - 3) + (a + 5) = 2a + a - 3 + 5$
$\qquad\qquad\qquad = 3a + 2$

31. $(7xy + 30) - (2xy + 5)$
$\qquad = 7xy + 30 - 2xy - 5 = 5xy + 25$

33. $(x^2 - 3x) + (-x^2 + 5x - 9)$
$\qquad = x^2 - x^2 - 3x + 5x - 9 = 2x - 9$

35. $(2x^3 - 4x - 3) - (x^2 - 2x + 5)$
$\qquad = 2x^3 - x^2 - 4x + 2x - 3 - 5$
$\qquad = 2x^3 - x^2 - 2x - 8$

37. Add the like terms to get $11x^2 - 2x - 9$.

39. $5x - 4x + 2 - (-3) = x + 5$.

41. Change the sign of every term on the bottom and add it to the appropriate like term on the top to get $-6x^2 + 5x + 2$.

43. Add the like terms to get $2x$.

45. Add exponents to get
$-3x^2 \cdot 5x^4 = -15x^6$.

47. $-1(3x - 2) = -3x + 2$

49. $5x^2y^3(3x^2y - 4x) = 15x^4y^4 - 20x^3y^3$

96

51. $(x-2)(x+2) = (x-2)x + (x-2)2$
$= x^2 - 2x + 2x - 4 = x^2 - 4$

53. $(x^2 + x + 2)(2x - 3)$
$= (x^2 + x + 2)2x + (x^2 + x + 2)(-3)$
$= 2x^3 + 2x^2 + 4x - 3x^2 - 3x - 6$
$= 2x^3 - x^2 + x - 6$

55. Multiply $-5x$ by $2x - 3$ to get
$-10x^2 + 15x$.

57.
$$\begin{array}{r} x+5 \\ \underline{x+5} \\ 5x + 25 \\ \underline{x^2 + 5x \quad\quad} \\ x^2 + 10x + 25 \end{array}$$

59.
$$\begin{array}{r} x + 6 \\ \underline{2x \; - \; 3} \\ -3x \; - \; 18 \\ \underline{2x^2 + 12x \quad\quad} \\ 2x^2 + 9x \; - \; 18 \end{array}$$

61.
$$\begin{array}{r} x^2 + xy + y^2 \\ \underline{x - y} \\ -x^2y - xy^2 - y^3 \\ \underline{x^3 + x^2y + xy^2 \quad\quad\quad} \\ x^3 \quad\quad\quad\quad\;\; - y^3 \end{array}$$

63. Add like terms to get $2x - 5$.

65. Add like terms: $4a^2 - 11a - 4$

67. $(w^2 - 7w - 2) - (w - 3w^2 + 5)$
$= w^2 - 7w - 2 - w + 3w^2 - 5$
$= 4w^2 - 8w - 7$

69. $(x - 2)(x^2 + 2x + 4)$
$= (x - 2)x^2 + (x - 2)2x + (x - 2)4$
$= x^3 - 2x^2 + 2x^2 - 4x + 4x - 8$
$= x^3 - 8$

71. $(x - w)(z + 2w)$
$= (x - w)z + (x - w)2w$
$= xz - wz + 2xw - 2w^2$

73. $(x^2 - x + 2)(x^2 + x - 2)$
$= (x^2 - x + 2)(x^2) + (x^2 - x + 2)(x)$
$\quad\quad\quad\quad + (x^2 - x + 2)(-2)$
$= x^4 - x^3 + 2x^2 + x^3 - x^2 + 2x$
$\quad\quad\quad - 2x^2 + 2x - 4$
$= x^4 - x^2 + 4x - 4$

75. $(2.31x - 5.4)6.25x + (2.31x - 5.4)1.8$
$= 14.4375x^2 - 29.592x - 9.72$

77. $(3.759 + 11.61)x^2 - 4.71x + 6.59x$
$\quad\quad\quad + 2.85 - 3.716$
$= 15.369x^2 + 1.88x - 0.866$

79. $\frac{2}{4}x + \frac{4}{2} + \frac{1}{4}x - \frac{1}{2} = \frac{3}{4}x + \frac{3}{2}$

81. $\frac{1}{2}x^2 + \frac{1}{3}x - \frac{1}{5} - \frac{2}{2}x^2 + \frac{2}{3}x + \frac{1}{5}$
$\quad\quad = -\frac{1}{2}x^2 + x$

83. Combine like terms within the first set of brackets:
$x^2 - 3 - x^2 - 5x + 4 = -5x + 1$
Combine like terms within the second set of brackets:
$x - 3x^2 + 15x = -3x^2 + 16x$
Now subtract these results:
$[-5x + 1] - [-3x^2 + 16x]$
$\quad = -5x + 1 + 3x^2 - 16x$
$\quad = 3x^2 - 21x + 1$

85. $[5x - 4(x - 3)][3x - 7(x + 2)]$
$\quad = [x + 12][-4x - 14]$
$\quad = [x + 12](-4x) + [x + 12](-14)$
$\quad = -4x^2 - 48x - 14x - 168$
$\quad = -4x^2 - 62x - 168$

87. $[x^2 - (m + 2)][x^2 + (m + 2)]$
$= [x^2 - m - 2][x^2 + m + 2]$
$= [x^2 - m - 2]x^2 + [x^2 - m - 2]m$
$\quad\quad\quad + [x^2 - m - 2]2$
$= x^4 - mx^2 - 2x^2 + mx^2 - m^2 - 2m$
$\quad\quad\quad + 2x^2 - 2m - 4$
$= x^4 - m^2 - 4m - 4$

89. $2x(5x - 4) - 3x[5x^2 - 3x(x - 7)]$
$\quad = 10x^2 - 8x - 3x[2x^2 + 21x]$
$\quad = 10x^2 - 8x - 6x^3 - 63x^2$
$\quad = -6x^3 - 53x^2 - 8x$

91. $(a^{2m} + 3a^m - 3) + (-5a^{2m} - 7a^m + 8)$
$\quad = -4a^{2m} - 4a^m + 5$

93. $(x^n - 1)(x^n + 3)$
$\quad = (x^n - 1)x^n + (x^n - 1)3$
$\quad = x^{2n} - x^n + 3x^n - 3$
$\quad = x^{2n} + 2x^n - 3$

95. $z^{3w} - z^{2w}(z^{1-w} - 4z^w)$
$\quad = z^{3w} - z^{1+w} + 4z^{3w} = 5z^{3w} - z^{1+w}$

97. $(x^{2r} + y)x^{4r} + (x^{2r} + y)(-x^{2r}y)$
$\quad\quad\quad + (x^{2r} + y)y^2$
$= x^{6r} + x^{4r}y - x^{4r}y - x^{2r}y^2 + x^{2r}y^2 + y^3$
$= x^{6r} + y^3$

99. $C(3) = 20(3) + 15 = \$75$

97

101. $C(3) = 50 \cdot 3 - 0.01(3)^4 = 149.19$
$C(2) = 50 \cdot 2 - 0.01(2)^4 = 99.84$
$C(3) - C(2) = 49.35$
Marginal cost of 3rd window is \$49.35.
$C(10) - C(9) = 400 - 384.39$
Marginal cost of 10th window is \$15.61.

103. a) $F(1950) - M(1950) \approx 6.2$ years
b) The lines are not parallel because the slopes are different.
c) If y is the year in which female life expectancy is 8 years larger then male life expectancy, then y satisfies the following equation.
$0.18268y - 284.98$
$$= 0.16252y - 251.91 + 8$$
$$0.02016y = 41.07$$
$$y \approx 2037$$

105. The binomial $x^4 + 1$ has degree 4.

107. There are 4 terms in the product $(a + b)(c + d)$. There are 8 terms in the product $(a + b)(c + d)(e + f)$. There are 16 terms in the product of four binomials with no like terms. There are 2^n terms in the product of n binomials with no like terms.

5.4 WARM-UPS

1. True, because the binomials are multiplied correctly.
2. True, because the binomials are multiplied correctly.
3. False, $(2 + 3)^2 = 5^2 = 25$ and $2^2 + 3^2 = 4 + 9 = 13$.
4. True, because the binomial is squared correctly.
5. False, $(8 - 3)^2 = 5^2 = 25$ and $64 - 9 = 55$.
6. True, because $(a + b)(a - b) = a^2 - b^2$.
7. True, $(60 - 1)(60 + 1) = 60^2 - 1^2$
$$= 3600 - 1.$$
8. True, because the binomial is squared correctly.
9. False, $(x - 3)^2 = x^2 - 6x + 9$ for any value of x.
10. False, it is a product of two monomials.

5.4 EXERCISES

1. The distributive property is used in multiplying binomials.
3. The purpose of FOIL is to provide a fast way to find the product of two binomials.
5. The square of a difference is the square of the first term minus twice the product of the two terms plus the square of the last term.
7. In general, $(a + b)^2 = a^2 + 2ab + b^2$.
9. $(x - 2)(x + 4) = x^2 + 4x - 2x - 8$
$$= x^2 + 2x - 8$$
11. $(1 + 2x)(3 + x) = 3 + x + 6x + 2x^2$
$$= 2x^2 + 7x + 3$$
13. $(-2a - 3)(-a + 5)$
$$= 2a^2 - 10a + 3a - 15$$
$$= 2a^2 - 7a - 15$$
15. $(2x^2 - 7)(2x^2 + 7)$
$$= 4x^2 + 14x^2 - 14x^2 - 49$$
$$= 4x^4 - 49$$
17. $(2x^3 - 1)(x^3 + 4) = 2x^6 + 8x^3 - x^3 - 4$
$$= 2x^6 + 7x^3 - 4$$
19. $(6z + w)(w - z)$
$$= 6zw - 6z^2 + w^2 - wz$$
$$= w^2 + 5wz - 6z^2$$
21. $(3k - 2t)(4t + 3k)$
$$= 12kt + 9k^2 - 8t^2 - 6tk$$
$$= 9k^2 + 6kt - 8t^2$$
23. $(x - 3)(y + w) = xy - 3y + xw - 3w$
25. $(m + 3)^2 = m^2 + 2(3)(m) + 3^2$
$$= m^2 + 6m + 9$$
27. $(4 - a)^2 = 4^2 - 2(a)(4) + a^2$
$$= a^2 - 8a + 16$$
29. $(2w + 1)^2 = (2w)^2 + 2(2w)(1) + 1^2$
$$= 4w^2 + 4w + 1$$
31. $(3t - 5u)^2 = (3t)^2 - 2(3t)(5u) + (5u)^2$
$$= 9t^2 - 30tu + 25u^2$$
33. $(-x - 1)^2 = (-x)^2 - 2(-x)(1) + 1^2$
$$= x^2 + 2x + 1$$
35. $(a - 3y^3)^2 = a^2 - 2(a)(3y^3) + (3y^3)^2$
$$= a^2 - 6ay^3 + 9y^6$$
37. $(w - 9)(w + 9) = w^2 - 9^2 = w^2 - 81$
39. $(w^3 - y)(w^3 + y) = (w^3)^2 - y^2$
$$= w^6 - y^2$$
41. $(7 - 2x)(7 + 2x) = (7)^2 - (2x)^2$
$$= 49 - 4x^2$$
43. $(3x^2 - 2)(3x^2 + 2) = (3x^2)^2 - 2^2$
$$= 9x^4 - 4$$

45. $(5a^3 - 2b)(5a^3 + 2b) = 25a^6 - 4b^2$

47. $[(m + t) + 5][(m + t) - 5]$
$$= (m + t)^2 - 25$$
$$= m^2 + 2mt + t^2 - 25$$

49. $[y - (r + 5)][y + (r + 5)]$
$$= y^2 - (r + 5)^2$$
$$= y^2 - r^2 - 10r - 25$$

51. $[(2y - t) + 3]^2$
$$= (2y - t)^2 + 2 \cdot 3(2y - t) + 3^2$$
$$= 4y^2 - 4yt + t^2 + 12y - 6t + 9$$

53. $[3h + (k - 1)]^2$
$$= 9h^2 + 2 \cdot 3h(k - 1) + (k - 1)^2$$
$$= 9h^2 + 6hk - 6h + k^2 - 2k + 1$$

55. $(x - 6)(x + 9) = x^2 + 9x - 6x - 54$
$$= x^2 + 3x - 54$$

57. $(5 - x)(5 + x) = 5^2 - x^2 = 25 - x^2$

59. $(3x - 4a)(2x + 5a)$
$$= 6x^2 + 15ax - 8ax - 20a^2$$
$$= 6x^2 + 7ax - 20a^2$$

61. $(2t - 3)(t + w) = 2t^2 + 2tw - 3t - 3w$

63. $(3x^2 + 2y^3)^2$
$$= (3x^2)^2 + 2(3x^2)(2y^3) + (2y^3)^2$$
$$= 9x^4 + 12x^2 y^3 + 4y^6$$

65. $(2 + 2y)(3y - 5) = 6y - 10 + 6y^2 - 10y$
$$= 6y^2 - 4y - 10$$

67. $(2m - 7)^2 = (2m)^2 - 2 \cdot 2m \cdot 7 + 7^2$
$$= 4m^2 - 28m + 49$$

69. $(3 + 7x)^2 = 3^2 + 2(3)(7x) + (7x)^2$
$$= 49x^2 + 42x + 9$$

71. $4y\left(3y + \frac{1}{2}\right)^2$
$$= 4y\left((3y)^2 + 2(3y)\frac{1}{2} + \left(\frac{1}{2}\right)^2\right)$$
$$= 4y\left(9y^2 + 3y + \frac{1}{4}\right)$$
$$= 36y^3 + 12y^2 + y$$

73. $(a + h)^2 - a^2 = a^2 + 2ah + h^2 - a^2$
$$= 2ah + h^2$$

75. $(x + 2)(x + 2)^2 = (x + 2)(x^2 + 4x + 4)$
$$= x(x^2 + 4x + 4) + 2(x^2 + 4x + 4)$$
$$= x^3 + 6x^2 + 12x + 8$$

77. $(y + 3)^3 = (y + 3)^2(y + 3)$
$$= (y^2 + 6y + 9)(y + 3)$$
$$= (y^2 + 6y + 9)y + (y^2 + 6y + 9)3$$
$$= y^3 + 6y^2 + 9y + 3y^2 + 18y + 27$$
$$= y^3 + 9y^2 + 27y + 27$$

79. $4x - 3(x - 5)^2$
$$= 4x - 3(x^2 - 10x + 25)$$

$$= 4x - 3x^2 + 30x - 75$$
$$= -3x^2 + 34x - 75$$

81. $(3.2x - 4.5)(5.1x + 3.9)$
$$= (3.2)(5.1)x^2 + (3.9)(3.2)x - (4.5)(5.1)x$$
$$- (4.5)(3.9)$$
$$= 16.32x^2 - 10.47x - 17.55$$

83. $(3.6y + 4.4)^2$
$$= (3.6y)^2 + 2(3.6)(4.4)y + (4.4)^2$$
$$= 12.96y^2 + 31.68y + 19.36$$

85. $(x^m + 2)(x^{2m} + 3)$
$$= x^m x^{2m} + 3x^m + 2x^{2m} + 6$$
$$= x^{3m} + 2x^{2m} + 3x^m + 6$$

87. $a^{n+1}(a^{2n} + a^n - 3)$
$$= a^{n+1+2n} + a^{n+1+n} - 3a^{n+1}$$
$$= a^{3n+1} + a^{2n+1} - 3a^{n+1}$$

89. $(a^m + a^n)^2 = (a^m)^2 + 2a^m a^n + (a^n)^2$
$$= a^{2m} + 2a^{m+n} + a^{2n}$$

91. $(5y^m + 8z^k)(3y^{2m} + 4z^{3-k})$
$$= 5y^m 3y^{2m} + 3 \cdot 8y^{2m} z^k + 4 \cdot 5y^m z^{3-k}$$
$$+ 32z^{k+3-k}$$
$$= 15y^{3m} + 24y^{2m} z^k + 20\, y^m z^{3-k} + 32z^3$$

93. Use $L = x + 3$ and $W = x + 1$ in the formula for the area of a rectangle, $A = LW$.
$A = (x + 3)(x + 1) = x^2 + 4x + 3$

95. $A = (8 - 2x)(10 - 2x) = 4x^2 - 36x + 80$
If $x = 0.4$, then $A = 4(0.4)^2 - 36(0.4) + 80$
$= 66.24$ km^2

97. Since x is taken off of both sides of the rectangle, the sides are $4 - 2x$ and $6 - 2x$
$V = x(4 - 2x)(6 - 2x)$
$$= x(4x^2 - 20x + 24)$$
$$= 4x^3 - 20x^2 + 24x$$
If $x = 4$ in. $= \frac{1}{3}$ ft, then
$$V = 4\left(\frac{1}{3}\right)^3 - 20\left(\frac{1}{3}\right)^2 + 24\left(\frac{1}{3}\right)$$
$$= 5\frac{25}{27} \text{ ft}^3 \approx 5.9 \text{ ft}^3$$

99. a) $(a + b)^3 = a^3 + 3a^2 b + 3ab^2 + b^3$
b) $(a + b)^4 = a^4 + 4a^3 b + 6a^2 b^2 + 4ab^3 + b^4$
$(a + b)^5 = a^5 + 5a^4 b + 10a^3 b^2 + 10a^2 b^3$
$$+ 5ab^4 + b^5$$

c) There are 4, 5, and 6 terms, respectively.
d) The number of terms in $(a + b)^n$ after combining like terms is $n + 1$.

101. The area of the four regions is a^2, ab, ab, and b^2. If you add these you get $a^2 + 2ab + b^2$. So we see geometrically that $(a + b)^2 = a^2 + 2ab + b^2$.

5.5 WARM-UPS

1. True, $3xy(x - 2y) = -3xy(-x + 2y)$.
2. True.
3. True, because $-2(2 - x)$
   $= -4 + 2x = 2x - 4$ for any value of x.
4. True, $a^2 - b^2 = (a - b)(a + b)$.
5. False, because $x^2 + 12x + 36$ is a perfect square trinomial.
6. False, because $y^2 + 8y + 16$ is a perfect square trinomial.
7. False, because $(3x + 7)^2 = 9x^2 + 42x + 49$.
8. True, $x^3 + 1 = (x + 1)(x^2 - x + 1)$.
9. False, $x^3 - 27 = (x - 3)(x^2 + 3x + 9)$.
10. False, $x^3 - 8 = (x - 2)(x^2 + 2x + 4)$.

5.5 EXERCISES

1. A prime number is a natural number greater than 1 that has no factors other than itself and 1.
3. The greatest common factor for the terms of a polynomial is a monomial that includes every number or variable that is a factor of all terms of the polynomial.
5. A linear polynomial is a polynomial of the form $ax + b$ with $a \neq 0$.
7. A prime polynomial is a polynomial that cannot be factored.
9. Since $48 = 2 \cdot 2 \cdot 2 \cdot 2 \cdot 3$ and $36x = 2 \cdot 2 \cdot 3 \cdot 3x$, the greatest common factor is $2 \cdot 2 \cdot 3$ or 12.
11. The gcf for 9, 21, and 15 is 3. Since there are no variables in common to all three terms, the gcf is 3.
13. Since $24 = 2 \cdot 2 \cdot 2 \cdot 3$, $42 = 2 \cdot 3 \cdot 7$, and $66 = 2 \cdot 3 \cdot 11$, the GCF of 24, 42, and 66 is $2 \cdot 3$ or 6. Using each common variable with the lowest power, we get the GCF $6xy$.
15. $x^3 - 5x = x \cdot x^2 - x \cdot 5 = x(x^2 - 5)$
17. $48wx + 36wy = 12w \cdot 4x + 12w \cdot 3y$
    $= 12w(4x + 3y)$
19. $2x^3 - 4x^2 + 6x$
    $= 2x \cdot x^2 - 2x \cdot 2x + 2x \cdot 3$
    $= 2x(x^2 - 2x + 3)$
21. $36a^3b^6 - 24a^4b^2 + 60a^5b^3$
    $= 12a^3b^2(3b^4 - 2a + 5a^2b)$

23. Factor out the quantity $x - 6$:
    $(x - 6)a + (x - 6)b = (x - 6)(a + b)$
25. Factor out $(y - 1)^2$:
    $(y - 1)^2y + (y - 1)^2z = (y - 1)^2(y + z)$
27. $2x - 2y = 2 \cdot x - 2 \cdot y = 2(x - y)$
    $2x - 2y = -2(-x) - 2 \cdot y = -2(-x + y)$
29. $6x^2 - 3x = 3x(2x - 1)$
    $6x^2 - 3x = -3x(-2x + 1)$
31. $-w^3 + 3w^2 = w^2 \cdot (-w) + w^2(3)$
    $= w^2(-w + 3)$
    $-w^3 + 3w^2 = -w^2 \cdot w - (-w^2) \cdot 3$
    $= -w^2(w - 3)$
33. $-a^3 + a^2 - 7a = a(-a^2 + a - 7)$
    $-a^3 + a^2 - 7a = -a(a^2 - a + 7)$
35. $ax + ay + 3x + 3y$
    $= a(x + y) + 3(x + y)$
    $= (x + y)(a + 3)$
37. $xy - 3y + x - 3$
    $= y(x - 3) + 1(x - 3)$
    $= (x - 3)(y + 1)$
39. $4a - 4b - ca + cb$
    $= 4(a - b) - c(a - b)$
    $= (a - b)(4 - c)$
41. $xy - y - 6x + 6$
    $= y(x - 1) - 6(x - 1)$
    $= (x - 1)(y - 6)$
43. The polynomial is a difference of two squares and so it factors as the product of a sum and a difference: $(x - 10)(x + 10)$
45. $4y^2 - 49 = (2y)^2 - 7^2$
    $= (2y - 7)(2y + 7)$
47. $9x^2 - 25a^2 = (3x)^2 - (5a)^2$
    $= (3x - 5a)(3x + 5a)$
49. $144w^2z^2 - h^2 = (12wz)^2 - h^2$
    $= (12wz - h)(12wz + h)$
51. This polynomial is a perfect square trinomial. The middle term is twice the product of the square roots of the first term and last term. $x^2 - 2(x)(10) + 10^2 = (x - 10)^2$
53. $4m^2 - 4m + 1 = (2m)^2 - 2 \cdot 2m + 1^2$
    $= (2m - 1)^2$
55. $w^2 - 2wt + t^2 = (w - t)^2$
57. $a^3 - 1 = a^3 - 1^3 = (a - 1)(a^2 + a + 1)$
59. $w^3 + 27 = w^3 + 3^3$
    $= (w + 3)(w^2 - 3w + 9)$
61. $8x^3 - 1 = (2x)^3 - 1^3$
    $= (2x - 1)(4x^2 + 2x + 1)$

100

63. $64x^3 + 125 = (4x)^3 + 5^3$
$$= (4x + 5)(16x^2 - 20x + 25)$$
65. $8a^3 - 27b^3 = (2a)^3 - (3b)^3$
$$= (2a - 3b)(4a^2 + 6ab + 9b^2)$$
67. $2x^2 - 8 = 2(x^2 - 4) = 2(x + 2)(x - 2)$
69. $x^3 + 10x^2 + 25x = x(x^2 + 10x + 25)$
$$= x(x + 5)^2$$
71. $4x^2 + 4x + 1 = (2x)^2 + 2(2x)(1) + 1^2$
$$= (2x + 1)^2$$
73. $(x + 3)x + (x + 3)7 = (x + 3)(x + 7)$
75. $6y^2 + 3y = 3y \cdot 2y + 3y \cdot 1 = 3y(2y + 1)$
77. $4x^2 - 20x + 25 = (2x)^2 - 2(2x)(5) + 5^2$
$$= (2x - 5)^2$$
79. $2m^4 - 2mn^3 = 2m(m^3 - n^3)$
$$= 2m(m - n)(m^2 + mn + n^2)$$
81. $(2x - 3)x - (2x - 3)2$
$$= (2x - 3)(x - 2)$$
83. $9a^3 - aw^2 = a(9a^2 - w^2)$
$$= a(3a + w)(3a - w)$$
85. $-5a^2 + 30a - 45 = -5(a^2 - 6a + 9)$
$$= -5(a - 3)^2$$
87. $16 - 54x^3 = 2(8 - 27x^3)$
$$= 2(2 - 3x)(4 + 6x + 9x^2)$$
89. $-3y^3 - 18y^2 - 27y = -3y(y^2 + 6y + 9)$
$$= -3y(y + 3)^2$$
91. $-7a^2b^2 + 7 = -7(a^2b^2 - 1)$
$$= -7(ab + 1)(ab - 1)$$
93. $7x - 7h - hx + h^2$
$$= 7(x - h) - h(x - h)$$
$$= (x - h)(7 - h)$$
95. $a^2x + 3a^2 - 4x - 12$
$$= a^2(x + 3) - 4(x + 3)$$
$$= (x + 3)(a^2 - 4)(x + 3)$$
$$= (x + 3)(a - 2)(a + 2)$$
97. If we choose $k = 9$, then $x^2 + 6x + k$ will be the perfect square $x^2 + 6x + 9$.
99. If we choose $k = 20$, then $4a^2 - ka + 25$ will be the perfect square $4a^2 - 20a + 25$.
101. If we choose $k = 16$, then $km^2 - 24m + 9$ will be the perfect square $16m^2 - 24m + 9$.
103. Choose $k = 100$ so that $81y^2 - 180y + k$ is the perfect square $81y^2 - 180y + 100$.
105. a) $b^3 - 6b^2 + 9b$
$$= b(b^2 - 6b + 9)$$
$$= b(b - 3)^2$$

Since $V = LWH$ and the height is b, the sides of the square bottom are each $b - 3$ cm.
b) $V = 18^3 - 6 \cdot 18^2 + 9 \cdot 18 = 4050$ in.^{3}
c) From the graph it appears that the height of a cage with a volume of 20,000 in.^{3} is 30 in.

5.6 WARM-UPS

1. True. Check by multiplying $(x + 3)(x + 6)$.
2. False, $y^2 + 2y - 35 = (y + 7)(y - 5)$.
3. False, $x^2 + 4$ cannot be factored. It is a sum of two squares.
4. False, $x^2 - 5x - 6 = (x - 6)(x + 1)$.
5. True. Check by multiplying $(x - 6)(x + 2)$.
6. False, $x^2 + 15x + 36 = (x + 12)(x + 3)$.
7. False, $3x^2 + 4x - 15 = (3x - 5)(x + 3)$.
8. False, $4x^2 + 4x - 3 = (2x + 3)(2x - 1)$.
9. True. Check by multiplying.
10. True. Check by multiplying.

5.6 EXERCISES

1. To factor $x^2 + bx + c$, find two integers whose sum is b and whose product is c.
3. Trial and error means to simply write down possible factors and then use FOIL to check until you get the correct factors.
5. Two numbers that have a product of 3 and a sum of 4 are 1 and 3:
$x^2 + 4x + 3 = (x + 1)(x + 3)$
7. Two numbers that have a product of 50 and a sum of 15 are 5 and 10:
$a^2 + 15a + 50 = (a + 10)(a + 5)$
9. Two numbers that have a product of -14 and a sum of -5 are -7 and 2:
$y^2 - 5y - 14 = (y - 7)(y + 2)$
11. Two numbers that have a product of 8 and a sum of -6 are -2 and -4:
$x^2 - 6x + 8 = (x - 2)(x - 4)$

13. Two numbers with product 27 and sum -12 are -9 and -3:
$a^2 - 12a + 27 = (a - 9)(a - 3)$

15. Two numbers that have a product of -30 and a sum of 7 are -3 and 10:
$a^2 + 7a - 30 = (a + 10)(a - 3)$

17. Two numbers with a product of $6 \cdot 1 = 6$ and a sum of 5 are 2 and 3:

$$6w^2 + 5w + 1 = 6w^2 + 3w + 2w + 1$$
$$= 3w(2w + 1) + 1(2w + 1)$$
$$= (3w + 1)(2w + 1)$$

19. Using the ac method we find two numbers that have a product of $2(-3) = -6$ and a sum of -5. These numbers are -6 and 1:

$$2x^2 - 5x - 3 = 2x^2 - 6x + 1x - 3$$
$$= 2x(x - 3) + 1(x - 3)$$
$$= (2x + 1)(x - 3)$$

21. Two numbers with a product of $4(15) = 60$ and a sum of 16 are 6 and 10:

$$4x^2 + 16x + 15 = 4x^2 + 6x + 10x + 15$$
$$= 2x(2x + 3) + 5(2x + 3)$$
$$= (2x + 5)(2x + 3)$$

23. Two numbers with a product of $6 \cdot 1 = 6$ and a sum of -5 are -2 and -3:

$$6x^2 - 5x + 1 = 6x^2 - 2x - 3x + 1$$
$$= 2x(3x - 1) - 1(3x - 1)$$
$$= (2x - 1)(3x - 1)$$

25. Two numbers with a product of $12(-1) = -12$ and a sum of 1 are 4 and -3:

$$12y^2 + y - 1 = 12y^2 - 3y + 4y - 1$$
$$= 3y(4y - 1) + 1(4y - 1)$$
$$= (3y + 1)(4y - 1)$$

27. Two numbers with a product of $-5(6) = -30$ and a sum of 1 are 6 and -5:

$$6a^2 + a - 5 = 6a^2 + 6a - 5a - 5$$
$$= 6a(a + 1) - 5(a + 1)$$
$$= (6a - 5)(a + 1)$$

29. $2x^2 + 15x - 8 = (2x - 1)(x + 8)$
31. $3b^2 - 16b - 35 = (3b + 5)(b - 7)$
33. $6w^2 + w - 12 = (3w - 4)(2w + 3)$
35. $4x^2 - 5x + 1 = (4x - 1)(x - 1)$
37. $5m^2 + 13m - 6 = (5m - 2)(m + 3)$
39. $6y^2 - 7y - 20 = (3y + 4)(2y - 5)$
41. If $a = x^5$ then

$$x^{10} - 9 = (x^5)^2 - 3^2 = a^2 - 3^2$$
$$= (a - 3)(a + 3) = (x^5 - 3)(x^5 + 3)$$

43. $z^{12} - 6z^6 + 9 = (z^6 - 3)^2$
45. $2x^7 + 8x^4 + 8x = 2x(x^6 + 4x^3 + 4)$
$$= 2x(x^3 + 2)^2$$
47. $4x^5 + 4x^3 + x = x(4x^4 + 4x^2 + 1)$
$$= x(2x^2 + 1)^2$$
49. $x^6 - 8 = (x^2)^3 - 2^3$
$$= (x^2 - 2)(x^4 + 2x^2 + 4)$$

51. $a^{2n} - 1 = (a^n)^2 - 1^2 = (a^n - 1)(a^n + 1)$
53. $a^{2r} + 6a^r + 9 = (a^r)^2 + 6a^r + 3^2$
$$= (a^r + 3)^2$$
55. $x^{3m} - 8 = (x^m)^3 - 2^3$
$$= (x^m - 2)(x^{2m} + 2x^m + 4)$$
57. $a^{3m} - b^3 = (a^m)^3 - b^3$
$$= (a^m - b)(a^{2m} + a^m b + b^2)$$
59. $k^{2w+1} - 10k^{w+1} + 25k$
$$= k(k^{2w} - 10k^w + 25) = k(k^w - 5)^2$$
61. $x^6 - 2x^3 - 35 = (x^3 + 5)(x^3 - 7)$
63. $a^{20} - 20a^{10} + 100 = (a^{10} - 10)^2$
65. $-12a^5 - 10a^3 - 2a$
$$= -2a(6a^4 + 5a^2 + 1)$$
$$= -2a(3a^2 + 1)(2a^2 + 1)$$
67. Two numbers with a product of -15 and a sum of 2 are 5 and -3:

$$x^{2a} + 2x^a - 15 = (x^a + 5)(x^a - 3)$$
69. $x^{2a} - y^{2b} = (x^a)^2 - (y^b)^2$
$$= (x^a - y^b)(x^a + y^b)$$
71. $x^8 - x^4 - 6 = (x^4 - 3)(x^4 + 2)$
73. $x^{a+2} - x^a = x^a(x^2 - 1)$
$$= x^a(x - 1)(x + 1)$$
75. $x^{2a} + 6x^a + 9 = (x^a)^2 + 2 \cdot x^a \cdot 3 + 3^2$
$$= (x^a + 3)^2$$
77. $2x^2 + 20x + 50 = 2(x^2 + 10x + 25)$
$$= 2(x + 5)^2$$
79. $a^3 - 36a = a(a^2 - 36)$
$$= a(a - 6)(a + 6)$$
81. $10a^2 + 55a - 30 = 5(2a^2 + 11a - 6)$
$$= 5(2a - 1)(a + 6)$$
83. $2x^2 - 128y^2 = 2(x^2 - 64y^2)$
$$= 2(x + 8y)(x - 8y)$$
85. $-9x^2 + 33x + 12 = -3(3x^2 - 11x - 4)$
$$= -3(3x + 1)(x - 4)$$
87. $m^5 + 20m^4 + 100m^3$
$$= m^3(m^2 + 20m + 100)$$
$$= m^3(m + 10)^2$$
89. Two numbers with a product of $6 \cdot 20 = 120$ and a sum of 23 are 8 and 15:

$$6x^2 + 23x + 20 = 6x^2 + 15x + 8x + 20$$
$$= 3x(2x + 5) + 4(2x + 5)$$
$$= (3x + 4)(2x + 5)$$
91. $9y^3 - 24y^2 + 16y = y(9y^2 - 24y + 16)$
$$= y(3y - 4)^2$$
93. $r^2 - 6rs + 8s^2 = (r - 4s)(r - 2s)$
95. $m^3 + 2m + 3m^2 = m(m^2 + 3m + 2)$
$$= m(m + 1)(m + 2)$$

97. $6m^3 - m^2n - 2mn^2$
$$= m(6m^2 - mn - 2n^2)$$
$$= m(2m + n)(3m - 2n)$$

99. $9m^2 - 25n^2 = (3m)^2 - (5n)^2$
$$= (3m - 5n)(3m + 5n)$$
101. $5a^2 + 20a - 60 = 5(a^2 + 4a - 12)$
$$= 5(a + 6)(a - 2)$$
103. $-2w^2 + 18w + 20$
$$= -2(w^2 - 9w - 10)$$
$$= -2(w - 10)(w + 1)$$
105. $w^2x^2 - 100x^2 = x^2(w^2 - 100)$
$$= x^2(w + 10)(w - 10)$$
107. $81x^2 - 9 = 9(9x^2 - 1) =$
$$= 9(3x + 1)(3x - 1)$$
109. Two numbers with a product of
$-15 \cdot 8 = -120$ and a sum of -2 are -12 and
10:
$$8x^2 - 2x - 15 = 8x^2 - 12x + 10x - 15$$
$$= 4x(2x - 3) + 5(2x - 3)$$
$$= (4x + 5)(2x - 3)$$
111. $3m^4 - 24m = 3m(m^3 - 8)$
$$= 3m(m - 2)(m^2 + 2m + 4)$$
113. The polynomials in (a) and (b) are not
perfect square trinomials.
115. a) $x^2 + 10x + 25 = (x + 5)^2$
b) $x^2 - 10x + 25 = (x - 5)^2$
c) $x^2 + 26x + 25 = (x + 25)(x + 1)$
d) $x^2 - 25 = (x + 5)(x - 5)$
e) $x^2 + 25$ is not factorable.

5.7 WARM-UPS

1. False, $x^2 - 9 = (x - 3)(x + 3)$ for any
value of x.
2. True, because $4x^2 + 12x + 9 = (2x + 3)^2$.
3. True.
4. False, because $x^2 - 4$ is not a prime
polynomial.
5. False, because
$y^3 - 27 = (y - 3)(y^2 + 3y + 9)$
for any value of y.
6. True, because $y^6 - 1 = (y^3)^2 - 1^2$.
7. False, because $2x - 4$ is not prime.
8. True, because it cannot be factored.
9. True, because $a^6 - 1 = (a^2)^3 - 1^3$.

10. False, because if we factor x out of the
first two terms and a out of the last two terms
we get
$x^2 + 3x - ax + 3a = x(x + 3) + a(-x + 3)$
and we cannot finish. If we factor x out of the
first two terms and $-a$ out of the last two
terms we get $x(x + 3) - a(x - 3)$ and again
we cannot finish the factoring.

5.7 EXERCISES

1. Always factor out the greatest common
factor first.
3. In factoring a trinomial look for the perfect
square trinomials.
5. Prime, because it is a sum of two squares.
7. Not prime, because
$-9w^2 - 9 = -9(w^2 + 1)$.

9. Not prime, $x^2 - 2x - 3 = (x - 3)(x + 1)$.
11. Prime, because no two integers have a
product of 3 and a sum of 2.
13. Prime, because no two numbers have a
product of -3 and a sum of -4.
15. Prime, because no two numbers have a
product of $6 \cdot (-4) = -24$ and a sum of 3.
17. Let $y = a^2$ in the polynomial:
$a^4 - 10a^2 + 25 = y^2 - 10y + 25$
$$= (y - 5)^2 = (a^2 - 5)^2$$
19. Let $y = x^2$ in the polynomial:
$x^4 - 6x^2 + 8 = y^2 - 6y + 8 = (y - 2)(y - 4)$
$= (x^2 - 2)(x^2 - 4) = (x^2 - 2)(x - 2)(x + 2)$
21. Let $y = 3x - 5$ in the polynomial:
$(3x - 5)^2 - 1 = y^2 - 1 = (y - 1)(y + 1)$
$$= (3x - 5 - 1)(3x - 5 + 1)$$
$$= (3x - 6)(3x - 4)$$
$$= 3(x - 2)(3x - 4)$$
23. If we let $a = y^3$, then $a^2 = (y^3)^2 = y^6$.
$2y^6 - 128 = 2(y^6 - 64) = 2(a^2 - 64)$
$= 2(a - 8)(a + 8) = 2(y^3 - 8)(y^3 + 8)$
$= 2(y - 2)(y^2 + 2y + 4)(y + 2)(y^2 - 2y + 4)$
25. $32a^4 - 18 = 2(16a^4 - 9)$
$$= 2(4a^2 + 3)(4a^2 - 3)$$

27. Let $a = x^2$ and $b = x - 6$:
$x^4 - (x - 6)^2 = a^2 - b^2 = (a + b)(a - b)$
$$= (x^2 + x - 6)(x^2 - [x - 6])$$
$$= (x + 3)(x - 2)(x^2 - x + 6)$$

29. Let $y = m + 2$ in the polynomial:
$(m + 2)^2 + 2(m + 2) - 3 = y^2 + 2y - 3$
$= (y + 3)(y - 1) = (m + 2 + 3)(m + 2 - 1)$
$= (m + 5)(m + 1)$

31. Let $w = y - 1$ in the polynomial:
$3(y - 1)^2 + 11(y - 1) - 20$
$\qquad = 3w^2 + 11w - 20$
$\qquad = (3w - 4)(w + 5)$
$\qquad = [3(y - 1) - 4][y - 1 + 5]$
$\qquad = (3y - 7)(y + 4)$

33. Let $a = y^2 - 3$:
$(y^2 - 3)^2 - 4(y^2 - 3) - 12 = a^2 - 4a - 12$
$\qquad = (a - 6)(a + 2)$
$\qquad = (y^2 - 3 - 6)(y^2 - 3 + 2)$
$\qquad = (y^2 - 9)(y^2 - 1)$
$\qquad = (y - 3)(y + 3)(y - 1)(y + 1)$

35. $x^2 - 2b - 2x + bx = x^2 - 2x + bx - 2b$
$\qquad = x(x - 2) + b(x - 2)$
$\qquad = (x - 2)(x + b)$

37. $x^2 - ay - xy + ax = x^2 - xy + ax - ay$
$\qquad = x(x - y) + a(x - y)$
$\qquad = (x - y)(x + a)$

39. $x^2 + 2x + 1 - a^2 = (x + 1)^2 - a^2$
$\qquad = (x + 1 - a)(x + 1 + a)$

41. $x^2 - 4x + 4 - w^2 = (x - 2)^2 - w^2$
$\qquad = (x - 2 - w)(x - 2 + w)$

43. $x^2 - z^2 + 4x + 4 = x^2 + 4x + 4 - z^2$
$\qquad = (x + 2)^2 - z^2$
$\qquad = (x + 2 - z)(x + 2 + z)$

45. This is a perfect square trinomial.
$9x^2 - 24x + 16 = (3x - 4)^2$

47. Two numbers whose product is 36 and whose sum is -13 are -4 and -9.
$12x^2 - 13x + 3 = 12x^2 - 4x - 9x + 3$
$\qquad = 4x(3x - 1) - 3(3x - 1)$
$\qquad = (3x - 1)(4x - 3)$

49. $3a^4 + 81a = 3a(a^3 + 27)$
$\qquad = 3a(a + 3)(a^2 - 3a + 9)$

51. $32 + 2x^2 = 2(x^2 + 16)$

53. Prime, because there are no two integers that have a product of 72 and a sum of -5.

55. Let $a = x + y$: $(x + y)^2 - 1 = a^2 - 1$
$= (a - 1)(a + 1) = (x + y - 1)(x + y + 1)$

57. $a^3b - ab^3 = ab(a^2 - b^2)$
$\qquad = ab(a - b)(a + b)$

59. $x^4 + 2x^3 - 8x - 16$
$\qquad = x^3(x + 2) - 8(x + 2)$
$\qquad = (x + 2)(x^3 - 8)$
$\qquad = (x + 2)(x - 2)(x^2 + 2x + 4)$

61. $m^2n + 2mn^2 + n^3$
$\qquad = n(m^2 + 2mn + n^2)$
$\qquad = n(m + n)^2$

63. $2m + wn + 2n + wm =$
$\qquad = 2m + 2n + wm + wn$
$\qquad = 2(m + n) + w(m + n)$
$\qquad = (m + n)(2 + w)$

65. Two numbers with a product of -12 and a sum of 4 are 6 and -2:
$4w^2 + 4w - 3 = 4w^2 - 2w + 6w - 3$
$\qquad = 2w(2w - 1) + 3(2w - 1)$
$\qquad = (2w + 3)(2w - 1)$

67. Let $a = t^2$:
$t^4 + 4t^2 - 21 = a^2 + 4a - 21$
$\qquad = (a + 7)(a - 3)$
$\qquad = (t^2 + 7)(t^2 - 3)$

69. $-a^3 - 7a^2 + 30a = -a(a^2 + 7a - 30)$
$\qquad = -a(a + 10)(a - 3)$

71. Let $a = y + 5$: $(y + 5)^2 - 2(y + 5) - 3$
$\qquad = a^2 - 2a - 3 = (a - 3)(a + 1)$
$\qquad = (y + 5 - 3)(y + 5 + 1)$
$\qquad = (y + 2)(y + 6)$

73. $-2w^4 + 1250 = -2(w^4 - 625)$
$\qquad = -2(w^2 - 25)(w^2 + 25)$
$\qquad = -2(w - 5)(w + 5)(w^2 + 25)$

75. $4a^2 + 16 = 4(a^2 + 4)$

77. $8a^3 + 8a = 8a(a^2 + 1)$

79. Let $y = w + 5$:
$(w + 5)^2 - 9 = y^2 - 9$
$\qquad = (y - 3)(y + 3)$
$\qquad = (w + 5 - 3)(w + 5 + 3)$
$\qquad = (w + 2)(w + 8)$

81. $4aw^2 - 12aw + 9a = a(4w^2 - 12w + 9)$
$\qquad = a(2w - 3)^2$

83. $x^2 - 6xy + 9y^2 = (x - 3y)^2$

85. $3x^4 - 75x^2 = 3x^2(x^2 - 25)$
$\qquad = 3x^2(x - 5)(x + 5)$

87. $m^3n - n = n(m^3 - 1)$
$\qquad = n(m - 1)(m^2 + m + 1)$

89. $12x^2 + 2x - 30 = 2(6x^2 + x - 15)$
$\qquad = 2[6x^2 - 9x + 10x - 15]$
$\qquad = 2[3x(2x - 3) + 5(2x - 3)]$
$\qquad = 2(3x + 5)(2x - 3)$

91. $2a^3 - 32 = 2(a^3 - 16)$

93. $a^{3m} - 1 = (a^m)^3 - 1^3$
$= (a^m - 1)(a^{2m} + a^m + 1)$

95. $a^{3w} - b^{6n} = (a^w)^3 - (b^{2n})^3$
$= (a^w - b^{2n})(a^{2w} + a^w b^{2n} + b^{4n})$

97. $t^{4n} - 16 = (t^{2n})^2 - 4^2$
$= (t^{2n} - 4)(t^{2n} + 4)$
$= (t^n - 2)(t^n + 2)(t^{2n} + 4)$

99. $a^{2n+1} - 2a^{n+1} - 15a$
$= a(a^{2n} - 2a^n - 15)$
$= a(a^n - 5)(a^n + 3)$

101. $a^{2n} - 3a^n + a^n b - 3b$
$= a^n(a^n - 3) + b(a^n - 3)$
$= (a^n - 3)(a^n + b)$

103. If you randomly write down trinomials, then most of them will be prime.

5.8 WARM-UPS

1. False, because 4 is a solution to $x - 1 = 3$, but $(4 - 1)(4 + 3) \neq 12$.
2. True, both 2 and 3 satisfy $(x - 2)(x - 3) = 0$.
3. True, because of the zero factor property.
4. False, $|x^2 + 4| = 5$ is equivalent to $x^2 + 4 = 5$ or $x^2 + 4 = -5$.
5. True, because of the zero factor property.
6. False, the Pythagorean theorem applies only to right triangles.
7. True, because the sum of the length and width of a rectangle is one-half of the perimeter.
8. True, $x + (8 - x) = 8$ for any number x.
9. False, the solution set also includes 0 because of the factor x.
10. False, because 3 is not a solution.

5.8 EXERCISES

1. The zero factor property says that if $a \cdot b = 0$ then either $a = 0$ or $b = 0$.
3. The hypotenuse of a right triangle is the side opposite the right angle.
5. The Pythagorean theorem says that a triangle is a right triangle if and only if the sum of the squares of the legs is equal to the square of the hypotenuse.

7. $(x - 5)(x + 4) = 0$
$x - 5 = 0$ or $x + 4 = 0$
$x = 5$ or $x = -4$
Solution set: $\{-4, 5\}$

9. $(2x - 5)(3x + 4) = 0$
$2x - 5 = 0$ or $3x + 4 = 0$
$2x = 5$ or $3x = -4$
$x = \frac{5}{2}$ or $x = -\frac{4}{3}$

Solution set: $\left\{ \frac{5}{2}, -\frac{4}{3} \right\}$

11. $4(x - 2)(x + 5) = 0$
Note that 4 is a factor, but 4 cannot cause the product to be zero. Only the factors containing variables could make the product zero.
$x - 2 = 0$ or $x + 5 = 0$
$x = 2$ or $x = -5$
Solution set: $\{-5, 2\}$

13. $x(x - 5)(x + 5) = 0$
$x = 0$ or $x - 5 = 0$ or $x + 5 = 0$
$x = 0$ or $x = 5$ or $x = -5$
Solution set: $\{-5, 0, 5\}$

15. $w^2 + 5w - 14 = 0$
$(w + 7)(w - 2) = 0$
$w + 7 = 0$ or $w - 2 = 0$
$w = -7$ or $w = 2$
Solution set: $\{-7, 2\}$

17. $m^2 - 7m = 0$
$m(m - 7) = 0$
$m = 0$ or $m - 7 = 0$
$m = 0$ or $m = 7$
Solution set: $\{0, 7\}$

19. $a^2 - a = 20$
$a^2 - a - 20 = 0$
$(a - 5)(a + 4) = 0$
$a - 5 = 0$ or $a + 4 = 0$
$a = 5$ or $a = -4$
Solution set: $\{-4, 5\}$

21. $2a^2 + 7a = 15$
$2a^2 + 7a - 15 = 0$
$(2a - 3)(a + 5) = 0$
$2a - 3 = 0$ or $a + 5 = 0$
$2a = 3$ or $a = -5$
$a = \frac{3}{2}$ or $a = -5$
Solution set: $\left\{ -5, \frac{3}{2} \right\}$

23.
$$10a^2 + 38a - 8 = 0$$
$$2(5a^2 + 19a - 4) = 0$$
$$2(5a - 1)(a + 4) = 0$$
$$5a - 1 = 0 \quad \text{or} \quad a + 4 = 0$$
$$5a = 1 \quad \text{or} \quad a = -4$$
$$a = \tfrac{1}{5} \quad \text{or} \quad a = -4$$
Solution set: $\left\{-4, \tfrac{1}{5}\right\}$

25.
$$3x^2 - 3x - 36 = 0$$
$$3(x^2 - x - 12) = 0$$
$$3(x - 4)(x + 3) = 0$$
$$x - 4 = 0 \quad \text{or} \quad x + 3 = 0$$
$$x = 4 \quad \text{or} \quad x = -3$$
Solution set: $\{-3, 4\}$

27.
$$z^2 + \tfrac{3}{2}z = 10$$
$$2\left(z^2 + \tfrac{3}{2}z\right) = 2(10)$$
$$2z^2 + 3z = 20$$
$$2z^2 + 3z - 20 = 0$$
$$(2z - 5)(z + 4) = 0$$
$$2z - 5 = 0 \quad \text{or} \quad z + 4 = 0$$
$$2z = 5 \quad \text{or} \quad z = -4$$
$$z = \tfrac{5}{2}$$
Solution set: $\left\{-4, \tfrac{5}{2}\right\}$

29.
$$x^3 - 4x = 0$$
$$x(x^2 - 4) = 0$$
$$x(x - 2)(x + 2) = 0$$
$$x = 0 \quad \text{or} \quad x - 2 = 0 \quad \text{or} \quad x + 2 = 0$$
$$x = 0 \quad \text{or} \quad x = 2 \quad \text{or} \quad x = -2$$
Solution set: $\{-2, 0, 2\}$

31.
$$-4x^3 + x = 3x^2$$
$$-4x^3 - 3x^2 + x = 0$$
$$-x(4x^2 + 3x - 1) = 0$$
$$-x(4x - 1)(x + 1) = 0$$
$$-x = 0 \text{ or } 4x - 1 = 0 \quad \text{or} \quad x + 1 = 0$$
$$x = 0 \quad \text{or} \quad x = \tfrac{1}{4} \quad \text{or} \quad x = -1$$
The solution set is $\left\{-1, 0, \tfrac{1}{4}\right\}$.

33.
$$w^3 + 4w^2 - 25w - 100 = 0$$
$$w^2(w + 4) - 25(w + 4) = 0$$
$$(w^2 - 25)(w + 4) = 0$$
$$(w - 5)(w + 5)(w + 4) = 0$$

$$w - 5 = 0 \quad \text{or } w + 5 = 0 \quad \text{or } w + 4 = 0$$
$$w = 5 \quad \text{or} \quad w = -5 \text{ or} \quad w = -4$$
Solution set: $\{-5, -4, 5\}$

35.
$$n^3 - 2n^2 - n + 2 = 0$$
$$n^2(n - 2) - 1(n - 2) = 0$$
$$(n^2 - 1)(n - 2) = 0$$
$$(n - 1)(n + 1)(n - 2) = 0$$
$$n - 1 = 0 \quad \text{or} \quad n + 1 = 0 \text{ or } n - 2 = 0$$
$$n = 1 \quad \text{or} \quad n = -1 \quad \text{or} \quad n = 2$$
Solution set: $\{-1, 1, 2\}$

37.
$$|x^2 - 5| = 4$$
$$x^2 - 5 = 4 \text{ or } x^2 - 5 = -4$$
$$x^2 - 9 = 0 \text{ or } x^2 - 1 = 0$$
$$(x - 3)(x + 3) = 0 \text{ or } (x - 1)(x + 1) = 0$$
$$x - 3 = 0 \text{ or } x + 3 = 0 \text{ or } x - 1 = 0$$
$$\text{or } x + 1 = 0$$
$$x = 3 \text{ or } x = -3 \quad \text{or} \quad x = 1 \text{ or } x = -1$$
Solution set: $\{-3, -1, 1, 3\}$

39.
$$|x^2 + 2x - 36| = 12$$
$$x^2 + 2x - 36 = 12 \text{ or } x^2 + 2x - 36 = -12$$
$$x^2 + 2x - 48 = 0 \text{ or } x^2 + 2x - 24 = 0$$
$$(x + 8)(x - 6) = 0 \text{ or } (x + 6)(x - 4) = 0$$
$$x = -8 \text{ or } x = 6 \text{ or } x = -6 \text{ or } x = 4$$
Solution set: $\{-8, -6, 4, 6\}$

41.
$$|x^2 + 4x + 2| = 2$$
$$x^2 + 4x + 2 = 2 \text{ or } x^2 + 4x + 2 = -2$$
$$x^2 + 4x = 0 \text{ or } x^2 + 4x + 4 = 0$$
$$x(x + 4) = 0 \quad \text{or} \quad (x + 2)^2 = 0$$
$$x = 0 \text{ or } x + 4 = 0 \text{ or } x + 2 = 0$$
$$x = 0 \text{ or } \quad x = -4 \text{ or } \quad x = -2$$
Solution set: $\{-4, -2, 0\}$

43.
$$|x^2 + 6x + 1| = 8$$
$$x^2 + 6x + 1 = 8 \text{ or } x^2 + 6x + 1 = -8$$
$$x^2 + 6x - 7 = 0 \text{ or } x^2 + 6x + 9 = 0$$
$$(x + 7)(x - 1) = 0 \quad \text{or} \quad (x + 3)^2 = 0$$
$$x + 7 = 0 \text{ or } x - 1 = 0 \text{ or } x + 3 = 0$$
$$x = -7 \text{ or } x = 1 \text{ or } \quad x = -3$$
Solution set: $\{-7, -3, 1\}$

45.
$$2x^2 - x = 6$$
$$2x^2 - x - 6 = 0$$
$$(2x + 3)(x - 2) = 0$$
$$2x + 3 = 0 \text{ or } x - 2 = 0$$
$$x = -\tfrac{3}{2} \text{ or } x = 2$$
Solution set: $\left\{-\tfrac{3}{2}, 2\right\}$

47.
$$|x^2 + 5x| = 6$$
$$x^2 + 5x = 6 \text{ or } x^2 + 5x = -6$$
$$x^2 + 5x - 6 = 0 \text{ or } x^2 + 5x + 6 = 0$$
$$(x+6)(x-1) = 0 \text{ or } (x+2)(x+3) = 0$$
$$x + 6 = 0 \text{ or } x - 1 = 0 \text{ or } x + 2 = 0$$
$$\text{or } x + 3 = 0$$
$$x = -6 \text{ or } x = 1 \text{ or } x = -2 \text{ or } x = -3$$
Solution set: $\{-6, -3, -2, 1\}$

49.
$$x^2 + 5x = 6$$
$$x^2 + 5x - 6 = 0$$
$$(x+6)(x-1) = 0$$
$$x + 6 = 0 \text{ or } x - 1 = 0$$
$$x = -6 \text{ or } x = 1$$
Solution set: $\{-6, 1\}$

51.
$$(x+2)(x+1) = 12$$
$$x^2 + 3x + 2 = 12$$
$$x^2 + 3x - 10 = 0$$
$$(x+5)(x-2) = 0$$
$$x + 5 = 0 \text{ or } x - 2 = 0$$
$$x = -5 \text{ or } x = 2$$
The solution set is $\{-5, 2\}$.

53.
$$y^3 + 9y^2 + 20y = 0$$
$$y(y^2 + 9y + 20) = 0$$
$$y(y+4)(y+5) = 0$$
$$y = 0 \text{ or } y + 4 = 0 \text{ or } y + 5 = 0$$
$$y = 0 \text{ or } y = -4 \text{ or } y = -5$$

Solution set: $\{-5, -4, 0\}$

55.
$$5a^3 = 45a$$
$$5a^3 - 45a = 0$$
$$5a(a^2 - 9) = 0$$
$$5a(a-3)(a+3) = 0$$
$$a = 0 \text{ or } a - 3 = 0 \text{ or } a + 3 = 0$$
$$a = 0 \text{ or } a = 3 \text{ or } a = -3$$
Solution set: $\{-3, 0, 3\}$

57. $(2x-1)(x^2 - 9) = 0$
$$(2x-1)(x-3)(x+3) = 0$$
$$2x - 1 = 0 \text{ or } x - 3 = 0 \text{ or } x + 3 = 0$$
$$x = \frac{1}{2} \text{ or } x = 3 \text{ or } x = -3$$
The solution set is $\left\{-3, \frac{1}{2}, 3\right\}$.

59. $(2x-1)(3x+1)(4x-1) = 0$
$$2x - 1 = 0 \text{ or } 3x + 1 = 0 \text{ or } 4x - 1 = 0$$
$$x = \frac{1}{2} \text{ or } x = -\frac{1}{3} \text{ or } x = \frac{1}{4}$$

The solution set is $\left\{-\frac{1}{3}, \frac{1}{4}, \frac{1}{2}\right\}$.

61.
$$4x^2 - 12x + 9 = 0$$
$$(2x-3)^2 = 0$$
$$2x - 3 = 0$$
$$x = \frac{3}{2}$$
The solution set is $\left\{\frac{3}{2}\right\}$.

63.
$$y^2 + by = 0$$
$$y(y+b) = 0$$
$$y = 0 \text{ or } y + b = 0$$
$$y = 0 \text{ or } y = -b$$
Solution set: $\{0, -b\}$

65.
$$a^2 y^2 - b^2 = 0$$
$$(ay - b)(ay + b) = 0$$
$$ay - b = 0 \text{ or } ay + b = 0$$
$$ay = b \text{ or } ay = -b$$
$$y = \frac{b}{a} \text{ or } y = -\frac{b}{a}$$
Solution set: $\left\{-\frac{b}{a}, \frac{b}{a}\right\}$

67.
$$4y^2 + 4by + b^2 = 0$$
$$(2y + b)^2 = 0$$
$$2y + b = 0$$
$$2y = -b$$
$$y = -\frac{b}{2}$$
Solution set: $\left\{-\frac{b}{2}\right\}$

69.
$$ay^2 + 3y - ay = 3$$
$$ay^2 + 3y - ay - 3 = 0$$
$$y(ay + 3) - 1(ay + 3) = 0$$
$$(y - 1)(ay + 3) = 0$$
$$y - 1 = 0 \text{ or } ay + 3 = 0$$
$$y = 1 \text{ or } y = -\frac{3}{a}$$
Solution set: $\left\{-\frac{3}{a}, 1\right\}$

71. Let $x =$ the width and $x + 2 =$ the length. Since $A = LW$, we can write the equation
$$x(x+2) = 24$$
$$x^2 + 2x = 24$$
$$x^2 + 2x - 24 = 0$$
$$(x+6)(x-4) = 0$$
$$x + 6 = 0 \text{ or } x - 4 = 0$$
$$x = -6 \text{ or } x = 4$$
$$x + 2 = 6$$
Since the width cannot be -6, we have a width of 4 inches and a length of 6 inches.

73. Let $x =$ one number and $13 - x =$ the other. Their product is 36:
$$x(13 - x) = 36$$
$$-x^2 + 13x - 36 = 0$$
$$x^2 - 13x + 36 = 0$$
$$(x - 4)(x - 9) = 0$$
$$x - 4 = 0 \quad \text{or} \quad x - 9 = 0$$
$$x = 4 \quad \text{or} \quad x = 9$$
$$13 - x = 9 \quad \text{or} \quad 13 - x = 4$$
The numbers are 4 and 9.

75. Let $x =$ the width and $x + 21 =$ the length.
$$x(x + 21) = 946$$
$$x^2 + 21x - 946 = 0$$
$$(x + 43)(x - 22) = 0$$
$$x + 43 = 0 \quad \text{or} \quad x - 22 = 0$$
$$x = -43 \quad \text{or} \quad x = 22$$
$$x + 21 = 43$$
The length is 43 in. and the width is 22 in.

77. a) From the graph it appears the ball is in the air for 4 seconds.

b) To find the time the ball is in the air we need the values of t that gives $h(t)$ a value of 0. We must solve the equation
$$-16t^2 + 64t = 0$$
$$-16t(t - 4) = 0$$
$$t = 0 \quad \text{or} \quad t - 4 = 0$$
$$t = 0 \quad \text{or} \quad t = 4$$
The ball is in the air for 4 seconds.

c) From the graph it appears that the maximum height is 64 feet.

d) The arrow reaches its maximum height at time $t = 2$ seconds.

79. a) $h(5) = -16 \cdot 5^2 + 325 \cdot 5$
$$= 1225 \text{ ft}$$

b) $-16t^2 + 325t = 0$
$$t(-16t + 325) = 0$$
$$t = 0 \quad \text{or} \quad -16t + 325 = 0$$
$$\text{or} \quad t = \frac{-325}{-16} = 20.3125$$
It takes 20.3125 sec to return to earth.

81. $\quad -16t^2 + 32t + 48 = 0$
$$-16(t^2 - 2t - 3) = 0$$
$$-16(t - 3)(t + 1) = 0$$
$$t - 3 = 0 \quad \text{or} \quad t + 1 = 0$$
$$t = 3 \quad \text{or} \quad t = -1$$
The ball strikes the earth after 3 sec.

83. Let $x =$ the width and $2x + 2 =$ the length. Using the Pythagorean theorem we can write
$$x^2 + (2x + 2)^2 = 13^2$$
$$x^2 + 4x^2 + 8x + 4 = 169$$
$$5x^2 + 8x - 165 = 0$$
$$(5x + 33)(x - 5) = 0$$
$$5x + 33 = 0 \quad \text{or} \quad x - 5 = 0$$
$$x = -\frac{33}{5} \quad \text{or} \quad x = 5$$
$$2x + 2 = 12$$
Since the dimensions must be positive numbers, the width is 5 feet and the length is 12 feet.

85. Since the perimeter is 34 feet the sum of the length and width is 17 feet. Let $x =$ the width and $17 - x =$ the length. Using the Pythagorean theorem we can write
$$x^2 + (17 - x)^2 = 13^2$$
$$x^2 + 289 - 34x + x^2 = 169$$
$$2x^2 - 34x + 120 = 0$$
$$x^2 - 17x + 60 = 0$$
$$(x - 5)(x - 12) = 0$$
$$x - 5 = 0 \quad \text{or} \quad x - 12 = 0$$
$$x = 5 \quad \text{or} \quad x = 12$$
$$17 - x = 12 \quad \text{or} \quad 17 - x = 5$$
The width is 5 feet and the length is 12 feet.

87. Let $x =$ the distance from the bottom of the ladder to the wall and $x + 3 =$ the distance from the top of the ladder to the ground.
$$x^2 + (x + 3)^2 = 15^2$$
$$x^2 + x^2 + 6x + 9 = 225$$
$$2x^2 + 6x - 216 = 0$$
$$x^2 + 3x - 108 = 0$$
$$(x + 12)(x - 9) = 0$$
$$x + 12 = 0 \quad \text{or} \quad x - 9 = 0$$
$$x = -12 \quad \text{or} \quad x = 9$$
$$x + 3 = 12$$
The distance from the top of the ladder to the ground is 12 feet.

89. Let $x =$ the number.
$$x^2 - x = 12$$
$$x^2 - x - 12 = 0$$
$$(x - 4)(x + 3) = 0$$
$$x - 4 = 0 \quad \text{or} \quad x + 3 = 0$$
$$x = 4 \quad \text{or} \quad x = -3$$
The number is either -3 or 4.

91. Let x = the first integer and $x + 1$ = the second integer. The sum of their squares is 25:
$$x^2 + (x + 1)^2 = 25$$
$$x^2 + x^2 + 2x + 1 = 25$$
$$2x^2 + 2x - 24 = 0$$
$$x^2 + x - 12 = 0$$
$$(x + 4)(x - 3) = 0$$
$$x + 4 = 0 \text{ or } x - 3 = 0$$
$$x = -4 \text{ or } \quad x = 3$$
$$x + 1 = -3 \text{ or } x + 1 = 4$$
The integers are 3 and 4, or -4 and -3.

93. Let x = the original length. Since the length times the width is 240 square feet, the width is $240/x$. The new length is $x - 4$ and the new width is $(240/x) + 3$. Since the area is still 240 square feet, we can write the following equation.
$$(x - 4)\left(\frac{240}{x} + 3\right) = 240$$
$$x \cdot \frac{240}{x} - 4 \cdot \frac{240}{x} + 3x - 12 = 240$$
$$240 - \frac{960}{x} + 3x - 252 = 0$$
$$-\frac{960}{x} + 3x - 12 = 0$$
$$x\left(-\frac{960}{x} + 3x - 12\right) = x \cdot 0$$
$$-960 + 3x^2 - 12x = 0$$
$$3x^2 - 12x - 960 = 0$$
$$x^2 - 4x - 320 = 0$$
$$(x - 20)(x + 16) = 0$$
$$x - 20 = 0 \text{ or } x + 16 = 0$$
$$x = 20 \text{ or } \quad x = -16$$
$$\frac{240}{x} = 12$$
The length is 20 feet and the width is 12 feet.

95. Dividing each side of an equation by a variable can cause you to lose a solution to the equation. Both 0 and 1 satisfy $x^2 = x$.

97. The 5 can be disregarded. If a product is equal to zero, it is because one of the factors is equal to zero and 5 is not 0.

Enriching Your Mathematical Word Power
1. d   2. b   3. c   4. d   5. b   6. a
7. c   8. a   9. a   10. d   11. c   12. a
13. c   14. b   15. c   16. a   17. c

CHAPTER 5 REVIEW

1. $2 \cdot 2 \cdot 2^{-1} = 4 \cdot \frac{1}{2} = 2$

3. $2^2 \cdot 3^2 = 4 \cdot 9 = 36$

5. $(-3)^{-3} = -\frac{1}{3^3} = -\frac{1}{27}$

7. $-(-1)^{-3} = -\frac{1}{(-1)^3} = -\frac{1}{-1} = 1$

9. $2x^3 \cdot 4x^{-6} = 8x^{-3} = \frac{8}{x^3}$

11. $\frac{y^{-5}}{y^{-3}} = y^{-2} = \frac{1}{y^2}$

13. $\frac{a^5 a^{-2}}{a^{-4}} = \frac{a^3}{a^{-4}} = a^7$

15. $\frac{6x^{-2}}{3x^2} = 2x^{-4} = \frac{2}{x^4}$

17. Move the decimal 6 places to the right: $8.36 \times 10^6 = 8,360,000$

19. Move the decimal point 4 places to the left: $5.7 \times 10^{-4} = 0.00057$

21. Move the decimal point 6 places to the left: $8,070,000 = 8.07 \times 10^6$

23. Move the decimal point 4 places to the right: $0.000709 = 7.09 \times 10^{-4}$

25. $\frac{(4 \times 10^9)(6 \times 10^{-7})}{(1.2 \times 10^{-5})(2 \times 10^6)} = \frac{24 \times 10^2}{2.4 \times 10^1}$
$$= 10 \times 10^1 = 1 \times 10^2$$

27. $(a^{-3})^{-2} \cdot a^{-7} = a^6 \cdot a^{-7} = a^{-1} = \frac{1}{a}$

29. $(m^2 n^3)^{-2}(m^{-3} n^2)^4 = m^{-4} n^{-6} m^{-12} n^8$
$$= m^{-16} n^2 = \frac{n^2}{m^{16}}$$

31. $\left(\frac{2}{3}\right)^{-4} = \left(\frac{3}{2}\right)^4 = \frac{3^4}{2^4} = \frac{81}{16}$

33. $\left(\frac{1}{2} + \frac{1}{3}\right)^2 = \left(\frac{5}{6}\right)^2 = \frac{25}{36}$

35. $\left(-\frac{3a}{4b^{-1}}\right)^{-1} = -\frac{4b^{-1}}{3a} = -\frac{4}{3ab}$

37. $\frac{(a^{-3}b)^4}{(ab^2)^{-5}} = \frac{a^{-12}b^4}{a^{-5}b^{-10}} = \frac{b^{14}}{a^7}$

39. $5^{2w} \cdot 5^{4w} \cdot 5^{-1} = 5^{6w-1}$

41. $\left(\frac{7^{3a}}{7^8}\right)^5 = \frac{7^{15a}}{7^{40}} = 7^{15a-40}$

43. $(2w - 3) + (6w + 5) = 8w + 2$

45. $(x^2 - 3x - 4) - (x^2 + 3x - 7)$
$$= x^2 - 3x - 4 - x^2 - 3x + 7 = -6x + 3$$

47. $(x^2 - 2x + 4)(x - 2)$
$$= (x^2 - 2x + 4)(x) - (x^2 - 2x + 4)(2)$$
$$= x^3 - 2x^2 + 4x - 2x^2 + 4x - 8$$
$$= x^3 - 4x^2 + 8x - 8$$

49. $xy + 7z - 5(xy - 3z)$
$$= xy + 7z - 5xy + 15z = -4xy + 22z$$

51. $m^2(5m^3 - m + 2) = 5m^5 - m^3 + 2m^2$

53. $(x - 3)(x + 7) = x^2 - 3x + 7x - 21$
$$= x^2 + 4x - 21$$

55. $(z - 5y)(z + 5y) = z^2 - (5y)^2$
$= z^2 - 25y^2$

57. $(m + 8)^2 = m^2 + 2 \cdot 8m + 8^2$
$= m^2 + 16m + 64$

59. $(w - 6x)(w - 4x)$
$= w^2 - 6xw - 4xw + 24x^2$
$= w^2 - 10xw + 24x^2$

61. $(k - 3)^2 = k^2 - 2 \cdot 3 \cdot k + 9$
$= k^2 - 6k + 9$

63. $(m^2 - 5)(m^2 + 5)$
$= m^4 - 5m^2 + 5m^2 - 25$
$= m^4 - 25$

65. $3x - 6 = 3(x - 2)$

67. $4a - 20 = -4(-a + 5)$

69. $3w - w^2 = -w(-3 + w) = -w(w - 3)$

71. $y^2 - 81 = (y - 9)(y + 9)$

73. $4x^2 + 28x + 49$
$= (2x)^2 + 2 \cdot 2x \cdot 7 + 7^2$
$= (2x + 7)^2$

75. $t^2 - 18t + 81$
$= t^2 - 2 \cdot 9t + 9^2 = (t - 9)^2$

77. $t^3 - 125 = (t - 5)(t^2 + 5t + 25)$

79. Two numbers with a product of -30 and a sum of -7 are -10 and 3.
$x^2 - 7x - 30 = (x - 10)(x + 3)$

81. Two numbers with a product of -28 and a sum of -3 are -7 and 4.
$w^2 - 3w - 28 = (w - 7)(w + 4)$

83. Two numbers with a product of -14 and a sum of 5 are -2 and 7.
$2m^2 + 5m - 7 = 2m^2 - 2m + 7m - 7$
$= 2m(m - 1) + 7(m - 1)$
$= (2m + 7)(m - 1)$

85. $m^7 - 3m^4 - 10m = m(m^6 - 3m^3 - 10)$
$= m(m^3 - 5)(m^3 + 2)$

87. $5x^3 + 40 = 5(x^3 + 8)$
$= 5(x + 2)(x^2 - 2x + 4)$

89. Two numbers with a product of 18 and a sum of 9 are 6 and 3.
$9x^2 + 9x + 2 = 9x^2 + 3x + 6x + 2$
$= 3x(3x + 1) + 2(3x + 1)$
$= (3x + 2)(3x + 1)$

91. $x^3 + x^2 - x - 1$
$= x^2(x + 1) - 1(x + 1)$
$= (x + 1)(x^2 - 1)$
$= (x + 1)(x + 1)(x - 1)$
$= (x + 1)^2(x - 1)$

93. $-x^2y + 16y = -y(x^2 - 16)$
$= -y(x - 4)(x + 4)$

95. $-a^3b^2 + 2a^2b^2 - ab^2$
$= -ab^2(a^2 - 2a + 1)$
$= -ab^2(a - 1)^2$

97. $x^3 - x^2 + 9x - 9$
$= x^2(x - 1) + 9(x - 1)$
$= (x - 1)(x^2 + 9)$

99. $x^4 - x^2 - 12 = (x^2 - 4)(x^2 + 3)$
$= (x - 2)(x + 2)(x^2 + 3)$

101. $a^6 - a^3 = a^3(a^3 - 1)$
$= a^3(a - 1)(a^2 + a + 1)$

103. $-8m^2 - 24m - 18$
$= -2(4m^2 + 12m + 9)$
$= -2(2m + 3)^2$

105. Let $y = 2x - 3$:
$(2x - 3)^2 - 16 = y^2 - 16$
$= (y - 4)(y + 4)$
$= (2x - 3 - 4)(2x - 3 + 4)$
$= (2x - 7)(2x + 1)$

107. $x^6 + 7x^3 - 8 = (x^3 + 8)(x^3 - 1)$
$= (x + 2)(x^2 - 2x + 4)(x - 1)(x^2 + x + 1)$

109. Let $y = a^2 - 9$:
$(a^2 - 9)^2 - 5(a^2 - 9) + 6$
$= y^2 - 5y + 6$
$= (y - 2)(y - 3)$
$= (a^2 - 9 - 2)(a^2 - 9 - 3)$
$= (a^2 - 11)(a^2 - 12)$

111. $x^{2k} - 49 = (x^k - 7)(x^k + 7)$

113. $m^{2a} - 2m^a - 3 = (m^a - 3)(m^a + 1)$

115. $9z^{2k} - 12z^k + 4 = (3z^k - 2)^2$

117. $y^{2a} - by^a + cy^a - bc$
$= y^a(y^a - b) + c(y^a - b) = (y^a - b)(y^a + c)$

119. $x^3 - 5x^2 = 0$
$x^2(x - 5) = 0$
$x^2 = 0$ or $x - 5 = 0$
$x = 0$ or $x = 5$
Solution set: $\{0, 5\}$

121. $(a - 2)(a - 3) = 6$
$a^2 - 5a + 6 = 6$
$a^2 - 5a = 0$
$a(a - 5) = 0$
$a = 0$ or $a - 5 = 0$
$a = 0$ or $a = 5$
Solution set: $\{0, 5\}$

123.
$$2m^2 - 9m - 5 = 0$$
$$(2m + 1)(m - 5) = 0$$
$$2m + 1 = 0 \quad \text{or} \quad m - 5 = 0$$
$$m = -1/2 \quad \text{or} \quad m = 5$$
Solution set: $\{-1/2, 5\}$

125.
$$w^3 + 5w^2 - w - 5 = 0$$
$$w^2(w + 5) - 1(w + 5) = 0$$
$$(w^2 - 1)(w + 5) = 0$$
$$(w - 1)(w + 1)(w + 5) = 0$$
$$w - 1 = 0 \text{ or } w + 1 = 0 \text{ or } w + 5 = 0$$
$$w = 1 \quad \text{or} \quad w = -1 \quad \text{or} \quad w = -5$$
Solution set: $\{-5, -1, 1\}$

127.
$$|x^2 - 5| = 4$$
$$x^2 - 5 = 4 \quad \text{or} \quad x^2 - 5 = -4$$
$$x^2 - 9 = 0 \quad \text{or} \quad x^2 - 1 = 0$$
$$(x - 3)(x + 3) = 0 \text{ or } (x - 1)(x + 1) = 0$$
$$x = 3 \text{ or } x = -3 \text{ or } x = 1 \text{ or } x = -1$$
Solution set: $\{-3, -1, 1, 3\}$

129. Let $x =$ the distance from the bottom of the ladder and the cactus and $x + 2 =$ the distance from the top of the ladder to the ground. Since the ladder is the hypotenuse of a right triangle, we can use the Pythagorean theorem:
$$x^2 + (x + 2)^2 = 10^2$$
$$x^2 + x^2 + 4x + 4 = 100$$
$$2x^2 + 4x - 96 = 0$$
$$x^2 + 2x - 48 = 0$$
$$(x + 8)(x - 6) = 0$$
$$x + 8 = 0 \quad \text{or } x - 6 = 0$$
$$x = -8 \text{ or } \quad x = 6$$
The ladder should be placed 6 feet from the cactus.

131. Since the perimeter is 32 meters, the sum of the length and width is 16 meters. So if x represents the length, then $16 - x$ represents the width. Now use the fact that the area is 63 square meters:
$$x(16 - x) = 63$$
$$-x^2 + 16x - 63 = 0$$
$$x^2 - 16x + 63 = 0$$
$$(x - 7)(x - 9) = 0$$
$$x - 7 = 0 \qquad \text{or} \qquad x - 9 = 0$$
$$x = 7 \qquad \text{or} \qquad x = 9$$
$$16 - x = 9 \qquad \text{or} \qquad 16 - x = 7$$
The width is 7 meters and the length is 9 meters.

133. Let x represent the width and $x + 17$ represent the length. The diagonal is 25.
$$x^2 + (x + 17)^2 = 25^2$$
$$x^2 + x^2 + 34x + 289 = 625$$
$$2x^2 + 34x - 336 = 0$$
$$x^2 + 17x - 168 = 0$$
$$(x + 24)(x - 7) = 0$$
$$x + 24 = 0 \qquad \text{or} \qquad x - 7 = 0$$
$$x = -24 \quad \text{or} \qquad x = 7$$
$$x + 17 = 24$$
The screen is 7 inches by 24 inches.

135. a) $L = 64.3(1.0033)^{20} \approx 68.7$ yr
b) From Exercise 95 Section 5.2, a 20-yr old white male is expected to live 75.1 yr. The White male is expected to live 6.5 years longer. To get this answer subtract before rounding each life expectancy.

137. a) The interest rate is approximately 15%.

b) $500,000 = R \cdot \dfrac{(1 + 0.07)^{20} - 1}{0.07}$

$R = 500,000 \div \dfrac{(1 + 0.07)^{20} - 1}{0.07}$

$\approx \$12,196.46$

CHAPTER 5 TEST

1. $3^{-2} = \dfrac{1}{3^2} = \dfrac{1}{9}$

2. $\dfrac{1}{6^{-2}} = 6^2 = 36$

3. $\left(\dfrac{1}{2}\right)^{-3} = 2^3 = 8$

4. $3x^4 \cdot 4x^3 = 3 \cdot 4x^4 x^3 = 12x^7$

5. $\dfrac{8y^9}{2y^{-3}} = 4y^{9-(-3)} = 4y^{12}$

6. $(4a^2b)^3 = 4^3(a^2)^3 b^3 = 64a^6 b^3$

7. $\left(\dfrac{x^2}{3}\right)^{-3} = \left(\dfrac{3}{x^2}\right)^3 = \dfrac{27}{x^6}$

8. $\dfrac{(2^{-1})^{-3}(a^2)^{-3}b^{-3}}{4a^{-9}} = \dfrac{2^3 a^{-6} b^{-3}}{4a^{-9}} = \dfrac{8a^3}{4b^3}$
$= \dfrac{2a^3}{b^3}$

9. $3.24 \times 10^9 = 3,240,000,000$

10. $8.673 \times 10^{-4} = 0.0008673$

11. $\dfrac{(8 \times 10^4)(6 \times 10^{-4})}{2 \times 10^6} = 24 \times 10^{-6}$
$= 2.4 \times 10^1 \times 10^{-6} = 2.4 \times 10^{-5}$

12. $\dfrac{(6 \times 10^{-5})^2 (5 \times 10^2)}{(3 \times 10^4)^2 (1 \times 10^{-2})}$
$= \dfrac{36 \times 10^{-10} \cdot 5 \times 10^2}{9 \times 10^8 \cdot 1 \times 10^{-2}}$

$$= 20 \times 10^{-14}$$
$$= 2 \times 10^{-13}$$

13. Add like terms to get $3x^3 + 3x^2 - 2x + 3$.

14. $(x^2 - 6x - 7) - (3x^2 + 2x - 4)$
$$= x^2 - 6x - 7 - 3x^2 - 2x + 4$$
$$= -2x^2 - 8x - 3$$

15. $(x^2 - 3x + 7)(x - 2)$
$$= (x^2 - 3x + 7)(x) - (x^2 - 3x + 7)(2)$$
$$= x^3 - 3x^2 + 7x - 2x^2 + 6x - 14$$
$$= x^3 - 5x^2 + 13x - 14$$

16. $(x - 2)^3 = (x - 2)^2(x - 2)$
$$= (x^2 - 4x + 4)(x - 2)$$
$$= (x^2 - 4x + 4)(x) - (x^2 - 4x + 4)(2)$$
$$= x^3 - 4x^2 + 4x - 2x^2 + 8x - 8$$
$$= x^3 - 6x^2 + 12x - 8$$

17. $(x - 7)(2x + 3) = 2x^2 - 14x + 3x - 21$
$$= 2x^2 - 11x - 21$$

18. $(x - 6)^2 = x^2 - 2 \cdot 6x + 36$
$$= x^2 - 12x + 36$$

19. $(2x + 5)^2 = 4x^2 + 20x + 25$

20. $(3y^2 - 5)(3y^2 + 5) = 9y^4 - 25$

21. Two numbers with a product of -24 and a sum of -2 are -6 and 4.
$$a^2 - 2a - 24 = (a - 6)(a + 4)$$

22. $4x^2 + 28x + 49$
$$= (2x)^2 + 2 \cdot 2x \cdot 7 + 7^2 = (2x + 7)^2$$

23. $3m^3 - 24 = 3(m^3 - 8)$
$$= 3(m - 2)(m^2 + 2m + 4)$$

24. $2x^2y - 32y = 2y(x^2 - 16)$
$$= 2y(x - 4)(x + 4)$$

25. $12m^2 + 28m + 15 = (2m + 3)(6m + 5)$

26. $3x^{10} + 5x^5 - 12 = (2x^5 - 3)(x^5 + 4)$

27. $2xa + 3a - 10x - 15$
$$= a(2x + 3) - 5(2x + 3)$$
$$= (2x + 3)(a - 5)$$

28. $x^4 + 3x^2 - 4 = (x^2 - 1)(x^2 + 4)$
$$= (x - 1)(x + 1)(x^2 + 4)$$

29. $2m^2 + 7m - 15 = 0$
$$(2m - 3)(m + 5) = 0$$
$$2m - 3 = 0 \text{ or } m + 5 = 0$$
$$m = \frac{3}{2} \text{ or } m = -5$$
Solution set: $\left\{ -5, \frac{3}{2} \right\}$

30. $\qquad x^3 - 4x = 0$
$$x(x^2 - 4) = 0$$
$$x(x - 2)(x + 2) = 0$$
$$x = 0 \text{ or } x - 2 = 0 \text{ or } x + 2 = 0$$
$$x = 0 \text{ or } \quad x = 2 \text{ or } \quad x = -2$$

Solution set: $\{-2, 0, 2\}$

31. $\qquad |x^2 + x - 9| = 3$
$$x^2 + x - 9 = 3 \text{ or } x^2 + x - 9 = -3$$
$$x^2 + x - 12 = 0 \text{ or } \quad x^2 + x - 6 = 0$$
$$(x + 4)(x - 3) = 0 \text{ or } (x + 3)(x - 2) = 0$$
$$x = -4 \text{ or } x = 3 \text{ or } x = -3 \text{ or } x = 2$$
Solution set: $\{-4, -3, 2, 3\}$

32. Let x = the height and $x + 2$ = the width. Using the Pythagorean theorem we can write
$$x^2 + (x + 2)^2 = 10^2$$
$$x^2 + x^2 + 4x + 4 = 100$$
$$2x^2 + 4x - 96 = 0$$
$$x^2 + 2x - 48 = 0$$
$$(x + 8)(x - 6) = 0$$
$$x = -8 \quad \text{or} \quad x = 6$$
$$x + 2 = -6 \quad \text{or} \quad x + 2 = 8$$
The width is 8 inches and the height is 6 inches.

33. $d = (1.8 \times 10^{28})(1.032)^{-1950} \approx 38.0$
$d = (1.8 \times 10^{28})(1.032)^{-1990} \approx 10.8$
$d = (1.8 \times 10^{28})(1.032)^{-2000} \approx 7.9$

34. a) $h(2) = -16 \cdot 2^2 + 80 \cdot 2 = 96$ ft
$h(3) = -16 \cdot 3^2 + 80 \cdot 3 = 96$ ft

b) $-16t^2 + 80t = 0$
$$-16t(t - 5) = 0$$
$$-16t = 0 \qquad \text{or} \qquad t - 5 = 0$$
$$t = 0 \qquad \text{or} \qquad t = 5$$
The stone is in the air for 5 seconds.

35. The sum of the length and width is 44 ft. So if x represents the width, then $44 - x$ represents the length.
$$x(44 - x) = 480$$
$$-x^2 + 44x - 480 = 0$$
$$x^2 - 44x + 480 = 0$$
$$(x - 20)(x - 24) = 0$$
$$x - 20 = 0 \qquad \text{or} \qquad x - 24 = 0$$
$$x = 20 \qquad \text{or} \qquad x = 24$$
$$44 - x = 24 \qquad \text{or} \qquad 44 - x = 20$$
So the width is 20 ft and the length is 24 ft.

Making Connections

Chapters 1 - 5

1. $4^2 = 4 \cdot 4 = 16$

2. $4(-2) = -8$

3. $4^{-2} = \frac{1}{4^2} = \frac{1}{16}$

4. $2^3 \cdot 4^{-1} = 8 \cdot \frac{1}{4} = 2$

5. $2^{-1} + 2^{-1} = \frac{1}{2} + \frac{1}{2} = 1$

6. $2^{-1} \cdot 3^{-1} = \frac{1}{2} \cdot \frac{1}{3} = \frac{1}{6}$

7. $3^{-1} - 2^{-2} = \frac{1}{3} - \frac{1}{4} = \frac{4}{12} - \frac{3}{12} = \frac{1}{12}$

8. $3^2 - 4(5)(-2) = 9 - (-40) = 49$

9. $2^7 - 2^6 = 128 - 64 = 64$

10. $0.08(32) + 0.08(68) = 0.08(32 + 68)$
$$= 0.08(100) = 8$$

11. $3 - 2\,|\,5 - 7 \cdot 3\,| = 3 - 2\,|-16\,|$
$$= 3 - 2 \cdot 16 = 3 - 32 = -29$$

12. $5^{-1} + 6^{-1} = \frac{1}{5} + \frac{1}{6} = \frac{6}{30} + \frac{5}{30} = \frac{11}{30}$

13. $0.05a - 0.04(a - 50) = 4$
$$0.05a - 0.04a + 2 = 4$$
$$0.01a = 2$$
$$a = 200$$
Solution set: $\{200\}$

14. $15b - 27 = 0$
$$15b = 27$$
$$b = \frac{27}{15} = \frac{9}{5}$$
Solution set: $\left\{\frac{9}{5}\right\}$

15. $2c^2 + 15c - 27 = 0$
$$(2c - 3)(c + 9) = 0$$
$$2c - 3 = 0 \ \text{or} \ c + 9 = 0$$
$$c = \frac{3}{2} \ \text{or} \ c = -9$$
Solution set: $\left\{-9, \frac{3}{2}\right\}$

16. $2t^2 + 15t = 0$
$$t(2t + 15) = 0$$
$$t = 0 \ \text{or} \ 2t + 15 = 0$$
$$t = 0 \ \text{or} \ t = -\frac{15}{2}$$
Solution set: $\left\{-\frac{15}{2}, 0\right\}$

17. $|\,15u - 27\,| = 3$
$$15u - 27 = 3 \ \text{or} \ 15u - 27 = -3$$
$$15u = 30 \ \text{or} \ 15u = 24$$
$$u = 2 \ \text{or} \ u = \frac{8}{5}$$
Solution set: $\left\{2, \frac{8}{5}\right\}$

18. $|\,15v - 27\,| = 0$
$$15v - 27 = 0$$
$$15v = 27$$
$$v = \frac{27}{15} = \frac{9}{5}$$
Solution set: $\left\{\frac{9}{5}\right\}$

19. Absolute value of any quantity is greater than or equal to 0. So the solution set is $\emptyset$.

20. $|\,x^2 + x - 4\,| = 2$
$$x^2 + x - 4 = 2 \ \text{or} \ x^2 + x - 4 = -2$$
$$x^2 + x - 6 = 0 \ \text{or} \ x^2 + x - 2 = 0$$
$$(x + 3)(x - 2) = 0 \ \text{or} \ (x + 2)(x - 1) = 0$$
$$x = -3 \ \text{or} \ x = 2 \ \text{or} \ x = -2 \ \text{or} \ x = 1$$
Solution set: $\{-3, -2, 1, 2\}$

21. $(2x - 1)(x + 5) = 0$
$$2x - 1 = 0 \ \text{or} \ x + 5 = 0$$
$$x = \frac{1}{2} \ \text{or} \ x = -5$$
Solution set: $\left\{-5, \frac{1}{2}\right\}$

22. $|\,3x - 1\,| + 6 = 9$
$$|\,3x - 1\,| = 3$$
$$3x - 1 = 3 \ \text{or} \ 3x - 1 = -3$$
$$3x = 4 \ \text{or} \ 3x = -2$$
$$x = \frac{4}{3} \ \text{or} \ x = -\frac{2}{3}$$
Solution set: $\left\{-\frac{2}{3}, \frac{4}{3}\right\}$

23. $(1.5 \times 10^{-4})w = 7 \times 10^6 + 5 \times 10^5$
$$w = \frac{7 \times 10^6 + 5 \times 10^5}{1.5 \times 10^{-4}}$$
$$= 5 \times 10^{10}$$

24. $(3 \times 10^7)(y - 5 \times 10^3) = 6 \times 10^{12}$
$$y - 5 \times 10^3 = \frac{6 \times 10^{12}}{3 \times 10^7}$$
$$y = \frac{6 \times 10^{12}}{3 \times 10^7} + 5 \times 10^3$$
$$y = 2.05 \times 10^5$$

25. a) If $E = \$100{,}000$, then
$$D = 0.75(100{,}000) + 5000$$
$$= \$80{,}000$$
$\$100{,}000 - \$80{,}000 = \$20{,}000$
So the tax is $\$20{,}000$.

b) If $E = \$10{,}000$, then
$$D = 0.75(10{,}000) + 5000 = 12{,}500$$
So Uncle Sam pays the person $\$2500$.

c) If $D = E$, then $E = 0.75E + 5000$ or $0.25E = 5000$, or $E = 20{,}000$.
So the lines intersect at an earned income of $\$20{,}000$ or at the point $(20{,}000, 20{,}000)$.

d) At the intersection the earned income is equal to the disposable income, which means that if you earn $\$20{,}000$ then you pay no tax.

6.1 WARM-UPS

1. True, because a number is a monomial.
2. True, because it is a ratio of two binomials.
3. False, because the domain is any number except 2.
4. True, because 9 and $-\frac{1}{2}$ both give a denominator of 0.
5. False, because the numerator may be zero in a rational expression.
6. False, because 5 is not a factor of the numerator.
7. False, because $x \cdot x = x^2$, but $(x-1)x = x^2 - x$.
8. True, because we can multiply the numerator and denominator by -1.
9. True, because it is a correctly reduced rational expression.
10. False, it reduces to $x + y$.

6.1 EXERCISES

1. A rational expression is a ratio of two polynomials with the denominator not zero.
3. The basic principle of rational numbers says that $(ab)/(ac) = b/c$ provided a and c are not 0.
5. We build up the denominator by multiplying the numerator and denominator by the same expression.
7. If $x = 1$, then $x - 1 = 0$. So the domain is $\{x \mid x \neq 1\}$ or $(-\infty, 1) \cup (1, \infty)$.
9. If $z = 0$ then $7z = 0$. So the domain is $\{z \mid z \neq 0\}$ or $(-\infty, 0) \cup (0, \infty)$.
11. If $y = \pm 2$, then $y^2 - 4 = 0$. So the domain is $\{y \mid y \neq -2 \text{ and } y \neq 2\}$ or $(-\infty, -2) \cup (-2, 2) \cup (2, \infty)$.
13. To find the domain solve
$$a^2 + 5a + 6 = 0$$
$$(a + 2)(a + 3) = 0$$
$$a = -2 \text{ or } a = -3$$
The domain is $\{a \mid a \neq -2 \text{ and } a \neq -3\}$ or $(-\infty, -3) \cup (-3, -2) \cup (-2, \infty)$.
15. To find the domain solve
$$x^2 + 4 = 0$$
$$x^2 = -4$$

Since $x^2 = -4$ has no solution there are no numbers that cause the denominator to be zero. Any real number can be used in place of x. The domain is $(-\infty, \infty)$.

17. Solve the equation
$$x^3 + x^2 - 6x = 0$$
$$x(x + 3)(x - 2) = 0$$
$$x = 0 \text{ or } x = -3 \text{ or } x = 2$$
So the domain is
$\{x \mid x \neq -3 \text{ and } x \neq 0 \text{ and } x \neq 2\}$ or
$(-\infty, -3) \cup (-3, 0) \cup (0, 2) \cup (2, \infty)$.

19. $\frac{6}{57} = \frac{3 \cdot 2}{3 \cdot 19} = \frac{2}{19}$

21. $\frac{42}{210} = \frac{42 \cdot 1}{42 \cdot 5} = \frac{1}{5}$

23. $\frac{2x + 2}{4} = \frac{2(x + 1)}{2 \cdot 2} = \frac{x + 1}{2}$

25. $\frac{3x - 6y}{10y - 5x} = \frac{3(x - 2y)}{-5(x - 2y)} = -\frac{3}{5}$

27. $\frac{ab^2}{a^3 b} = \frac{b}{a^2}$

29. $\frac{-2w^2 x^3 y}{6wx^5 y^2} = \frac{-w}{3x^2 y}$

31. $\frac{a^3 b^2}{a^3 + a^4} = \frac{a^3 \cdot b^2}{a^3(1 + a)} = \frac{b^2}{1 + a}$

33. $\frac{a - b}{2b - 2a} = \frac{1(a - b)}{-2(a - b)} = -\frac{1}{2}$

35. $\frac{3x + 6}{3x} = \frac{3(x + 2)}{3x} = \frac{x + 2}{x}$

37. $\frac{a^3 - b^3}{a - b} = \frac{(a - b)(a^2 + ab + b^2)}{a - b}$
$= a^2 + ab + b^2$

39. $\frac{4x^2 - 4}{4x^2 + 4} = \frac{4(x^2 - 1)}{4(x^2 + 1)} = \frac{x^2 - 1}{x^2 + 1}$

41. $\frac{12x^2 - 26x - 10}{4x^2 - 25}$
$= \frac{2(3x + 1)(2x - 5)}{(2x - 5)(2x + 5)} = \frac{6x + 2}{2x + 5}$

43. $\frac{x^3 + 7x^2 - 4x}{x^3 - 16x} = \frac{x(x^2 + 7x - 4)}{x(x - 4)(x + 4)}$
$= \frac{x^2 + 7x - 4}{x^2 - 16}$

45. $\dfrac{2ab + 2by + 3a + 3y}{2b^2 - 7b - 15} = \dfrac{(a+y)(2b+3)}{(2b+3)(b-5)}$

$= \dfrac{a+y}{b-5}$

47. $\dfrac{4x^2 - 10x - 6}{2x^2 + 11x + 5} = \dfrac{2(2x+1)(x-3)}{(2x+1)(x+5)}$

$= \dfrac{2x-6}{x+5}$

49. $\dfrac{2a^2 + 5ab + 3b^2}{4a^2 + 12ab + 9b^2} = \dfrac{(a+b)(2a+3b)}{(2a+3b)(2a+3b)}$

$= \dfrac{a+b}{2a+3b}$

51. $\dfrac{1}{5} = \dfrac{1 \cdot 10}{5 \cdot 10} = \dfrac{10}{50}$

53. $\dfrac{1}{xy} = \dfrac{1 \cdot 3xy^2}{xy \cdot 3xy^2} = \dfrac{3xy^2}{3x^2y^3}$

55. $\dfrac{5}{x-1} = \dfrac{5(x-1)}{(x-1)(x-1)} = \dfrac{5x-5}{x^2 - 2x + 1}$

57. $\dfrac{3}{2x-5} = \dfrac{3(2x+5)}{(2x-5)(2x+5)}$

$= \dfrac{6x+15}{4x^2 - 25}$

59. $\dfrac{1}{2x+2} = \dfrac{1(-3)}{(2x+2)(-3)} = \dfrac{-3}{-6x-6}$

61. $5 = \dfrac{5 \cdot a}{1 \cdot a} = \dfrac{5a}{a}$

63. $\dfrac{x+2}{x+3} = \dfrac{(x+2)(x-1)}{(x+3)(x-1)} = \dfrac{x^2 + x - 2}{x^2 + 2x - 3}$

65. $\dfrac{7}{x-1} = \dfrac{7(-1)}{(x-1)(-1)} = \dfrac{-7}{1-x}$

67. $\dfrac{3}{x+2} = \dfrac{3(x^2 - 2x + 4)}{(x+2)(x^2 - 2x + 4)}$

$= \dfrac{3x^2 - 6x + 12}{x^3 + 8}$

69. $\dfrac{x+2}{3x-1} = \dfrac{(x+2)(2x+5)}{(3x-1)(2x+5)}$

$= \dfrac{2x^2 + 9x + 10}{6x^2 + 13x - 5}$

71. $R(3) = \dfrac{3 \cdot 3 - 5}{3 + 4} = \dfrac{4}{7}$

73. $H(-2) = \dfrac{(-2)^2 - 5}{3(-2) - 4} = \dfrac{-1}{-10} = \dfrac{1}{10}$

75. $W(-2) = \dfrac{4(-2)^3 - 1}{(-2)^2 - (-2) - 6} = \dfrac{-33}{0}$

$W(-2)$ is undefined.

77. $\dfrac{1}{3} = \dfrac{1 \cdot 7}{3 \cdot 7} = \dfrac{7}{21}$

79. $5 = \dfrac{5 \cdot 2}{1 \cdot 2} = \dfrac{10}{2}$

81. $\dfrac{3}{a} = \dfrac{3a}{a \cdot a} = \dfrac{3a}{a^2}$

83. $\dfrac{2}{a-b} = \dfrac{2(-1)}{(a-b)(-1)} = \dfrac{-2}{b-a}$

85. $\dfrac{2}{x-1} = \dfrac{2(x+1)}{(x-1)(x+1)} = \dfrac{2x+2}{x^2 - 1}$

87. $\dfrac{2}{w-3} = \dfrac{2(-1)}{(w-3)(-1)} = \dfrac{-2}{3-w}$

89. $\dfrac{2x+4}{6} = \dfrac{2(x+2)}{2 \cdot 3} = \dfrac{x+2}{3}$

91. $\dfrac{3a+3}{3a} = \dfrac{3(a+1)}{3a} = \dfrac{a+1}{a}$

93. $\dfrac{1}{x-1} = \dfrac{1(x^2 + x + 1)}{(x-1)(x^2 + x + 1)}$

$= \dfrac{x^2 + x + 1}{x^3 - 1}$

95. $\dfrac{x^{2a} - 4}{x^a + 2} = \dfrac{(x^a - 2)(x^a + 2)}{x^a + 2} = x^a - 2$

97. $\dfrac{x^a + m + wx^a + wm}{x^{2a} - m^2}$

$= \dfrac{(x^a + m)(1 + w)}{(x^a - m)(x^a + m)}$

$= \dfrac{1 + w}{x^a - m}$

99. $\dfrac{x^{3b+1} - x}{x^{2b+1} - x} = \dfrac{x(x^b - 1)(x^{2b} + x^b + 1)}{x(x^b - 1)(x^b + 1)}$

$= \dfrac{x^{2b} + x^b + 1}{x^b + 1}$

101. Since $D = RT$, his rate is $\dfrac{500}{2x}$ or $\dfrac{250}{x}$ mph.

103. a) Average cost per invitation is

$A(n) = \dfrac{0.50n + 45}{n}$ dollars.

b) $A(200) = \dfrac{0.50(200) + 45}{200} = \$0.725$

$A(300) = \dfrac{0.50(300) + 45}{300} = \$0.65$

It costs 7.5 cents less per invitation to print 300 rather than 200 invitations.

c) As the number of invitations increases, the average cost per invitation decreases.

d) As the number of invitations increases, the total cost of the invitations increases.

105. a) $p(n) = \dfrac{0.053n^2 - 0.64n + 6.71}{3.43n + 87.24}$

b) $p(0) = \dfrac{0.053(0)^2 - 0.64(0) + 6.71}{3.43(0) + 87.24}$

$= 7.7\%$

$p(30) = \dfrac{0.053(30)^2 - 0.64(30) + 6.71}{3.43(30) + 87.24}$

$= 18.5\%$

$p(50) = \dfrac{0.053(50)^2 - 0.64(50) + 6.71}{3.43(50) + 87.24}$

$= 41.4\%$

107. The value of $R(x)$ gets closer and closer to $1/2$.

6.2 WARM-UPS

1. False, because we can multiply any two fractions.

2. False, $\frac{2}{7} \cdot \frac{3}{7} = \frac{6}{49}$.

3. True.

4. False, $a \div b = a \cdot \frac{1}{b}$.

5. True, because the expressions are correctly multiplied.

6. True, $\frac{1}{2} \cdot \frac{1}{3} = \frac{1}{6}$.

7. True, $\frac{1}{3} \div \frac{1}{2} = \frac{1}{3} \cdot \frac{2}{1} = \frac{2}{3}$.

8. True, $\frac{w-z}{z-w} = -1$.

9. True, $\frac{x}{3} \div 2 = \frac{x}{3} \cdot \frac{1}{2} = \frac{x}{6}$.

10. False, $\frac{a}{b} \div \frac{b}{a} = \frac{a}{b} \cdot \frac{a}{b} = \frac{a^2}{b^2}$.

6.2 EXERCISES

1. To multiply rational numbers, multiply the numerators and the denominators.

3. The expressions $a - b$ and $b - a$ are opposites.

5. $\frac{12}{42} \cdot \frac{35}{22} = \frac{2 \cdot 2 \cdot 3}{2 \cdot 3 \cdot 7} \cdot \frac{5 \cdot 7}{2 \cdot 11} = \frac{5}{11}$

7. $\frac{3a}{10b} \cdot \frac{5b^2}{6} = \frac{3a}{2 \cdot 5b} \cdot \frac{5b^2}{2 \cdot 3} = \frac{ab}{4}$

9. $\frac{3x - 3}{6} \cdot \frac{x}{x^2 - x}$
$= \frac{3(x - 1)}{2 \cdot 3} \cdot \frac{x}{x(x - 1)} = \frac{1}{2}$

11. $\frac{10x + 5}{5x^2 + 5} \cdot \frac{2x^2 + x - 1}{4x^2 - 1}$
$= \frac{5(2x + 1)}{5(x^2 + 1)} \cdot \frac{(2x - 1)(x + 1)}{(2x - 1)(2x + 1)} = \frac{x + 1}{x^2 + 1}$

13. $\frac{ax + aw + bx + bw}{x^2 - w^2} \cdot \frac{x - w}{a^2 - b^2}$
$= \frac{(a + b)(x + w)}{(x - w)(x + w)} \cdot \frac{x - w}{(a - b)(a + b)} = \frac{1}{a - b}$

15. $\frac{a^2 - 2a + 4}{a^3 + 8} \cdot \frac{(a + 2)^3}{2a + 4}$
$= \frac{a^2 - 2a + 4}{(a + 2)(a^2 - 2a + 4)} \cdot \frac{(a + 2)^3}{2(a + 2)}$
$= \frac{a + 2}{2}$

17. $\frac{x - 9}{12y} \cdot \frac{8y}{9 - x}$
$= \frac{x - 9}{4 \cdot 3y} \cdot \frac{4 \cdot 2y}{-1(x - 9)} = \frac{2}{-3} = -\frac{2}{3}$

19. $(a^2 - 4) \cdot \frac{7}{2 - a}$
$= (a - 2)(a + 2) \cdot \frac{7}{-1(a - 2)}$

21. $\frac{(3x + 1)^3}{2x - 1} \cdot \frac{4x^2 - 4x + 1}{9x^2 + 6x + 1}$
$= \frac{(3x + 1)^3}{2x - 1} \cdot \frac{(2x - 1)^2}{(3x + 1)^2}$
$= (3x + 1)(2x - 1)$
$= 6x^2 - x - 1$

$= -7(a + 2) = -7a - 14$

23. $\frac{15}{17} \div \frac{10}{17} = \frac{3 \cdot 5}{17} \cdot \frac{17}{2 \cdot 5} = \frac{3}{2}$

25. $\frac{36x}{5y} \div \frac{20x}{35y} = \frac{2^2 3^2 x}{5y} \cdot \frac{5 \cdot 7y}{2^2 5x} = \frac{3^2 7}{5}$
$= \frac{63}{5}$

27. $\frac{24a^5 b^2}{5c^3} \div (4a^5 bc^5)$
$= \frac{2^3 \cdot 3a^5 b^2}{5c^3} \cdot \frac{1}{2^2 a^5 bc^5} = \frac{6b}{5c^8}$

29. $(w + 1) \div \frac{w^2 - 1}{w} = (w + 1) \cdot \frac{w}{w^2 - 1}$
$= (w + 1) \cdot \frac{w}{(w - 1)(w + 1)} = \frac{w}{w - 1}$

31. $\frac{x - y}{5} \div \frac{x^2 - 2xy + y^2}{10}$
$= \frac{x - y}{5} \cdot \frac{2 \cdot 5}{(x - y)^2}$
$= \frac{2}{x - y}$

33. $\frac{4x - 2}{x^2 - 5x} \div \frac{2x^2 + 9x - 5}{x^2 - 25}$
$= \frac{2(2x - 1)}{x(x - 5)} \cdot \frac{(x - 5)(x + 5)}{(2x - 1)(x + 5)} = \frac{2}{x}$

35. $\frac{x - y}{3} \div \frac{1}{6} = \frac{x - y}{3} \cdot \frac{6}{1} = 2x - 2y$

37. $\frac{x^2 - 25}{3} \div \frac{x - 5}{6}$
$= \frac{(x - 5)(x + 5)}{3} \cdot \frac{2 \cdot 3}{x - 5} = 2x + 10$

39. $\frac{a - b}{2} \div 3 = \frac{a - b}{2} \cdot \frac{1}{3} = \frac{a - b}{6}$

41. $(a^2 - b^2) \div \frac{a + b}{3}$
$(a - b)(a + b) \cdot \frac{3}{a + b} = 3a - 3b$

43. $\frac{5x}{2} \div 3 = \frac{5x}{2} \cdot \frac{1}{3} = \frac{5x}{6}$

45. $\frac{3}{4} \div \frac{1}{4} = \frac{3}{4} \cdot \frac{4}{1} = 3$

47. $\frac{1}{2} \cdot \frac{1}{6} = \frac{1}{12}$

49. $\frac{1}{2} \cdot \frac{4x}{3} = \frac{2x}{3}$

51. $\frac{a - b}{b - a} = \frac{-1(b - a)}{b - a} = -1$

53. $\frac{x - y}{3} \cdot \frac{6}{y - x} = \frac{-1(y - x)}{3} \cdot \frac{2 \cdot 3}{y - x}$
$= -2$

55. $\frac{2a + 2b}{a} \cdot \frac{1}{2} = \frac{2(a + b)}{a} \cdot \frac{1}{2} = \frac{a + b}{a}$

57. $-1\left(\dfrac{9-x}{2}\right) = \dfrac{-1(9-x)}{2} = \dfrac{x-9}{2}$

59. $\dfrac{4}{y-7} \div \dfrac{2}{7-y} = \dfrac{2 \cdot 2}{y-7} \cdot \dfrac{-1(y-7)}{2}$
$$= -2$$

61. $(a+b) \div \dfrac{1}{2} = (a+b) \cdot 2 = 2a + 2b$

63. $\dfrac{3x}{5} \div y = \dfrac{3x}{5} \cdot \dfrac{1}{y} = \dfrac{3x}{5y}$

65. $\dfrac{3a}{5b} \div 2 = \dfrac{3a}{5b} \cdot \dfrac{1}{2} = \dfrac{3a}{10b}$

67. $\dfrac{3x^2 + 13x - 10}{x} \cdot \dfrac{x^3}{9x^2 - 4} \cdot \dfrac{7x - 35}{x^2 - 25}$
$$= \dfrac{(3x-2)(x+5)}{x} \cdot \dfrac{x^3}{(3x-2)(3x+2)}$$
$$\cdot \dfrac{7(x-5)}{(x-5)(x+5)}$$
$$= \dfrac{7x^2}{3x+2}$$

69. $\dfrac{(a^2 b^3 c)^2}{(-2ab^2 c)^3} \cdot \dfrac{(a^3 b^2 c)^3}{(abc)^4} = \dfrac{a^4 b^6 c^2 a^9 b^6 c^3}{-8a^3 b^6 c^3 a^4 b^4 c^4}$
$$= -\dfrac{a^6 b^2}{8c^2}$$

71. $\dfrac{(2mn)^3}{6mn^2} \div \dfrac{2m^2 n^3}{(m^2 n)^4} = \dfrac{(2mn)^3}{6mn^2} \cdot \dfrac{(m^2 n)^4}{2m^2 n^3}$
$$= \dfrac{8m^3 n^3 m^8 n^4}{12m^3 n^5} = \dfrac{2m^8 n^2}{3}$$

73. $\dfrac{2x^2 + 7x - 15}{4x^2 - 100} \cdot \dfrac{2x^2 - 9x - 5}{4x^2 - 1}$
$$= \dfrac{(2x-3)(x+5)}{4(x-5)(x+5)} \cdot \dfrac{(2x+1)(x-5)}{(2x-1)(2x+1)}$$
$$= \dfrac{2x-3}{8x-4}$$

75. $\dfrac{2h^2 - 5h - 3}{5h^2 - 4h - 1} \div \dfrac{2h^2 + 7h + 3}{h^2 + 2h - 3}$
$$= \dfrac{(2h+1)(h-3)}{(5h+1)(h-1)} \cdot \dfrac{(h+3)(h-1)}{(2h+1)(h+3)}$$
$$= \dfrac{h-3}{5h+1}$$

77. $\dfrac{9a - 3}{1 - 9a^2} \cdot \dfrac{9a^2 + 6a + 1}{6}$
$$= \dfrac{-3(1-3a)}{(1-3a)(1+3a)} \cdot \dfrac{(3a+1)^2}{6}$$
$$= \dfrac{-3a-1}{2}$$

79. We can factor $k^2 - 2km + m^2$ as $(k-m)^2$ or $(m-k)^2$.
$$\dfrac{k^2 + 2km + m^2}{k^2 - 2km + m^2} \cdot \dfrac{m^2 + 3m - mk - 3k}{m^2 + mk + 3m + 3k}$$
$$= \dfrac{(k+m)^2}{(m-k)^2} \cdot \dfrac{(m-k)(m+3)}{(m+3)(m+k)} = \dfrac{k+m}{m-k}$$

81. $\dfrac{x^a}{y^2} \cdot \dfrac{y^{b+2}}{x^{2a}} = \dfrac{y^b}{x^a}$

83. $\dfrac{x^{2a} + x^a - 6}{x^{2a} + 6x^a + 9} \div \dfrac{x^{2a} - 4}{x^{2a} + 2x^a - 3}$

$$= \dfrac{(x^a + 3)(x^a - 2)}{(x^a + 3)^2} \cdot \dfrac{(x^a + 3)(x^a - 1)}{(x^a - 2)(x^a + 2)}$$
$$= \dfrac{x^a - 1}{x^a + 2}$$

85. $\dfrac{m^k v^k + 3v^k - 2m^k - 6}{m^{2k} - 9}$
$$\cdot \dfrac{m^{2k} - 2m^k - 3}{v^k m^k - 2m^k + 2v^k - 4}$$
$$= \dfrac{(m^k + 3)(v^k - 2)}{(m^k - 3)(m^k + 3)} \cdot \dfrac{(m^k - 3)(m^k + 1)}{(v^k - 2)(m^k + 2)}$$
$$= \dfrac{m^k + 1}{m^k + 2}$$

87. $\dfrac{1}{50} \div \dfrac{7}{25} = \dfrac{1}{50} \cdot \dfrac{25}{7} = \dfrac{1}{14} \approx 7.1\%$

89. Her rate is $\dfrac{100}{x}$ miles/hour. Since $D = RT$, in $\dfrac{3}{4}$ hr she traveled $\dfrac{100}{x} \cdot \dfrac{3}{4}$ or $\dfrac{75}{x}$ miles.

91. Because $x \div \dfrac{y}{z} = x \cdot \dfrac{z}{y} = \dfrac{xz}{y}$, only (e) is not equivalent to the given expression.

6.3 WARM-UPS

1. False, the LCM is 30.
2. False, the LCM is $24a^2 b^3$.
3. True, because $x - 1$ is a factor of $x^2 - 1$.
4. False, the LCD is $x(x+1)$.
5. False, $\dfrac{1}{2} + \dfrac{2}{3} = \dfrac{7}{6}$.
6. False, $5 + \dfrac{1}{x} = \dfrac{5x}{x} + \dfrac{1}{x} = \dfrac{5x+1}{x}$ for any nonzero x.
7. True, because the expressions are added correctly.
8. True, because the expressions are subtracted correctly.
9. True, because $\dfrac{8}{12} + \dfrac{9}{12} = \dfrac{17}{12}$.
10. False, he uses x reams per day

6.3 EXERCISES

1. The sum of a/b and c/b is $(a+c)/b$.
3. The least common multiple of some numbers is the smallest number that is a multiple of all of the numbers.
5. To add rational expressions with different denominators, you must build-up the expressions to equivalent expressions with the same denominator.
7. $\dfrac{3x}{2} + \dfrac{5x}{2} = \dfrac{8x}{2} = 4x$

9. $\dfrac{x-3}{2x} - \dfrac{3x-5}{2x} = \dfrac{x-3-3x+5}{2x}$

$= \dfrac{-2x+2}{2x} = \dfrac{2(-x+1)}{2x} = \dfrac{-x+1}{x}$

11. $\dfrac{3x-4}{2x-4} + \dfrac{2x-6}{2x-4} = \dfrac{5x-10}{2x-4} = \dfrac{5(x-2)}{2(x-2)}$

$= \dfrac{5}{2}$

13. $\dfrac{x^2+4x-6}{x^2-9} - \dfrac{x^2+2x-12}{x^2-9} = \dfrac{2x+6}{x^2-9}$

$= \dfrac{2(x+3)}{(x-3)(x+3)} = \dfrac{2}{x-3}$

15. $\dfrac{2x^2-8x-4}{2x^2+7x+3} + \dfrac{4x^2+x-1}{2x^2+7x+3}$

$= \dfrac{6x^2-7x-5}{2x^2+7x+3}$

$= \dfrac{(3x-5)(2x+1)}{(2x+1)(x+3)} = \dfrac{3x-5}{x+3}$

$= \dfrac{3x-5}{x+3}$

17. Since $24 = 2^3 \cdot 3$ and $20 = 2^2 \cdot 5$, the LCM $= 2^3 \cdot 3 \cdot 5 = 120$.

19. The LCM for 10 and 15 is 30. So the LCM is $30x^3y$.

21. The highest power of a is 3, the highest power of b is 5, and the highest power of c is 2. So the LCM $= a^3b^5c^2$.

23. Since the three polynomials have no common factors, the LCM is the product of the polynomials: $x(x+2)(x-2)$

25. $4a+8 = 4(a+2), 6a+12 = 6(a+2)$

LCM $= 12(a+2) = 12a+24$

27. $x^2-1 = (x-1)(x+1)$

$x^2+2x+1 = (x+1)^2$

LCM $= (x-1)(x+1)^2$

29. $x^2-4x = x(x-4)$

$x^2-16 = (x-4)(x+4)$

$x^2+6x+8 = (x+4)(x+2)$

LCM $= x(x-4)(x+4)(x+2)$

31. $6x^2+17x+12 = (2x+3)(3x+4)$

$9x^2-16 = (3x-4)(3x+4)$

LCM $= (2x+3)(3x+4)(3x-4)$

33. $\dfrac{1}{28} + \dfrac{3}{35} = \dfrac{1 \cdot 5}{28 \cdot 5} + \dfrac{3 \cdot 4}{35 \cdot 4}$

$= \dfrac{5}{140} + \dfrac{12}{140} = \dfrac{17}{140}$

35. $\dfrac{7}{48} - \dfrac{5}{36} = \dfrac{7 \cdot 3}{48 \cdot 3} - \dfrac{5 \cdot 4}{36 \cdot 4}$

$= \dfrac{21}{144} - \dfrac{20}{144} = \dfrac{1}{144}$

37. $\dfrac{3}{wz^2} + \dfrac{5}{w^2z} = \dfrac{3 \cdot w}{wz^2 \cdot w} + \dfrac{5 \cdot z}{w^2z \cdot z}$

$= \dfrac{3w}{w^2z^2} + \dfrac{5z}{w^2z^2} = \dfrac{3w+5z}{w^2z^2}$

39. $\dfrac{2x-3}{8} - \dfrac{x-2}{6}$

$= \dfrac{(2x-3)3}{8 \cdot 3} - \dfrac{(x-2)4}{6 \cdot 4}$

$= \dfrac{6x-9}{24} - \dfrac{4x-8}{24} = \dfrac{2x-1}{24}$

41. $\dfrac{xa^3}{2a^4} + \dfrac{21x^2}{35ax} = \dfrac{x}{2a} + \dfrac{3x}{5a}$

$= \dfrac{5x}{10a} + \dfrac{6x}{10a} = \dfrac{11x}{10a}$

43. $\dfrac{9}{4y} - x = \dfrac{9}{4y} - \dfrac{x \cdot 4y}{4y} = \dfrac{9}{4y} - \dfrac{4xy}{4y}$

$= \dfrac{9-4xy}{4y}$

45. $\dfrac{5}{a+2} - \dfrac{7}{a} = \dfrac{5 \cdot a}{(a+2)a} - \dfrac{7(a+2)}{a(a+2)}$

$= \dfrac{5a}{a(a+2)} - \dfrac{7a+14}{a(a+2)}$

$= \dfrac{5a-(7a+14)}{a(a+2)} = \dfrac{-2a-14}{a(a+2)}$

47. $\dfrac{1}{a-b} + \dfrac{2}{a+b}$

$= \dfrac{1(a+b)}{(a-b)(a+b)} + \dfrac{2(a-b)}{(a+b)(a-b)}$

$= \dfrac{a+b}{(a-b)(a+b)} + \dfrac{2a-2b}{(a-b)(a+b)}$

$= \dfrac{3a-b}{(a-b)(a+b)}$

49. $\dfrac{2x^2}{2x^3-18x} + \dfrac{15}{5x-15}$

$= \dfrac{2x^2}{2x(x+3)(x-3)} + \dfrac{3 \cdot 5}{5(x-3)}$

$= \dfrac{x}{(x+3)(x-3)} + \dfrac{3}{x-3}$

$= \dfrac{x}{(x+3)(x-3)} + \dfrac{3(x+3)}{(x-3)(x+3)}$

$= \dfrac{4x+9}{(x+3)(x-3)}$

51. $\dfrac{1}{a-b} + \dfrac{1(-1)}{(b-a)(-1)} = \dfrac{1}{a-b} + \dfrac{-1}{a-b}$

$= \dfrac{0}{a-b} = 0$

53. $\dfrac{10}{4x-8} - \dfrac{15}{10-5x}$

$= \dfrac{10}{4(x-2)} - \dfrac{15}{-5(x-2)}$

$= \dfrac{5}{2(x-2)} + \dfrac{3}{(x-2)}$

$= \dfrac{5}{2(x-2)} + \dfrac{3 \cdot 2}{(x-2)2}$

$= \dfrac{5}{2x-4} + \dfrac{6}{2x-4} = \dfrac{11}{2x-4}$

55. $\dfrac{5}{x^2 + x + 2} - \dfrac{6}{x^2 + 2x - 3}$

$= \dfrac{5(x+3)}{(x+2)(x-1)(x+3)}$

$\qquad - \dfrac{6(x+2)}{(x+3)(x-1)(x+2)}$

$= \dfrac{5x + 15}{(x-1)(x+2)(x+3)}$

$\qquad - \dfrac{6x + 12}{(x-1)(x+2)(x+3)}$

$= \dfrac{-x + 3}{(x-1)(x+2)(x+3)}$

57. $\dfrac{x}{2x^2 + x - 1} + \dfrac{3}{3x^2 + 2x - 1}$

$= \dfrac{x(3x-1)}{(2x-1)(x+1)(3x-1)}$

$\qquad + \dfrac{3(2x-1)}{(3x-1)(x+1)(2x-1)}$

$= \dfrac{3x^2 - x}{(x+1)(2x-1)(3x-1)}$

$\qquad + \dfrac{6x - 3}{(x+1)(2x-1)(3x-1)}$

$= \dfrac{3x^2 + 5x - 3}{(x+1)(2x-1)(3x-1)}$

59. $\dfrac{1}{x} + \dfrac{2}{x-1} - \dfrac{3}{x+2}$

$= \dfrac{1(x-1)(x+2)}{x(x-1)(x+2)} + \dfrac{2x(x+2)}{(x-1)x(x+2)}$

$\qquad - \dfrac{3x(x-1)}{(x+2)x(x-1)}$

$= \dfrac{x^2 + x - 2}{x(x-1)(x+2)} + \dfrac{2x^2 + 4x}{x(x-1)(x+2)}$

$\qquad - \dfrac{3x^2 - 3x}{x(x-1)(x+2)}$

$= \dfrac{8x - 2}{x(x-1)(x+2)}$

61. $\dfrac{1}{3} + \dfrac{1}{4} = \dfrac{1 \cdot 4 + 1 \cdot 3}{12} = \dfrac{7}{12}$

63. $\dfrac{1}{8} - \dfrac{3}{5} = \dfrac{1 \cdot 5 - 8 \cdot 3}{40} = -\dfrac{19}{40}$

65. $\dfrac{x}{3} + \dfrac{x}{2} = \dfrac{x \cdot 2 + x \cdot 3}{6} = \dfrac{5x}{6}$

67. $\dfrac{a}{b} - \dfrac{2}{3} = \dfrac{3 \cdot a - 2 \cdot b}{3b} = \dfrac{3a - 2b}{3b}$

69. $a + \dfrac{2}{3} = \dfrac{a \cdot 3 + 1 \cdot 2}{3 \cdot 1} = \dfrac{3a + 2}{3}$

71. $\dfrac{3}{a} + 1 = \dfrac{a \cdot 1 + 3 \cdot 1}{a \cdot 1} = \dfrac{a + 3}{a}$

73. $\dfrac{3+x}{x} - 1 = \dfrac{3 + x - x}{x} = \dfrac{3}{x}$

75. $\dfrac{2}{3} + \dfrac{1}{4x} = \dfrac{2 \cdot 4x + 1 \cdot 3}{12x} = \dfrac{8x + 3}{12x}$

77. $\dfrac{w^2 - 3w + 6}{w - 5} + \dfrac{9 - w^2}{w - 5} = \dfrac{-3w + 15}{w - 5}$

$\qquad = \dfrac{-3(w - 5)}{w - 5} = -3$

79. $\dfrac{1}{3x - 6} - \dfrac{6}{5x - 10}$

$= \dfrac{1}{3(x-2)} - \dfrac{6}{5(x-2)}$

$= \dfrac{5}{15(x-2)} - \dfrac{18}{15(x-2)}$

$= \dfrac{-13}{15(x-2)}$

81. $\dfrac{x-1}{2x^2 + 3x + 1} - \dfrac{x+1}{2x^2 - x - 1}$

$= \dfrac{x-1}{(2x+1)(x+1)} - \dfrac{x+1}{(2x+1)(x-1)}$

$= \dfrac{(x-1)(x-1)}{(2x+1)(x+1)(x-1)}$

$\qquad - \dfrac{(x+1)(x+1)}{(2x+1)(x-1)(x+1)}$

$= \dfrac{x^2 - 2x + 1}{(x+1)(x-1)(2x+1)}$

$\qquad - \dfrac{x^2 + 2x + 1}{(x+1)(x-1)(2x+1)}$

$= \dfrac{-4x}{(x+1)(x-1)(2x+1)}$

83. $\dfrac{(a^2 b^3)^4}{(ab^4)^3} \cdot \dfrac{(ab)^3}{(a^4 b)^2}$

$= \dfrac{a^8 b^{12}}{a^3 b^{12}} \cdot \dfrac{a^3 b^3}{a^8 b^2} = \dfrac{a^{11} b^{15}}{a^{11} b^{14}} = b$

85. $\dfrac{x^2 + 4}{25x^2 - 20x + 4} + \dfrac{10x + 4}{25x^2 - 4}$

$= \dfrac{x^2 + 4}{(5x - 2)^2} + \dfrac{10x + 4}{(5x - 2)(5x + 2)}$

$= \dfrac{x^2 + 4}{(5x - 2)^2} + \dfrac{2(5x + 2)}{(5x - 2)(5x + 2)}$

$= \dfrac{x^2 + 4}{(5x - 2)^2} + \dfrac{2}{5x - 2}$

$= \dfrac{x^2 + 4}{(5x - 2)^2} + \dfrac{2(5x - 2)}{(5x - 2)(5x - 2)}$

$= \dfrac{x^2 + 4}{(5x - 2)^2} + \dfrac{10x - 4}{(5x - 2)^2}$

$= \dfrac{x^2 + 10x}{(5x - 2)^2}$

87. $\dfrac{4x^2 + 9}{4x^2 - 9} \cdot \dfrac{4x^2 + 12x + 9}{2x^2 + 3x}$

$= \dfrac{4x^2 + 9}{(2x - 3)(2x + 3)} \cdot \dfrac{(2x + 3)^2}{x(2x + 3)}$

$= \dfrac{4x^2 + 9}{x(2x - 3)}$

89. $\dfrac{w^2 - 3}{3w^3 + 81} - \dfrac{2}{6w + 18} - \dfrac{w - 4}{w^2 - 3w + 9}$

$= \dfrac{w^2 - 3}{3w^3 + 81} - \dfrac{1}{3(w + 3)} - \dfrac{w - 4}{w^2 - 3w + 9}$

$= \dfrac{w^2 - 3}{3(w + 3)(w^2 - 3w + 9)}$

$\qquad - \dfrac{1(w^2 - 3w + 9)}{3(w + 3)(w^2 - 3w + 9)}$

$\qquad - \dfrac{3(w - 4)(w + 3)}{3(w + 3)(w^2 - 3w + 9)}$

$$= \frac{w^2 - 3 - w^2 + 3w - 9 - 3(w^2 - w - 12)}{3(w+3)(w^2 - 3w + 9)}$$

$$= \frac{-3w^2 + 6w + 24}{3(w+3)(w^2 - 3w + 9)}$$

$$= \frac{-w^2 + 2w + 8}{(w+3)(w^2 - 3w + 9)}$$

91. $\dfrac{a^2 - 6a + 9}{a^3 + 8} \div \dfrac{a^2 - a - 6}{a^2 - 4}$

$$= \frac{(a-3)^2}{(a-2)(a^2 + 2a + 4)} \cdot \frac{(a-2)(a+2)}{(a-3)(a+2)}$$

$$= \frac{a-3}{a^2 + 2a + 4}$$

93. $\dfrac{w^2 + 3}{w^3 - 8} - \dfrac{2w}{w^2 - 4}$

$$= \frac{(w^2 + 3)(w+2)}{(w-2)(w^2 + 2w + 4)(w+2)}$$

$$- \frac{2w(w^2 + 2w + 4)}{(w-2)(w+2)(w^2 + 2w + 4)}$$

$$= \frac{w^3 + 3w + 2w^2 + 6 - (2w^3 + 4w^2 + 8w)}{(w-2)(w+2)(w^2 + 2w + 4)}$$

$$= \frac{-w^3 - 2w^2 - 5w + 6}{(w-2)(w+2)(w^2 + 2w + 4)}$$

95. $\dfrac{1}{x^3 - 1} - \dfrac{1}{x^2 - 1} + \dfrac{1}{x - 1}$

$$= \frac{x+1}{(x-1)(x^2 + x + 1)(x+1)}$$

$$- \frac{x^2 + x + 1}{(x-1)(x+1)(x^2 + x + 1)}$$

$$+ \frac{(x+1)(x^2 + x + 1)}{(x-1)(x+1)(x^2 + x + 1)}$$

$$= \frac{x + 1 - x^2 - x - 1 + x^3 + 2x^2 + 2x + 1}{(x-1)(x+1)(x^2 + x + 1)}$$

$$= \frac{x^3 + x^2 + 2x + 1}{(x-1)(x+1)(x^2 + x + 1)}$$

97. Joe processes $\frac{1}{x}$ claims/hr while Ellen processes $\frac{1}{x+1}$ claims/hr. Together they process $\frac{1}{x} + \frac{1}{x+1}$ or $\frac{2x+1}{x(x+1)}$ claims/hr. In 8 hours they process $\frac{8(2x+1)}{x(x+1)}$ or $\frac{16x+8}{x^2+x}$ claims.

99. George's rate is $\frac{1}{20}$ mag/min while Theresa's rate is $\frac{1}{x}$ mag/min. Their rate together is $\frac{1}{x} + \frac{1}{20}$ or $\frac{x+20}{20x}$ mag/min. In 60

min they will sell $60 \cdot \dfrac{x+20}{20x}$ or $\dfrac{3x+60}{x}$ magazines.

101. Let x be Joan's original speed and $x + 5$ be her increased speed. Since $T = \dfrac{D}{R}$, her total travel time is $\dfrac{100}{x} + \dfrac{200}{x+5}$ or $\dfrac{300x + 500}{x^2 + 5x}$ hours.

103. By definition, $a/b + c/b = (a + c)/b$ provided $b \neq 0$. By definition of fraction multiplication, common denominators are not needed.

6.4 WARM-UPS

1. False, the LCM is $6x^2$.
2. True, since $2b - 2a = -2(a - b)$.
3. True.
4. True, because $\frac{1}{2} + \frac{1}{3} = \frac{5}{6}$ and $1 + \frac{1}{2} = \frac{3}{2}$.
5. False, because $2^{-1} + 3^{-1} = \frac{1}{2} + \frac{1}{3} = \frac{5}{6}$ and $(2 + 3)^{-1} = \frac{1}{5}$.
6. False, because the left side is the reciprocal of $\frac{5}{6}$ and the right side is 5.
7. False, because $2 + 3^{-1} = \frac{7}{3}$ and $5^{-1} = \frac{1}{5}$.
8. False, because $x + 2^{-1} = x + \frac{1}{2} = \frac{2x+1}{2}$ for any value of x.
9. True, because ab is the LCD.
10. True, because multiplying by $a^5 b^2$ will eliminate all negative exponents.

6.4 EXERCISES

1. A complex fraction is a fraction that contains fractions in the numerator, denominator, or both.

3. $\dfrac{\left(\frac{1}{2} - \frac{1}{3}\right)60}{\left(\frac{1}{4} - \frac{1}{5}\right)60} = \dfrac{30 - 20}{15 - 12} = \dfrac{10}{3}$

5. $\dfrac{\frac{2}{3} + \frac{5}{6} - \frac{1}{2}}{\frac{1}{8} - \frac{1}{3} + \frac{1}{12}}$

$$= \frac{24 \cdot \frac{2}{3} + 24 \cdot \frac{5}{6} - 24 \cdot \frac{1}{2}}{24 \cdot \frac{1}{8} - 24 \cdot \frac{1}{3} + 24 \cdot \frac{1}{12}} = \frac{16 + 20 - 12}{3 - 8 + 2}$$

$$= -8$$

120

7. $\dfrac{\frac{a+b}{b}}{\frac{a-b}{ab}} = \dfrac{ab\left(\frac{a+b}{b}\right)}{ab\left(\frac{a-b}{ab}\right)} = \dfrac{a^2+ab}{a-b}$

9. $\dfrac{a+\frac{3}{b}}{\frac{b}{a}+\frac{1}{b}} = \dfrac{ab\cdot a + ab\cdot\frac{3}{b}}{ab\cdot\frac{b}{a}+ab\cdot\frac{1}{b}} = \dfrac{a^2 b + 3a}{a+b^2}$

11. $\dfrac{\frac{x-3y}{xy}}{\frac{1}{x}+\frac{1}{y}} = \dfrac{xy\left(\frac{x-3y}{xy}\right)}{xy\cdot\frac{1}{x}+xy\cdot\frac{1}{y}}$

$= \dfrac{x-3y}{x+y}$

13. $\dfrac{3-\frac{m-2}{6}}{\frac{4}{9}+\frac{2}{m}} = \dfrac{18m\cdot 3 - 18m\cdot\frac{m-2}{6}}{18m\cdot\frac{4}{9}+18m\cdot\frac{2}{m}}$

$= \dfrac{54m-3m^2+6m}{8m+36}$

$= \dfrac{60m-3m^2}{8m+36}$

15. $\dfrac{\frac{a^2-b^2}{a^2 b^3}}{\frac{a+b}{a^3 b}} = \dfrac{a^3 b^3\left(\frac{a^2-b^2}{a^2 b^3}\right)}{a^3 b^3\left(\frac{a+b}{a^3 b}\right)}$

$= \dfrac{a(a^2-b^2)}{b^2(a+b)} = \dfrac{a(a-b)(a+b)}{b^2(a+b)} = \dfrac{a^2-ab}{b^2}$

17. $\dfrac{\frac{1}{x^2 y^2}+\frac{1}{xy^3}}{\frac{1}{x^3 y}-\frac{1}{xy}} = \dfrac{\left(\frac{1}{x^2 y^2}+\frac{1}{xy^3}\right)x^3 y^3}{\left(\frac{1}{x^3 y}-\frac{1}{xy}\right)x^3 y^3}$

$= \dfrac{xy+x^2}{y^2-x^2 y^2}$

19. $\dfrac{x+\frac{4}{x+4}}{x-\frac{4x+4}{x+4}}$

$= \dfrac{(x+4)x+(x+4)\cdot\frac{4}{x+4}}{(x+4)x-(x+4)\cdot\frac{4x+4}{x+4}}$

$= \dfrac{x^2+4x+4}{x^2+4x-4x-4} = \dfrac{(x+2)^2}{(x-2)(x+2)}$

$= \dfrac{x+2}{x-2}$

21.

$\dfrac{1-\frac{1}{y-1}}{3+\frac{1}{y+1}} = \dfrac{(y-1)(y+1)\left(1-\frac{1}{y-1}\right)}{(y-1)(y+1)\left(3+\frac{1}{y+1}\right)}$

$= \dfrac{y^2-1-(y+1)}{3(y^2-1)+y-1} = \dfrac{y^2-y-2}{3y^2+y-4}$

$= \dfrac{y^2-y-2}{(y-1)(3y+4)}$

23. $\dfrac{\frac{2}{3-x}-4}{\frac{1}{x-3}-1} = \dfrac{(x-3)\cdot\frac{2}{3-x}-(x-3)4}{(x-3)\cdot\frac{1}{x-3}-(x-3)1}$

$= \dfrac{-2-4x+12}{1-x+3} = \dfrac{-4x+10}{-x+4} = \dfrac{4x-10}{x-4}$

25. $\dfrac{\frac{w+2}{w-1}-\frac{w-3}{w}}{\frac{w+4}{w}+\frac{w-2}{w-1}}$

$= \dfrac{w(w-1)\left(\frac{w+2}{w-1}-\frac{w-3}{w}\right)}{w(w-1)\left(\frac{w+4}{w}+\frac{w-2}{w-1}\right)}$

$= \dfrac{w^2+2w-(w-1)(w-3)}{(w-1)(w+4)+w^2-2w}$

$= \dfrac{w^2+2w-w^2+4w-3}{w^2+3w-4+w^2-2w}$

$= \dfrac{6w-3}{2w^2+w-4}$

27. $\dfrac{\frac{1}{a-b}-\frac{3}{a+b}}{\frac{2}{b-a}+\frac{4}{a+b}}$

$= \dfrac{(a-b)(a+b)\left(\frac{1}{a-b}-\frac{3}{a+b}\right)}{(a-b)(a+b)\left(\frac{2}{b-a}+\frac{4}{a+b}\right)}$

$= \dfrac{a+b-3(a-b)}{-2(a+b)+4(a-b)} = \dfrac{-2a+4b}{2a-6b}$

$= \dfrac{2(2b-a)}{2(a-3b)} = \dfrac{2b-a}{a-3b}$

29. $\dfrac{\frac{4}{y}-\frac{y+4}{y-3}}{\frac{2}{y-3}+\frac{y+1}{y}}$

$= \dfrac{\left(\frac{4}{y}-\frac{y+4}{y-3}\right)(y)(y-3)}{\left(\frac{2}{y-3}+\frac{y+1}{y}\right)(y)(y-3)}$

$= \dfrac{4(y-3)-y(y+4)}{2y+(y+1)(y-3)}$

$= \dfrac{4y-12-y^2-4y}{2y+y^2-2y-3} = \dfrac{-y^2-12}{y^2-3}$

31. $\dfrac{3 - \dfrac{4}{a-1}}{5 - \dfrac{3}{1-a}} = \dfrac{(a-1)3 - (a-1)\cdot \dfrac{4}{a-1}}{(a-1)5 - (a-1)\cdot \dfrac{3}{1-a}}$

$\qquad = \dfrac{3a - 3 - 4}{5a - 5 + 3} = \dfrac{3a - 7}{5a - 2}$

33. $\dfrac{\dfrac{2}{m-3} + \dfrac{4}{m}}{\dfrac{3}{m-2} + \dfrac{1}{m}}$

$= \dfrac{m(m-3)(m-2)\left(\dfrac{2}{m-3} + \dfrac{4}{m}\right)}{m(m-3)(m-2)\left(\dfrac{3}{m-2} + \dfrac{1}{m}\right)}$

$= \dfrac{2m(m-2) + 4(m-3)(m-2)}{3m(m-3) + (m-3)(m-2)}$

$= \dfrac{2m^2 - 4m + 4m^2 - 20m + 24}{3m^2 - 9m + m^2 - 5m + 6}$

$= \dfrac{6m^2 - 24m + 24}{4m^2 - 14m + 6} = \dfrac{3m^2 - 12m + 12}{2m^2 - 7m + 3}$

$= \dfrac{3m^2 - 12m + 12}{(m-3)(2m-1)}$

35. $\dfrac{\dfrac{3}{x^2-1} - \dfrac{x-2}{x^3-1}}{\dfrac{3}{x^2+x+1} + \dfrac{x-3}{x^3-1}}$

$= \dfrac{(x-1)(x+1)(x^2+x+1)\left(\frac{3}{x^2-1} - \frac{x-2}{x^3-1}\right)}{(x-1)(x+1)(x^2+x+1)\left(\frac{3}{x^2+x+1} + \frac{x-3}{x^3-1}\right)}$

$= \dfrac{3x^2 + 3x + 3 - (x+1)(x-2)}{3x^2 - 3 + (x+1)(x-3)}$

$= \dfrac{3x^2 + 3x + 3 - x^2 + x + 2}{3x^2 - 3 + x^2 - 2x - 3}$

$= \dfrac{2x^2 + 4x + 5}{4x^2 - 2x - 6}$

37. $\dfrac{w^{-1} + y^{-1}}{z^{-1} + y^{-1}} = \dfrac{wyz(w^{-1} + y^{-1})}{wyz(z^{-1} + y^{-1})}$

$\qquad = \dfrac{yz + wz}{wy + wz}$

39. $\dfrac{1 - x^{-1}}{1 - x^{-2}} = \dfrac{x^2(1 - x^{-1})}{x^2(1 - x^{-2})} = \dfrac{x^2 - x}{x^2 - 1}$

$\qquad = \dfrac{x(x-1)}{(x-1)(x+1)} = \dfrac{x}{x+1}$

41. $\dfrac{a^{-2} + b^{-2}}{a^{-1}b} = \dfrac{a^2 b^2 (a^{-2} + b^{-2})}{a^2 b^2 (a^{-1}b)}$

$\qquad = \dfrac{a^2 + b^2}{ab^3}$

43. $1 - a^{-1} = \dfrac{a(1 - a^{-1})}{a} = \dfrac{a-1}{a}$

45. $\dfrac{x^{-1} + x^{-2}}{x + x^{-2}} = \dfrac{x^2(x^{-1} + x^{-2})}{x^2(x + x^{-2})}$

$\qquad = \dfrac{x+1}{x^3 + 1} = \dfrac{x+1}{(x+1)(x^2 - x + 1)}$

$\qquad = \dfrac{1}{x^2 - x + 1}$

47. $\dfrac{2m^{-1} - 3m^{-2}}{m^{-2}} = \dfrac{m^2(2m^{-1} - 3m^{-2})}{m^2(m^{-2})}$

$\qquad = 2m - 3$

49. $\dfrac{a^{-1} - b^{-1}}{a - b} = \dfrac{ab(a^{-1} - b^{-1})}{ab(a-b)}$

$\qquad = \dfrac{b - a}{-ab(b-a)} = -\dfrac{1}{ab}$

51. $\dfrac{x^3 - y^3}{x^{-3} - y^{-3}} = \dfrac{(x^3 - y^3)x^3 y^3}{(x^{-3} - y^{-3})x^3 y^3}$

$\qquad = \dfrac{(x^3 - y^3)x^3 y^3}{y^3 - x^3} = (-1)x^3 y^3 = -x^3 y^3$

53. $\dfrac{1 - 8x^{-3}}{x^{-1} + 2x^{-2} + 4x^{-3}}$

$= \dfrac{x^3(1 - 8x^{-3})}{x^3(x^{-1} + 2x^{-2} + 4x^{-3})}$

$= \dfrac{x^3 - 8}{x^2 + 2x + 4} = \dfrac{(x-2)(x^2 + 2x + 4)}{x^2 + 2x + 4}$

$\qquad = x - 2$

55. $(x^{-1} + y^{-1})^{-1} = \dfrac{1}{x^{-1} + y^{-1}}$

$\qquad = \dfrac{xy \cdot 1}{xy(x^{-1} + y^{-1})} = \dfrac{xy}{x + y}$

57. $\dfrac{\frac{5}{3} - \frac{4}{5}}{\frac{1}{3} - \frac{5}{6}} \approx -1.7333 \text{ or } -26/15$

59. $\dfrac{4^{-1} - 9^{-1}}{2^{-1} + 3^{-1}} \approx 0.1667 \text{ or } 1/6$

61. If $x =$ the number of students at Central, then $\frac{1}{2}x =$ the number at Northside and $\frac{2}{3}x =$ the number at Southside. To find the percentage of black students among the city's elementary students we divide the number of

122

black students by the total number of students:

$$\frac{\frac{1}{3} \cdot \frac{1}{2}x + \frac{3}{4} \cdot x + \frac{1}{6} \cdot \frac{2}{3}x}{\frac{1}{2}x + x + \frac{2}{3}x} = \frac{\frac{1}{6}x + \frac{3}{4}x + \frac{1}{9}x}{\frac{1}{2}x + x + \frac{2}{3}x}$$

$$= \frac{36\left(\frac{1}{6}x + \frac{3}{4}x + \frac{1}{9}x\right)}{36\left(\frac{1}{2}x + x + \frac{2}{3}x\right)} = \frac{6x + 27x + 4x}{18x + 36x + 24x}$$

$$= \frac{37x}{78x} = \frac{37}{78} = 47.4\%$$

63. Let x = the distance and $\frac{x}{45}$ = the time from Clarksville to Leesville. We can also say that x = the distance and $\frac{x}{55}$ = the time for the return trip. To find the average speed for any trip we divide the total distance by the total time. In this case the total distance is $2x$ and the total time is $\frac{x}{45} + \frac{x}{55}$.

$$\frac{2x}{\frac{x}{45} + \frac{x}{55}} = \frac{2x \cdot 5 \cdot 9 \cdot 11}{\left(\frac{x}{45} + \frac{x}{55}\right)5 \cdot 9 \cdot 11} = \frac{990x}{11x + 9x}$$

$$= \frac{990x}{20x} = \frac{99}{2} = 49.5$$

Her average speed for the trip is 49.5 mph.

67. The complex fraction simplifies to $\frac{2x+1}{3x+2}$. The complex fraction is undefined for $x = 0, -1, -1/2, -2/3$.

6.5 WARM-UPS

1. False, c is the quotient.
2. False, (quotient)(divisor) + remainder = dividend.
3. True, because
$(x + 2)(x + 3) = x^2 + 5x + 6$ and
$(x + 2)(x + 3) + 1 = x^2 + 5x + 7$.
4. True. From warm-up number 3 we can see that if $x + 3$ is the divisor then $x + 2$ is the quotient.
5. True. From warm-up number 3 we can see that if $x + 2$ is the divisor then $x + 3$ is the quotient and 1 is the remainder.
6. False, to divide by $x - c$ with synthetic division we use c.
7. False, synthetic division is used only for dividing by $x - c$.

8. True, when dividing by $x - c$ the degree of the quotient is 1 less than the degree of the dividend.
9. True, because if remainder is zero, then (quotient)(divisor) = dividend.
10. True, because if remainder is zero, then (quotient)(divisor) = dividend.

6.5 EXERCISES

1. If $a \div b = c$, then the divisor is b, the dividend is a, and the quotient is c.
3. If the term x^n is missing in the dividend, insert the term $0 \cdot x^n$ for the missing term.
5. Synthetic division is used only for dividing by a binomial of the form $x - c$.
7. $\frac{36x^7}{3x^3} = 12x^4$
9. $\frac{16x^2}{-8x^2} = -2$
11. $\frac{6b - 9}{3} = \frac{6b}{3} - \frac{9}{3} = 2b - 3$
13. $\frac{3x^2 + 3x}{3x} = \frac{3x^2}{3x} + \frac{3x}{3x} = x + 1$

15. $\frac{10x^4 - 8x^3 + 6x^2}{-2x^2}$
$$= \frac{10x^4}{-2x^2} - \frac{8x^3}{-2x^2} + \frac{6x^2}{-2x^2}$$
$$= -5x^2 + 4x - 3$$
17. $\frac{7x^3 - 4x^2}{2x} = \frac{7x^3}{2x} - \frac{4x^2}{2x} = \frac{7}{2}x^2 - 2x$

19. $(8x - 3) \div (4x) = \frac{8x}{4x} + \frac{-3}{4x}$
$$= 2 + \frac{-3}{4x}$$
The quotient is 2 and the remainder is -3.

21. $(2x^3 + x^2 + 4x - 3) \div (3x^2)$
$$= \frac{2x^3}{3x^2} + \frac{x^2}{3x^2} + \frac{4x}{3x^2} + \frac{-3}{3x^2}$$
$$= \frac{2}{3}x + \frac{1}{3} + \frac{4x - 3}{3x^2}$$
The quotient is $\frac{2}{3}x + \frac{1}{3}$ and the remainder is $4x - 3$.

23. $(-10x^4 - 5x^3 - 6x - 7) \div (-5x^2)$
$$= \frac{-10x^4}{-5x^2} - \frac{5x^3}{-5x^2} + \frac{-6x - 7}{-5x^2}$$
$$= 2x^2 + x + \frac{-6x - 7}{-5x^2}$$
The quotient is $2x^2 + x$ and the remainder is $-6x - 7$.

25.
$$
x + 3 \overline{\smash{\big)}\ x^2 + 8x + 13} \quad \overset{x+5}{}
$$

$$
\begin{array}{r}
x + 5 \\
x + 3 \overline{\smash{\big)}\ x^2 + 8x + 13} \\
\underline{x^2 + 3x} \\
5x + 13 \\
\underline{5x + 15} \\
-2
\end{array}
$$

Quotient: $x + 5$ Remainder: -2

27.
$$
\begin{array}{r}
x - 4 \\
x + 2 \overline{\smash{\big)}\ x^2 - 2x} \\
\underline{x^2 + 2x} \\
-4x \\
\underline{-4x - 8} \\
8
\end{array}
$$

Quotient: $x - 4$ Remainder: 8

29.
$$
\begin{array}{r}
x^2 - 2x + 4 \\
x + 2 \overline{\smash{\big)}\ x^3 + 0x^2 + 0x + 8} \\
\underline{x^3 + 2x^2} \\
-2x^2 + 0x \\
\underline{-2x^2 - 4x} \\
4x + 8 \\
\underline{4x + 8} \\
0
\end{array}
$$

Quotient: $x^2 - 2x + 4$ Remainder: 0

31.
$$
\begin{array}{r}
a^2 + 2a + 8 \\
a - 2 \overline{\smash{\big)}\ a^3 + 0a^2 + 4a - 5} \\
\underline{a^3 - 2a^2} \\
2a^2 + 4a \\
\underline{2a^2 - 4a} \\
8a - 5 \\
\underline{8a - 16} \\
11
\end{array}
$$

Quotient: $a^2 + 2a + 8$ Remainder: 11

33.
$$
\begin{array}{r}
x^2 - 2x + 3 \\
x + 1 \overline{\smash{\big)}\ x^3 - x^2 + x - 3} \\
\underline{x^3 + x^2} \\
-2x^2 + x \\
\underline{-2x^2 - 2x} \\
3x - 3 \\
\underline{3x + 3} \\
-6
\end{array}
$$

Quotient: $x^2 - 2x + 3$ Remainder: -6

35.
$$
\begin{array}{r}
x^3 + 3x^2 + 6x + 11 \\
x - 2 \overline{\smash{\big)}\ x^4 + x^3 + 0x^2 - x - 1} \\
\underline{x^4 - 2x^3} \\
3x^3 + 0x^2 \\
\underline{3x^3 - 6x^2} \\
6x^2 - x \\
\underline{6x^2 - 12x} \\
11x - 1 \\
\underline{11x - 22} \\
21
\end{array}
$$

Quotient: $x^3 + 3x^2 + 6x + 11$ Remainder: 21

37.
$$
\begin{array}{r}
-3x^2 - 1 \\
x^2 - 2 \overline{\smash{\big)}\ -3x^4 + 0x^3 + 5x^2 + x - 2} \\
\underline{-3x^4 + 6x^2} \\
-x^2 + x - 2 \\
\underline{-x^2 + 2} \\
x - 4
\end{array}
$$

Quotient: $-3x^2 - 1$ Remainder: $x - 4$

39.
$$
\begin{array}{r}
3x + 5 \\
2x - 3 \overline{\smash{\big)}\ 6x^2 + x - 16} \\
\underline{6x^2 - 9x} \\
10x - 16 \\
\underline{10x - 15} \\
-1
\end{array}
$$

Quotient: $3x + 5$ Remainder: -1

41.
$$
\begin{array}{r}
2b - 5 \\
5b + 4 \overline{\smash{\big)}\ 10b^2 - 17b - 22} \\
\underline{10b^2 + 8b} \\
-25b - 22 \\
\underline{-25b - 20} \\
-2
\end{array}
$$

Quotient: $2b - 5$ Remainder: -2

43.
$$2x + 1 \overline{\smash{\big)}\ 2x^3 + 3x^2 - 3x - 2} \quad \overset{\textstyle x^2 + x - 2}{}$$

$$\underline{2x^3 + x^2}$$
$$2x^2 - 3x$$
$$\underline{2x^2 + x}$$
$$-4x - 2$$
$$\underline{-4x - 2}$$
$$0$$

Quotient: $x^2 + x - 2$ Remainder: 0

45.
$$x^2 + x + 2 \overline{\smash{\big)}\ x^3 - 4x^2 - 3x - 10} \quad \overset{\textstyle x - 5}{}$$

$$\underline{x^3 + x^2 + 2x}$$
$$-5x^2 - 5x - 10$$
$$\underline{-5x^2 - 5x - 10}$$
$$0$$

Quotient: $x - 5$ Remainder: 0

47.
$$x - 5 \overline{\smash{\big)}\ 2x + 0} \quad \overset{\textstyle 2}{}$$

$$\underline{2x - 10}$$
$$10$$

$$\frac{2x}{x - 5} = 2 + \frac{10}{x - 5}$$

49.
$$2x + 1 \overline{\smash{\big)}\ 2x^2 - x + 0} \quad \overset{\textstyle x - 1}{}$$

$$\underline{2x^2 + x}$$
$$-2x + 0$$
$$\underline{-2x - 1}$$
$$1$$

$$\frac{2x^2 - x}{2x + 1} = x - 1 + \frac{1}{2x + 1}$$

51.
$$x + 2 \overline{\smash{\big)}\ x^3 + 0x^2 + 0x + 0} \quad \overset{\textstyle x^2 - 2x + 4}{}$$

$$\underline{x^3 + 2x^2}$$
$$-2x^2 + 0x$$
$$\underline{-2x^2 - 4x}$$
$$4x + 0$$
$$\underline{4x + 8}$$
$$-8$$

$$\frac{x^3}{x + 2} = x^2 - 2x + 4 + \frac{-8}{x + 2}$$

53. $\dfrac{x^3 + 2x}{x^2} = \dfrac{x^3}{x^2} + \dfrac{2x}{x^2} = x + \dfrac{2}{x}$

55.
$$2x + 1 \overline{\smash{\big)}\ 2x^2 - 11x - 4} \quad \overset{\textstyle x - 6}{}$$

$$\underline{2x^2 + x}$$
$$-12x - 4$$
$$\underline{-12x - 6}$$
$$2$$

$$\frac{2x^2 - 11x - 4}{2x + 1} = x - 6 + \frac{2}{2x + 1}$$

57.
$$x - 1 \overline{\smash{\big)}\ 3x^3 - 4x^2 + 0x + 7} \quad \overset{\textstyle 3x^2 - x - 1}{}$$

$$\underline{3x^3 - 3x^2}$$
$$-x^2 + 0x$$
$$\underline{-x^2 + x}$$
$$-x + 7$$
$$\underline{-x + 1}$$
$$6$$

$$\frac{3x^3 - 4x^2 + 7}{x - 1} = 3x^2 - x - 1 + \frac{6}{x - 1}$$

59.
$$x - 2 \overline{\smash{\big)}\ 6x^3 + 0x^2 - 4x + 5} \quad \overset{\textstyle 6x^2 + 12x + 20}{}$$

$$\underline{6x^3 - 12x^2}$$
$$12x^2 - 4x$$
$$\underline{12x^2 - 24x}$$
$$20x + 5$$
$$\underline{20x - 40}$$
$$45$$

$$\frac{6x^3 - 4x + 5}{x - 2} = 6x^2 + 12x + 20 + \frac{45}{x - 2}$$

61.
$$x + 1 \overline{\smash{\big)}\ x^3 + 0x^2 + x + 0} \quad \overset{\textstyle x^2 - x + 2}{}$$

$$\underline{x^3 + x^2}$$
$$-x^2 + x$$
$$\underline{-x^2 - x}$$
$$2x + 0$$
$$\underline{2x + 2}$$
$$-2$$

$$\frac{x^3 + x}{x + 1} = x^2 - x + 2 + \frac{-2}{x + 1}$$

63.

$$\begin{array}{r} 2 \enspace \begin{array}{rrrr} 1 & -5 & 6 & -3 \\ & 2 & -6 & 0 \\ \hline 1 & -3 & 0 & -3 \end{array} \end{array}$$

The quotient is $x^2 - 3x$ and the remainder is -3.

65.

$$\begin{array}{r} -1 \enspace \begin{array}{rrr} 2 & -4 & 5 \\ & -2 & 6 \\ \hline 2 & -6 & 11 \end{array} \end{array}$$

The quotient is $2x - 6$ and the remainder is 11.

67.

$$\begin{array}{r} 3 \enspace \begin{array}{rrrrr} 3 & 0 & -15 & 7 & -9 \\ & 9 & 27 & 36 & 129 \\ \hline 3 & 9 & 12 & 43 & 120 \end{array} \end{array}$$

The quotient is $3x^3 + 9x^2 + 12x + 43$ and the remainder is 120.

69.

$$\begin{array}{r} 1 \enspace \begin{array}{rrrrrr} 1 & 0 & 0 & 0 & 0 & -1 \\ & 1 & 1 & 1 & 1 & 1 \\ \hline 1 & 1 & 1 & 1 & 1 & 0 \end{array} \end{array}$$

The quotient is $x^4 + x^3 + x^2 + x + 1$ and the remainder is 0.

71.

$$\begin{array}{r} -2 \enspace \begin{array}{rrrr} 1 & 0 & -5 & 6 \\ & -2 & 4 & 2 \\ \hline 1 & -2 & -1 & 8 \end{array} \end{array}$$

Quotient is $x^2 - 2x - 1$ and remainder is 8.

73.

$$\begin{array}{r} 5 \enspace \begin{array}{rrrr} 1 & -3 & 0 & 5 \\ & 5 & 10 & 50 \\ \hline 1 & 2 & 10 & 55 \end{array} \end{array}$$

The quotient is $x^2 + 2x + 10$ and the remainder is 55.

75. Divide $x^3 + x^2 - 11x + 8$ by $x + 4$:

$$\begin{array}{r} -4 \enspace \begin{array}{rrrr} 1 & 1 & -11 & 8 \\ & -4 & 12 & -4 \\ \hline 1 & -3 & 1 & 4 \end{array} \end{array}$$

Since the remainder is not 0, $x + 4$ is not a factor of $x^3 + x^2 - 11x + 8$.

77. Divide $x^2 - 6x + 8$ by $x - 4$:

$$\begin{array}{r} 4 \enspace \begin{array}{rrr} 1 & -6 & 8 \\ & 4 & -8 \\ \hline 1 & -2 & 0 \end{array} \end{array}$$

Since the remainder is 0, $x - 4$ is a factor of $x^2 - 6x + 8$: $x^2 - 6x + 8 = (x - 4)(x - 2)$.

79. Divide $w^3 - 27$ by $w - 3$:

$$\begin{array}{r} 3 \enspace \begin{array}{rrrr} 1 & 0 & 0 & -27 \\ & 3 & 9 & 27 \\ \hline 1 & 3 & 9 & 0 \end{array} \end{array}$$

Since the remainder is zero, $w - 3$ is a factor of $w^2 - 27$: $w^3 - 27 = (w - 3)(w^2 + 3w + 9)$.

81. Divide $2x^3 - 3x^2 - 4x + 7$ by $2x - 3$ using long division to get a quotient of $x^2 - 2$ and a remainder of 1. By the remainder theorem, $2x - 3$ is not a factor of $2x^3 - 3x^2 - 4x + 7$.

83. Divide $y^3 - 4y^2 + 6y - 4$ by $y - 2$:

$$\begin{array}{r} 2 \enspace \begin{array}{rrrr} 1 & -4 & 6 & -4 \\ & 2 & -4 & 4 \\ \hline 1 & -2 & 2 & 0 \end{array} \end{array}$$

Since the remainder is 0, $y - 2$ is a factor of $y^3 - 4y^2 + 6y - 4$ and
$y^3 - 4y^2 + 6y - 4 = (y - 2)(y^2 - 2y + 2)$.

85. Divide $x^2 - 5x - 9$ by $x - 3$:

$$\begin{array}{r} 3 \enspace \begin{array}{rrr} 1 & -5 & -9 \\ & 3 & -6 \\ \hline 1 & -2 & -15 \end{array} \end{array}$$

By the remainder theorem $P(3) = -15$.

87. Divide $4y^3 - 6y + 7$ by $y + 1$:

$$\begin{array}{r} -1 \enspace \begin{array}{rrrr} 4 & 0 & -6 & 7 \\ & -4 & 4 & 2 \\ \hline 4 & -4 & -2 & 9 \end{array} \end{array}$$

By the remainder theorem $P(-1) = 9$.

89. Divide $-w^3 + 5w^2 + 3w$ by $w - 4$:

$$\begin{array}{r} 4 \enspace \begin{array}{rrrr} -1 & 5 & 3 & 0 \\ & -4 & 4 & 28 \\ \hline -1 & 1 & 7 & 28 \end{array} \end{array}$$

By the remainder theorem $P(4) = 28$.

91. a) Divide $0.03x^2 + 300x$ by x to get $AC(x) = 0.03x + 300$.
b) $AC(x)$ is not constant, it is a linear function.
c) Because $AC(x)$ is very close to 300 for x less than 15, the graph looks horizontal.

93. Divide $x^2 - 1$ by $x + 1$ to get $x - 1$. The width is $x - 1$ ft.

95. $V = \dfrac{10(30^3 - 20^3)}{3(30 - 20)} \approx 6{,}333.3 \text{ m}^3$

97. The answer is wrong because the degree of the remainder $9x + 7$ is not less than the degree of the divisor $x - 3$.

6.6 WARM-UPS

1. True, because that will eliminate all denominators.

2. False, we should multiply each side by $6x$.

3. False, extraneous roots are real numbers that are not roots to the equation.

4. True, $x^2 - 4$ is the LCD.

5. False, $-\frac{1}{2}$ cannot be a solution because it causes $2x + 1$ to be 0.

6. True, because the equation is equivalent to $2x = 15$.

7. False, because the extremes-means property is only applied to equations of the type $\frac{a}{b} = \frac{c}{d}$.

8. False, because $x^2 = x$ has two solutions, 0 and 1.

9. False, the solution set is $\left\{ \frac{3}{2}, -\frac{4}{3} \right\}$.

10. True, because of the extremes-means property.

6.6 EXERCISES

1. The first step is to multiply each side of the equation by the LCD.

3. A proportion is an equation expressing equality of two rational expressions.

5. In $a/b = c/d$ the extremes are a and d.

7.
$$\frac{1}{x} + \frac{1}{6} = \frac{1}{8}$$
$$24x\left(\frac{1}{x} + \frac{1}{6}\right) = 24x\left(\frac{1}{8}\right)$$
$$24 + 4x = 3x$$
$$x = -24$$
Solution set: $\{-24\}$

9.
$$\frac{2}{3x} + \frac{1}{15x} = \frac{1}{2}$$
$$30x\left(\frac{2}{3x} + \frac{1}{15x}\right) = 30x\left(\frac{1}{2}\right)$$
$$20 + 2 = 15x$$
$$22 = 15x$$
$$\frac{22}{15} = x$$
Solution set: $\left\{ \frac{22}{15} \right\}$

11.
$$\frac{3}{x-2} + \frac{5}{x} = \frac{10}{x}$$
$$x(x-2)\left(\frac{3}{x-2} + \frac{5}{x}\right) = x(x-2)\left(\frac{10}{x}\right)$$
$$3x + 5x - 10 = 10x - 20$$
$$8x - 10 = 10x - 20$$

$$10 = 2x$$
$$5 = x$$
Solution set: $\{5\}$

13.
$$\frac{x}{x-2} + \frac{3}{x} = 2$$
$$x(x-2)\left(\frac{x}{x-2} + \frac{3}{x}\right) = x(x-2)2$$
$$x^2 + 3x - 6 = 2x^2 - 4x$$
$$-x^2 + 7x - 6 = 0$$
$$x^2 - 7x + 6 = 0$$
$$(x-6)(x-1) = 0$$
$$x - 6 = 0 \quad \text{or} \quad x - 1 = 0$$
$$x = 6 \quad \text{or} \quad x = 1$$
Solution set: $\{1, 6\}$

15.
$$\frac{100}{x} = \frac{150}{x+5} - 1$$
$$x(x+5)\left(\frac{100}{x}\right) = x(x+5)\left(\frac{150}{x+5} - 1\right)$$
$$100x + 500 = 150x - x(x+5)$$
$$100x + 500 = 150x - x^2 - 5x$$
$$x^2 - 45x + 500 = 0$$
$$(x-25)(x-20) = 0$$
$$x = 25 \quad \text{or} \quad x = 20$$
Solution set: $\{20, 25\}$

17.
$$\frac{3x-5}{x-1} = 2 - \frac{2x}{x-1}$$
$$(x-1)\left(\frac{3x-5}{x-1}\right) = (x-1)\left(2 - \frac{2x}{x-1}\right)$$
$$3x - 5 = 2x - 2 - 2x$$
$$3x - 5 = -2$$
$$3x = 3$$
$$x = 1$$
Since replacing x by 1 gives 0 in the denominator, 1 is an extraneous root. The solution set is $\emptyset$.

19.
$$x + 1 + \frac{2x-5}{x-5} = \frac{x}{x-5}$$
$$(x-5)\left(x + 1 + \frac{2x-5}{x-5}\right) = (x-5)\frac{x}{x-5}$$
$$x^2 - 4x - 5 + 2x - 5 = x$$
$$x^2 - 3x - 10 = 0$$
$$(x-5)(x+2) = 0$$
$$x = 5 \quad \text{or} \quad x = -2$$
Since $x - 5$ is in the denominator, 5 is an extraneous root. The solution set is $\{-2\}$

21.
$$5 + \frac{9}{x-2} = 2 + \frac{x+7}{x-2}$$
$$(x-2)\left(5 + \frac{9}{x-2}\right) = (x-2)\left(2 + \frac{x+7}{x-2}\right)$$
$$5x - 10 + 9 = 2x - 4 + x + 7$$
$$2x = 4$$
$$x = 2$$

Since replacing x by 2 gives 0 in the denominator, 2 is an extraneous root. The solution set is $\emptyset$.

23. $\dfrac{2}{x+2} + \dfrac{x}{x-3} + \dfrac{1}{x^2-x-6} = 0$

Multiply each side by $(x+2)(x-3)$.

$(x-3)2 + (x+2)x + 1 = 0$

$2x - 6 + x^2 + 2x + 1 = 0$

$x^2 + 4x - 5 = 0$

$(x+5)(x-1) = 0$

$x = -5 \text{ or } x = 1$

The solution set is $\{-5, 1\}$.

25. $\dfrac{2}{x} = \dfrac{3}{4}$

$3x = 8$

$x = \dfrac{8}{3}$

Solution set: $\left\{\dfrac{8}{3}\right\}$

27. $\dfrac{a}{3} = \dfrac{-1}{4}$

$4a = -3$

$a = -\dfrac{3}{4}$

Solution set: $\left\{-\dfrac{3}{4}\right\}$

29. $\dfrac{-5}{7} = \dfrac{2}{x}$

$-5x = 14$

$x = -\dfrac{14}{5}$

Solution set: $\left\{-\dfrac{14}{5}\right\}$

31. $\dfrac{10}{x} = \dfrac{20}{x+20}$

$10x + 200 = 20x$

$-10x = -200$

$x = 20$

Solution set: $\{20\}$

33. $\dfrac{2}{x+1} = \dfrac{x-1}{4}$

$x^2 - 1 = 8$

$x^2 - 9 = 0$

$(x-3)(x+3) = 0$

$x = 3 \text{ or } x = -3$

Solution set: $\{-3, 3\}$

35. $\dfrac{x}{6} = \dfrac{5}{x-1}$

$x^2 - x = 30$

$x^2 - x - 30 = 0$

$(x-6)(x+5) = 0$

$x = 6 \text{ or } x = -5$

Solution set: $\{-5, 6\}$

37. $\dfrac{x+7}{x+4} = \dfrac{x+1}{x-2}$

$x^2 + 5x - 14 = x^2 + 5x + 4$

$-14 = 4$

Solution set: $\emptyset$

39. $\dfrac{x-2}{x-3} = \dfrac{x+5}{x+2}$

$(x-2)(x+2) = (x-3)(x+5)$

$x^2 - 4 = x^2 + 2x - 15$

$-2x = -11$

$x = \dfrac{11}{2}$

Solution set: $\left\{\dfrac{11}{2}\right\}$

41. $\dfrac{3w}{3w-5} = \dfrac{w}{w+2}$

$3w^2 + 6w = 3w^2 - 5w$

$6w = 5w$

$w = 0$

The solution set is $\{0\}$.

43. $\dfrac{a}{9} = \dfrac{4}{a}$

$a^2 = 36$

$a^2 - 36 = 0$

$(a-6)(a+6) = 0$

$a = 6 \text{ or } a = -6$

Solution set: $\{-6, 6\}$

45. $\dfrac{1}{2x-4} + \dfrac{1}{x-2} = \dfrac{1}{4}$

$4(x-2)\left(\dfrac{1}{2(x-2)} + \dfrac{1}{x-2}\right)$
$= 4(x-2) \cdot \dfrac{1}{4}$

$2 + 4 = x - 2$

$8 = x$

Solution set: $\{8\}$

47. $\dfrac{x-2}{4} = \dfrac{x-2}{x}$

$x^2 - 2x = 4x - 8$

$x^2 - 6x + 8 = 0$

$(x-2)(x-4) = 0$

$x = 2 \text{ or } x = 4$

Solution set: $\{2, 4\}$

49. $\dfrac{5}{2x+4} - \dfrac{1}{x-1} = \dfrac{3}{x+2}$

$2(x+2)(x-1)\left(\dfrac{5}{2(x+2)} - \dfrac{1}{x-1}\right)$
$= 2(x+2)(x-1)\left(\dfrac{3}{x+2}\right)$

$5x - 5 - 2(x+2) = 6(x-1)$

$5x - 5 - 2x - 4 = 6x - 6$

$3x - 9 = 6x - 6$

$$-3x = 3$$
$$x = -1$$

Solution set: $\{-1\}$

51.
$$\frac{5}{x-3} = \frac{x}{x-3}$$
$$5x - 15 = x^2 - 3x$$
$$-x^2 + 8x - 15 = 0$$
$$x^2 - 8x + 15 = 0$$
$$(x-3)(x-5) = 0$$
$$x = 3 \text{ or } x = 5$$

Since replacing x by 3 causes 0 to appear in a denominator, 3 is an extraneous root. The solution set is $\{5\}$.

53.
$$\frac{w}{6} = \frac{3}{2w}$$
$$2w^2 = 18$$
$$w^2 = 9$$
$$w^2 - 9 = 0$$
$$(w-3)(w+3) = 0$$
$$w = 3 \text{ or } w = -3$$

Solution set: $\{-3, 3\}$

55.
$$\frac{5}{4x-2} - \frac{1}{1-2x} = \frac{7}{3x+6}$$

$$6(2x-1)(x+2)\left(\frac{5}{2(2x-1)} - \frac{-1}{2x-1}\right)$$

$$= 6(2x-1)(x+2)\left(\frac{7}{3(x+2)}\right)$$
$$15(x+2) + 6(x+2) = 14(2x-1)$$
$$15x + 30 + 6x + 12 = 28x - 14$$
$$21x + 42 = 28x - 14$$
$$-7x = -56$$
$$x = 8$$

Solution set: $\{8\}$

57.
$$\frac{5}{x} = \frac{2}{5}$$
$$2x = 25$$
$$x = \frac{25}{2}$$

Solution set: $\left\{\frac{25}{2}\right\}$

59.
$$\frac{x}{x-2} - \frac{x+2}{x^2 - 2x} = \frac{1}{x}$$
$$x(x-2)\left(\frac{x}{x-2} - \frac{x+2}{x(x-2)}\right)$$
$$= x(x-2) \cdot \frac{1}{x}$$
$$x^2 - x - 2 = x - 2$$
$$x^2 - 2x = 0$$
$$x(x-2) = 0$$
$$x = 0 \text{ or } x = 2$$

Both solutions are extraneous. The solution set is $\emptyset$.

61.
$$\frac{5}{x^2-9} + \frac{2}{x+3} = \frac{1}{x-3}$$
$$(x-3)(x+3)\left(\frac{5}{x^2-9} + \frac{2}{x+3}\right)$$
$$= (x-3)(x+3)\left(\frac{1}{x-3}\right)$$
$$5 + 2x - 6 = x + 3$$
$$x = 4$$

Solution set: $\{4\}$

63.
$$\frac{9}{x^3-1} - \frac{1}{x-1} = \frac{2}{x^2+x+1}$$

$$(x-1)(x^2+x+1)\left(\frac{9}{x^3-1} - \frac{1}{x-1}\right)$$
$$= (x-1)(x^2+x+1)\left(\frac{2}{x^2+x+1}\right)$$
$$9 - (x^2+x+1) = 2(x-1)$$
$$9 - x^2 - x - 1 = 2x - 2$$
$$-x^2 - 3x + 10 = 0$$
$$x^2 + 3x - 10 = 0$$
$$(x+5)(x-2) = 0$$
$$x = -5 \text{ or } x = 2$$

Solution set: $\{-5, 2\}$

65.
$$\frac{4}{x} = \frac{3}{4}$$
$$3x = 16$$
$$x = \frac{16}{3}$$

The solution set is $\left\{\frac{16}{3}\right\}$.

67. $\frac{4}{x} + \frac{3}{4} = \frac{4 \cdot 4}{x \cdot 4} + \frac{3x}{4x} = \frac{3x+16}{4x}$

69.
$$\frac{2}{x} - \frac{3}{4} = \frac{1}{2}$$
$$4x \cdot \frac{2}{x} - 4x \cdot \frac{3}{4} = 4x \cdot \frac{1}{2}$$
$$8 - 3x = 2x$$
$$8 = 5x$$
$$x = \frac{8}{5}$$

The solution set is $\left\{\frac{8}{5}\right\}$.

71. $\frac{2}{x} - \frac{3}{4} - \frac{1}{2} = \frac{2 \cdot 4}{x \cdot 4} - \frac{3x}{4x} - \frac{1 \cdot 2x}{2 \cdot 2x}$

$$= \frac{8 - 3x - 2x}{4x} = \frac{-5x+8}{4x}$$

73. $\frac{300,000}{250,000} = \frac{200,000}{a}$
$$300,000a = 200,000 \cdot 250,000$$
$$a = \frac{200,000 \cdot 250,000}{300,000}$$
$$a = \$166,666.67$$

75. $\dfrac{2}{23} = \dfrac{12}{p}$

$$2p = 12 \cdot 23$$
$$p = \dfrac{12 \cdot 23}{2} = 138$$

77. Let w = the width and $w + 22$ = the length.

$$\dfrac{7}{6} = \dfrac{w + 22}{w}$$
$$7w = 6w + 132$$
$$w = 132$$
$$w + 22 = 154$$

The width is 132 cm and the length is 154 cm.

79. $1{,}000{,}000 = \dfrac{4{,}000{,}000p}{100 - p}$

$$100{,}000{,}000 - 1{,}000{,}000p = 4{,}000{,}000p$$
$$100{,}000{,}000 = 5{,}000{,}000p$$
$$p = 20$$

For \$1 million, 20% of the pollution can be cleaned up. For \$100 million, 96% of the pollution can be cleaned up.

81. a) Let s = the amount invested in stocks and $s - 20{,}000$ = the amount invested in bonds. For a wealth-building portfolio, the ratio of stocks to bonds should be 65 to 30:

$$\dfrac{65}{30} = \dfrac{s}{s - 20{,}000}$$
$$65s - 1{,}300{,}000 = 30s$$
$$35s = 1{,}300{,}000$$
$$s = \$37{,}142.86$$
$$s - 20{,}000 = \$17{,}142.86$$

She invested \$17,142.86 in bonds.

b) Since the stock investment is 65% of her bonus, b, we have $0.65b = 37{,}142.86$ or $b = 37{,}142.86/0.65 = \$57{,}142.86$.

83. To solve the equation, multiply each side by the LCD. To find the sum, build up each rational expression so that its denominator is the LCD.

6.7 WARM-UPS

1. False, t must not appear on both sides of the formula.

2. True, pqs is the LCD.

3. False, the price is $\dfrac{x}{50}$ dollars per pound.

4. True, rate is distance divided by time.

5. True, because this is his rate.

6. True, this is the rate at which he is working.

7. False, because $y - 1 = x$.

8. True, multiply by B and divide by m.

9. False, if $a = \dfrac{x}{y}$ then $y = \dfrac{x}{a}$.

10. False, the correct equation is $x - 3 = y$.

6.7 EXERCISES

1. $\dfrac{y - 3}{x - 2} = 5$

$$(x - 2)\left(\dfrac{y - 3}{x - 2}\right) = (x - 2)(5)$$
$$y - 3 = 5x - 10$$
$$y = 5x - 7$$

3. $\dfrac{y + 1}{x - 6} = -\dfrac{1}{3}$

$$(x - 6)\left(\dfrac{y + 1}{x - 6}\right) = -\dfrac{1}{3}(x - 6)$$
$$y + 1 = -\dfrac{1}{3}x + 2$$
$$y = -\dfrac{1}{3}x + 1$$

5. $\dfrac{y - a}{x - b} = m$

$$(x - b)\left(\dfrac{y - a}{x - b}\right) = m(x - b)$$
$$y - a = mx - mb$$
$$y = mx - bm + a$$

7. $\dfrac{y - 2}{x + 5} = -\dfrac{7}{3}$

$$(x + 5)\left(\dfrac{y - 2}{x + 5}\right) = -\dfrac{7}{3}(x + 5)$$
$$y - 2 = -\dfrac{7}{3}x - \dfrac{35}{3}$$
$$y = -\dfrac{7}{3}x - \dfrac{35}{3} + \dfrac{6}{3}$$
$$y = -\dfrac{7}{3}x - \dfrac{29}{3}$$

9. $M = \dfrac{F}{f}$

$$f \cdot M = f \cdot \dfrac{F}{f}$$
$$fM = F$$
$$f = \dfrac{F}{M}$$

11. $A = \dfrac{\pi}{4} \cdot D^2$

$$4 \cdot A = 4 \cdot \dfrac{\pi}{4} \cdot D^2$$
$$4A = \pi D^2$$
$$D^2 = \dfrac{4A}{\pi}$$

13. $F = k\dfrac{m_1 m_2}{r^2}$

$$Fr^2 = km_1 m_2$$
$$m_1 = \dfrac{Fr^2}{km_2}$$

15.
$$\frac{1}{p} + \frac{1}{q} = \frac{1}{f}$$
$$pqf\left(\frac{1}{p} + \frac{1}{q}\right) = pqf \cdot \frac{1}{f}$$
$$qf + pf = pq$$
$$qf - pq = -pf$$
$$q(f - p) = -pf$$
$$q = \frac{-pf}{f - p}$$
$$q = \frac{pf}{p - f}$$

17.
$$e^2 = 1 - \frac{b^2}{a^2}$$
$$a^2\left(e^2\right) = a^2\left(1 - \frac{b^2}{a^2}\right)$$
$$a^2 e^2 = a^2 - b^2$$
$$a^2 e^2 - a^2 = -b^2$$
$$a^2(e^2 - 1) = -b^2$$
$$a^2 = \frac{-b^2}{e^2 - 1} = \frac{b^2}{1 - e^2}$$

19.
$$\frac{P_1 V_1}{T_1} = \frac{P_2 V_2}{T_2}$$
$$T_1 T_2\left(\frac{P_1 V_1}{T_1}\right) = T_1 T_2\left(\frac{P_2 V_2}{T_2}\right)$$
$$T_2 P_1 V_1 = T_1 P_2 V_2$$
$$\frac{T_2 P_1 V_1}{P_2 V_2} = \frac{T_1 P_2 V_2}{P_2 V_2}$$
$$T_1 = \frac{P_1 V_1 T_2}{P_2 V_2}$$

21.
$$V = \frac{4}{3}\pi r^2 h$$
$$3V = 3 \cdot \frac{4}{3}\pi r^2 h$$
$$3V = 4\pi r^2 h$$
$$\frac{3V}{4\pi r^2} = \frac{4\pi r^2 h}{4\pi r^2}$$
$$h = \frac{3V}{4\pi r^2}$$

23.
$$10 = \frac{5}{f}$$
$$10f = 5$$
$$f = \frac{1}{2}$$

25.
$$6\pi = \frac{\pi}{4}D^2$$
$$\frac{4}{\pi} \cdot 6\pi = \frac{4}{\pi} \cdot \frac{\pi}{4}D^2$$
$$24 = D^2$$

27.
$$32 = k\frac{6 \cdot 8}{4^2}$$

$$32 = 3k$$
$$k = \frac{32}{3}$$

29.
$$\frac{1}{p} + \frac{1}{1.7} = \frac{1}{2.3}$$
$$\frac{1}{p} \approx -0.15345$$
$$p \approx \frac{1}{-0.15345} \approx -6.517$$

31.
$$(0.62)^2 = 1 - \frac{(3.5)^2}{a^2}$$

$$-0.6156 = -\frac{12.25}{a^2}$$
$$a^2 = \frac{12.25}{0.6156} \approx 19.899$$

33.
$$25.6 = \frac{4}{3}\pi r^2(3.2)$$
$$r^2 = \frac{25.6(3)}{4\pi(3.2)} \approx 1.910$$

35. Let x = her walking speed and $x + 10$ = her riding speed. Since $T = D/R$, her time to school is $7/(x + 10)$ and her time to the post office is $2/x$. Since the times are equal, we can write
$$\frac{2}{x} = \frac{7}{x + 10}$$
$$7x = 2x + 20$$
$$5x = 20$$
$$x = 4$$
She walks 4 mph.

37. Let x = the speed of each. Since $T = D/R$, Patrick's time is $\frac{40}{x}$ and Guy's time is $\frac{60}{x}$. Since Guy's time is $\frac{1}{5}$ hr longer than Patrick's we can write the equation

$$\frac{60}{x} - \frac{1}{5} = \frac{40}{x}$$
$$5x\left(\frac{60}{x} - \frac{1}{5}\right) = 5x \cdot \frac{40}{x}$$
$$300 - x = 200$$
$$100 = x$$

They are both driving 100 mph. Patrick takes $\frac{40}{100} = \frac{2}{5}$ hr = 24 minutes to get to work and Guy takes 12 minutes longer or 36 minutes. If Guy's claim is correct, then they are both driving 100 mph. So it is not likely that Guy's claim is correct.

131

39. Let $x =$ his walking speed and $x + 6 =$ his running speed. His time walking was $\frac{1}{x}$ hours and his time running was $\frac{5}{x+6}$ hours. Since his total time was 3/4 of an hour, we can write the equation

$$\frac{1}{x} + \frac{5}{x+6} = \frac{3}{4}$$
$$4x(x+6)\left(\frac{1}{x} + \frac{5}{x+6}\right) = 4x(x+6)\frac{3}{4}$$
$$4x + 24 + 20x = 3x^2 + 18x$$
$$-3x^2 + 6x + 24 = 0$$
$$x^2 - 2x - 8 = 0$$
$$(x-4)(x+2) = 0$$
$$x = 4 \text{ or } x = -2$$

Disregard the negative solution. Use $x = 4$ to find that his running speed was 10 mph.

41. Let $x =$ the number of hours for the smaller pump to drain the pool working alone. In one hour of operation, the larger pump drains $\frac{1}{3}$ of the pool, the smaller pump drains $\frac{1}{x}$ of the pool, and together $\frac{1}{2}$ of the pool is emptied.

$$\frac{1}{3} + \frac{1}{x} = \frac{1}{2}$$
$$6x\left(\frac{1}{3} + \frac{1}{x}\right) = 6x\left(\frac{1}{2}\right)$$
$$2x + 6 = 3x$$
$$6 = x$$

The smaller pump would take 6 hours to drain the pool by itself.

43. Let $x =$ the number of minutes to fill the tub with the drain left open. In one minute, the faucet fills $\frac{1}{10}$ of the tub, the drain takes $\frac{1}{12}$ of the tub, but the tub is $\frac{1}{x}$ full.

$$\frac{1}{10} - \frac{1}{12} = \frac{1}{x}$$
$$60x\left(\frac{1}{10} - \frac{1}{12}\right) = 60x \cdot \frac{1}{x}$$
$$6x - 5x = 60$$
$$x = 60$$

The tub is filled in 60 minutes.

45. Let $x =$ their time working together. Since Gina takes 90 minutes and Hilda works twice as fast, Hilda can do the job alone in 45 minutes. Hilda does $\frac{1}{45}$ of the job per minute and Gina does $\frac{1}{90}$ or the job per minute. Since together they do $\frac{1}{x}$ of the job per minute, we can write the following equation.

$$\frac{1}{45} + \frac{1}{90} = \frac{1}{x}$$
$$90x\left(\frac{1}{45} + \frac{1}{90}\right) = 90x \cdot \frac{1}{x}$$
$$2x + x = 90$$
$$3x = 90$$
$$x = 30$$

It will take them 30 minutes working together.

47. Let $x =$ the number of pounds of apples and $x + 2 =$ the number of pounds of oranges. The apples sell for $\frac{8.80}{x}$ dollars per pound and the oranges sell for $\frac{5.28}{x+2}$. The apples cost twice as much per pound is expressed as

$$\frac{8.80}{x} = 2\left(\frac{5.28}{x+2}\right)$$
$$\frac{8.80}{x} = \frac{10.56}{x+2}$$
$$8.80x + 17.60 = 10.56x$$
$$17.60 = 1.76x$$
$$10 = x$$

She bought 10 pounds of apples and 12 pounds of oranges.

49.
$$\frac{1}{2} = \frac{1}{3} + \frac{1}{R_2}$$
$$6R_2\left(\frac{1}{2}\right) = 6R_2\left(\frac{1}{3} + \frac{1}{R_2}\right)$$
$$3R_2 = 2R_2 + 6$$
$$R_2 = 6 \text{ ohms}$$

51. a) Let $x =$ the number in the original group. Their cost is $\frac{24,000}{x}$ per person. For 40 more people, the cost is $\frac{24,000}{x+40}$. We have

$$\frac{24,000}{x} = \frac{24,000}{x+40} + 100$$
$$x(x+40)\frac{24,000}{x} = x(x+40)\left(\frac{24,000}{x+40} + 100\right)$$
$$24,000x + 960,000$$
$$\qquad = 24,000x + 100x^2 + 4000x$$
$$100x^2 + 4000x - 960,000 = 0$$
$$x^2 + 40x - 9600 = 0$$
$$(x - 80)(x + 120) = 0$$
$$x = 80 \quad \text{or} \quad x = -120$$

There are 80 people in the initial group.

b) $C(n) = \frac{24,000}{n}$ where n is the number of people sharing the cost and $C(n)$ is the cost per person in dollars.

53. Let $x =$ the number of days for both dogs to eat a 50 pound bag together. Muffy eats at a rate of $\frac{25}{28}$ pounds per day, Missy eats at the rate of $\frac{25}{23}$ pounds per day, and together they eat at a rate of $\frac{50}{x}$ pounds per day.

$$\frac{25}{28} + \frac{25}{23} = \frac{50}{x}$$
$$\frac{1}{28} + \frac{1}{23} = \frac{2}{x}$$
$$28 \cdot 23x\left(\frac{1}{28} + \frac{1}{23}\right) = 28 \cdot 23x\left(\frac{2}{x}\right)$$
$$23x + 28x = 1288$$
$$51x = 1288$$
$$x = \frac{1288}{51} \approx 25.255$$

It would take 25.255 days for them to eat 50 pounds of dog food together.

Enriching Your Mathematical Word Power
1. b   **2.** d   **3.** b   **4.** d   **5.** b
6. a   **7.** a   **8.** d   **9.** a   **10.** b
11. d   **12.** d   **13.** a   **14.** c   **15.** d

CHAPTER 6 REVIEW

1. If $x = 1$, then $3x - 3 = 0$. So the domain is $\{x \mid x \neq 1\}$ or $(-\infty, 1) \cup (1, \infty)$.
3. Solve $x^2 - x - 2 = 0$
$$(x - 2)(x + 1) = 0$$
$$x - 2 = 0 \text{ or } x + 1 = 0$$
$$x = 2 \text{ or } \quad x = -1$$
The domain is $\{x \mid x \neq -1 \text{ and } x \neq 2\}$ or $(-\infty, -1) \cup (-1, 2) \cup (2, \infty)$.
5. $\dfrac{a^3bc^3}{a^5b^2c} = \dfrac{c^2}{a^2b}$

7. $\dfrac{68x^3}{51xy} = \dfrac{2 \cdot 2 \cdot 17x^3}{3 \cdot 17xy} = \dfrac{4x^2}{3y}$

9. $\dfrac{a^3b^2}{b^3a} \cdot \dfrac{ab - b^2}{ab - a^2} = \dfrac{a^3b^2b(a - b)}{b^3a(-a)(a - b)} = -a$

11. $\dfrac{w - 4}{3w} \div \dfrac{2w - 8}{9w} = \dfrac{w - 4}{3w} \cdot \dfrac{3 \cdot 3w}{2(w - 4)}$
$$= \dfrac{3}{2}$$

13. Since $6x = 2 \cdot 3x$, $3x - 6 = 3(x - 2)$, and $x^2 - 2x = x(x - 2)$, the LCM $= 6x(x - 2)$.

15. Since $6ab^3 = 2 \cdot 3ab^3$ and $4a^5b^2 = 2 \cdot 2a^5b^2$, the LCM $= 2 \cdot 2 \cdot 3a^5b^3 = 12a^5b^3$.

17. $\dfrac{3}{2x - 6} + \dfrac{1}{x^2 - 9}$

$$= \dfrac{3(x + 3)}{2(x - 3)(x + 3)} + \dfrac{1 \cdot 2}{(x - 3)(x + 3)2}$$

$$= \dfrac{3x + 9}{2(x - 3)(x + 3)} + \dfrac{2}{2(x - 3)(x + 3)}$$

$$= \dfrac{3x + 11}{2(x - 3)(x + 3)}$$

19. $\dfrac{w}{ab^2} - \dfrac{5}{a^2b} = \dfrac{w \cdot a}{ab^2 \cdot a} - \dfrac{5 \cdot b}{a^2b \cdot b}$
$$= \dfrac{aw - 5b}{a^2b^2}$$

21. $\dfrac{\frac{3}{2x} - \frac{4}{5x}}{\frac{1}{3} - \frac{2}{x}} = \dfrac{30x\left(\frac{3}{2x} - \frac{4}{5x}\right)}{30x\left(\frac{1}{3} - \frac{2}{x}\right)}$

$$= \dfrac{45 - 24}{10x - 60} = \dfrac{21}{10x - 60}$$

23. $\dfrac{\frac{1}{y - 2} - 3}{\frac{5}{y - 2} + 4} = \dfrac{(y - 2)\left(\frac{1}{y - 2} - 3\right)}{(y - 2)\left(\frac{5}{y - 2} + 4\right)}$

$$= \dfrac{1 - 3(y - 2)}{5 + 4(y - 2)} = \dfrac{7 - 3y}{4y - 3}$$

25. $\dfrac{a^{-2} - b^{-3}}{a^{-1}b^{-2}} = \dfrac{a^2b^3(a^{-2} - b^{-3})}{a^2b^3(a^{-1}b^{-2})}$
$$= \dfrac{b^3 - a^2}{ab}$$

27.
$$\begin{array}{r}
x^2 + 3x - 5 \\
x - 2 \overline{)\, x^3 + x^2 - 11x + 10} \\
\underline{x^3 - 2x^2} \\
3x^2 - 11x \\
\underline{3x^2 - 6x} \\
-5x + 10 \\
\underline{-5x + 10} \\
0
\end{array}$$

Quotient: $x^2 + 3x - 5$ Remainder: 0

29.

$$m+1 \overline{\smash{\big)}\, m^4 + 0m^3 + 0m^2 + 0m - 1}$$

$$m^3 - m^2 + m - 1$$

$$\underline{m^4 + m^3}$$
$$-m^3 + 0m^2$$
$$\underline{-m^3 - m^2}$$
$$m^2 + 0m$$
$$\underline{m^2 + m}$$
$$-m - 1$$
$$\underline{-m - 1}$$
$$0$$

Quotient: $m^3 - m^2 + m - 1$ Remainder: 0

31.

$$a^3 - 2 \overline{\smash{\big)}\, a^9 + 0a^6 + 0a^3 - 8}$$

$$a^6 + 2a^3 + 4$$

$$\underline{a^9 - 2a^6}$$
$$2a^6 + 0a^3$$
$$\underline{2a^6 - 4a^3}$$
$$4a^3 - 8$$
$$\underline{4a^3 - 8}$$
$$0$$

Quotient: $a^6 + 2a^3 + 4$ Remainder: 0

33. $\dfrac{3m^3 + 6m^2 - 18m}{3m}$

$$= \frac{3m^3}{3m} + \frac{6m^2}{3m} - \frac{18m}{3m}$$
$$= m^2 + 2m - 6$$

Quotient: $m^2 + 2m - 6$ Remainder: 0

35.

$$\begin{array}{c|ccc} 1 & 1 & 0 & -5 \\ & & 1 & 1 \\ \hline & 1 & 1 & -4 \end{array}$$

Quotient is $x + 1$ and remainder is -4.

$$\frac{x^2 - 5}{x - 1} = x + 1 + \frac{-4}{x - 1}$$

37.

$$\begin{array}{c|cc} 2 & 3 & 0 \\ & & 6 \\ \hline & 3 & 6 \end{array}$$

Quotient is 3 and remainder is 6.

$$\frac{3x}{x - 2} = 3 + \frac{6}{x - 2}$$

39. When $x^3 - 2x^2 + 3x + 22$ is divided by $x + 2$ the quotient is $x^2 - 4x + 11$ and the remainder is 0. So the first polynomial is a factor of the second.

41. If $x^3 - x - 120$ is divided by $x - 5$ the quotient is $x^2 + 5x + 24$ and remainder is 0. So the first polynomial is a factor of the second.

43. When $x^3 + x^2 - 3$ is divided by $x - 1$ the quotient is $x^2 + 2x + 2$ and the remainder is -1. So the first polynomial is not a factor of the second.

45. When $x^4 + x^3 + 5x^2 + 2x + 6$ is divided by $x^2 + 2$ the quotient is $x^2 + x + 3$ and the remainder is 0. So the first polynomial is a factor of the second.

47.

$$\frac{-3}{8} = \frac{2}{x}$$
$$-3x = 16$$
$$x = -\frac{16}{3}$$

Solution set: $\left\{ -\dfrac{16}{3} \right\}$

49.

$$5 + \frac{x + 1}{x - 1} = 3 + \frac{5x - 3}{x - 1}$$
$$(x - 1)\left(5 + \frac{x + 1}{x - 1} \right) = (x - 1)\left(3 + \frac{5x - 3}{x - 1} \right)$$
$$5x - 5 + x + 1 = 3x - 3 + 5x - 3$$
$$6x - 4 = 8x - 6$$
$$2 = 2x$$
$$1 = x$$

The solution set is the empty set, $\varnothing$.

51.

$$\frac{15}{a^2 - 25} + \frac{1}{a - 5} = \frac{6}{a + 5}$$
$$(a - 5)(a + 5)\left(\frac{15}{a^2 - 25} + \frac{1}{a - 5} \right)$$
$$= (a - 5)(a + 5)\left(\frac{6}{a + 5} \right)$$
$$15 + a + 5 = 6a - 30$$
$$-5a = -50$$
$$a = 10$$

Solution set: $\{10\}$

53.

$$\frac{y - b}{m} = x$$
$$m\left(\frac{y - b}{m} \right) = mx$$
$$y - b = mx$$
$$y = mx + b$$

55.

$$F = \frac{mv^2}{r}$$
$$r \cdot F = r\left(\frac{mv^2}{r} \right)$$
$$rF = mv^2$$
$$m = \frac{Fr}{v^2}$$

57.
$$A = \tfrac{2}{3}\pi rh$$
$$3 \cdot A = 3 \cdot \tfrac{2}{3}\pi rh$$
$$3A = 2\pi rh$$
$$\frac{3A}{2\pi h} = \frac{2\pi rh}{2\pi h}$$
$$r = \frac{3A}{2\pi h}$$

59.
$$\frac{y+3}{x-7} = 2$$
$$(x-7)\left(\frac{y+3}{x-7}\right) = (x-7)2$$
$$y+3 = 2x - 14$$
$$y = 2x - 17$$

61.
$$\frac{5x}{3x^2 y} + \frac{7a^2}{6a^2 x} = \frac{5}{3xy} + \frac{7}{6x}$$
$$= \frac{5 \cdot 2}{3xy \cdot 2} + \frac{7 \cdot y}{6x \cdot y} = \frac{10 + 7y}{6xy}$$

63.
$$\frac{5}{a-5} - \frac{3}{-a-5}$$
$$= \frac{5}{a-5} - \frac{3}{-1(a+5)}$$
$$= \frac{5}{a-5} + \frac{3}{a+5}$$
$$= \frac{5(a+5)}{(a-5)(a+5)} + \frac{3(a-5)}{(a+5)(a-5)}$$
$$= \frac{8a+10}{(a-5)(a+5)}$$

65.
$$\frac{1}{x-2} - \frac{1}{x+2} = \frac{1}{15}$$
$$15(x-2)(x+2)\left(\frac{1}{x-2} - \frac{1}{x+2}\right)$$
$$= 15(x-2)(x+2)\left(\frac{1}{15}\right)$$
$$15x + 30 - 15(x-2) = x^2 - 4$$
$$60 = x^2 - 4$$
$$x^2 - 64 = 0$$
$$(x-8)(x+8) = 0$$
$$x = 8 \text{ or } x = -8$$
Solution set: $\{-8, 8\}$

67.
$$\frac{-3}{x+2} \cdot \frac{5x+10}{10} = \frac{-3}{x+2} \cdot \frac{5(x+2)}{2 \cdot 5}$$
$$= -\frac{3}{2}$$

69.
$$\frac{x}{-3} = \frac{-27}{x}$$
$$x^2 = 81$$
$$x^2 - 81 = 0$$
$$(x-9)(x+9) = 0$$
$$x = 9 \text{ or } x = -9$$
Solution set: $\{-9, 9\}$

71.
$$\frac{wx + wm + 3x + 3m}{w^2 - 9} \div \frac{x^2 - m^2}{w - 3}$$
$$= \frac{(w+3)(x+m)}{(w-3)(w+3)} \cdot \frac{w-3}{(x-m)(x+m)}$$
$$= \frac{1}{x-m}$$

73.
$$\frac{5}{a^2 - 25} + \frac{3}{a^2 - 4a - 5}$$
$$= \frac{5(a+1)}{(a-5)(a+5)(a+1)}$$
$$+ \frac{3(a+5)}{(a-5)(a+1)(a+5)}$$
$$= \frac{8a + 20}{(a-5)(a+5)(a+1)}$$

75.
$$\frac{-7}{2a^2 - 18} - \frac{4}{a^2 + 5a + 6}$$
$$= \frac{-7(a+2)}{2(a-3)(a+3)(a+2)}$$
$$- \frac{4 \cdot 2(a-3)}{(a+2)(a+3)2(a-3)}$$
$$= \frac{-7a - 14 - (8a - 24)}{2(a+2)(a+3)(a-3)}$$
$$= \frac{-15a + 10}{2(a+2)(a+3)(a-3)}$$

77.
$$\frac{7}{a^2 - 1} + \frac{2}{1-a} = \frac{1}{a+1}$$
$$= (a-1)(a+1)\left(\frac{7}{a^2-1} + \frac{2}{1-a}\right)$$
$$= (a-1)(a+1)\left(\frac{1}{a+1}\right)$$
$$7 - 2(a+1) = a - 1$$
$$-2a + 5 = a - 1$$
$$-3a = -6$$
$$a = 2$$
Solution set: $\{2\}$

79.
$$\frac{2x}{x-3} + \frac{3}{x-2} = \frac{6}{(x-2)(x-3)}$$
$$(x-2)(x-3)\left(\frac{2x}{x-3} + \frac{3}{x-2}\right)$$
$$= (x-2)(x-3)\frac{6}{(x-2)(x-3)}$$
$$2x(x-2) + 3(x-3) = 6$$
$$2x^2 - x - 15 = 0$$
$$(2x + 5)(x - 3) = 0$$
$$2x + 5 = 0 \text{ or } x - 3 = 0$$
$$x = -\frac{5}{2} \text{ or } x = 3$$
Since 3 is an extraneous root, the solution set is $\left\{-\frac{5}{2}\right\}$.

81. $\dfrac{x-2}{6} \div \dfrac{2-x}{2} = \dfrac{x-2}{2\cdot 3}\cdot\dfrac{2}{2-x} = -\dfrac{1}{3}$

83. $\dfrac{x-3}{x^2+3x+2}\cdot\dfrac{x^2-4}{3x-9}$

$= \dfrac{x-3}{(x+2)(x+1)}\cdot\dfrac{(x-2)(x+2)}{3(x-3)}$

$\qquad = \dfrac{x-2}{3x+3}$

85. $\dfrac{a+4}{a^3-8} - \dfrac{3}{2-a}$

$= \dfrac{a+4}{(a-2)(a^2+2a+4)} - \dfrac{-3}{a-2}$

$= \dfrac{a+4}{(a-2)(a^2+2a+4)}$

$\qquad - \dfrac{-3(a^2+2a+4)}{(a-2)(a^2+2a+4)}$

$= \dfrac{a+4+3a^2+6a+12}{a^3-8} = \dfrac{3a^2+7a+16}{a^3-8}$

87. $\dfrac{x^3-9x}{1-x^2} \div \dfrac{x^3+6x^2+9x}{x-1}$

$= \dfrac{x(x-3)(x+3)}{(1-x)(1+x)}\cdot\dfrac{x-1}{x(x+3)^2}$

$= \dfrac{3-x}{(x+1)(x+3)}$

89. $\dfrac{a^2+3a+3w+aw}{a^2+6a+8}\cdot\dfrac{a^2-aw-2w+2a}{a^2+3a-3w-aw}$

$= \dfrac{(a+w)(a+3)}{(a+2)(a+4)}\cdot\dfrac{(a+2)(a-w)}{(a-w)(a+3)}$

$= \dfrac{a+w}{a+4}$

91. $\dfrac{5}{x} - \dfrac{4}{x+2} = \dfrac{1}{5} + \dfrac{1}{5x}$

$5x(x+2)\left(\dfrac{5}{x} - \dfrac{4}{x+2}\right)$

$\qquad = 5x(x+2)\left(\dfrac{1}{5} + \dfrac{1}{5x}\right)$

$25x + 50 - 20x = x^2 + 2x + x + 2$

$\qquad 5x + 50 = x^2 + 3x + 2$

$\qquad -x^2 + 2x + 48 = 0$

$\qquad x^2 - 2x - 48 = 0$

$\qquad (x-8)(x+6) = 0$

$\qquad x = 8 \ \text{ or } \ x = -6$

Solution set: $\{-6, 8\}$

93. $\dfrac{6}{x} = \dfrac{6\cdot 3}{x\cdot 3} = \dfrac{18}{3x}$

95. $\dfrac{3}{a-b} = \dfrac{3(-1)}{(a-b)(-1)} = \dfrac{-3}{b-a}$

97. $4 = 4\cdot\dfrac{x}{x} = \dfrac{4x}{x}$

99. $5x \div \dfrac{1}{2} = 5x\cdot 2 = 10x$

101. $4a \div \dfrac{1}{3} = 4a\cdot 3 = 12a$

103. $\dfrac{a-3}{a^2-9} = \dfrac{1(a-3)}{(a-3)(a+3)} = \dfrac{1}{a+3}$

105. $\dfrac{1}{2} - \dfrac{1}{5} = \dfrac{5}{10} - \dfrac{2}{10} = \dfrac{3}{10}$

107. $\dfrac{a}{3} + \dfrac{a}{2} = \dfrac{2a}{6} + \dfrac{3a}{6} = \dfrac{5a}{6}$

109. Let $x =$ the number of reported female AIDs cases and $x + 16,000 =$ the number of reported male AIDs cases.

$$\dfrac{7}{3} = \dfrac{x+16,000}{x}$$

$$7x = 3x + 48,000$$

$$4x = 48,000$$

$$x \approx 12,000$$

$x + 16,000 = 28,000$

There were 28,000 reported male AIDs cases.

111. If $x =$ the time east of Louisville, then we also have $x =$ the time west of Louisville. His speed east of Louisville was $\dfrac{310}{x}$ and his speed west of Louisville was $\dfrac{360}{x}$. We can write an equation expressing the fact that his speed west of Louisville was 10 mph greater than east of Louisville:

$$\dfrac{360}{x} - 10 = \dfrac{310}{x}$$

$$x\left(\dfrac{360}{x} - 10\right) = x\left(\dfrac{310}{x}\right)$$

$$360 - 10x = 310$$

$$-10x = -50$$

$$x = 5$$

His journey took a total of 10 hours.

113. Let $x =$ the time for all three to make the quilt working together. Debbie makes $\dfrac{1}{2000}$ of the quilt per hour, Pat makes $\dfrac{1}{1000}$ of the quilt per hour, and Cheryl makes $\dfrac{1}{1000}$ of the quilt per hour. Working together, $\dfrac{1}{x}$ of the quilt gets done per hour:

$$\dfrac{1}{1000} + \dfrac{1}{1000} + \dfrac{1}{2000} = \dfrac{1}{x}$$

$$2000x\left(\dfrac{1}{1000} + \dfrac{1}{1000} + \dfrac{1}{2000}\right) = 2000x\cdot\dfrac{1}{x}$$

$$2x + 2x + x = 2000$$

$$5x = 2000$$

$$x = 400$$

It will take them 400 hours to make the quilt together.

CHAPTER 6 TEST

1. The solution to $4 - 3x = 0$ is $\{4/3\}$. So the domain of the rational expression is $\left\{x \mid x \neq \frac{4}{3}\right\}$ or $(-\infty, 4/3) \cup (4/3, \infty)$.

2. The solution to $x^2 - 9 = 0$ is $\{-3, 3\}$. So the domain of the rational expression is $\{x \mid x \neq 3 \text{ and } x \neq -3\}$ or $(-\infty, -3) \cup (-3, 3) \cup (3, \infty)$.

3. There is no solution to $x^2 + 9 = 0$ in the real numbers. So the domain is the set of all real numbers, $(-\infty, \infty)$.

4. $\dfrac{12a^9 b^8}{(2a^2 b^3)^3} = \dfrac{12a^9 b^8}{8a^6 b^9} = \dfrac{3a^3}{2b}$

5. $\dfrac{y^2 - x^2}{2x^2 - 4xy + 2y^2} = \dfrac{(y-x)(y+x)}{2(x-y)^2}$

$= \dfrac{-1(y+x)}{2(x-y)} = \dfrac{-x-y}{2x-2y}$

6. $\dfrac{5y}{12y} - \dfrac{4x}{9x} = \dfrac{5}{12} - \dfrac{4}{9} = \dfrac{15}{36} - \dfrac{16}{36} = -\dfrac{1}{36}$

7. $\dfrac{3}{y} + 7y = \dfrac{3}{y} + \dfrac{7y^2}{y} = \dfrac{7y^2 + 3}{y}$

8. $\dfrac{4}{a-9} - \dfrac{1}{(9-a)} = \dfrac{4}{a-9} - \dfrac{1(-1)}{(9-a)(-1)}$

$= \dfrac{4}{a-9} + \dfrac{1}{a-9} = \dfrac{5}{a-9}$

9. $\dfrac{1}{6ab^2} + \dfrac{1}{8a^2 b} = \dfrac{1 \cdot 4a}{6ab^2 \cdot 4a} + \dfrac{1 \cdot 3b}{8a^2 b \cdot 3b}$

$= \dfrac{4a}{24a^2 b^2} + \dfrac{3b}{24a^2 b^2} = \dfrac{4a + 3b}{24a^2 b^2}$

10. $\dfrac{3a^3 b}{20ab} \cdot \dfrac{2a^2 b}{9ab^3} = \dfrac{3a^3 b}{2 \cdot 2 \cdot 5ab} \cdot \dfrac{2a^2 b}{3 \cdot 3ab^3}$

$= \dfrac{a^3}{30b^2}$

11. $\dfrac{a-b}{7} \div \dfrac{b^2 - a^2}{21}$

$= \dfrac{a-b}{7} \cdot \dfrac{3 \cdot 7}{(b-a)(b+a)} = -\dfrac{3}{a+b}$

12. $\dfrac{x-3}{x-1} \div (x^2 - 2x - 3)$

$= \dfrac{x-3}{x-1} \cdot \dfrac{1}{(x-3)(x+1)} = \dfrac{1}{x^2 - 1}$

13. $\dfrac{2}{x^2 - 4} - \dfrac{6}{x^2 - 3x - 10}$

$= \dfrac{2}{(x-2)(x+2)} - \dfrac{6}{(x-5)(x+2)}$

$= \dfrac{2(x-5)}{(x-2)(x+2)(x-5)}$

$\qquad - \dfrac{6(x-2)}{(x-5)(x+2)(x-2)}$

$= \dfrac{2x - 10 - (6x - 12)}{(x+2)(x-2)(x-5)}$

$= \dfrac{-4x + 2}{(x+2)(x-2)(x-5)}$

14. $\dfrac{m^3 - 1}{(m-1)^2} \cdot \dfrac{m^2 - 1}{3m^2 + 3m + 3}$

$= \dfrac{(m-1)(m^2 + m + 1)}{(m-1)^2} \cdot \dfrac{(m-1)(m+1)}{3(m^2 + m + 1)}$

$= \dfrac{m+1}{3}$

15. $\dfrac{3}{x} = \dfrac{7}{4}$

$7x = 12$

$x = \dfrac{12}{7}$

The solution set is $\left\{\dfrac{12}{7}\right\}$.

16. $\dfrac{x}{x-2} - \dfrac{5}{x} = \dfrac{3}{4}$

$4x(x-2)\left(\dfrac{x}{x-2} - \dfrac{5}{x}\right) = 4x(x-2)\left(\dfrac{3}{4}\right)$

$4x^2 - 5(4x - 8) = 3x^2 - 6x$

$x^2 - 14x + 40 = 0$

$(x-4)(x-10) = 0$

$x = 4 \quad \text{or} \quad x = 10$

Solution set: $\{4, 10\}$

17. $\dfrac{3m}{2} = \dfrac{6}{m}$

$3m^2 = 12$

$m^2 = 4$

$m^2 - 4 = 0$

$(m-4)(m+4) = 0$

$m = 2 \quad \text{or} \quad m = -2$

Solution set: $\{-2, 2\}$

18. $W = \dfrac{a^2}{t}$

$tW = a^2$

$t = \dfrac{a^2}{W}$

19. $\dfrac{1}{a} + \dfrac{1}{b} = \dfrac{1}{2}$

$2ab\left(\dfrac{1}{a} + \dfrac{1}{b}\right) = 2ab \cdot \dfrac{1}{2}$

$2b + 2a = ab$

$2b - ab = -2a$

$b(2 - a) = -2a$

$b = \dfrac{-2a}{2 - a}$

$b = \dfrac{2a}{a - 2}$

20. $\dfrac{\frac{1}{x}+\frac{1}{3x}}{\frac{3}{4x}-\frac{1}{2}} = \dfrac{12x\left(\frac{1}{x}+\frac{1}{3x}\right)}{12x\left(\frac{3}{4x}-\frac{1}{2}\right)}$

$$= \dfrac{12+4}{9-6x} = \dfrac{16}{9-6x}$$

21. $\dfrac{m^{-2}-w^{-2}}{m^{-2}w^{-1}+m^{-1}w^{-2}}$

$$= \dfrac{m^2 w^2(m^{-2}-w^{-2})}{m^2 w^2(m^{-2}w^{-1}+m^{-1}w^{-2})}$$

$$= \dfrac{w^2-m^2}{w+m} = \dfrac{(w-m)(w+m)}{w+m} = w-m$$

22. $\dfrac{a^2 b^3}{4a} \div \dfrac{ab^3}{6a^2} = \dfrac{a^2 b^3}{2\cdot 2a} \cdot \dfrac{2\cdot 3a^2}{ab^3} = \dfrac{3a^2}{2}$

23. $\begin{array}{r} 3x+2 \\ 2x+1\,\overline{\smash{)}\,6x^2+7x-6} \\ \underline{6x^2+3x} \\ 4x-6 \\ \underline{4x+2} \\ -8 \end{array}$

Quotient: $3x+2$ Remainder: -8

24. $(x-3)\div(3-x) = \dfrac{x-3}{3-x}$

$$= \dfrac{-1(3-x)}{3-x} = -1$$

Quotient: -1 Remainder: 0

25. $\begin{array}{r} 5 \\ x+3\,\overline{\smash{)}\,5x+0} \\ \underline{5x+15} \\ -15 \end{array}$

$\dfrac{5x}{x+3} = 5 + \dfrac{-15}{x+3}$

26. $\begin{array}{r} x+5 \\ x-2\,\overline{\smash{)}\,x^2+3x-6} \\ \underline{x^2-2x} \\ 5x-6 \\ \underline{5x-10} \\ 4 \end{array}$

$\dfrac{x^2+3x-6}{x-2} = x+5 + \dfrac{4}{x-2}$

27. Let $x=$ the number of minutes to fill the

leaky pool. In one minute, the hose is supposed to fill $\frac{1}{6}$ of the pool, the leak removes $\frac{1}{8}$ of the pool, but together $\frac{1}{x}$ of the pool actually gets filled.

$$\frac{1}{6}-\frac{1}{8}=\frac{1}{x}$$
$$24x\left(\frac{1}{6}-\frac{1}{8}\right) = 24x\cdot\frac{1}{x}$$
$$4x-3x = 24$$
$$x = 24$$

The leaky pool will be filled in 24 minutes.

28. Let $x=$ the number of miles hiked in one day. Milton's time is $\frac{x}{4}$ and Bonnie's time is $\frac{x}{3}$ hours. Since Milton's time is 2.5 hours less than Bonnie's, we can write the equation

$$\frac{x}{4}+\frac{5}{2}=\frac{x}{3}$$
$$12\left(\frac{x}{4}+\frac{5}{2}\right) = 12\cdot\frac{x}{3}$$
$$3x+30 = 4x$$
$$30 = x$$

They hiked 30 miles that day.

29. Let $x=$ the number of sailors in the original group and $x+3=$ the number in the larger group. The cost is $\dfrac{72{,}000}{x}$ dollars per person for x people and $\dfrac{72{,}000}{x+3}$ dollars per person for $x+3$ people.

$$\frac{72{,}000}{x} = \frac{72{,}000}{x+3}+2000$$
$$x(x+3)\frac{72{,}000}{x} = x(x+3)\left(\frac{72{,}000}{x+3}+2000\right)$$
$$72{,}000x+216{,}000$$
$$= 72{,}000x+2000x^2+6000x$$
$$2000x^2+6000x-216{,}000 = 0$$
$$x^2+3x-108 = 0$$
$$(x-9)(x+12) = 0$$
$$x=9 \quad\text{or}\quad x=-12$$

There are 9 sailors in the original group.

Making Connections

Chapters 1 - 6

1. $\dfrac{3}{x}=\dfrac{4}{5}$
$$4x = 15$$
$$x = \frac{15}{4}$$

Solution set: $\left\{\dfrac{15}{4}\right\}$

2.
$$\frac{2}{x} = \frac{x}{8}$$
$$x^2 = 16$$
$$x^2 - 16 = 0$$
$$(x-4)(x+4) = 0$$
$$x = 4 \text{ or } x = -4$$
Solution set: $\{-4, 4\}$

3.
$$\frac{x}{3} = \frac{4}{5}$$
$$5x = 12$$
$$x = \frac{12}{5}$$
Solution set: $\left\{ \frac{12}{5} \right\}$

4.
$$\frac{3}{x} = \frac{x+3}{6}$$
$$x^2 + 3x = 18$$
$$x^2 + 3x - 18 = 0$$
$$(x+6)(x-3) = 0$$
$$x = -6 \text{ or } x = 3$$
Solution set: $\{-6, 3\}$

5.
$$\frac{1}{x} = 4$$
$$4x = 1$$
$$x = \frac{1}{4}$$
Solution set: $\left\{ \frac{1}{4} \right\}$

6.
$$\frac{2}{3}x = 4$$
$$2x = 12$$
$$x = 6$$
Solution set: $\{6\}$

7.
$$2x + 3 = 4$$
$$2x = 1$$
$$x = \frac{1}{2}$$
Solution set: $\left\{ \frac{1}{2} \right\}$

8.
$$2x + 3 = 4x$$
$$3 = 2x$$
$$\frac{3}{2} = x$$
Solution set: $\left\{ \frac{3}{2} \right\}$

9.
$$\frac{2a}{3} = \frac{6}{a}$$
$$2a^2 = 18$$
$$a^2 = 9$$
$$a^2 - 9 = 0$$
$$(a-3)(a+3) = 0$$
$$a = 3 \text{ or } a = -3$$

Solution set: $\{-3, 3\}$

10. $\frac{12}{x} - \frac{14}{x+1} = \frac{1}{2}$

$$2x(x+1)\left(\frac{12}{x} - \frac{14}{x+1} \right) = 2x(x+1) \cdot \frac{1}{2}$$
$$24x + 24 - 28x = x^2 + x$$
$$-x^2 - 5x + 24 = 0$$
$$x^2 + 5x - 24 = 0$$
$$(x+8)(x-3) = 0$$
$$x = -8 \text{ or } x = 3$$
Solution set: $\{-8, 3\}$

11.
$$|6x - 3| = 1$$
$$6x - 3 = 1 \text{ or } 6x - 3 = -1$$
$$6x = 4 \text{ or } \qquad 6x = 2$$
$$x = \frac{2}{3} \text{ or } \qquad x = \frac{1}{3}$$
Solution set: $\left\{ \frac{1}{3}, \frac{2}{3} \right\}$

12.
$$\frac{x}{2x+9} = \frac{3}{x}$$
$$x^2 = 6x + 27$$
$$x^2 - 6x - 27 = 0$$
$$(x-9)(x+3) = 0$$
$$x = 9 \text{ or } x = -3$$
Solution set: $\{-3, 9\}$

13.
$$4(6x - 3)(2x + 9) = 0$$
$$6x - 3 = 0 \text{ or } 2x + 9 = 0$$
$$6x = 3 \text{ or } \qquad 2x = -9$$
$$x = \frac{1}{2} \text{ or } \qquad x = -\frac{9}{2}$$
Solution set: $\left\{ -\frac{9}{2}, \frac{1}{2} \right\}$

14. $\frac{x-1}{x+2} - \frac{1}{5(x+2)} = 1$

$$5(x+2)\left(\frac{x-1}{x+2} - \frac{1}{5(x+2)} \right)$$
$$= 5(x+2) \cdot 1$$
$$5x - 5 - 1 = 5x + 10$$
$$-6 = 10$$
Solution set: $\emptyset$

15.
$$Ax + By = C$$
$$By = C - Ax$$
$$y = \frac{C - Ax}{B}$$

16.
$$\frac{y-3}{x+5} = -\frac{1}{3}$$
$$y - 3 = -\frac{1}{3}(x+5)$$
$$y - 3 = -\frac{1}{3}x - \frac{5}{3}$$
$$y = -\frac{1}{3}x - \frac{5}{3} + \frac{9}{3}$$
$$y = -\frac{1}{3}x + \frac{4}{3}$$

17.
$$Ay = By + C$$
$$Ay - By = C$$
$$y(A - B) = C$$
$$y = \frac{C}{A - B}$$

18.
$$\frac{A}{y} = \frac{y}{A}$$
$$y^2 = A^2$$
$$y^2 - A^2 = 0$$
$$(y - A)(y + A) = 0$$
$$y - A = 0 \quad \text{or} \quad y + A = 0$$
$$y = A \quad \text{or} \quad y = -A$$

19.
$$\frac{A}{y} - \frac{1}{2} = \frac{B}{y}$$
$$2y\left(\frac{A}{y} - \frac{1}{2}\right) = 2y \cdot \frac{B}{y}$$
$$2A - y = 2B$$
$$2A - 2B = y$$
$$y = 2A - 2B$$

20.
$$\frac{A}{y} - \frac{1}{2} = \frac{B}{C}$$
$$2Cy\left(\frac{A}{y} - \frac{1}{2}\right) = 2Cy \cdot \frac{B}{C}$$
$$2AC - Cy = 2yB$$
$$2AC = Cy + 2yB$$
$$2AC = y(C + 2B)$$
$$\frac{2AC}{C + 2B} = y$$
$$y = \frac{2AC}{2B + C}$$

21.
$$3x - 4y = 6$$
$$-4y = -3x + 6$$
$$y = \frac{-3x + 6}{-4}$$
$$y = \frac{3}{4}x - \frac{3}{2}$$

22.
$$y^2 - 2y - Ay + 2A = 0$$
$$y(y - 2) - A(y - 2) = 0$$
$$(y - A)(y - 2) = 0$$
$$y - A = 0 \quad \text{or} \quad y - 2 = 0$$
$$y = A \quad \text{or} \quad y = 2$$

23.
$$A = \frac{1}{2}B(C + y)$$
$$2A = B(C + y)$$
$$2A = BC + By$$
$$2A - BC = By$$
$$\frac{2A - BC}{B} = y$$
$$y = \frac{2A - BC}{B}$$

24.
$$y^2 + Cy = BC + By$$
$$y^2 + Cy - BC - By = 0$$
$$y(y + C) - B(y + C) = 0$$
$$(y - B)(y + C) = 0$$
$$y - B = 0 \quad \text{or} \quad y + C = 0$$
$$y = B \quad \text{or} \quad y = -C$$

25. $3x^5 \cdot 4x^8 = 12x^{13}$

26. $3x^2(x^3 + 5x^6) = 3x^5 + 15x^8$

27. $(5x^6)^2 = 25x^{12}$

28. $(3a^3b^2)^3 = 27a^9b^6$

29. $\dfrac{12a^9b^4}{-3a^3b^{-2}} = -4a^6b^6$

30. $\left(\dfrac{x^{-2}}{2}\right)^5 = \dfrac{x^{-10}}{2^5} = \dfrac{1}{32x^{10}}$

31. $\left(\dfrac{2x^{-4}}{3y^5}\right)^{-3} = \dfrac{2^{-3}x^{12}}{3^{-3}y^{-15}} = \dfrac{27x^{12}y^{15}}{8}$

32. $(-2a^{-1}b^3c)^{-2} = (-2)^{-2}a^2b^{-6}c^{-2}$
$$= \dfrac{a^2}{4b^6c^2}$$

33. $\dfrac{a^{-1} + b^3}{a^{-2} + b^{-1}} = \dfrac{a^2b(a^{-1} + b^3)}{a^2b(a^{-2} + b^{-1})}$
$$= \dfrac{ab + a^2b^4}{b + a^2}$$

34. $\dfrac{(a + b)^{-1}}{(a + b)^{-2}} = a + b$

35. a) $B = 655 + 9.56(292/2.2)$
$$+ 1.85(86 \cdot 2.54) - 4.68(30)$$
$$= 2188 \text{ calories}$$

b) The graph shows the energy requirement increasing as the weight increases.

c) $B = 655 + 9.56W + 1.85(86 \cdot 2.54)$
$$- 4.68(30)$$
$$B = 9.56W + 918.7$$

d) $B = 655 + 9.56(292/2.2)$
$$+ 1.85(86 \cdot 2.54) - 4.68(A)$$
$$B = 2328 - 4.68A$$

7.1 WARM-UPS

1. True, because of the definition of square root.

2. False, because $\sqrt[3]{2} \cdot \sqrt[3]{2} \cdot \sqrt[3]{2} = 2$.

3. True, because $(-3)^3 = -27$.

4. False, because $(-5)^2 = 25$.

5. True, because $2^4 = 16$.

6. False, because $\sqrt{9} = 3$.

7. False, because $(2^3)^2 = 2^6$.

8. False, because $\dfrac{\sqrt{10}}{\sqrt{2}} = \sqrt{5}$.

9. True, because $\left(\dfrac{1}{2}\right)^2 = \dfrac{1}{4}$.

10. True, because $\dfrac{\sqrt{6}}{\sqrt{3}} = \sqrt{\dfrac{6}{3}} = \sqrt{2}$

7.1 EXERCISES

1. If $b^n = a$ then b is an nth root of a.

3. If $b^n = a$ then b is an even root of a provided n is even or an odd root provided n is odd.

5. The product rule for radicals says that $\sqrt[n]{a} \cdot \sqrt[n]{b} = \sqrt[n]{ab}$ provided all of these roots are real.

7. Because $6^2 = 36$, $\sqrt{36} = 6$.

9. Because $10^2 = 100$, $\sqrt{100} = 10$.

11. Because $3^2 = 9$, $\sqrt{9} = 3$ and $-\sqrt{9} = -3$.

13. Because $2^3 = 8$, $\sqrt[3]{8} = 2$.

15. Because $(-2)^3 = -8$, $\sqrt[3]{-8} = -2$.

17. Because $2^5 = 32$, $\sqrt[5]{32} = 2$.

19. Because $10^3 = 1000$, $\sqrt[3]{1000} = 10$.

21. The expression $\sqrt[4]{-16}$ is not a real number since it is an even root of a negative number.

23. Since $(-2)^5 = -32$, $\sqrt[5]{-32} = -2$.

25. Since $m \cdot m = m^2$, $\sqrt{m^2} = m$.

27. Since $(x^8)^2 = x^{16}$, $\sqrt{x^{16}} = x^8$.

29. Since $(y^3)^5 = y^{15}$, $\sqrt[5]{y^{15}} = y^3$.

31. Since $(y^5)^3 = y^{15}$, $\sqrt[3]{y^{15}} = y^5$.

33. Since $(m)^3 = m^3$, $\sqrt[3]{m^3} = m$.

35. Since $(w^3)^4 = w^{12}$, $\sqrt[4]{w^{12}} = w^3$.

37. $\sqrt{9y} = \sqrt{9}\sqrt{y} = 3\sqrt{y}$

39. $\sqrt{4a^2} = \sqrt{4}\sqrt{a^2} = 2a$

41. $\sqrt{x^4 y^2} = \sqrt{x^4}\sqrt{y^2} = x^2 y$

43. $\sqrt{5m^{12}} = \sqrt{m^{12}}\sqrt{5} = m^6\sqrt{5}$

45. $\sqrt[3]{8y} = \sqrt[3]{8}\sqrt[3]{y} = 2\sqrt[3]{y}$

47. $\sqrt[3]{3a^6} = \sqrt[3]{a^6}\sqrt[3]{3} = a^2\sqrt[3]{3}$

49. $\sqrt{20} = \sqrt{4} \cdot \sqrt{5} = 2\sqrt{5}$

51. $\sqrt{50} = \sqrt{25} \cdot \sqrt{2} = 5\sqrt{2}$

53. $\sqrt{72} = \sqrt{36} \cdot \sqrt{2} = 6\sqrt{2}$

55. $\sqrt[3]{40} = \sqrt[3]{8} \cdot \sqrt[3]{5} = 2\sqrt[3]{5}$

57. $\sqrt[3]{81} = \sqrt[3]{27} \cdot \sqrt[3]{3} = 3\sqrt[3]{3}$

59. $\sqrt[4]{48} = \sqrt[4]{16} \cdot \sqrt[4]{3} = 2\sqrt[4]{3}$

61. $\sqrt[5]{96} = \sqrt[5]{32} \cdot \sqrt[5]{3} = 2\sqrt[5]{3}$

63. $\sqrt{a^3} = \sqrt{a^2}\sqrt{a} = a\sqrt{a}$

65. $\sqrt{18a^6} = \sqrt{9a^6}\sqrt{2} = 3a^3\sqrt{2}$

67. $\sqrt{20x^5 y} = \sqrt{4x^4}\sqrt{5xy} = 2x^2\sqrt{5xy}$

69. $\sqrt[3]{24m^4} = \sqrt[3]{8m^3} \cdot \sqrt[3]{3m} = 2m\sqrt[3]{3m}$

71. $\sqrt[4]{32a^5} = \sqrt[4]{16a^4} \cdot \sqrt[4]{2a} = 2a\sqrt[4]{2a}$

73. $\sqrt[5]{64x^6} = \sqrt[5]{32x^5} \cdot \sqrt[5]{2x} = 2x\sqrt[5]{2x}$

75. $\sqrt{48x^3 y^8 z^7} = \sqrt{16x^2 y^8 z^6} \cdot \sqrt{3xz}$
$= 4xy^4 z^3 \sqrt{3xz}$

77. $\sqrt{\dfrac{t}{4}} = \dfrac{\sqrt{t}}{\sqrt{4}} = \dfrac{\sqrt{t}}{2}$

79. $\sqrt{\dfrac{625}{16}} = \dfrac{\sqrt{625}}{\sqrt{16}} = \dfrac{25}{4}$

81. $\dfrac{\sqrt{30}}{\sqrt{3}} = \sqrt{\dfrac{30}{3}} = \sqrt{10}$

83. $\sqrt[3]{\dfrac{t}{8}} = \dfrac{\sqrt[3]{t}}{\sqrt[3]{8}} = \dfrac{\sqrt[3]{t}}{2}$

85. $\sqrt[3]{\dfrac{-8x^6}{y^3}} = \dfrac{\sqrt[3]{-8x^6}}{\sqrt[3]{y^3}} = \dfrac{-2x^2}{y}$

87. $\sqrt{\dfrac{4a^6}{9}} = \dfrac{\sqrt{4a^6}}{\sqrt{9}} = \dfrac{2a^3}{3}$

89. $\sqrt{\dfrac{12}{25}} = \dfrac{\sqrt{12}}{\sqrt{25}} = \dfrac{\sqrt{4}\sqrt{3}}{5} = \dfrac{2\sqrt{3}}{5}$

91. $\sqrt{\dfrac{27}{16}} = \dfrac{\sqrt{27}}{\sqrt{16}} = \dfrac{\sqrt{9}\sqrt{3}}{4} = \dfrac{3\sqrt{3}}{4}$

93. $\sqrt[3]{\dfrac{a^4}{125}} = \dfrac{\sqrt[3]{a^4}}{\sqrt[3]{125}} = \dfrac{\sqrt[3]{a^3} \cdot \sqrt[3]{a}}{5}$
$= \dfrac{a \cdot \sqrt[3]{a}}{5}$

95. $\sqrt[3]{\dfrac{81}{8b^3}} = \dfrac{\sqrt[3]{81}}{\sqrt[3]{8b^3}} = \dfrac{\sqrt[3]{27} \cdot \sqrt[3]{3}}{2b} = \dfrac{3\sqrt[3]{3}}{2b}$

97. $\sqrt[4]{\dfrac{x^7}{y^8}} = \dfrac{\sqrt[4]{x^7}}{\sqrt[4]{y^8}} = \dfrac{\sqrt[4]{x^4} \cdot \sqrt[4]{x^3}}{y^2} = \dfrac{x\sqrt[4]{x^3}}{y^2}$

99. $\sqrt[4]{\dfrac{a^5}{16b^{12}}} = \dfrac{\sqrt[4]{a^5}}{\sqrt[4]{16b^{12}}} = \dfrac{\sqrt[4]{a^4} \cdot \sqrt[4]{a}}{2b^3}$
$= \dfrac{a\sqrt[4]{a}}{2b^3}$

101. If $f(x) = \sqrt{x-2}$, then
$f(3) = \sqrt{3-2} = \sqrt{1} = 1$.
In $f(x) = \sqrt{x-2}$ we must have $x - 2 \geq 0$ or
$x \geq 2$. So the domain of f is the interval
$[2, \infty)$.
103. If $f(x) = \sqrt{2x+8}$, then
$f(3) = \sqrt{2(3)+8} = \sqrt{14}$.
In $f(x) = \sqrt{2x+8}$ we must have $2x + 8 \geq 0$,
or $2x \geq -8$, or $x \geq -4$. So the domain of f is
the interval $[-4, \infty)$.
105. If $f(x) = \sqrt{3-3x}$, then
$f(3) = \sqrt{3-3(3)} = \sqrt{-6}$. So there is no
$f(3)$. In $f(x) = \sqrt{3-3x}$ we must have
$3 - 3x \geq 0$, or $-3x \geq -3$, or $x \leq 1$. So the
domain of f is the interval $(-\infty, 1]$.
107. $w = 91.4 - \frac{(10.5 + 6.7\sqrt{20} - 0.45 \cdot 20)(457 - 5 \cdot 25)}{110}$
$w \approx -4°$F
If the air temp is 25°F and wind is 30 mph,
then from the graph w is approximately
$-10°$F.
109. a) $t = \sqrt{\dfrac{h}{16}} = \dfrac{\sqrt{h}}{4}$

b) $t = \dfrac{\sqrt{40}}{4} = \dfrac{2\sqrt{10}}{4} = \dfrac{\sqrt{10}}{2}$ sec

c) From the graph it appears that if a diver
takes 2.5 seconds then the height is 100 feet.
111. $M = 1.3\sqrt{20} \approx 5.8$ knots

113. $V = \sqrt{\dfrac{841 \cdot 8700}{2.81 \cdot 200}} \approx 114.1$ ft/sec

$114.1 \dfrac{\text{ft}}{\text{sec}} \cdot \dfrac{1 \text{ mi}}{5280 \text{ ft}} \cdot \dfrac{3600 \text{ sec}}{1 \text{ hr}} \approx 77.8$ mph

115. The equations in (a), (c), and (d) are
identities because the radical symbol always
indicates a positive root. In (b), $\sqrt[3]{x^3}$ is
negative if x is negative, but $|x|$ is positive if x
is negative. So (b) is not an identity.
117. The arithmetic mean of 80 and 100 is 90.
Solve $80/h = h/100$ to get $h^2 = 8000$ or
$h = \sqrt{8000} \approx 89.4$. So you are better of with
the arithmetic mean.

7.2 WARM-UPS

1. True, by definition of exponent 1/3.
2. False, because $8^{5/3} = \sqrt[3]{8^5}$.

3. False, because $-16^{1/2} = -4$, while
$(-16)^{1/2}$ is not a real number.
4. True, $9^{-3/2} = \dfrac{1}{(\sqrt{9})^3} = \dfrac{1}{3^3} = \dfrac{1}{27}$.
5. True, $\dfrac{\sqrt{6}}{6} = \dfrac{6^{1/2}}{6^1} = 6^{1/2-1} = 6^{-1/2}$.
6. True, because $\dfrac{2^1}{2^{1/2}} = 2^{1-1/2} = 2^{1/2}$.
7. True, because $2^{1/2} \cdot 2^{1/2} = 2^1 = 2$ and
$4^{1/2} = 2$.
8. False, because $16^{-1/4} = \dfrac{1}{16^{1/4}} = \dfrac{1}{2}$.
9. True, $6^{1/6} \cdot 6^{1/6} = 6^{1/6+1/6} = 6^{2/6} = 6^{1/3}$.
10. True, $(2^8)^{3/4} = 2^{24/4} = 2^6$.

7.2 EXERCISES

1. The nth root of a is $a^{1/n}$.
3. The expression $a^{-m/n}$ means $\dfrac{1}{a^{m/n}}$.
5. The operations can be performed in any
order, but the easiest is usually root, power,
and then reciprocal.
7. $\sqrt[4]{7} = 7^{1/4}$
9. $\sqrt{5x} = (5x)^{1/2}$
11. $9^{1/5} = \sqrt[5]{9}$
13. $a^{1/2} = \sqrt{a}$
15. $25^{1/2} = \sqrt{25} = 5$
17. $(-125)^{1/3} = \sqrt[3]{-125} = -5$
19. $16^{1/4} = \sqrt[4]{16} = 2$
21. $(-4)^{1/2}$ is not a real number because it is
$\sqrt{-4}$.
23. $\sqrt[3]{w^7} = w^{7/3}$
25. $\dfrac{1}{\sqrt[3]{2^{10}}} = 2^{-10/3}$
27. $w^{-3/4} = \sqrt[4]{\dfrac{1}{w^3}}$
29. $(ab)^{3/2} = \sqrt{(ab)^3}$

31. $125^{2/3} = (\sqrt[3]{125})^2 = 5^2 = 25$
33. $25^{3/2} = (\sqrt{25})^3 = 5^3 = 125$
35. $27^{-4/3} = \dfrac{1}{(\sqrt[3]{27})^4} = \dfrac{1}{3^4} = \dfrac{1}{81}$

37. $16^{-3/2} = \dfrac{1}{(\sqrt{16})^3} = \dfrac{1}{4^3} = \dfrac{1}{64}$

39. $(-27)^{-1/3} = \dfrac{1}{(-27)^{1/3}} = \dfrac{1}{-3} = -\dfrac{1}{3}$

41. $(-16)^{-1/4}$ is not a real number because it is a fourth root of a negative number.

43. $3^{1/3} \cdot 3^{1/4} = 3^{\frac{1}{3}+\frac{1}{4}} = 3^{7/12}$

45. $3^{1/3} \cdot 3^{-1/3} = 3^0 = 1$

47. $\dfrac{8^{1/3}}{8^{2/3}} = 8^{1/3-2/3} = 8^{-1/3} = \dfrac{1}{8^{1/3}} = \dfrac{1}{2}$

49. $4^{3/4} \div 4^{1/4} = 4^{3/4-1/4} = 4^{1/2} = 2$

51. $18^{1/2} \cdot 2^{1/2} = 36^{1/2} = 6$

53. $(2^6)^{1/3} = 2^{6/3} = 2^2 = 4$

55. $(3^8)^{1/2} = 3^{8/2} = 3^4 = 81$

57. $(2^{-4})^{1/2} = 2^{-2} = \dfrac{1}{4}$

59. $\left(\dfrac{3^4}{2^6}\right)^{1/2} = \dfrac{3^2}{2^3} = \dfrac{9}{8}$

61. $(x^4)^{1/4} = |x|$

63. $(a^8)^{1/2} = a^4$

65. $(y^3)^{1/3} = y$

67. $(9x^6 y^2)^{1/2} = |3x^3 y|$

69. $\left(\dfrac{81x^{12}}{y^{20}}\right)^{1/4} = \left|\dfrac{3x^3}{y^5}\right|$

71. $x^{1/2} x^{1/4} = x^{2/4+1/4} = x^{3/4}$

73. $(x^{1/2} y)(x^{-3/4} y^{1/2}) = x^{-1/4} y^{3/2} = \dfrac{y^{3/2}}{x^{1/4}}$

75. $\dfrac{w^{1/3}}{w^3} = w^{\frac{1}{3}-3} = w^{-8/3} = \dfrac{1}{w^{8/3}}$

77. $(144x^{16})^{1/2} = 12x^8$

79. $\left(\dfrac{a^{-1/2}}{b^{-1/4}}\right)^{-4} = \dfrac{a^{4/2}}{b^{4/4}} = \dfrac{a^2}{b}$

81. $\left(\dfrac{2w^{1/3}}{w^{-3/4}}\right)^3 = \dfrac{8w}{w^{-9/4}} = 8w^{1-(-9/4)}$
$= 8w^{13/4}$

83. $\dfrac{9^{1/4} h^{1/2} k^{3/2}}{9^{3/4} h^{1/3} k^2} = \dfrac{h^{1/6}}{9^{1/2} k^{1/2}} = \dfrac{h^{1/6}}{3k^{1/2}}$

85. $(9^2)^{1/2} = 9$

87. $-16^{-3/4} = -\dfrac{1}{2^3} = -\dfrac{1}{8}$

89. $125^{-4/3} = \dfrac{1}{5^4} = \dfrac{1}{625}$

91. $2^{1/2} \cdot 2^{-1/4} = 2^{2/4-1/4} = 2^{1/4}$

93. $3^{0.26} 3^{0.74} = 3^{0.26+0.74} = 3^1 = 3$

95. $3^{1/4} \cdot 27^{1/4} = (3 \cdot 27)^{1/4} = 81^{1/4} = 3$

97. $\left(-\dfrac{8}{27}\right)^{2/3} = \dfrac{(-8)^{2/3}}{27^{2/3}} = \dfrac{4}{9}$

99. Not a real number, because the fourth root of $-1/16$ is not real.

101. $\left(\dfrac{9}{16}\right)^{-1/2} = \left(\dfrac{16}{9}\right)^{1/2} = \dfrac{4}{3}$

103. $-\left(\dfrac{25}{36}\right)^{-3/2} = -\left(\dfrac{36}{25}\right)^{3/2} = -\dfrac{216}{125}$

105. $(9x^9)^{1/2} = 9^{1/2} x^{9/2} = 3x^{9/2}$

107. $(3a^{-2/3})^{-3} = 3^{-3} a^2 = \dfrac{a^2}{27}$

109. $(a^{1/2} b)^{1/2} (ab^{1/2}) = a^{1/4} b^{1/2} a^1 b^{1/2}$
$= a^{5/4} b$

111. $(km^{1/2})^3 (k^3 m^5)^{1/2} = k^3 m^{3/2} k^{3/2} m^{5/2}$
$= k^{9/2} m^4$

113. $2^{1/3} \approx 2^{0.33333333} \approx 1.2599$

115. $-2^{1/2} = -(2^{0.5}) \approx -1.4142$

117. $1024^{1/10} = 2$

119. $\left(\dfrac{64}{15,625}\right)^{-1/6} = 2.5$

121. $a^{m/2} \cdot a^{m/4} = a^{m/2+m/4} = a^{3m/4}$

123. $\dfrac{a^{-m/5}}{a^{-m/3}} = a^{-m/5+m/3} = a^{2m/15}$

125. $\left(a^{-1/m} b^{-1/n}\right)^{-mn} = a^n b^m$

127. $\left(\dfrac{a^{-3m} b^{-6n}}{a^{9m}}\right)^{-1/3} = \dfrac{a^m b^{2n}}{a^{-3m}} = a^{4m} b^{2n}$

129. $D = (12^2 + 4^2 + 3^2)^{1/2}$
$= (144 + 16 + 9)^{1/2}$
$= 169^{1/2} = 13$ inches

131. $S = (13.0368 + 7.84(18.42)^{1/3}$
$- 0.8(21.45))^2$
$S \approx 274.96$ m^2

133. $r = \left(\dfrac{20,130}{10,000}\right)^{1/3} - 1 \approx 0.263$
$= 26.3\%$

135. $r = \left(\dfrac{141,600,000,000}{450,000}\right)^{1/213} - 1$
$\approx 0.061 = 6.1\%$

137. The expression $(-1)^{1/2}$ is not a real number. The rules of exponents hold only for expressions that represent real numbers. So we should not be using either rule of exponents to perform these computations. Using complex numbers from Section 7.6 we can compute $(-1)^{1/2} \cdot (-1)^{1/2} = i \cdot i = -1$.

7.3 WARM-UPS

1. False, because $\sqrt{3} + \sqrt{3} = 2\sqrt{3}$.

2. True, because
$\sqrt{8} + \sqrt{2} = 2\sqrt{2} + \sqrt{2} = 3\sqrt{2}$.

3. False, because $2\sqrt{3} \cdot 3\sqrt{3} = 6 \cdot 3 = 18$.

4. False, because $\sqrt[3]{2} \cdot \sqrt[3]{2} = \sqrt[3]{4}$.

5. True, because $\sqrt{5} \cdot \sqrt{2} = \sqrt{10}$.

6. False, because $2\sqrt{5} + 3\sqrt{5} = 5\sqrt{5}$.

7. True, because $\sqrt{2}\sqrt{3} = \sqrt{6}$ and $\sqrt{2}\sqrt{2} = 2$.

8. False, because $\sqrt{12} = \sqrt{4}\sqrt{3} = 2\sqrt{3}$.

9. False, because $(\sqrt{2} + \sqrt{3})^2$
$= 2 + 2\sqrt{2}\sqrt{3} + 3 = 5 + 2\sqrt{6}$.

10. True,
$(\sqrt{3} - \sqrt{2})(\sqrt{3} + \sqrt{2}) = 3 - 2 = 1$.

7.3 EXERCISES

1. Like radicals are radicals with the same index and same radicand.

3. In the product rule the radicals must have the same index, but do not have to have the same radicand.

5. $\sqrt{3} - 2\sqrt{3} = 1\sqrt{3} - 2\sqrt{3}$
$$= -1\sqrt{3} = -\sqrt{3}$$

7. $5\sqrt{7x} + 4\sqrt{7x} = (5 + 4)\sqrt{7x} = 9\sqrt{7x}$

9. $2 \cdot \sqrt[3]{2} + 3 \cdot \sqrt[3]{2} = (2 + 3)\sqrt[3]{2} = 5\sqrt[3]{2}$

11. $\sqrt{3} - \sqrt{5} + 3\sqrt{3} - \sqrt{5}$
$= \sqrt{3} + 3\sqrt{3} - \sqrt{5} - \sqrt{5} = 4\sqrt{3} - 2\sqrt{5}$

13. $\sqrt[3]{2} + \sqrt[3]{x} - \sqrt[3]{2} + 4\sqrt[3]{x}$
$= \sqrt[3]{2} - \sqrt[3]{2} + \sqrt[3]{x} + 4\sqrt[3]{x} = 5\sqrt[3]{x}$

15. $\sqrt[3]{x} - \sqrt{2x} + \sqrt[3]{x} = \sqrt[3]{x} + \sqrt[3]{x} - \sqrt{2x}$
$$= 2\sqrt[3]{x} - \sqrt{2x}$$

17. $\sqrt{8} + \sqrt{28} = \sqrt{4}\sqrt{2} + \sqrt{4}\sqrt{7}$
$$= 2\sqrt{2} + 2\sqrt{7}$$

19. $\sqrt{8} + \sqrt{18} = \sqrt{4}\sqrt{2} + \sqrt{9}\sqrt{2}$
$$= 2\sqrt{2} + 3\sqrt{2} = 5\sqrt{2}$$

21. $2\sqrt{45} - 3\sqrt{20} = 2\sqrt{9}\sqrt{5} - 3\sqrt{4}\sqrt{5}$
$$= 6\sqrt{5} - 6\sqrt{5} = 0$$

23. $\sqrt{2} - \sqrt{8} = \sqrt{2} - 2\sqrt{2} = -\sqrt{2}$

25. $\sqrt{45x^3} - \sqrt{18x^2} + \sqrt{50x^2} - \sqrt{20x^3}$
$= 3x\sqrt{5x} - 3x\sqrt{2} + 5x\sqrt{2} - 2x\sqrt{5x}$
$= x\sqrt{5x} + 2x\sqrt{2}$

27. $2\sqrt[3]{24} + \sqrt[3]{81} = 2\sqrt[3]{8} \cdot \sqrt[3]{3} + \sqrt[3]{27} \cdot \sqrt[3]{3}$
$= 4\sqrt[3]{3} + 3 \cdot \sqrt[3]{3} = 7\sqrt[3]{3}$

29. $\sqrt[4]{48} - 2\sqrt[4]{243} = \sqrt[4]{16 \cdot 3} - 2\sqrt[4]{81 \cdot 3}$
$$= 2\sqrt[4]{3} - 6\sqrt[4]{3}$$
$$= -4\sqrt[4]{3}$$

31. $\sqrt[3]{54t^4y^3} - \sqrt[3]{16t^4y^3}$
$= \sqrt[3]{27t^3y^3} \cdot \sqrt[3]{2t} - \sqrt[3]{8t^3y^3} \cdot \sqrt[3]{2t}$
$= 3ty \cdot \sqrt[3]{2t} - 2ty \cdot \sqrt[3]{2t} = ty\sqrt[3]{2t}$

33. $\sqrt{3}\sqrt{5} = \sqrt{3 \cdot 5} = \sqrt{15}$

35. $(2\sqrt{5})(3\sqrt{10}) = 6\sqrt{50} = 6\sqrt{25}\sqrt{2}$
$$= 6 \cdot 5\sqrt{2} = 30\sqrt{2}$$

37. $(2\sqrt{7a})(3\sqrt{2a}) = 6\sqrt{14a^2} = 6\sqrt{a^2}\sqrt{14}$
$$= 6a\sqrt{14}$$

39. $(\sqrt[4]{9})(\sqrt[4]{27}) = \sqrt[4]{243} = \sqrt[4]{81} \cdot \sqrt[4]{3}$
$$= 3\sqrt[4]{3}$$

41. $(2\sqrt{3})^2 = 4 \cdot 3 = 12$

43. $2\sqrt{3}(\sqrt{6} + 3\sqrt{3}) = 2\sqrt{18} + 18$
$$= 2\sqrt{9}\sqrt{2} + 18 = 6\sqrt{2} + 18$$

45. $\sqrt{5}(\sqrt{10} - 2) = \sqrt{50} - 2\sqrt{5}$
$$= \sqrt{25}\sqrt{2} - 2\sqrt{5} = 5\sqrt{2} - 2\sqrt{5}$$

47. $\sqrt[3]{3t}\left(\sqrt[3]{9t} - \sqrt[3]{t^2}\right) = \sqrt[3]{27t^2} - \sqrt[3]{3t^3}$
$$= 3\sqrt[3]{t^2} - t\sqrt[3]{3}$$

49. $\left(\sqrt{3} + 2\right)\left(\sqrt{3} - 5\right)$
$= 3 + 2\sqrt{3} - 5\sqrt{3} - 10 = -7 - 3\sqrt{3}$

51. $(\sqrt{11} - 3)(\sqrt{11} + 3) = 11 - 9 = 2$

53. $(2\sqrt{5} - 7)(2\sqrt{5} + 4)$
$$= 20 - 14\sqrt{5} + 8\sqrt{5} - 28$$
$$= -8 - 6\sqrt{5}$$

55. $(2\sqrt{3} - \sqrt{6})(\sqrt{3} + 2\sqrt{6})$
$= 6 - \sqrt{18} + 4\sqrt{18} - 12$
$= -6 - 3\sqrt{2} + 4 \cdot 3\sqrt{2} = -6 + 9\sqrt{2}$

57. $\sqrt[3]{3} \cdot \sqrt{3} = 3^{1/3}3^{1/2} = 3^{5/6} = \sqrt[6]{3^5}$
$$= \sqrt[6]{243}$$

59. $\sqrt[3]{5} \cdot \sqrt[4]{5} = 5^{1/3}5^{1/4} = 5^{7/12} = \sqrt[12]{5^7}$

61. $\sqrt[3]{2} \cdot \sqrt{5} = 2^{1/3}5^{1/2} = 2^{2/6}5^{3/6}$
$$= \sqrt[6]{2^25^3} = \sqrt[6]{500}$$

63. $\sqrt[3]{2} \cdot \sqrt[4]{3} = 2^{1/3}3^{1/4} = 2^{4/12}3^{3/12}$
$$= \sqrt[12]{2^43^3} = \sqrt[12]{432}$$

65. $(\sqrt{3} - 2)(\sqrt{3} + 2) = 3 - 4 = -1$

67. $(\sqrt{5} + \sqrt{2})(\sqrt{5} - \sqrt{2}) = 5 - 2 = 3$

69. $(2\sqrt{5} + 1)(2\sqrt{5} - 1) = 4 \cdot 5 - 1 = 19$

71. $(3\sqrt{2} + \sqrt{5})(3\sqrt{2} - \sqrt{5}) = 9 \cdot 2 - 5$
$$= 13$$

73. $(5 - 3\sqrt{x})(5 + 3\sqrt{x}) = 25 - 9x$

75. $\sqrt{300} + \sqrt{3} = 10\sqrt{3} + \sqrt{3} = 11\sqrt{3}$

77. $2\sqrt{5} \cdot 5\sqrt{6} = 10\sqrt{30}$

79. $(3 + 2\sqrt{7})(\sqrt{7} - 2)$
$$= 3\sqrt{7} - 6 + 2 \cdot 7 - 4\sqrt{7}$$
$$= 8 - \sqrt{7}$$

81. $4\sqrt{w} \cdot 4\sqrt{w} = 16(\sqrt{w})^2 = 16w$

83. $\sqrt{3x^3} \cdot \sqrt{6x^2} = \sqrt{18x^5} = \sqrt{9x^4} \cdot \sqrt{2x}$
$$= 3x^2\sqrt{2x}$$

85. $(2\sqrt{5} + \sqrt{2})(3\sqrt{5} - \sqrt{2})$
$= 30 + 3\sqrt{10} - 2\sqrt{10} - 2 = 28 + \sqrt{10}$

144

87. $\dfrac{\sqrt{2}}{3} + \dfrac{\sqrt{2}}{5} = \dfrac{5\sqrt{2}}{5\cdot3} + \dfrac{3\sqrt{2}}{3\cdot5} = \dfrac{8\sqrt{2}}{15}$

89. $(5 + 2\sqrt{2})(5 - 2\sqrt{2}) = 25 - 4\cdot2 = 17$

91. $(3 + \sqrt{x})^2 = 9 + 2\cdot3\sqrt{x} + x$
$\qquad\qquad = 9 + 6\sqrt{x} + x$

93. $(5\sqrt{x} - 3)^2 = 25x - 2\cdot3\cdot5\sqrt{x} + 9$
$\qquad\qquad = 25x - 30\sqrt{x} + 9$

95. $(1 + \sqrt{x+2})^2 = 1 + 2\sqrt{x+2} + x + 2$
$\qquad\qquad = x + 3 + 2\sqrt{x+2}$

97. $\sqrt{4w} - \sqrt{9w} = 2\sqrt{w} - 3\sqrt{w}$
$\qquad\qquad = -\sqrt{w}$

99. $2\sqrt{a^3} + 3\sqrt{a^3} - 2a\sqrt{4a}$
$\qquad = 2a\sqrt{a} + 3a\sqrt{a} - 4a\sqrt{a} = a\sqrt{a}$

101. $\sqrt{x^5} + 2x\sqrt{x^3}$
$\qquad = \sqrt{x^4}\sqrt{x} + 2x\sqrt{x^2}\sqrt{x}$
$\qquad = x^2\sqrt{x} + 2x^2\sqrt{x} = 3x^2\sqrt{x}$

103. $\sqrt[3]{-16x^4} + 5x\sqrt[3]{54x}$
$= -2x\sqrt[3]{2x} + 5x\cdot3\cdot\sqrt[3]{2x} = 13x\sqrt[3]{2x}$

105. $\sqrt[3]{2x}\cdot\sqrt{2x} = (2x)^{1/3}(2x)^{1/2} = (2x)^{5/6}$
$\qquad\qquad = \sqrt[6]{32x^5}$

107. $A = LW = \sqrt{6}\cdot\sqrt{3} = \sqrt{18}$
$\qquad\qquad = 3\sqrt{2}\text{ ft}^2$

109. $A = \frac{1}{2}h(b_1 + b_2)$
$\qquad = \frac{1}{2}\sqrt{6}\left(\sqrt{3} + \sqrt{12}\right)$
$\qquad = \frac{1}{2}\sqrt{6}\left(\sqrt{3} + 2\sqrt{3}\right)$
$\qquad = \frac{1}{2}\sqrt{6}\cdot3\sqrt{3} = \dfrac{3\sqrt{18}}{2}$
$\qquad = \dfrac{9\sqrt{2}}{2}\text{ ft}^2$

111. No because $\sqrt{9} + \sqrt{16} = 3 + 4 = 7$ and $\sqrt{9+16} = \sqrt{25} = 5$.

113. a) $y^2 - 3 = \left(y + \sqrt{3}\right)\left(y - \sqrt{3}\right)$

$2a^2 - 7 = \left(\sqrt{2}a + \sqrt{7}\right)\left(\sqrt{2}a - \sqrt{7}\right)$

b) $\qquad\qquad x^2 - 8 = 0$
$\qquad\left(x + \sqrt{8}\right)\left(x - \sqrt{8}\right) = 0$
$x + \sqrt{8} = 0 \quad$ or $\quad x - \sqrt{8} = 0$
$\qquad x = -\sqrt{8}\quad$ or $\qquad x = \sqrt{8}$
$\qquad x = -2\sqrt{2}\quad$ or $\quad x = 2\sqrt{2}$

The solution set is $\left\{\pm2\sqrt{2}\right\}$.

c) $\qquad x^2 - a = 0$
$\qquad\left(x + \sqrt{a}\right)\left(x - \sqrt{a}\right) = 0$
$x + \sqrt{a} = 0 \quad$ or $\quad x - \sqrt{a} = 0$
$\qquad x = -\sqrt{a}\quad$ or $\qquad x = \sqrt{a}$

The solution set is $\left\{\pm\sqrt{a}\right\}$.

7.4 WARM-UPS

1. True, because $\sqrt{3}\sqrt{2} = \sqrt{6}$.

2. True, because
$$\dfrac{2}{\sqrt{2}} = \dfrac{2\sqrt{2}}{\sqrt{2}\sqrt{2}} = \dfrac{2\sqrt{2}}{2} = \sqrt{2}.$$

3. False, because $\dfrac{4 - \sqrt{10}}{2} = 2 - \dfrac{\sqrt{10}}{2}$.

4. True, because $\dfrac{1}{\sqrt{3}} = \dfrac{1\cdot\sqrt{3}}{\sqrt{3}\sqrt{3}} = \dfrac{\sqrt{3}}{3}$.

5. False, because $\dfrac{8\sqrt{7}}{2\sqrt{7}} = 4$.

6. True, because
$(2 - \sqrt{3})(2 + \sqrt{3}) = 4 - 3 = 1$.

7. False, because $\dfrac{\sqrt{12}}{3} = \dfrac{2\sqrt{3}}{3}$.

8. True, because
$$\dfrac{\sqrt{20}}{\sqrt{5}} = \dfrac{\sqrt{4}\sqrt{5}}{\sqrt{5}} = \sqrt{4} = 2.$$

9. True, because $(2\sqrt{4})^2 = 4\cdot4 = 16$.

10. True, because
$(3\sqrt{5})^3 = 3^3\cdot\sqrt{5^3} = 27\sqrt{125}$.

7.4 EXERCISES

1. $\dfrac{2}{\sqrt{5}} = \dfrac{2\sqrt{5}}{\sqrt{5}\sqrt{5}} = \dfrac{2\sqrt{5}}{5}$

3. $\dfrac{\sqrt{3}}{\sqrt{7}} = \dfrac{\sqrt{3}\sqrt{7}}{\sqrt{7}\sqrt{7}} = \dfrac{\sqrt{21}}{\sqrt{49}} = \dfrac{\sqrt{21}}{7}$

5. $\dfrac{1}{\sqrt[3]{4}} = \dfrac{1\cdot\sqrt[3]{2}}{\sqrt[3]{4}\cdot\sqrt[3]{2}} = \dfrac{\sqrt[3]{2}}{\sqrt[3]{8}} = \dfrac{\sqrt[3]{2}}{2}$

7. $\dfrac{\sqrt[3]{6}}{\sqrt[3]{5}} = \dfrac{\sqrt[3]{6}\cdot\sqrt[3]{25}}{\sqrt[3]{5}\cdot\sqrt[3]{25}} = \dfrac{\sqrt[3]{150}}{\sqrt[3]{125}} = \dfrac{\sqrt[3]{150}}{5}$

9. $\dfrac{\sqrt{5}}{\sqrt{12}} = \dfrac{\sqrt{5}\sqrt{3}}{\sqrt{12}\sqrt{3}} = \dfrac{\sqrt{15}}{\sqrt{36}} = \dfrac{\sqrt{15}}{6}$

11. $\dfrac{\sqrt{3}}{\sqrt{12}} = \dfrac{\sqrt{3}}{\sqrt{4}\sqrt{3}} = \dfrac{1}{\sqrt{4}} = \dfrac{1}{2}$

13. $\sqrt{\dfrac{1}{2}} = \dfrac{1}{\sqrt{2}} = \dfrac{1\cdot\sqrt{2}}{\sqrt{2}\sqrt{2}} = \dfrac{\sqrt{2}}{2}$

15. $\sqrt[3]{\dfrac{2}{3}} = \dfrac{\sqrt[3]{2}\cdot\sqrt[3]{9}}{\sqrt[3]{3}\cdot\sqrt[3]{9}} = \dfrac{\sqrt[3]{18}}{3}$

17. $\sqrt[3]{\dfrac{7}{4}} = \dfrac{\sqrt[3]{7}\cdot\sqrt[3]{2}}{\sqrt[3]{4}\cdot\sqrt[3]{2}} = \dfrac{\sqrt[3]{14}}{2}$

19. $\sqrt{\dfrac{x}{y}} = \dfrac{\sqrt{x}\sqrt{y}}{\sqrt{y}\sqrt{y}} = \dfrac{\sqrt{xy}}{y}$

21. $\dfrac{\sqrt{a^3}}{\sqrt{b^7}} = \dfrac{\sqrt{a^2}\sqrt{a}}{\sqrt{b^6}\sqrt{b}} = \dfrac{a\sqrt{a}\sqrt{b}}{b^3\sqrt{b}\sqrt{b}} = \dfrac{a\sqrt{ab}}{b^4}$

23. $\sqrt{\dfrac{a}{3b}} = \dfrac{\sqrt{a}\sqrt{3b}}{\sqrt{3b}\sqrt{3b}} = \dfrac{\sqrt{3ab}}{3b}$

25. $\sqrt[3]{\dfrac{a}{b}} = \dfrac{\sqrt[3]{a}\cdot\sqrt[3]{b^2}}{\sqrt[3]{b}\cdot\sqrt[3]{b^2}} = \dfrac{\sqrt[3]{ab^2}}{b}$

27. $\sqrt[3]{\dfrac{5}{2b^2}} = \dfrac{\sqrt[3]{5}\cdot\sqrt[3]{4b}}{\sqrt[3]{2b^2}\cdot\sqrt[3]{4b}} = \dfrac{\sqrt[3]{20b}}{2b}$

29. $\sqrt{15} \div \sqrt{5} = \sqrt{\dfrac{15}{5}} = \sqrt{3}$

31. $\sqrt{3} \div \sqrt{5} = \dfrac{\sqrt{3}}{\sqrt{5}} = \dfrac{\sqrt{3}\sqrt{5}}{\sqrt{5}\sqrt{5}} = \dfrac{\sqrt{15}}{5}$

33. $(3\sqrt{3}) \div (5\sqrt{6}) = \dfrac{3\sqrt{3}}{5\sqrt{6}} = \dfrac{3\sqrt{3}}{5\sqrt{2}\sqrt{3}}$

$= \dfrac{3}{5\sqrt{2}} = \dfrac{3\sqrt{2}}{5\sqrt{2}\sqrt{2}} = \dfrac{3\sqrt{2}}{10}$

35. $(2\sqrt{3}) \div (3\sqrt{6}) = \dfrac{2\sqrt{3}}{3\sqrt{6}} = \dfrac{2}{3\sqrt{2}}$

$= \dfrac{2\sqrt{2}}{3\sqrt{2}\sqrt{2}} = \dfrac{2\sqrt{2}}{3\cdot 2} = \dfrac{\sqrt{2}}{3}$

37. $(\sqrt{24a^2}) \div (\sqrt{72a}) = \dfrac{\sqrt{24a^2}}{\sqrt{72a}} = \dfrac{2a\sqrt{6}}{6\sqrt{2a}}$

$= \dfrac{a\sqrt{3}}{3\sqrt{a}} = \dfrac{a\sqrt{3}\sqrt{a}}{3\sqrt{a}\sqrt{a}} = \dfrac{a\sqrt{3a}}{3a} = \dfrac{\sqrt{3a}}{3}$

39. $\sqrt[3]{20} \div \sqrt[3]{2} = \dfrac{\sqrt[3]{20}}{\sqrt[3]{2}} = \sqrt[3]{\dfrac{20}{2}} = \sqrt[3]{10}$

41. $\sqrt[4]{48} \div \sqrt[4]{3} = \dfrac{\sqrt[4]{48}}{\sqrt[4]{3}} = \dfrac{\sqrt[4]{16}\cdot\sqrt[4]{3}}{\sqrt[4]{3}} = 2$

43. $\sqrt[4]{16w} \div \sqrt[4]{w^5} = \dfrac{\sqrt[4]{16w}}{\sqrt[4]{w^5}} = \dfrac{2\sqrt[4]{w}}{w\sqrt[4]{w}}$

$= \dfrac{2}{w}$

45. $\dfrac{6+\sqrt{45}}{3} = \dfrac{6+3\sqrt{5}}{3} = 2+\sqrt{5}$

47. $\dfrac{-2+\sqrt{12}}{-2} = \dfrac{-2+2\sqrt{3}}{-2} = 1-\sqrt{3}$

49. $\dfrac{4}{2+\sqrt{8}} = \dfrac{4(2-\sqrt{8})}{(2+\sqrt{8})(2-\sqrt{8})}$

$= \dfrac{8-4\sqrt{8}}{4-8} = \dfrac{8-8\sqrt{2}}{-4} = -2+2\sqrt{2}$

$= 2\sqrt{2}-2$

51. $\dfrac{3}{\sqrt{11}-\sqrt{5}}$

$= \dfrac{3(\sqrt{11}+\sqrt{5})}{(\sqrt{11}-\sqrt{5})(\sqrt{11}+\sqrt{5})}$

$= \dfrac{3\sqrt{11}+3\sqrt{5}}{11-5} = \dfrac{3\sqrt{11}+3\sqrt{5}}{6}$

$= \dfrac{\sqrt{11}+\sqrt{5}}{2}$

53. $\dfrac{1+\sqrt{2}}{\sqrt{3}-1} = \dfrac{(1+\sqrt{2})(\sqrt{3}+1)}{(\sqrt{3}-1)(\sqrt{3}+1)}$

$= \dfrac{1+\sqrt{6}+\sqrt{2}+\sqrt{3}}{2}$

55. $\dfrac{\sqrt{2}}{\sqrt{6}+\sqrt{3}} = \dfrac{\sqrt{2}(\sqrt{6}-\sqrt{3})}{(\sqrt{6}+\sqrt{3})(\sqrt{6}-\sqrt{3})}$

$= \dfrac{\sqrt{12}-\sqrt{6}}{6-3} = \dfrac{2\sqrt{3}-\sqrt{6}}{3}$

57. $\dfrac{2\sqrt{3}}{3\sqrt{2}-\sqrt{5}}$

$= \dfrac{2\sqrt{3}(3\sqrt{2}+\sqrt{5})}{(3\sqrt{2}-\sqrt{5})(3\sqrt{2}+\sqrt{5})}$

$= \dfrac{6\sqrt{6}+2\sqrt{15}}{18-5} = \dfrac{6\sqrt{6}+2\sqrt{15}}{13}$

59. $(2\sqrt{2})^5 = 2^5 \cdot \sqrt{2^5} = 32\sqrt{16}\sqrt{2}$

$= 128\sqrt{2}$

61. $(\sqrt{x})^5 = \sqrt{x^5} = \sqrt{x^4}\sqrt{x} = x^2\sqrt{x}$

63. $(-3\sqrt{x^3})^3 = -27\sqrt{x^9} = -27\sqrt{x^8}\sqrt{x}$

$= -27x^4\sqrt{x}$

65. $(2x\sqrt[3]{x^2})^3 = 8x^3\sqrt[3]{x^6} = 8x^3\cdot x^2 = 8x^5$

67. $(-2\sqrt[3]{5})^2 = (-2)^2\cdot\sqrt[3]{5^2} = 4\sqrt[3]{25}$

69. $(\sqrt[3]{x^2})^6 = (x^2)^{6/3} = (x^2)^2 = x^4$

71. $\dfrac{\sqrt{3}}{\sqrt{2}} + \dfrac{2}{\sqrt{2}} = \dfrac{\sqrt{3}\sqrt{2}}{\sqrt{2}\sqrt{2}} + \dfrac{2\sqrt{2}}{\sqrt{2}\sqrt{2}}$

$= \dfrac{\sqrt{6}}{2} + \dfrac{2\sqrt{2}}{2} = \dfrac{\sqrt{6}+2\sqrt{2}}{2}$

73. $\dfrac{\sqrt{3}}{\sqrt{2}} + \dfrac{3\sqrt{6}}{2} = \dfrac{\sqrt{3}\sqrt{2}}{\sqrt{2}\sqrt{2}} + \dfrac{3\sqrt{6}}{2}$

$= \dfrac{\sqrt{6}}{2} + \dfrac{3\sqrt{6}}{2} = \dfrac{4\sqrt{6}}{2} = 2\sqrt{6}$

75. $\dfrac{\sqrt{6}}{2}\cdot\dfrac{1}{\sqrt{3}} = \dfrac{\sqrt{2}\sqrt{3}}{2\sqrt{3}} = \dfrac{\sqrt{2}}{2}$

77. $(2\sqrt{w}) \div (3\sqrt{w}) = \dfrac{2\sqrt{w}}{3\sqrt{w}} = \dfrac{2}{3}$

79. $\dfrac{8-\sqrt{32}}{20} = \dfrac{8-4\sqrt{2}}{20} = \dfrac{4(2-\sqrt{2})}{4\cdot 5}$

$= \dfrac{2-\sqrt{2}}{5}$

81. $\dfrac{5+\sqrt{75}}{10} = \dfrac{5+5\sqrt{3}}{10} = \dfrac{5(1+\sqrt{3})}{5\cdot 2}$

$= \dfrac{1+\sqrt{3}}{2}$

83. $\sqrt{a}(\sqrt{a} - 3) = a - 3\sqrt{a}$

85. $4\sqrt{a}(a + \sqrt{a}) = 4a\sqrt{a} + 4a$

87. $(2\sqrt{3m})^2 = 4 \cdot 3m = 12m$

89. $\left(-2\sqrt{xy^2z}\right)^2 = (-2)^2xy^2z = 4xy^2z$

91. $\sqrt[3]{m}\left(\sqrt[3]{m^2} - \sqrt[3]{m^5}\right) = \sqrt[3]{m^3} - \sqrt[3]{m^6}$
$$= m - m^2$$

93. $\sqrt[3]{8x^4} + \sqrt[3]{27x^4}$
$$= \sqrt[3]{8x^3} \cdot \sqrt[3]{x} + \sqrt[3]{27x^3} \cdot \sqrt[3]{x}$$
$$= 2x \cdot \sqrt[3]{x} + 3x \cdot \sqrt[3]{x} = 5x\sqrt[3]{x}$$

95. $\left(2m\sqrt[4]{2m^2}\right)^3 = 8m^3\sqrt[4]{8m^6}$
$$= 8m^3\sqrt[4]{m^4}\sqrt[4]{8m^2}$$
$$= 8m^3 \cdot m \cdot \sqrt[4]{8m^2} = 8m^4\sqrt[4]{8m^2}$$

97. $\dfrac{x-9}{\sqrt{x}-3} = \dfrac{(x-9)(\sqrt{x}+3)}{(\sqrt{x}-3)(\sqrt{x}+3)}$
$$= \dfrac{(x-9)(\sqrt{x}+3)}{x-9} = \sqrt{x}+3$$

99. $\dfrac{3\sqrt{k}}{\sqrt{k}+\sqrt{7}} = \dfrac{3\sqrt{k}(\sqrt{k}-\sqrt{7})}{(\sqrt{k}+\sqrt{7})(\sqrt{k}-\sqrt{7})}$
$$= \dfrac{3k - 3\sqrt{7k}}{k-7}$$

101. $\dfrac{5}{\sqrt{2}-1} + \dfrac{3}{\sqrt{2}+1}$
$$= \dfrac{5(\sqrt{2}+1)}{(\sqrt{2}-1)(\sqrt{2}+1)} + \dfrac{3(\sqrt{2}-1)}{(\sqrt{2}+1)(\sqrt{2}-1)}$$
$$= \dfrac{5\sqrt{2}+5}{2-1} + \dfrac{3\sqrt{2}-3}{2-1} = 2 + 8\sqrt{2}$$

103. $\dfrac{1}{\sqrt{2}} + \dfrac{1}{\sqrt{3}} = \dfrac{\sqrt{2}}{2} + \dfrac{\sqrt{3}}{3}$
$$= \dfrac{3\sqrt{2}}{3 \cdot 2} + \dfrac{2\sqrt{3}}{2 \cdot 3} = \dfrac{3\sqrt{2}+2\sqrt{3}}{6}$$

105. $\dfrac{3}{\sqrt{2}-1} + \dfrac{4}{\sqrt{2}+1}$
$$= \dfrac{3(\sqrt{2}+1)}{(\sqrt{2}-1)(\sqrt{2}+1)} + \dfrac{4(\sqrt{2}-1)}{(\sqrt{2}+1)(\sqrt{2}-1)}$$
$$= \dfrac{3\sqrt{2}+3}{1} + \dfrac{4\sqrt{2}-4}{1} = 7\sqrt{2}-1$$

107. $\dfrac{\sqrt{x}}{\sqrt{x}+2} + \dfrac{3\sqrt{x}}{\sqrt{x}-2}$
$$= \dfrac{\sqrt{x}(\sqrt{x}-2)}{(\sqrt{x}+2)(\sqrt{x}-2)}$$
$$+ \dfrac{3\sqrt{x}(\sqrt{x}+2)}{(\sqrt{x}-2)(\sqrt{x}+2)}$$
$$= \dfrac{x-2\sqrt{x}}{x-4} + \dfrac{3x+6\sqrt{x}}{x-4} = \dfrac{4x+4\sqrt{x}}{x-4}$$

109. $\dfrac{1}{\sqrt{x}} + \dfrac{1}{1-\sqrt{x}}$
$$= \dfrac{1(1-\sqrt{x})}{\sqrt{x}(1-\sqrt{x})} + \dfrac{1\sqrt{x}}{(1-\sqrt{x})\sqrt{x}}$$
$$= \dfrac{1-\sqrt{x}}{\sqrt{x}-x} + \dfrac{\sqrt{x}}{\sqrt{x}-x} = \dfrac{1}{\sqrt{x}-x}$$
$$= \dfrac{1(\sqrt{x}+x)}{(\sqrt{x}-x)(\sqrt{x}+x)} = \dfrac{x+\sqrt{x}}{x-x^2}$$

111. a) Use $(a-b)(a^2+ab+b^2) = a^3 - b^3$ with $a = x$ and $b = \sqrt[3]{2}$ to get $x^3 - 2$.

b) Use $a^3 + b^3 = (a+b)(a^2 - ab + b^2)$ to get
$$x^3 + 5 = x^3 + \left(\sqrt[3]{5}\right)^3$$
$$= (x + \sqrt[3]{5})(x^2 - \sqrt[3]{5}x + \sqrt[3]{25})$$

c) Use $(a-b)(a^2+ab+b^2) = a^3 - b^3$ with $a = \sqrt[3]{5}$ and $b = \sqrt[3]{2}$ to get
$$\left(\sqrt[3]{5} - \sqrt[3]{2}\right)\left(\sqrt[3]{25} + \sqrt[3]{10} + \sqrt[3]{4}\right)$$
$$= 5 - 2 = 3$$

d)
$$a+b = \left(\sqrt[3]{a} + \sqrt[3]{b}\right)\left(\sqrt[3]{a^2} - \sqrt[3]{ab} + \sqrt[3]{b^2}\right)$$
$$a-b = \left(\sqrt[3]{a} - \sqrt[3]{b}\right)\left(\sqrt[3]{a^2} + \sqrt[3]{ab} + \sqrt[3]{b^2}\right)$$

7.5 WARM-UPS

1. False, because $x^2 = 4$ is equivalent to the compound equation $x = 2$ or $x = -2$.

2. True, because the square of any real number is nonnegative.

3. False, 0 is a solution.

4. False, because -2 is not a solution to $x^3 = 8$.

5. True, because if x is positive the left side of the equation is negative and the right side is positive.

6. False, we should square each side.

7. False, extraneous roots are found but they do not satisfy the equation.

8. True, because we get $x = 49$, and 49 does not satisfy $\sqrt{x} = -7$.

9. True, because both square roots of 6 satisfy $x^2 - 6 = 0$.

10. True, we get extraneous roots only by raising each side to an even power.

7.5 EXERCISES

1. The odd-root property says that if n is an odd positive integer then $x^n = k$ is equivalent to $x = \sqrt[n]{k}$ for any real number k.

3. An extraneous solution is a solution that appears when solving an equation, but does not satisfy the original equation.

5. $x^3 = -1000$
$$x = \sqrt[3]{-1000} = -10$$
Solution set: $\{-10\}$

7. $32m^5 - 1 = 0$
$$32m^5 = 1$$
$$m^5 = \frac{1}{32}$$
$$m = \sqrt[5]{\frac{1}{32}} = \frac{1}{2}$$
Solution set: $\left\{\frac{1}{2}\right\}$

9. $(y-3)^3 = -8$
$$y - 3 = \sqrt[3]{-8}$$
$$y - 3 = -2$$
$$y = 1$$
Solution set: $\{1\}$

11. $\frac{1}{2}x^3 + 4 = 0$
$$\frac{1}{2}x^3 = -4$$
$$x^3 = -8$$
$$x = \sqrt[3]{-8} = -2$$
Solution set: $\{-2\}$

13. $x^2 = 25$
$$x = \pm 5$$
Solution set: $\{-5, 5\}$

15. $x^2 - 20 = 0$
$$x^2 = 20$$
$$x = \pm\sqrt{20} = \pm 2\sqrt{5}$$
Solution set: $\{-2\sqrt{5}, 2\sqrt{5}\,\}$

17. $x^2 = -9$
Since an even root of a negative number is not a real number, there is no real solution.

19. $(x-3)^2 = 16$
$$x - 3 = \pm 4$$
$$x = 3 \pm 4$$
$$x = 3 + 4 \quad \text{or} \quad x = 3 - 4$$
$$x = 7 \quad \text{or} \quad x = -1$$
Solution set: $\{-1, 7\}$

21. $(x+1)^2 - 8 = 0$
$$(x+1)^2 = 8$$

$$x + 1 = \pm\sqrt{8}$$
$$x = -1 \pm 2\sqrt{2}$$
Solution set: $\left\{-1 - 2\sqrt{2}, -1 + 2\sqrt{2}\right\}$

23. $\frac{1}{2}x^2 = 5$
$$x^2 = 10$$
$$x = \pm\sqrt{10}$$
Solution set: $\left\{-\sqrt{10}, \sqrt{10}\right\}$

25. $(y-3)^4 = 0$
$$y - 3 = \sqrt[4]{0} = 0$$
$$y = 3$$
Solution set: $\{3\}$

27. $2x^6 = 128$
$$x^6 = 64$$
$$x = \pm\sqrt[6]{64}$$
$$x = \pm 2$$
Solution set: $\{-2, 2\}$

29. $\sqrt{x-3} - 3 = 4$
$$\sqrt{x-3} = 7$$
$$(\sqrt{x-3})^2 = 7^2$$
$$x - 3 = 49$$
$$x = 52$$
Check 52 in the original equation.
Solution set: $\{52\}$

31. $2\sqrt{w+4} = 5$
$$(2\sqrt{w+4})^2 = 5^2$$
$$4(w+4) = 25$$
$$4w + 16 = 25$$
$$4w = 9$$
$$w = \frac{9}{4}$$
Check $\frac{9}{4}$ in the original equation.
Solution set: $\left\{\frac{9}{4}\right\}$

33. $\sqrt[3]{2x+3} = \sqrt[3]{x+12}$
$$\left(\sqrt[3]{2x+3}\right)^3 = \left(\sqrt[3]{x+12}\right)^3$$
$$2x + 3 = x + 12$$
$$x = 9$$
Check 9 in the original equation.
Solution set: $\{9\}$

35. $\sqrt{2t-4} = \sqrt{t-1}$
$$(\sqrt{2t-4})^2 = (\sqrt{t-1})^2$$
$$2t - 4 = t - 1$$
$$t = 3$$
Check: $\sqrt{2(3)-4} = \sqrt{3-1}$
Since each side of the equation is $\sqrt{2}$, the solution set is $\{3\}$.

148

37.
$$\sqrt{4x^2 + x - 3} = 2x$$
$$\left(\sqrt{4x^2 + x - 3}\right)^2 = (2x)^2$$
$$4x^2 + x - 3 = 4x^2$$
$$x - 3 = 0$$
$$x = 3$$
Check 3 in the original equation.
Solution set: $\{3\}$

39.
$$\sqrt{x^2 + 2x - 6} = 3$$
$$\left(\sqrt{x^2 + 2x - 6}\right)^2 = 3^2$$
$$x^2 + 2x - 6 = 9$$
$$x^2 + 2x - 15 = 0$$
$$(x + 5)(x - 3) = 0$$
$$x + 5 = 0 \text{ or } x - 3 = 0$$
$$x = -5 \text{ or } x = 3$$
Check 3 and -5 in the original equation.
Solution set: $\{-5, 3\}$

41.
$$\sqrt{2x^2 - 1} = x$$
$$\left(\sqrt{2x^2 - 1}\right)^2 = x^2$$
$$2x^2 - 1 = x^2$$
$$x^2 = 1$$
$$x = \pm 1$$
Checking in the original we find that if $x = -1$ we get $\sqrt{1} = -1$, which is incorrect. So the solution set is $\{1\}$.

43.
$$\sqrt{2x^2 + 5x + 6} = x$$
$$\left(\sqrt{2x^2 + 5x + 6}\right)^2 = x^2$$
$$2x^2 + 5x + 6 = x^2$$
$$x^2 + 5x + 6 = 0$$
$$(x + 2)(x + 3) = 0$$
$$x = -2 \text{ or } x = -3$$
If we use $x = -2$ in the original equation we get $\sqrt{4} = -2$, and if $x = -3$ in the original we get $\sqrt{9} = -3$. Since both of these equations are incorrect, the solution set is $\emptyset$.

45.
$$\sqrt{x} + \sqrt{x - 3} = 3$$
$$\sqrt{x} = 3 - \sqrt{x - 3}$$
$$\left(\sqrt{x}\right)^2 = \left(3 - \sqrt{x - 3}\right)^2$$
$$x = 9 - 6\sqrt{x - 3} + x - 3$$
$$-6 = -6\sqrt{x - 3}$$
$$1 = \sqrt{x - 3}$$

$$1^2 = \left(\sqrt{x - 3}\right)^2$$
$$1 = x - 3$$
$$4 = x$$
Check 4 in the original equation. The solution set is $\{4\}$.

47.
$$\sqrt{x + 2} + \sqrt{x - 1} = 3$$
$$\sqrt{x + 2} = 3 - \sqrt{x - 1}$$
$$\left(\sqrt{x + 2}\right)^2 = \left(3 - \sqrt{x - 1}\right)^2$$
$$x + 2 = 9 - 6\sqrt{x - 1} + x - 1$$
$$-6 = -6\sqrt{x - 1}$$
$$1 = \sqrt{x - 1}$$
$$1^2 = \left(\sqrt{x - 1}\right)^2$$
$$1 = x - 1$$
$$2 = x$$
Check 2 in the original equation. The solution set is $\{2\}$.

49.
$$\sqrt{x + 3} - \sqrt{x - 2} = 1$$
$$\sqrt{x + 3} = 1 + \sqrt{x - 2}$$
$$\left(\sqrt{x + 3}\right)^2 = \left(1 + \sqrt{x - 2}\right)^2$$
$$x + 3 = 1 + 2\sqrt{x - 2} + x - 2$$
$$4 = 2\sqrt{x - 2}$$
$$2 = \sqrt{x - 2}$$
$$2^2 = \left(\sqrt{x - 2}\right)^2$$
$$4 = x - 2$$
$$6 = x$$
Check 6 in the original equation. The solution set is $\{6\}$.

51.
$$\sqrt{2x + 2} - \sqrt{x - 3} = 2$$
$$\sqrt{2x + 2} = 2 + \sqrt{x - 3}$$
$$\left(\sqrt{2x + 2}\right)^2 = \left(2 + \sqrt{x - 3}\right)^2$$
$$2x + 2 = 4 + 4\sqrt{x - 3} + x - 3$$
$$x + 1 = 4\sqrt{x - 3}$$
$$(x + 1)^2 = \left(4\sqrt{x - 3}\right)^2$$
$$x^2 + 2x + 1 = 16(x - 3)$$
$$x^2 - 14x + 49 = 0$$
$$(x - 7)^2 = 0$$
$$x = 7$$
Check 7 in the original equation.
Solution set: $\{7\}$

53.
$$\sqrt{4 - x} - \sqrt{x + 6} = 2$$
$$\sqrt{4 - x} = 2 + \sqrt{x + 6}$$
$$\left(\sqrt{4 - x}\right)^2 = \left(2 + \sqrt{x + 6}\right)^2$$

$$4 - x = 4 + 4\sqrt{x+6} + x + 6$$
$$-6 - 2x = 4\sqrt{x+6}$$
$$-3 - x = 2\sqrt{x+6}$$
$$(-3-x)^2 = \left(2\sqrt{x+6}\right)^2$$
$$9 + 6x + x^2 = 4(x+6)$$
$$x^2 + 2x - 15 = 0$$
$$(x+5)(x-3) = 0$$
$$x = -5 \quad \text{or} \quad x = 3$$

If $x = 3$ in the original equation we get $1 - 3 = 2$, which is incorrect. If $x = -5$ we get $3 - 1 = 2$. So the solution set is $\{-5\}$.

55.
$$\sqrt{x-5} - \sqrt{x} = 3$$
$$\sqrt{x-5} = 3 + \sqrt{x}$$
$$\left(\sqrt{x-5}\right)^2 = (3+\sqrt{x})^2$$
$$x - 5 = 9 + 6\sqrt{x} + x$$
$$-14 = 6\sqrt{x}$$
$$-\frac{7}{3} = \sqrt{x}$$
$$x = \left(-\frac{7}{3}\right)^2 = \frac{49}{9}$$

Since $49/9$ does not satisfy the original equation, the solution set is $\emptyset$.

57.
$$\sqrt{3x+1} + \sqrt{2x+4} = 3$$
$$\sqrt{3x+1} = 3 - \sqrt{2x+4}$$
$$\left(\sqrt{3x+1}\right)^2 = (3 - \sqrt{2x+4})^2$$
$$3x + 1 = 9 - 6\sqrt{2x+4} + 2x + 4$$
$$x - 12 = -6\sqrt{2x+4}$$
$$(x-12)^2 = (-6\sqrt{2x+4})^2$$
$$x^2 - 24x + 144 = 36(2x+4)$$
$$x^2 - 96x = 0$$
$$(x)(x-96) = 0$$
$$x = 0 \quad \text{or} \quad x - 96 = 0$$
$$x = 6 \quad \text{or} \qquad x = 96$$

Since 96 does not satisfy the original equation, the solution set is $\{0\}$.

59.
$$x^{2/3} = 3$$
$$\left(x^{2/3}\right)^3 = 3^3$$
$$x^2 = 27$$
$$x = \pm\sqrt{27} = \pm 3\sqrt{3}$$
Solution set: $\left\{-3\sqrt{3}, 3\sqrt{3}\right\}$

61.
$$y^{-2/3} = 9$$
$$\left(y^{-2/3}\right)^{-3} = (9)^{-3}$$
$$y^2 = \frac{1}{729}$$
$$y = \pm\sqrt{\frac{1}{729}} = \pm\frac{1}{27}$$

Solution set: $\left\{-\frac{1}{27}, \frac{1}{27}\right\}$

63.
$$w^{1/3} = 8$$
$$\left(w^{1/3}\right)^3 = 8^3$$
$$w = 512$$
Solution set: $\{512\}$

65.
$$t^{-1/2} = 9$$
$$\left(t^{-1/2}\right)^{-2} = (9)^{-2}$$
$$t = \frac{1}{81}$$
Solution set: $\left\{\frac{1}{81}\right\}$

67.
$$\left((3a-1)^{-2/5}\right)^{-5} = 1^{-5}$$
$$(3a-1)^2 = 1$$
$$3a - 1 = \pm 1$$
$$3a - 1 = 1 \quad \text{or} \quad 3a - 1 = -1$$
$$3a = 2 \quad \text{or} \qquad 3a = 0$$
$$a = \frac{2}{3} \quad \text{or} \qquad a = 0$$
Solution set: $\left\{0, \frac{2}{3}\right\}$

69. $(t-1)^{-2/3} = 2$
$$\left((t-1)^{-2/3}\right)^{-3} = 2^{-3}$$
$$(t-1)^2 = \frac{1}{8}$$
$$t - 1 = \pm\sqrt{\frac{1}{8}} = \pm\frac{\sqrt{2}}{4}$$
$$t = 1 \pm \frac{\sqrt{2}}{4} = \frac{4}{4} \pm \frac{\sqrt{2}}{4} = \frac{4 \pm \sqrt{2}}{4}$$
Solution set: $\left\{\frac{4-\sqrt{2}}{4}, \frac{4+\sqrt{2}}{4}\right\}$

71. $(x-3)^{2/3} = -4$
$$\left((x-3)^{2/3}\right)^3 = (-4)^3$$
$$(x-3)^2 = -64$$
Because the square of any real number is nonnegative, the solution set is $\emptyset$. There is no real solution.

73.
$$2x^2 + 3 = 7$$
$$2x^2 = 4$$
$$x^2 = 2$$
$$x = \pm\sqrt{2}$$
Solution set: $\left\{-\sqrt{2}, \sqrt{2}\right\}$

75.
$$\sqrt[3]{2w+3} = \sqrt[3]{w-2}$$
$$\left(\sqrt[3]{2w+3}\right)^3 = \left(\sqrt[3]{w-2}\right)^3$$
$$2w + 3 = w - 2$$
$$w = -5$$
Solution set: $\{-5\}$

150

77. $(w+1)^{2/3} = -3$

$$\left((w+1)^{2/3}\right)^3 = (-3)^3$$
$$(w+1)^2 = -27$$

This equation has no real solution by the even root property. The solution set is $\emptyset$.

79. $(a+1)^{1/3} = -2$

$$\left((a+1)^{1/3}\right)^3 = (-2)^3$$
$$a+1 = -8$$
$$a = -9$$

Solution set: $\{-9\}$

81. $(4y-5)^7 = 0$

$$4y - 5 = \sqrt[7]{0} = 0$$
$$4y = 5$$
$$y = \frac{5}{4}$$

Solution set: $\left\{\frac{5}{4}\right\}$

83. $\sqrt{5x^2+4x+1} - x = 0$

$$\sqrt{5x^2+4x+1} = x$$
$$5x^2+4x+1 = x^2$$
$$4x^2+4x+1 = 0$$
$$(2x+1)^2 = 0$$
$$2x+1 = 0$$
$$x = -\frac{1}{2}$$

Since $-1/2$ does not satisfy the original equation, the solution set is $\emptyset$.

85. $\sqrt{4x^2} = x+2$

$$\left(\sqrt{4x^2}\right)^2 = (x+2)^2$$
$$4x^2 = x^2 + 4x + 4$$
$$3x^2 - 4x - 4 = 0$$
$$(3x+2)(x-2) = 0$$
$$3x+2 = 0 \quad \text{or} \quad x-2 = 0$$
$$x = -\frac{2}{3} \quad \text{or} \quad x = 2$$

Solution set: $\left\{-\frac{2}{3}, 2\right\}$

87. $(t+2)^4 = 32$

$$t+2 = \pm\sqrt[4]{32} = \pm 2 \cdot \sqrt[4]{2}$$
$$t = -2 \pm 2 \cdot \sqrt[4]{2}$$

Solution set: $\left\{-2 - 2\sqrt[4]{2}, -2 + 2\sqrt[4]{2}\right\}$

89. $\sqrt{x^2-3x} = x$

$$\left(\sqrt{x^2-3x}\right)^2 = x^2$$
$$x^2 - 3x = x^2$$
$$-3x = 0$$
$$x = 0$$

Solution set: $\{0\}$

91. $x^{-3} = 8$

$$\left(x^{-3}\right)^{-1} = 8^{-1}$$
$$x^3 = \frac{1}{8}$$
$$x = \sqrt[3]{\frac{1}{8}} = \frac{1}{2}$$

Solution set: $\left\{\frac{1}{2}\right\}$

93. Let $x =$ the length of a side. Two sides and the diagonal of a square form a right triangle. By the Pythagorean theorem we can write the equation

$$x^2 + x^2 = 8^2$$
$$2x^2 = 64$$
$$x^2 = 32$$
$$x = \pm\sqrt{32} = \pm 4\sqrt{2}$$

The length of the side is not a negative number. So the side is $4\sqrt{2}$ feet in length.

95. Let $s =$ the length of the side of the square. Since $A = s^2$ for a square, we can write the equation

$$s^2 = 50$$
$$s = \pm\sqrt{50} = \pm 5\sqrt{2}$$

Since the side of a square is not negative, the length of the side is $5\sqrt{2}$ feet.

97. Let $d =$ the length of a diagonal of the rectangle whose sides are 30 and 40 feet. By the Pythagorean theorem we can write the equation

$$d^2 = 30^2 + 40^2$$
$$d^2 = 2500$$
$$d = \pm\sqrt{2500} = \pm 50$$

Since the diagonal is not negative, the length of the diagonal is 50 feet.

99. a) $C = 4(23{,}245)^{-1/3}(13.5) \approx 1.89$

b)
$$C = 4d^{-1/3}b$$
$$d^{-1/3} = \frac{C}{4b}$$
$$(d^{-1/3})^{-3} = \left(\frac{C}{4b}\right)^{-3}$$
$$d = \frac{64b^3}{C^3}$$

c) The capsize screening value is less than 2 when $d > 19{,}683$ pounds.

101. If the volume of the cube is 2, each side of the cube has length $\sqrt[3]{2}$, because $V = s^3$. Let $d =$ the length of the diagonal of a side. The diagonal of a side is the diagonal of a square with sides of length $\sqrt[3]{2}$. By the Pythagorean theorem we can write

$$d^2 = \left(\sqrt[3]{2}\right)^2 + \left(\sqrt[3]{2}\right)^2$$
$$d^2 = \sqrt[3]{4} + \sqrt[3]{4}$$
$$d^2 = 2 \cdot \sqrt[3]{4} = \sqrt[3]{8} \cdot \sqrt[3]{4} = \sqrt[3]{32}$$
$$d = \left(\sqrt[3]{32}\right)^{1/2} = \left(32^{1/3}\right)^{1/2} = 32^{1/6}$$

The length of the diagonal is $\sqrt[6]{32}$ meters.

103. Let $x =$ the third side to the triangle whose given sides are 3 and 5. By the Pythagorean theorem we can find x:
$$x^2 + 3^2 = 5^2$$
$$x^2 = 16$$
$$x = 4$$

Since $x = 4$, the base of length 12 is divided into 2 parts, one of length 4 and the other of length 8. The side marked a is the hypotenuse of a right triangle with legs 3 and 8:
$$a^2 = 3^2 + 8^2$$
$$a^2 = 73$$
$$a = \sqrt{73} \text{ km}$$

105. $\quad r = \left(\dfrac{S}{P}\right)^{1/n} - 1$
$$1 + r = \left(\frac{S}{P}\right)^{1/n}$$
$$\frac{S}{P} = (1+r)^n$$
$$S = P(1+r)^n$$

Solve for P:
$$P = \frac{S}{(1+r)^n}$$
$$P = S(1+r)^{-n}$$

107. $\quad \dfrac{11.86^2}{5.2^3} = \dfrac{29.46^2}{R^3}$
$$11.86^2 R^3 = 5.2^3 \cdot 29.46^2$$
$$R^3 = \frac{5.2^3 \cdot 29.46^2}{11.86^2}$$
$$R = \sqrt[3]{\frac{5.2^3 \cdot 29.46^2}{11.86^2}} \approx 9.5 \text{ AU}$$

109. $\quad x^2 = 3.24$
$$x = \pm\sqrt{3.24} = \pm 1.8$$
Solution set: $\{-1.8, 1.8\}$

111. $\quad \sqrt{x-2} = 1.73$
$$x - 2 = (1.73)^2$$
$$x = 2 + (1.73)^2 \approx 4.993$$
Solution set: $\{4.993\}$

113. $\quad x^{2/3} = 8.86$
$$\left(x^{2/3}\right)^3 = (8.86)^3$$
$$x^2 = 695.506$$

$$x = \pm\sqrt{695.506} \approx \pm 26.372$$
Solution set: $\{-26.372, 26.372\}$

115. Since $V = s^3$, $s = \sqrt[3]{V}$.

7.6 WARM-UPS

1. True, because every real number is a complex number.

2. False, because $2 - \sqrt{-6} = 2 - i\sqrt{6}$.

3. False, because $\sqrt{-9} = 3i$.

4. True, because $(\pm 3i)^2 = -9$.

5. True, because we subtract complex numbers just like we subtract binomials.

6. True, because $i^4 = i^2 \cdot i^2 = (-1)(-1) = 1$.

7. True, because
$(2-i)(2+i) = 4 - i^2 = 4 - (-1) = 5$.

8. False, $i^3 = i^2 \cdot i = -1 \cdot i = -i$.

9. True, $i^{48} = (i^4)^{12} = 1^{12} = 1$.

10. False, $x^2 = 0$ has only one solution.

7.6 EXERCISES

1. A complex number is a number of the form $a + bi$ where a and b are real numbers.

3. The union of the real numbers and the imaginary numbers is the set of complex numbers.

5. The conjugate of $a + bi$ is $a - bi$.

7. $(2+3i) + (-4+5i) = -2+8i$

9. $(2-3i) - (6-7i) = 2 - 3i - 6 + 7i$
$$= -4 + 4i$$

11. $(-1+i) + (-1-i) = -2$

13. $(-2-3i) - (6-i) = -2 - 3i - 6 + i$
$$= -8 - 2i$$

15. $3(2+5i) = 3 \cdot 2 + 3 \cdot 5i = 6 + 15i$

17. $2i(i-5) = 2i^2 - 10i = 2(-1) - 10i$
$$= -2 - 10i$$

19. $-4i(3-i) = -12i + 4i^2$
$$= -12i + 4(-1) = -4 - 12i$$

21. $(2+3i)(4+6i) = 8 + 24i + 18i^2$
$$= 8 + 24i + 18(-1) = -10 + 24i$$

23. $(-1+i)(2-i) = -2 + 3i - i^2$
$$= -2 + 3i - (-1) = -1 + 3i$$

25. $(-1-2i)(2+i) = -2 - 5i - 2i^2$
$$= -2 - 5i - 2(-1) = -5i$$

27. $(5-2i)(5+2i) = 25 - 4i^2$
$$= 25 - 4(-1) = 29$$

29. $(1-i)(1+i) = 1 - i^2 = 1 - (-1) = 2$

31. $(4 + 2i)(4 - 2i) = 16 - 4i^2$
$$= 16 - 4(-1) = 20$$

33. $(3i)^2 = 9i^2 = 9(-1) = -9$

35. $(-5i)^2 = (-5)^2 i^2 = 25(-1) = -25$

37. $(2i)^4 = 2^4 i^4 = 16(1) = 16$

39. $i^9 = (i^4)^2 \cdot i = 1^2 \cdot i = i$

41. $(3 + 5i)(3 - 5i) = 9 - 25i^2 = 9 + 25$
$$= 34$$

43. $(1 - 2i)(1 + 2i) = 1 - 4i^2 = 1 - 4(-1)$
$$= 5$$

45. $(-2 + i)(-2 - i) = 4 - i^2 = 4 - (-1)$
$$= 5$$

47. $(2 - i\sqrt{3})(2 + i\sqrt{3}) = 4 - 3i^2$
$$= 4 - 3(-1) = 7$$

49. $\dfrac{3}{4 + i} = \dfrac{3(4 - i)}{(4 + i)(4 - i)} = \dfrac{12 - 3i}{16 - i^2}$
$$= \dfrac{12 - 3i}{17} = \dfrac{12}{17} - \dfrac{3}{17}i$$

51. $\dfrac{2 + i}{3 - 2i} = \dfrac{(2 + i)(3 + 2i)}{(3 - 2i)(3 + 2i)}$
$$= \dfrac{6 + 7i + 2i^2}{9 - 4i^2} = \dfrac{4 + 7i}{13} = \dfrac{4}{13} + \dfrac{7}{13}i$$

53. $\dfrac{4 + 3i}{i} = \dfrac{(4 + 3i)(-i)}{(i)(-i)} = \dfrac{-4i - 3i^2}{-i^2}$
$$= \dfrac{-4i + 3}{1} = 3 - 4i$$

55. $\dfrac{2 + 6i}{2} = \dfrac{2}{2} + \dfrac{6i}{2} = 1 + 3i$

57. $\dfrac{1 + i}{3i - 2} = \dfrac{(1 + i)(3i + 2)}{(3i - 2)(3i + 2)}$
$$= \dfrac{2 + 5i + 3i^2}{9i^2 - 4} = \dfrac{-1 + 5i}{-13} = \dfrac{1}{13} - \dfrac{5}{13}i$$

59. $\dfrac{6}{3i} = \dfrac{6 \cdot (-3i)}{(3i)(-3i)} = \dfrac{-18i}{-9i^2} = \dfrac{-18i}{9} = -2i$

61. $2 + \sqrt{-4} = 2 + i\sqrt{4} = 2 + 2i$

63. $2\sqrt{-9} + 5 = 2i\sqrt{9} + 5 = 2i \cdot 3 + 5$
$$= 5 + 6i$$

65. $7 - \sqrt{-6} = 7 - i\sqrt{6}$

67. $\sqrt{-8} + \sqrt{-18} = i\sqrt{4}\sqrt{2} + i\sqrt{9}\sqrt{2}$
$$= 2i\sqrt{2} + 3i\sqrt{2} = 5i\sqrt{2}$$

69. $\dfrac{2 + \sqrt{-12}}{2} = \dfrac{2 + i\sqrt{4}\sqrt{3}}{2} = 1 + i\sqrt{3}$

71. $\dfrac{-4 - \sqrt{-24}}{4} = \dfrac{-4 - i\sqrt{4}\sqrt{6}}{4}$
$$= \dfrac{-4}{4} - \dfrac{2i\sqrt{6}}{4} = -1 - \dfrac{1}{2}i\sqrt{6}$$

73. $\sqrt{-2} \cdot \sqrt{-6} = i\sqrt{2} \cdot i\sqrt{6} = i^2\sqrt{12}$
$$= -1 \cdot 2\sqrt{3} = -2\sqrt{3}$$

75. $\sqrt{-3} \cdot \sqrt{-27} = i\sqrt{3} \cdot i\sqrt{27} = i^2\sqrt{81}$
$$= -1 \cdot 9 = -9$$

77. $\dfrac{\sqrt{8}}{\sqrt{-4}} = \dfrac{\sqrt{8}}{i\sqrt{4}} = \dfrac{\sqrt{2}}{i} = \dfrac{\sqrt{2} \cdot (-i)}{i(-i)}$
$$= \dfrac{-i\sqrt{2}}{-i^2} = \dfrac{-i\sqrt{2}}{1} = -i\sqrt{2}$$

79. $x^2 = -36$
$$x = \pm\sqrt{-36} = \pm 6i$$
Solution set: $\{\pm 6i\}$

81. $x^2 = -12$
$$x = \pm\sqrt{-12} = \pm i\sqrt{4}\sqrt{3} = \pm 2i\sqrt{3}$$
Solution set: $\{\pm 2i\sqrt{3}\}$

83. $2x^2 + 5 = 0$
$$x^2 = -\dfrac{5}{2}$$
$$x = \pm\sqrt{-\dfrac{5}{2}} = \pm\dfrac{i\sqrt{5}}{\sqrt{2}} = \pm\dfrac{i\sqrt{10}}{2}$$
Solution set: $\left\{\pm\dfrac{i\sqrt{10}}{2}\right\}$

85. $3x^2 + 6 = 0$
$$x^2 = -2$$
$$x = \pm\sqrt{-2} = \pm i\sqrt{2}$$
Solution set: $\{\pm i\sqrt{2}\}$

87. $(2 - 3i)(3 + 4i) = 6 - i - 12i^2$
$$= 6 - i - 12(-1) = 18 - i$$

89. $(2 - 3i) + (3 + 4i) = 5 + i$

91. $\dfrac{2 - 3i}{3 + 4i} = \dfrac{(2 - 3i)(3 - 4i)}{(3 + 4i)(3 - 4i)}$
$$= \dfrac{6 - 17i + 12i^2}{9 - 16i^2} = \dfrac{6 - 17i + 12(-1)}{9 - 16(-1)}$$
$$= \dfrac{-6 - 17i}{25} = -\dfrac{6}{25} - \dfrac{17}{25}i$$

93. $i(2 - 3i) = 2i - 3i^2 = 2i + 3 = 3 + 2i$

95. $(-3i)^2 = 9i^2 = -9$

97. $\sqrt{-12} + \sqrt{-3} = 2i\sqrt{3} + i\sqrt{3} = 3i\sqrt{3}$

99. $(2 - 3i)^2 = 4 - 12i + 9i^2 = -5 - 12i$

101. $\dfrac{-4 + \sqrt{-32}}{2} = \dfrac{-4 + 4i\sqrt{2}}{2}$
$$= -2 + 2i\sqrt{2}$$

103. If $x = 2 - i$ then
$$x^2 - 4x + 5 = (2 - i)^2 - 4(2 - i) + 5$$
$$= 4 - 4i + i^2 - 8 + 4i + 5$$
$$= 1 + i^2 = 0$$
So $2 - i$ is a solution to $x^2 - 4x + 5 = 0$.

105. The product rule for radicals is valid only for radicals that represent real numbers. So the product of 4 is not correct. The correct product is found as follows:
$$\sqrt{-4} \cdot \sqrt{-4} = 2i \cdot 2i = 4i^2 = -4$$

Enriching Your Mathematical Word Power
1. d **2.** b **3.** b **4.** b **5.** d
6. b **7.** c **8.** a **9.** d **10.** c
11. a **12.** c **13.** d **14.** b

CHAPTER 7 REVIEW

1. $\sqrt[5]{32} = 2$ because $2^5 = 32$.

3. $\sqrt[3]{1000} = 10$ because $10^3 = 1000$.

5. $\sqrt{72} = \sqrt{36}\sqrt{2} = 6\sqrt{2}$

7. $\sqrt{x^{12}} = x^6$ because $\left(x^6\right)^2 = x^{12}$.

9. $\sqrt[3]{x^6} = x^2$ because $\left(x^2\right)^3 = x^6$.

11. $\sqrt{2x^9} = \sqrt{x^8}\sqrt{2x} = x^4\sqrt{2x}$

13. $\sqrt{8w^5} = \sqrt{4w^4}\sqrt{2w} = 2w^2\sqrt{2w}$

15. $\sqrt[3]{16x^4} = \sqrt[3]{8x^3}\sqrt[3]{2x} = 2x\sqrt[3]{2x}$

17. $\sqrt[4]{a^9 b^5} = \sqrt[4]{a^8 b^4}\sqrt[4]{ab} = a^2 b\sqrt[4]{ab}$

19. $\sqrt{\dfrac{x^3}{16}} = \dfrac{\sqrt{x^2}\sqrt{x}}{\sqrt{16}} = \dfrac{x\sqrt{x}}{4}$

21. $(-27)^{-2/3} = \dfrac{1}{(-3)^2} = \dfrac{1}{9}$

23. $(2^6)^{1/3} = 2^{6/3} = 2^2 = 4$

25. $100^{-3/2} = \dfrac{1}{10^3} = \dfrac{1}{1000}$

27. $\dfrac{3x^{-1/2}}{3^{-2}x^{-1}} = 3^3 x^{-1/2+1} = 27x^{1/2}$

29. $(a^{1/2}b)^3(ab^{1/4})^2 = a^{3/2}b^3 a^2 b^{2/4}$
$$= a^{7/2}b^{7/2}$$

31. $(x^{1/2}y^{1/4})(x^{1/4}y) = x^{1/2+1/4}y^{1/4+1}$
$$= x^{3/4}y^{5/4}$$

33. $\sqrt{13}\sqrt{13} = 13$

35. $\sqrt{27} + \sqrt{45} - \sqrt{75}$
$$= 3\sqrt{3} + 3\sqrt{5} - 5\sqrt{3} = 3\sqrt{5} - 2\sqrt{3}$$

37. $3\sqrt{2}(5\sqrt{2} - 7\sqrt{3}) = 15\cdot 2 - 21\sqrt{6}$
$$= 30 - 21\sqrt{6}$$

39. $(2 - \sqrt{3})(3 + \sqrt{2})$
$$= 6 - 3\sqrt{3} + 2\sqrt{2} - \sqrt{6}$$

41. $5 \div \sqrt{2} = \dfrac{5}{\sqrt{2}} = \dfrac{5\sqrt{2}}{\sqrt{2}\sqrt{2}} = \dfrac{5\sqrt{2}}{2}$

43. $\sqrt{\dfrac{2}{5}} = \dfrac{\sqrt{2}\sqrt{5}}{\sqrt{5}\sqrt{5}} = \dfrac{\sqrt{10}}{5}$

45. $\sqrt[3]{\dfrac{2}{3}} = \dfrac{\sqrt[3]{2}\cdot\sqrt[3]{9}}{\sqrt[3]{3}\cdot\sqrt[3]{9}} = \dfrac{\sqrt[3]{18}}{\sqrt[3]{27}} = \dfrac{\sqrt[3]{18}}{3}$

47. $\dfrac{2}{\sqrt{3x}} = \dfrac{2\sqrt{3x}}{\sqrt{3x}\sqrt{3x}} = \dfrac{2\sqrt{3x}}{3x}$

49. $\dfrac{\sqrt{10y^3}}{\sqrt{6}} = \dfrac{\sqrt{2}\sqrt{5y}\sqrt{y^2}}{\sqrt{2}\sqrt{3}} = \dfrac{y\sqrt{5y}}{\sqrt{3}}$

$$= \dfrac{y\sqrt{5y}\sqrt{3}}{\sqrt{3}\sqrt{3}} = \dfrac{y\sqrt{15y}}{3}$$

51. $\dfrac{3}{\sqrt[3]{2a}} = \dfrac{3\sqrt[3]{4a^2}}{\sqrt[3]{2a}\sqrt[3]{4a^2}} = \dfrac{3\sqrt[3]{4a^2}}{2a}$

53. $\dfrac{5}{\sqrt[4]{3x^2}} = \dfrac{5\sqrt[4]{27x^2}}{\sqrt[4]{3x^2}\cdot\sqrt[4]{27x^2}}$
$$= \dfrac{5\sqrt[4]{27x^2}}{\sqrt[4]{81x^4}} = \dfrac{5\sqrt[4]{27x^2}}{3x}$$

55. $(\sqrt{3})^4 = 3^{4/2} = 3^2 = 9$

57. $\dfrac{2 - \sqrt{8}}{2} = \dfrac{2 - 2\sqrt{2}}{2} = \dfrac{2(1 - \sqrt{2})}{2}$
$$= 1 - \sqrt{2}$$

59. $\dfrac{\sqrt{6}}{1 - \sqrt{3}} = \dfrac{\sqrt{6}(1 + \sqrt{3})}{(1 - \sqrt{3})(1 + \sqrt{3})}$
$$= \dfrac{\sqrt{6} + \sqrt{18}}{1 - 3} = \dfrac{\sqrt{6} + 3\sqrt{2}}{-2} = \dfrac{-\sqrt{6} - 3\sqrt{2}}{2}$$

61. $\dfrac{2\sqrt{3}}{3\sqrt{6} - \sqrt{12}} = \dfrac{2\sqrt{3}}{3\sqrt{6} - 2\sqrt{3}}$
$$= \dfrac{2\sqrt{3}(3\sqrt{6} + 2\sqrt{3})}{(3\sqrt{6} - 2\sqrt{3})(3\sqrt{6} + 2\sqrt{3})}$$

$$= \dfrac{6\sqrt{18} + 12}{54 - 12} = \dfrac{6\cdot 3\sqrt{2} + 12}{42} = \dfrac{6\cdot 3\sqrt{2} + 6\cdot 2}{6\cdot 7}$$

$$= \dfrac{3\sqrt{2} + 2}{7}$$

63. $\left(2w\sqrt[3]{2w^2}\right)^6 = 2^6 w^6 \sqrt[3]{2^6 w^{12}}$
$$= 2^6 2^2 w^6 w^4 = 256w^{10}$$

65. $x^2 = 16$
$$x = \pm 4$$
Solution set: $\{-4, 4\}$

67. $(a - 5)^2 = 4$
$$a - 5 = \pm 2$$
$$a = 5 \pm 2$$
$a = 5 + 2$ or $a = 5 - 2$
$a = 7$ or $a = 3$
Solution set: $\{3, 7\}$

69. $(a + 1)^2 = 5$
$$a + 1 = \pm\sqrt{5}$$
$$a = -1 \pm\sqrt{5}$$
Solution set: $\{-1 - \sqrt{5}, -1 + \sqrt{5}\}$

71. $(m+1)^2 = -8$

Since the square root of -8 is not a real number, there is no real solution. The solution set is $\emptyset$.

73.
$$\sqrt{m-1} = 3$$
$$(\sqrt{m-1})^2 = 3^2$$
$$m-1 = 9$$
$$m = 10$$

Solution set: $\{10\}$

75.
$$\sqrt[3]{2x+9} = 3$$
$$(\sqrt[3]{2x+9})^3 = 3^3$$
$$2x+9 = 27$$
$$2x = 18$$
$$x = 9$$

Solution set: $\{9\}$

77.
$$w^{2/3} = 4$$
$$(w^{2/3})^3 = 4^3$$
$$w^2 = 64$$
$$w = \pm 8$$

Solution set: $\{-8, 8\}$

79.
$$(m+1)^{1/3} = 5$$
$$((m+1)^{1/3})^3 = 5^3$$
$$m+1 = 125$$
$$m = 124$$

Solution set: $\{124\}$

81.
$$\sqrt{x-3} = \sqrt{x+2} - 1$$
$$\left(\sqrt{x-3}\right)^2 = \left(\sqrt{x+2} - 1\right)^2$$
$$x-3 = x+2 - 2\sqrt{x+2} + 1$$
$$-6 = -2\sqrt{x+2}$$
$$3 = \sqrt{x+2}$$
$$3^2 = (\sqrt{x+2})^2$$
$$9 = x+2$$
$$7 = x$$

Solution set: $\{7\}$

83.
$$\sqrt{5x-x^2} = \sqrt{6}$$
$$\left(\sqrt{5x-x^2}\right)^2 = (\sqrt{6})^2$$
$$5x-x^2 = 6$$
$$-x^2 + 5x - 6 = 0$$
$$x^2 - 5x + 6 = 0$$
$$(x-2)(x-3) = 0$$
$$x = 2 \text{ or } x = 3$$

Solution set: $\{2, 3\}$

85.
$$\sqrt{x+7} - 2\sqrt{x} = -2$$
$$\sqrt{x+7} = 2\sqrt{x} - 2$$
$$\left(\sqrt{x+7}\right)^2 = \left(2\sqrt{x} - 2\right)^2$$

$$x+7 = 4x - 8\sqrt{x} + 4$$
$$8\sqrt{x} = 3x - 3$$
$$\left(8\sqrt{x}\right)^2 = \left(3x - 3\right)^2$$
$$64x = 9x^2 - 18x + 9$$
$$0 = 9x^2 - 82x + 9$$
$$0 = (9x - 1)(x - 9)$$
$$x = \frac{1}{9} \text{ or } x = 9$$

Since $\sqrt{\frac{1}{9} + 7} - 2\sqrt{\frac{1}{9}} = \frac{8}{3} - \frac{2}{3} = 2$, $1/9$ is an extraneous root. The solution set is $\{9\}$.

87.
$$2\sqrt{x} - \sqrt{x-3} = 3$$
$$2\sqrt{x} = \sqrt{x-3} + 3$$
$$\left(2\sqrt{x}\right)^2 = \left(\sqrt{x-3} + 3\right)^2$$
$$4x = x - 3 + 6\sqrt{x-3} + 9$$
$$3x - 6 = 6\sqrt{x-3}$$
$$x - 2 = 2\sqrt{x-3}$$
$$(x-2)^2 = (2\sqrt{x-3})^2$$
$$x^2 - 4x + 4 = 4(x-3)$$
$$x^2 - 8x + 16 = 0$$
$$(x-4)^2 = 0$$
$$x = 4$$

Solution set: $\{4\}$

89. $(2-3i)(-5+5i) = -10 + 25i - 15i^2$
$$= -10 + 25i + 15 = 5 + 25i$$

91. $(2+i) + (5-4i) = 7 - 3i$

93. $(1-i) - (2-3i) = 1 - i - 2 + 3i$
$$= -1 + 2i$$

95. $\frac{6+3i}{3} = \frac{6}{3} + \frac{3i}{3} = 2 + i$

97. $\frac{4-\sqrt{-12}}{2} = \frac{4 - 2i\sqrt{3}}{2} = 2 - i\sqrt{3}$

99. $\frac{2-3i}{4+i} = \frac{(2-3i)(4-i)}{(4+i)(4-i)}$
$$= \frac{5-14i}{17} = \frac{5}{17} - \frac{14}{17}i$$

101. $x^2 + 100 = 0$
$$x^2 = -100$$
$$x = \pm\sqrt{-100} = \pm 10i$$

The solution set is $\{\pm 10i\}$.

103. $2b^2 + 9 = 0$
$$2b^2 = -9$$
$$b^2 = -\frac{9}{2}$$
$$b = \sqrt{-\frac{9}{2}} = \pm\frac{3i}{\sqrt{2}} = \pm\frac{3i\sqrt{2}}{2}$$

The solution set is $\left\{\pm\frac{3i\sqrt{2}}{2}\right\}$.

105. False, because $2^3 \cdot 3^2 = 8 \cdot 9 = 72$.
107. True, because $(\sqrt{2})^3 = \sqrt{8} = 2\sqrt{2}$.
109. True, because
$$8^{200}8^{200} = (8 \cdot 8)^{200} = 64^{200}.$$
111. False, because $4^{1/2} = \sqrt{4} = 2$.
113. False, because $5^2 \cdot 5^2 = 5^4 = 625$.
115. False, because $\sqrt{w^{10}} = |w^5|$.
117. False, because $\sqrt{x^6} = |x^3|$.
119. True, $\sqrt{x^8} = x^{8/2} = x^4$.
121. False, $\sqrt{16} = 4$.
123. True, $2^{600} = (2^2)^{300} = 4^{300}$.
125. False, $\dfrac{2 + \sqrt{6}}{2} = 1 + \dfrac{\sqrt{6}}{2}$.
127. False, $\sqrt{\dfrac{4}{6}} = \sqrt{\dfrac{2}{3}} = \dfrac{\sqrt{2}\sqrt{3}}{\sqrt{3}\sqrt{3}} = \dfrac{\sqrt{6}}{3}$.
129. True, $81^{2/4} = 3^2 = 9 = \sqrt{81}$.
131. True, because $(a^4b^2)^{1/2}$ is nonnegative and a^2b could be negative, absolute value symbols are necessary.
133. To find the time for which $s = 12,000$, solve the equation
$$16t^2 = 12,000$$
$$t^2 = 750$$
$$t = \sqrt{750} = 5\sqrt{30}$$
The time is $5\sqrt{30}$ seconds.
135. The guy wire of length 40 is the hypotenuse of a right triangle where one leg is length 30 and the other is length x. By the Pythagorean theorem we can write
$$x^2 + 30^2 = 40^2$$
$$x^2 = 700$$
$$x = \sqrt{700} = 10\sqrt{7}$$
The wire is attached to the ground $10\sqrt{7}$ feet from the base of the antenna.
137. Let $x =$ the length of the guy wire. The height of the antenna is 200 feet and the distance from the base of the antenna to the point on the ground where the guy wire is attached is 200 feet. The guy wire is the hypotenuse of a right triangle whose legs each have length 200. By the Pythagorean theorem we can write
$$x^2 = 200^2 + 200^2$$
$$x^2 = 80,000$$
$$x = \sqrt{80,000} = \sqrt{40,000}\sqrt{2} = 200\sqrt{2}$$
The length of the guy wires should be $200\sqrt{2}$ feet.

139. If the volume is 40 ft^3, then each side is $\sqrt[3]{40}$ ft in length. The surface area of the six square sides is $6(\sqrt[3]{40})^2$ ft^2. Multiply by 1.1 to get $1.1 \cdot 6(\sqrt[3]{40})^2$ ft^2 as the amount of cardboard needed to make the box. Simplify:
$$6.6\left(2\sqrt[3]{5}\right)^2 = 26.4\sqrt[3]{25} \text{ ft}^2$$
This is the exact answer. You can find an approximate answer with a calculator.
141. a) $1392.7 = 993.3(1 + r)^7$
$$(1 + r)^7 = \frac{1392.7}{993.3}$$
$$1 + r = \left(\frac{1392.7}{993.3}\right)^{1/7}$$
$$r = \left(\frac{1392.7}{993.3}\right)^{1/7} - 1$$
$$r \approx 0.049 = 4.9\%$$
b) From the graph it appears that the annual cost of health care in 2005 will be \$1600 billion.
143. $V = \sqrt{\dfrac{841L}{CS}} = \dfrac{\sqrt{841} \cdot \sqrt{L} \cdot \sqrt{CS}}{\sqrt{CS} \cdot \sqrt{CS}}$
$$V = \frac{29\sqrt{LCS}}{CS}$$

CHAPTER 7   TEST

1. $8^{2/3} = 2^2 = 4$
2. $4^{-3/2} = \dfrac{1}{2^3} = \dfrac{1}{8}$
3. $\sqrt{21} \div \sqrt{7} = \dfrac{\sqrt{21}}{\sqrt{7}} = \sqrt{\dfrac{21}{7}} = \sqrt{3}$
4. $2\sqrt{5} \cdot 3\sqrt{5} = 6 \cdot 5 = 30$
5. $\sqrt{20} + \sqrt{5} = 2\sqrt{5} + \sqrt{5} = 3\sqrt{5}$
6. $\sqrt{5} + \dfrac{1\sqrt{5}}{\sqrt{5}\sqrt{5}} = \sqrt{5} + \dfrac{\sqrt{5}}{5}$
$$= \frac{5\sqrt{5}}{5} + \frac{\sqrt{5}}{5} = \frac{6\sqrt{5}}{5}$$
7. $2^{1/2} \cdot 2^{1/2} = 2^1 = 2$
8. $\sqrt{72} = \sqrt{36}\sqrt{2} = 6\sqrt{2}$
9. $\sqrt{\dfrac{5}{12}} = \dfrac{\sqrt{5}\sqrt{3}}{\sqrt{12}\sqrt{3}} = \dfrac{\sqrt{15}}{\sqrt{36}} = \dfrac{\sqrt{15}}{6}$
10. $\dfrac{6 + \sqrt{18}}{6} = \dfrac{6 + 3\sqrt{2}}{6} = \dfrac{3(2 + \sqrt{2})}{3 \cdot 2}$
$$= \frac{2 + \sqrt{2}}{2}$$
11. $(2\sqrt{3} + 1)(\sqrt{3} - 2)$
$$= 6 + \sqrt{3} - 4\sqrt{3} - 2 = 4 - 3\sqrt{3}$$

12. $\sqrt[4]{32a^5y^8} = \sqrt[4]{16a^4y^8} \cdot \sqrt[4]{2a}$
$\qquad\qquad = 2ay^2 \sqrt[4]{2a}$

13. $\dfrac{1}{\sqrt[3]{2x^2}} = \dfrac{1 \cdot \sqrt[3]{4x}}{\sqrt[3]{2x^2} \cdot \sqrt[3]{4x}} = \dfrac{\sqrt[3]{4x}}{\sqrt[3]{8x^3}}$
$\qquad\qquad\qquad\qquad = \dfrac{\sqrt[3]{4x}}{2x}$

14. $\sqrt{\dfrac{8a^9}{b^3}} = \dfrac{2a^4 \sqrt{2a}}{b\sqrt{b}} = \dfrac{2a^4 \sqrt{2a}\sqrt{b}}{b\sqrt{b}\sqrt{b}}$
$\qquad\qquad\qquad\qquad = \dfrac{2a^4 \sqrt{2ab}}{b^2}$

15. $\sqrt[3]{-27x^9} = -3x^{9/3} = -3x^3$

16. $\sqrt{20m^3} = \sqrt{4m^2}\,\sqrt{5m} = 2m\sqrt{5m}$

17. $x^{1/2}x^{1/4} = x^{1/2+1/4} = x^{3/4}$

18. $(9y^4x^{1/2})^{1/2} = 3y^2x^{1/4}$

19. $\sqrt[3]{40x^7} = \sqrt[3]{8x^6} \cdot \sqrt[3]{5x} = 2x^2\sqrt[3]{5x}$

20. $(4+\sqrt{3})^2 = 16 + 8\sqrt{3} + 3 = 19 + 8\sqrt{3}$

21. $\dfrac{2}{5-\sqrt{3}} = \dfrac{2(5+\sqrt{3})}{(5-\sqrt{3})(5+\sqrt{3})}$
$\qquad = \dfrac{2(5+\sqrt{3})}{25-3} = \dfrac{2(5+\sqrt{3})}{22} = \dfrac{5+\sqrt{3}}{11}$

22. $\dfrac{\sqrt{6}}{4\sqrt{3}+\sqrt{2}} = \dfrac{\sqrt{6}(4\sqrt{3}-\sqrt{2})}{(4\sqrt{3}+\sqrt{2})(4\sqrt{3}-\sqrt{2})}$
$\qquad = \dfrac{4\sqrt{18}-\sqrt{12}}{48-2} = \dfrac{12\sqrt{2}-2\sqrt{3}}{46}$
$\qquad = \dfrac{6\sqrt{2}-\sqrt{3}}{23}$

23. $(3-2i)(4+5i) = 12 + 7i - 10i^2$
$\qquad\qquad\qquad = 22 + 7i$

24. $i^4 - i^5 = 1 - i^4 \cdot i = 1 - i$

25. $\dfrac{3-i}{1+2i} = \dfrac{(3-i)(1-2i)}{(1+2i)(1-2i)}$
$\qquad = \dfrac{3-7i+2i^2}{1-4i^2} = \dfrac{1-7i}{5} = \dfrac{1}{5} - \dfrac{7}{5}i$

26. $\dfrac{-6+\sqrt{-12}}{8} = \dfrac{-6+2i\sqrt{3}}{8}$
$\qquad\qquad = \dfrac{-3+i\sqrt{3}}{4} = -\dfrac{3}{4} + \dfrac{1}{4}i\sqrt{3}$

27. $(x-2)^2 = 49$
$\qquad x - 2 = \pm 7$
$\qquad\quad x = 2 \pm 7$
$\quad x = 2 + 7 \quad\text{or}\quad x = 2 - 7$
$\quad x = 9 \quad\quad\text{or}\quad\quad x = -5$
Solution set: $\{-5, 9\}$

28. $2\sqrt{x+4} = 3$
$\qquad \left(2\sqrt{x+4}\right)^2 = (3)^2$

$\qquad\qquad 4(x+4) = 9$
$\qquad\qquad\qquad 4x = -7$
$\qquad\qquad\qquad\quad x = -\dfrac{7}{4}$
Solution set: $\left\{-\dfrac{7}{4}\right\}$

29. $w^{2/3} = 4$
$\qquad (w^{2/3})^3 = 4^3$
$\qquad\qquad w^2 = 64$
$\qquad\qquad\; w = \pm 8$
Solution set: $\{-8, 8\}$

30. $9y^2 + 16 = 0$
$\qquad\quad 9y^2 = -16$
$\qquad\quad\; y^2 = -\dfrac{16}{9}$
$\qquad\qquad y = \pm\sqrt{-\dfrac{16}{9}} = \pm\dfrac{4}{3}i$
The solution set is $\left\{\pm\dfrac{4}{3}i\right\}$.

31. $\sqrt{2x^2+x-12} = x$
$\qquad \left(\sqrt{2x^2+x-12}\right)^2 = x^2$
$\qquad\qquad 2x^2+x-12 = x^2$
$\qquad\qquad\quad x^2+x-12 = 0$
$\qquad\qquad (x+4)(x-3) = 0$
$\quad x+4 = 0 \quad\text{or}\quad x-3 = 0$
$\qquad x = -4 \;\text{or}\qquad x = 3$
Since -4 does not check in the original equation, the solution set is $\{3\}$.

32. $\sqrt{x-1} + \sqrt{x+4} = 5$
$\qquad\quad \sqrt{x-1} = 5 - \sqrt{x+4}$
$\qquad \left(\sqrt{x-1}\right)^2 = \left(5 - \sqrt{x+4}\right)^2$
$\qquad\qquad x - 1 = 25 - 10\sqrt{x+4} + x + 4$
$\qquad 10\sqrt{x+4} = 30$
$\qquad \left(\sqrt{x+4}\right)^2 = (3)^2$
$\qquad\qquad x + 4 = 9$
$\qquad\qquad\quad x = 5$
Solution set: $\{5\}$

33. Let x = the length of the side. Since the diagonal is the hypotenuse of a right triangle, we can write the equation
$\qquad x^2 + x^2 = 3^2$
$\qquad\quad 2x^2 = 9$
$\qquad\quad\; x^2 = \dfrac{9}{2}$
$\qquad\qquad x = \sqrt{\dfrac{9}{2}} = \dfrac{3\sqrt{2}}{\sqrt{2}\sqrt{2}} = \dfrac{3\sqrt{2}}{2}$

The length of each side is $\dfrac{3\sqrt{2}}{2}$ feet.

34. Let $x =$ one number and $x + 11 =$ the other. Since their square roots differ by 1, we can write

$$\sqrt{x+11} - \sqrt{x} = 1$$
$$\sqrt{x+11} = \sqrt{x} + 1$$
$$\left(\sqrt{x+11}\right)^2 = \left(\sqrt{x}+1\right)^2$$
$$x + 11 = x + 2\sqrt{x} + 1$$
$$10 = 2\sqrt{x}$$
$$10^2 = (2\sqrt{x})^2$$
$$4x = 100$$
$$x = 25$$
$$x + 11 = 36$$

The numbers are 25 and 36.

35. If the perimeter is 20, the sum of the length and width is 10. Let $x =$ the length and $10 - x =$ the width. Use the Pythagorean theorem to write the equation

$$x^2 + (10-x)^2 = \left(2\sqrt{13}\right)^2$$
$$x^2 + 100 - 20x + x^2 = 52$$
$$2x^2 - 20x + 48 = 0$$
$$x^2 - 10x + 24 = 0$$
$$(x-4)(x-6) = 0$$
$$x = 4 \quad \text{or} \quad x = 6$$
$$10 - x = 6 \quad \text{or} \quad 10 - x = 4$$

The length and width are 4 feet and 6 feet.

36. $R = (248.530)^{2/3} \approx 39.53$ AU

$$30.08 = T^{2/3}$$
$$T = 30.08^{3/2} = 164.97 \text{ years}$$

Making Connections

Chapters 1 - 7

1.
$$3(x-2) + 5 = 7 - 4(x+3)$$
$$3x - 6 + 5 = 7 - 4x - 12$$
$$3x - 1 = -4x - 5$$
$$7x = -4$$
$$x = -\frac{4}{7}$$
Solution set: $\left\{-\frac{4}{7}\right\}$

2.
$$\sqrt{6x+7} = 4$$
$$\left(\sqrt{6x+7}\right)^2 = (4)^2$$

$$6x + 7 = 16$$
$$6x = 9$$
$$x = \frac{3}{2}$$
Solution set: $\left\{\frac{3}{2}\right\}$

3.
$$|2x+5| > 1$$
$$2x + 5 > 1 \quad \text{or} \quad 2x + 5 < -1$$
$$2x > -4 \quad \text{or} \quad 2x < -6$$
$$x > -2 \quad \text{or} \quad x < -3$$
$$(-\infty, -3) \cup (-2, \infty)$$

4.
$$8x^3 - 27 = 0$$
$$x^3 = \frac{27}{8}$$
$$x = \sqrt[3]{\frac{27}{8}} = \frac{3}{2}$$
Solution set: $\left\{\frac{3}{2}\right\}$

5.
$$2x - 3 > 3x - 4$$
$$1 > x$$
$$(-\infty, 1)$$

6.
$$\sqrt{2x-3} - \sqrt{3x+4} = 0$$
$$\left(\sqrt{2x-3}\right)^2 = \left(\sqrt{3x+4}\right)^2$$
$$2x - 3 = 3x + 4$$
$$-7 = x$$

Checking -7 gives us a square root of a negative number. So the solution set is $\emptyset$.

7.
$$\frac{w}{3} + \frac{w-4}{2} = \frac{11}{2}$$
$$6\left(\frac{w}{3} + \frac{w-4}{2}\right) = 6\left(\frac{11}{2}\right)$$
$$2w + 3w - 12 = 33$$
$$5w = 45$$
$$w = 9$$

Solution set: $\{9\}$

8.
$$2(x+7) - 4 = x - (10-x)$$
$$2x + 14 - 4 = x - 10 + x$$
$$2x + 10 = 2x - 10$$
$$10 = -10$$
Solution set: $\emptyset$

9.
$$(x+7)^2 = 25$$
$$x + 7 = \pm 5$$
$$x = -7 \pm 5$$
Solution set: $\{-12, -2\}$

158

10.
$$a^{-1/2} = 4$$
$$\left(a^{-1/2}\right)^{-2} = (4)^{-2}$$
$$a = \frac{1}{16}$$
Solution set: $\left\{\frac{1}{16}\right\}$

11. $x - 3 > 2$ or $x < 2x + 6$
$$x > 5 \text{ or } -x < 6$$
$$x > 5 \text{ or } x > -6$$
$(-6, \infty)$

12.
$$a^{-2/3} = 16$$
$$\left(a^{-2/3}\right)^{-3} = (16)^{-3}$$
$$a^2 = 2^{-12}$$
$$a = \pm\sqrt{2^{-12}} = \pm 2^{-6} = \pm\frac{1}{64}$$
Solution set: $\left\{-\frac{1}{64}, \frac{1}{64}\right\}$

13.
$$3x^2 - 1 = 0$$
$$x^2 = \frac{1}{3}$$
$$x = \pm\sqrt{\frac{1}{3}} = \pm\frac{\sqrt{3}}{3}$$
Solution set: $\left\{-\frac{\sqrt{3}}{3}, \frac{\sqrt{3}}{3}\right\}$

14. $5 - 2(x - 2) = 3x - 5(x - 2) - 1$
$$5 - 2x + 4 = 3x - 5x + 10 - 1$$
$$-2x + 9 = -2x + 9$$
Solution set is all real numbers R.

15. $|3x - 4| < 5$
$$-5 < 3x - 4 < 5$$
$$-1 < 3x < 9$$
$$-\frac{1}{3} < x < 3$$
$\left(-\frac{1}{3}, 3\right)$

16.
$$3x - 1 = 0$$
$$x = \frac{1}{3}$$
Solution set: $\left\{\frac{1}{3}\right\}$

17.
$$\sqrt{y - 1} = 9$$
$$\left(\sqrt{y - 1}\right)^2 = 9^2$$

$$y - 1 = 81$$
$$y = 82$$
Solution set: $\{82\}$

18. $|5(x - 2) + 1| = 3$
$$|5x - 10 + 1| = 3$$
$$|5x - 9| = 3$$
$$5x - 9 = 3 \text{ or } 5x - 9 = -3$$
$$5x = 12 \qquad 5x = 6$$
$$x = \frac{12}{5} \text{ or } \quad x = \frac{6}{5}$$
Solution set: $\left\{\frac{6}{5}, \frac{12}{5}\right\}$

19. $0.06x - 0.04(x - 20) = 2.8$
$$0.06x - 0.04x + 0.8 = 2.8$$
$$0.02x = 2.0$$
$$x = 100$$
Solution set: $\{100\}$

20. $|3x - 1| > -2$
Since absolute value of any quantity is greater than or equal to zero, any real number satisfies this inequality. Solution set: R

21.
$$\frac{3\sqrt{2}}{x} = \frac{\sqrt{3}}{4\sqrt{5}}$$
$$\sqrt{3}x = 12\sqrt{10}$$
$$x = \frac{12\sqrt{10}}{\sqrt{3}} = \frac{12\sqrt{10}\sqrt{3}}{\sqrt{3}\sqrt{3}} = 4\sqrt{30}$$
Solution set: $\left\{4\sqrt{30}\right\}$

22.
$$\frac{\sqrt{x} - 4}{x} = \frac{1}{\sqrt{x} + 5}$$
$$x = (\sqrt{x} - 4)(\sqrt{x} + 5)$$
$$x = x + \sqrt{x} - 20$$
$$20 = \sqrt{x}$$
$$400 = x$$
Solution set: $\{400\}$

23.
$$\frac{3\sqrt{2} + 4}{\sqrt{2}} = \frac{x\sqrt{18}}{3\sqrt{2} + 2}$$
$$6x = (3\sqrt{2} + 4)(3\sqrt{2} + 2)$$
$$6x = 18 + 18\sqrt{2} + 8$$
$$x = \frac{26 + 18\sqrt{2}}{6} = \frac{13 + 9\sqrt{2}}{3}$$
Solution set: $\left\{\frac{13 + 9\sqrt{2}}{3}\right\}$

24. $\dfrac{x}{2\sqrt{5}-\sqrt{2}} = \dfrac{2\sqrt{5}+\sqrt{2}}{x}$

$x^2 = 20 - 2$

$x = \pm\sqrt{18} = \pm 3\sqrt{2}$

Solution set: $\{-3\sqrt{2}, 3\sqrt{2}\}$

25. $\dfrac{\sqrt{2x}-5}{x} = \dfrac{-3}{\sqrt{2x}+5}$

$-3x = 2x - 25$

$-5x = -25$

$x = 5$

Solution set: $\{5\}$

26. $\dfrac{\sqrt{6}+2}{x} = \dfrac{2}{\sqrt{6}+4}$

$2x = (\sqrt{6}+2)(\sqrt{6}+4)$

$2x = 14 + 6\sqrt{6}$

$x = 7 + 3\sqrt{6}$

Solution set: $\{7 + 3\sqrt{6}\}$.

27. $\dfrac{x-1}{\sqrt{6}} = \dfrac{\sqrt{6}}{x}$

$x^2 - x = 6$

$x^2 - x - 6 = 0$

$(x-3)(x+2) = 0$

$x = 3 \text{ or } x = -2$

Solution set: $\{-2, 3\}$

28. $\dfrac{x+3}{\sqrt{10}} = \dfrac{\sqrt{10}}{x}$

$x^2 + 3x = 10$

$x^2 + 3x - 10 = 0$

$(x+5)(x-2) = 0$

$x = -5 \text{ or } x = 2$

Solution set: $\{-5, 2\}$

29. $\dfrac{1}{x} - \dfrac{1}{x-1} = -\dfrac{1}{6}$

$6x(x-1)(\dfrac{1}{x} - \dfrac{1}{x-1}) = 6x(x-1)(-\dfrac{1}{6})$

$6x - 6 - 6x = -x^2 + x$

$x^2 - x - 6 = 0$

$(x-3)(x+2) = 0$

$x - 3 = 0 \text{ or } x + 2 = 0$

$x = 3 \text{ or } \quad x = -2$

Solution set: $\{-2, 3\}$

30. $\dfrac{1}{x^2 - 2x} + \dfrac{1}{x} = \dfrac{2}{3}$

$3x(x-2)(\dfrac{1}{x^2-2x} + \dfrac{1}{x}) = 3x(x-2)\dfrac{2}{3}$

$3 + 3x - 6 = 2x^2 - 4x$

$-2x^2 + 7x - 3 = 0$

$2x^2 - 7x + 3 = 0$

$(2x-1)(x-3) = 0$

$2x - 1 = 0 \text{ or } x - 3 = 0$

$x = \dfrac{1}{2} \text{ or } \quad x = 3$

Solution set: $\left\{\dfrac{1}{2}, 3\right\}$

31. $\dfrac{-2+\sqrt{2^2 - 4(1)(-15)}}{2(1)}$

$= \dfrac{-2+\sqrt{64}}{2} = 3$

32. $\dfrac{-8+\sqrt{8^2 - 4(1)(12)}}{2(1)}$

$= \dfrac{-8+\sqrt{16}}{2} = -2$

33. $\dfrac{-5+\sqrt{5^2 - 4(2)(-3)}}{2(2)}$

$= \dfrac{-5+\sqrt{49}}{4} = \dfrac{1}{2}$

34. $\dfrac{-7+\sqrt{7^2 - 4(6)(-3)}}{2(6)}$

$= \dfrac{-7+\sqrt{121}}{12} = \dfrac{1}{3}$

35. a) $v = -94.8 + 21.4x - 0.761x^2$

$v = -94.8 + 21.4(11) - 0.761(11)^2$

$\approx 48.5 \, \text{cm}^3$

b) From the graph it appears that a moisture content of 14% will produce the maximum volume of popped corn.

c) From the graph it appears that the maximum volume is about 56 cm³.

8.1 WARM-UPS

1. False, completing the square involves perfect square trinomials.
2. False, it is equivalent to $x - 3 = \pm 2\sqrt{3}$.
3. False, because some quadratic polynomials cannot be factored.
4. False, because one-half of 4/3 is 2/3 and 2/3 squared is 4/9.
5. True.
6. False, we must first divide each side of the equation by 2.
7. False, $x = 3/2$ or $x = -5/3$.
8. True, one-half of 3 is 3/2 and 3/2 squared is 9/4.
9. False, $x = \pm 2i\sqrt{2}$.
10. False, $(x - 3)^2 = 0$ is a quadratic equation with only one solution.

8.1 EXERCISES

1. In this section, quadratic equations are solved by factoring, the even root property, and completing the square.
3. The last term is the square of one-half the coefficient of the middle term.
5.
$$x^2 - x - 6 = 0$$
$$(x - 3)(x + 2) = 0$$
$$x - 3 = 0 \quad \text{or} \quad x + 2 = 0$$
$$x = 3 \quad \text{or} \quad x = -2$$
Solution set: $\{-2, 3\}$
7.
$$a^2 + 2a = 15$$
$$a^2 + 2a - 15 = 0$$
$$(a + 5)(a - 3) = 0$$
$$a + 5 = 0 \quad \text{or} \quad a - 3 = 0$$
$$a = -5 \quad \text{or} \quad a = 3$$
Solution set: $\{-5, 3\}$
9.
$$2x^2 - x - 3 = 0$$
$$(2x - 3)(x + 1) = 0$$
$$2x - 3 = 0 \quad \text{or} \quad x + 1 = 0$$
$$x = \frac{3}{2} \quad \text{or} \quad x = -1$$
Solution set: $\left\{-1, \frac{3}{2}\right\}$
11.
$$y^2 + 14y + 49 = 0$$
$$(y + 7)^2 = 0$$
$$y + 7 = 0$$
$$y = -7$$
Solution set: $\{-7\}$

13.
$$a^2 - 16 = 0$$
$$(a - 4)(a + 4) = 0$$
$$a - 4 = 0 \quad \text{or} \quad a + 4 = 0$$
$$a = 4 \quad \text{or} \quad a = -4$$
Solution set: $\{-4, 4\}$
15.
$$x^2 = 81$$
$$x = \pm\sqrt{81} = \pm 9$$
Solution set: $\{-9, 9\}$
17.
$$x^2 = \frac{16}{9}$$
$$x = \pm\sqrt{\frac{16}{9}} = \pm\frac{4}{3}$$
Solution set: $\left\{-\frac{4}{3}, \frac{4}{3}\right\}$
19.
$$(x - 3)^2 = 16$$
$$x - 3 = \pm\sqrt{16}$$
$$x = 3 \pm 4$$
$$x = 3 + 4 \quad \text{or} \quad x = 3 - 4$$
$$x = 7 \quad \text{or} \quad x = -1$$
Solution set: $\{-1, 7\}$
21.
$$(z + 1)^2 = 5$$
$$z + 1 = \pm\sqrt{5}$$
$$z = -1 \pm\sqrt{5}$$
Solution set: $\{-1 - \sqrt{5}, -1 + \sqrt{5}\}$
23.
$$\left(w - \frac{3}{2}\right)^2 = \frac{7}{4}$$
$$w - \frac{3}{2} = \pm\sqrt{\frac{7}{4}}$$
$$w = \frac{3}{2} \pm \frac{\sqrt{7}}{2} = \frac{3 \pm \sqrt{7}}{2}$$
Solution set: $\left\{\frac{3 - \sqrt{7}}{2}, \frac{3 + \sqrt{7}}{2}\right\}$
25. One-half of 2 is 1, and 1 squared is 1:
$x^2 + 2x + 1$
27. One-half of -3 is $-3/2$, and $-3/2$ squared is 9/4: $x^2 - 3x + \frac{9}{4}$
29. One-half of 1/4 is 1/8, and 1/8 squared is 1/64: $y^2 + \frac{1}{4}y + \frac{1}{64}$
31. One-half of 2/3 is 1/3, and 1/3 squared is 1/9: $x^2 + \frac{2}{3}x + \frac{1}{9}$
33. $x^2 + 8x + 16 = (x + 4)^2$
35. $y^2 - 5y + \frac{25}{4} = \left(y - \frac{5}{2}\right)^2$
37. $z^2 - \frac{4}{7}z + \frac{4}{49} = \left(z - \frac{2}{7}\right)^2$
39. $t^2 + \frac{3}{5}t + \frac{9}{100} = \left(t + \frac{3}{10}\right)^2$
41. $x^2 - 2x - 15 = 0$
$$x^2 - 2x = 15$$
$$x^2 - 2x + 1 = 15 + 1$$

$$(x-1)^2 = 16$$
$$x - 1 = \pm 4$$
$$x = 1 \pm 4$$
$$x = 5 \quad \text{or} \quad x = -3$$
Solution set: $\{-3, 5\}$

43.
$$2x^2 - 4x = 70$$
$$x^2 - 2x = 35$$
$$x^2 - 2x + 1 = 35 + 1$$
$$(x-1)^2 = 36$$
$$x - 1 = \pm 6$$
$$x = 1 \pm 6$$
$$x = 7 \quad \text{or} \quad x = -5$$
Solution set: $\{-5, 7\}$

45.
$$w^2 - w - 20 = 0$$
$$w^2 - w = 20$$
$$w^2 - w + \frac{1}{4} = 20 + \frac{1}{4}$$
$$\left(w - \frac{1}{2}\right)^2 = \frac{81}{4}$$
$$w - \frac{1}{2} = \pm \frac{9}{2}$$
$$w = \frac{1}{2} \pm \frac{9}{2}$$
$$x = \frac{10}{2} = 5 \quad \text{or} \quad x = \frac{-8}{2} = -4$$
Solution set: $\{-4, 5\}$

47.
$$q^2 + 5q = 14$$
$$q^2 + 5q + \frac{25}{4} = 14 + \frac{25}{4}$$
$$\left(q + \frac{5}{2}\right)^2 = \frac{81}{4}$$
$$q + \frac{5}{2} = \pm \frac{9}{2}$$
$$q = -\frac{5}{2} \pm \frac{9}{2}$$
$$q = \frac{4}{2} = 2 \quad \text{or} \quad q = \frac{-14}{2} = -7$$
Solution set: $\{-7, 2\}$

49.
$$2h^2 - h - 3 = 0$$
$$2h^2 - h = 3$$
$$h^2 - \frac{1}{2}h = \frac{3}{2}$$
$$h^2 - \frac{1}{2}h + \frac{1}{16} = \frac{3}{2} + \frac{1}{16}$$
$$\left(h - \frac{1}{4}\right)^2 = \frac{25}{16}$$
$$h - \frac{1}{4} = \pm \frac{5}{4}$$
$$h = \frac{1}{4} \pm \frac{5}{4}$$
Solution set: $\left\{-1, \frac{3}{2}\right\}$

51.
$$x^2 + 4x = 6$$
$$x^2 + 4x + 4 = 6 + 4$$
$$(x+2)^2 = 10$$
$$x + 2 = \pm \sqrt{10}$$
$$x = -2 \pm \sqrt{10}$$

Solution set: $\{-2 - \sqrt{10}, -2 + \sqrt{10}\}$

53.
$$x^2 + 8x - 4 = 0$$
$$x^2 + 8x = 4$$
$$x^2 + 8x + 16 = 4 + 16$$
$$(x+4)^2 = 20$$
$$x + 4 = \pm \sqrt{20}$$
$$x = -4 \pm 2\sqrt{5}\tfrac{1}{2}$$
Solution set: $\{-4 - 2\sqrt{5}, -4 + 2\sqrt{5}\}$

55.
$$4x^2 - 4x - 1 = 0$$
$$x^2 - x - \frac{1}{4} = 0$$
$$x^2 - x = \frac{1}{4}$$
$$x^2 - x + \frac{1}{4} = \frac{1}{4} + \frac{1}{4}$$
$$\left(x - \frac{1}{2}\right)^2 = \frac{1}{2}$$
$$x - \frac{1}{2} = \pm \sqrt{\frac{1}{2}}$$
$$x = \frac{1}{2} \pm \frac{\sqrt{2}}{2}$$
$$x = \frac{1 \pm \sqrt{2}}{2}$$
Solution set: $\left\{\frac{1 \pm \sqrt{2}}{2}\right\}$

57.
$$2x^2 + 3x - 4 = 0$$
$$2x^2 + 3x = 4$$
$$x^2 + \frac{3}{2}x = 2$$
$$x^2 + \frac{3}{2}x + \frac{9}{16} = 2 + \frac{9}{16}$$
$$\left(x + \frac{3}{4}\right)^2 = \frac{41}{16}$$
$$x + \frac{3}{4} = \pm \frac{\sqrt{41}}{4}$$
$$x = -\frac{3}{4} \pm \frac{\sqrt{41}}{4}$$
Solution set: $\left\{\frac{-3 - \sqrt{41}}{4}, \frac{-3 + \sqrt{41}}{4}\right\}$

59.
$$\sqrt{2x+1} = x - 1$$
$$(\sqrt{2x+1})^2 = (x-1)^2$$
$$2x + 1 = x^2 - 2x + 1$$
$$0 = x^2 - 4x$$
$$x(x-4) = 0$$
$$x = 0 \quad \text{or} \quad x = 4$$
Since 0 is an extraneous root, the solution set is $\{4\}$.

61.
$$w = \frac{\sqrt{w+1}}{2}$$
$$2w = \sqrt{w+1}$$
$$(2w)^2 = (\sqrt{w+1})^2$$
$$4w^2 = w + 1$$
$$4w^2 - w = 1$$

162

$$w^2 - \tfrac{1}{4}w = \tfrac{1}{4}$$
$$w^2 - \tfrac{1}{4}w + \tfrac{1}{64} = \tfrac{1}{4} + \tfrac{1}{64}$$
$$\left(w - \tfrac{1}{8}\right)^2 = \tfrac{17}{64}$$
$$w - \tfrac{1}{8} = \pm \tfrac{\sqrt{17}}{8}$$
$$w = \tfrac{1}{8} \pm \tfrac{\sqrt{17}}{8}$$

The number $\dfrac{1 - \sqrt{17}}{8}$ is a negative number. No negative number can be a solution to the original equation because the left side would be negative and the right side is a principal square root. The solution set is $\left\{ \dfrac{1 + \sqrt{17}}{8} \right\}$.

63.
$$\frac{t}{t - 2} = \frac{2t - 3}{t}$$
$$t^2 = (t - 2)(2t - 3)$$
$$t^2 = 2t^2 - 7t + 6$$
$$-t^2 + 7t - 6 = 0$$
$$t^2 - 7t + 6 = 0$$
$$(t - 6)(t - 1) = 0$$
$$t = 6 \quad \text{or} \quad t = 1$$
Solution set: $\{1, 6\}$

65.
$$\frac{2}{x^2} + \frac{4}{x} + 1 = 0$$
$$x^2\left(\frac{2}{x^2} + \frac{4}{x} + 1\right) = x^2(0)$$
$$2 + 4x + x^2 = 0$$
$$x^2 + 4x = -2$$
$$x^2 + 4x + 4 = -2 + 4$$
$$(x + 2)^2 = 2$$
$$x + 2 = \pm\sqrt{2}$$
$$x = -2 \pm \sqrt{2}$$
Solution set: $\{-2 - \sqrt{2}, -2 + \sqrt{2}\}$

67. $x^2 + 2x + 5 = 0$
$$x^2 + 2x + 1 = -5 + 1$$
$$(x + 1)^2 = -4$$
$$x + 1 = \pm\sqrt{-4}$$
$$x = -1 \pm 2i$$
The solution set is $\{-1 - 2i, -1 + 2i\}$.

69. $x^2 - 6x + 11 = 0$
$$x^2 - 6x + 9 = -11 + 9$$
$$(x - 3)^2 = -2$$
$$x - 3 = \pm\sqrt{-2}$$
$$x = 3 \pm i\sqrt{2}$$
The solution set is $\left\{3 - i\sqrt{2}, 3 + i\sqrt{2}\right\}$.

71. $x^2 = -\tfrac{1}{2}$
$$x = \pm\sqrt{-\tfrac{1}{2}} = \pm i\tfrac{\sqrt{2}}{2}$$

Solution set: $\left\{ \pm i\dfrac{\sqrt{2}}{2} \right\}$

73. $x^2 + 12 = 0$
$$x^2 = -12$$
$$x = \pm\sqrt{-12}$$
$$x = \pm 2i\sqrt{3}$$
The solution set is $\{-2i\sqrt{3}, 2i\sqrt{3}\}$

75. $5z^2 - 4z + 1 = 0$
$$z^2 - \tfrac{4}{5}z + \tfrac{4}{25} = -\tfrac{1}{5} + \tfrac{4}{25}$$
$$\left(z - \tfrac{2}{5}\right)^2 = \tfrac{-1}{25}$$
$$z - \tfrac{2}{5} = \pm\sqrt{\tfrac{-1}{25}}$$
$$z = \tfrac{2}{5} \pm \sqrt{\tfrac{-1}{25}}$$
$$z = \tfrac{2}{5} \pm \tfrac{i}{5} = \tfrac{2 \pm i}{5}$$
The solution set is $\left\{ \dfrac{2 \pm i}{5} \right\}$.

77. $x^2 = -121$
$$x = \pm\sqrt{-121} = \pm 11i$$
Solution set: $\{ \pm 11i\}$

79. $4x^2 + 25 = 0$
$$4x^2 = -25$$
$$x^2 = -\tfrac{25}{4}$$
$$x = \pm\sqrt{-\tfrac{25}{4}} = \pm \tfrac{5}{2}i$$
Solution set: $\left\{ -\tfrac{5}{2}i, \tfrac{5}{2}i \right\}$

81. $\left(p + \tfrac{1}{2}\right)^2 = \tfrac{9}{4}$
$$p + \tfrac{1}{2} = \pm\tfrac{3}{2}$$
$$p = -\tfrac{1}{2} \pm \tfrac{3}{2}$$
The solution set is $\left\{ -2, 1 \right\}$

83. $5t^2 + 4t - 3 = 0$
$$5t^2 + 4t = 3$$
$$t^2 + \tfrac{4}{5}t = \tfrac{3}{5}$$
$$t^2 + \tfrac{4}{5}t + \tfrac{4}{25} = \tfrac{3}{5} + \tfrac{4}{25}$$
$$\left(t + \tfrac{2}{5}\right)^2 = \tfrac{19}{25}$$
$$t + \tfrac{2}{5} = \pm\tfrac{\sqrt{19}}{5}$$
$$t = -\tfrac{2}{5} \pm \tfrac{\sqrt{19}}{5}$$
Solution set: $\left\{ \dfrac{-2 - \sqrt{19}}{5}, \dfrac{-2 + \sqrt{19}}{5} \right\}$

85. $m^2 + 2m - 24 = 0$
$$m^2 + 2m = 24$$
$$m^2 + 2m + 1 = 24 + 1$$
$$(m + 1)^2 = 25$$

$$m + 1 = \pm 5$$
$$m = -1 \pm 5$$
$$m = -1 - 5 = -6 \text{ or } m = -1 + 5 = 4$$
The solution set is $\{-6, 4\}$.

87. $(x - 2)^2 = -9$
$$x - 2 = \pm\sqrt{-9}$$
$$x = 2 \pm 3i$$
Solution set: $\{2 \pm 3i\}$

89. $-x^2 + x + 6 = 0$
$$-x^2 + x = -6$$
$$x^2 - x = 6$$
$$x^2 - x + \frac{1}{4} = 6 + \frac{1}{4}$$
$$\left(x - \frac{1}{2}\right)^2 = \frac{25}{4}$$
$$x - \frac{1}{2} = \pm\frac{5}{2}$$
$$x = \frac{1}{2} \pm \frac{5}{2}$$
Solution set: $\{-2, 3\}$

91. $x^2 - 6x + 10 = 0$
$$x^2 - 6x + 9 = -10 + 9$$
$$(x - 3)^2 = -1$$
$$x - 3 = \pm\sqrt{-1} = \pm i$$
$$x = 3 \pm i$$
The solution set is $\{3 - i, 3 + i\}$.

93. $2x - 5 = \sqrt{7x + 7}$
$$(2x - 5)^2 = (\sqrt{7x + 7})^2$$
$$4x^2 - 20x + 25 = 7x + 7$$
$$4x^2 - 27x + 18 = 0$$
$$(4x - 3)(x - 6) = 0$$
$$4x - 3 = 0 \text{ or } x - 6 = 0$$
$$x = \frac{3}{4} \text{ or } \quad x = 6$$
Since 3/4 does not check, the solution set is $\{6\}$.

95. $\frac{1}{x} + \frac{1}{x - 1} = \frac{1}{4}$
$$4x(x - 1)\left(\frac{1}{x} + \frac{1}{x - 1}\right)$$
$$= 4x(x - 1)\left(\frac{1}{4}\right)$$
$$4x - 4 + 4x = x^2 - x$$
$$-x^2 + 9x - 4 = 0$$
$$x^2 - 9x + 4 = 0$$
$$x^2 - 9x + \frac{81}{4} = -4 + \frac{81}{4}$$
$$\left(x - \frac{9}{2}\right)^2 = \frac{65}{4}$$
$$x - \frac{9}{2} = \pm\frac{\sqrt{65}}{2}$$
$$x = \frac{9}{2} \pm \frac{\sqrt{65}}{2}$$

Solution set: $\left\{\frac{9 - \sqrt{65}}{2}, \frac{9 + \sqrt{65}}{2}\right\}$

97. $(2 + \sqrt{3})^2 - 4(2 + \sqrt{3}) + 1$
$$= 4 + 4\sqrt{3} + 3 - 8 - 4\sqrt{3} + 1 = 0$$

$(2 - \sqrt{3})^2 - 4(2 - \sqrt{3}) + 1$
$$= 4 - 4\sqrt{3} + 3 - 8 + 4\sqrt{3} + 1 = 0$$

99. $(i + 1)^2 - 2(i + 1) + 2$
$$= -1 + 2i + 1 - 2i - 2 + 2 = 0$$

$(1 - i)^2 - 2(1 - i) + 2$
$$= 1 - 2i - 1 - 2 + 2i + 2 = 0$$

101. $1211.1 \cdot 8700 = 2.81A^2 \cdot 200$
$$A^2 = \frac{1211.1 \cdot 8700}{2.81 \cdot 200}$$
$$A = \sqrt{\frac{1211.1 \cdot 8700}{2.81 \cdot 200}} \approx 136.9 \text{ ft/sec}$$

103. $17{,}568 = 1500x - 3x^2$
$$3x^2 - 1500x = -17{,}568$$
$$x^2 - 500x = -5856$$
$$x^2 - 500x + 62500 = -5856 + 62500$$
$$(x - 250)^2 = 56644$$
$$x - 250 = \pm 238$$
$$x = 250 \pm 238$$
$$x = 12 \quad \text{or} \quad x = 488$$
Since x is less than 25, the answer is 12.

105. Equation (c) is not quadratic because there is no x^2 term.

109. $\{4.56, 2.74\}$

111. $\{3.53\}$

8.2 WARM-UPS

1. True.

2. False, before identifying a, b, and c, we must write the equation as $3x^2 - 4x + 7 = 0$.

3. True, this is just the quadratic formula with $a = d$, $b = e$, and $c = f$.

4. False, the quadratic formula works to solve any quadratic equation.

5. True, because
$(-3)^2 - 4(2)(-4) = 9 + 32 = 41$.

6. True, if the discriminant is 0, then the quadratic equation has one real solution.

7. True.

8. True, because we can write the equation in the form $-1x^2 + 2x + 0 = 0$.

9. False, x and $6 - x$ have a sum of 6.

10. False, there can be one real solution, 2 real solutions, or 2 imaginary solutions.

8.2 EXERCISES

1. The quadratic formula can be used to solve any quadratic equation.

3. Factoring is used when the quadratic polynomial is simple enough to factor.

5. The discriminant is $b^2 - 4ac$.

7. $x^2 + 5x + 6 = 0$

$a = 1, b = 5, \ c = 6$

$$x = \frac{-5 \pm \sqrt{5^2 - 4(1)(6)}}{2(1)} = \frac{-5 \pm \sqrt{1}}{2}$$

$$= \frac{-5 \pm 1}{2} = \frac{-4}{2}, \frac{-6}{2}$$

Solution set: $\{-3, -2\}$

9. $y^2 + y - 6 = 0$

$a = 1, b = 1, c = -6$

$$y = \frac{-1 \pm \sqrt{1^2 - 4(1)(-6)}}{2(1)} = \frac{-1 \pm \sqrt{25}}{2}$$

$$= \frac{-1 \pm 5}{2} = \frac{4}{2}, \frac{-6}{2}$$

Solution set: $\{-3, 2\}$

11. $-6z^2 + 7z + 3 = 0$

$a = -6, b = 7, c = 3$

$$z = \frac{-7 \pm \sqrt{(7)^2 - 4(-6)(3)}}{2(-6)} = \frac{-7 \pm \sqrt{121}}{-12}$$

$$= \frac{-7 \pm 11}{-12} = \frac{-18}{-12}, \frac{4}{-12}$$

Solution set: $\left\{ -\frac{1}{3}, \frac{3}{2} \right\}$

13. $4x^2 - 4x + 1 = 0$

$a = 4, \ b = -4, \ c = 1$

$$x = \frac{4 \pm \sqrt{16 - 4(4)(1)}}{2(4)} = \frac{4 \pm 0}{8} = \frac{1}{2}$$

Solution set: $\left\{ \frac{1}{2} \right\}$

15. $-9x^2 + 6x - 1 = 0$

$a = -9, b = 6, c = -1$

$$x = \frac{-6 \pm \sqrt{36 - 4(-9)(-1)}}{2(-9)} = \frac{-6 \pm \sqrt{0}}{-18}$$

$$= \frac{1}{3}$$

Solution set: $\left\{ \frac{1}{3} \right\}$

17. $16x^2 + 24x + 9 = 0$

$a = 16, b = 24, c = 9$

$$x = \frac{-24 \pm \sqrt{576 - 4(16)(9)}}{2(16)} = \frac{-24 \pm \sqrt{0}}{32}$$

$$= -\frac{3}{4}$$

Solution set: $\left\{ -\frac{3}{4} \right\}$

19. $v^2 + 8v + 6 = 0$

$a = 1, b = 8, c = 6$

$$v = \frac{-8 \pm \sqrt{8^2 - 4(1)(6)}}{2(1)} = \frac{-8 \pm \sqrt{40}}{2}$$

$$= \frac{-8 \pm 2\sqrt{10}}{2} = -4 \pm \sqrt{10}$$

Solution set: $\{-4 \pm \sqrt{10}\,\}$

21. $x^2 + 5x - 1 = 0$: $a = 1, b = 5, c = -1$

$$x = \frac{-5 \pm \sqrt{5^2 - 4(1)(-1)}}{2(1)} = \frac{-5 \pm \sqrt{29}}{2}$$

Solution set: $\left\{ \dfrac{-5 \pm \sqrt{29}}{2} \right\}$

23. $2t^2 - 6t + 1 = 0$

$a = 2, b = -6, c = 1$

$$t = \frac{6 \pm \sqrt{36 - 4(2)(1)}}{2(2)} = \frac{6 \pm \sqrt{28}}{4} = \frac{6 \pm 2\sqrt{7}}{4}$$

$$= \frac{2(3 \pm \sqrt{7})}{2(2)} = \frac{3 \pm \sqrt{7}}{2}$$

Solution set: $\left\{ \dfrac{3 \pm \sqrt{7}}{2} \right\}$

25. $2t^2 - 6t + 5 = 0$

$$t = \frac{6 \pm \sqrt{36 - 4(2)(5)}}{2(2)} = \frac{6 \pm \sqrt{-4}}{4} = \frac{6 \pm 2i}{4}$$

$$= \frac{3 \pm i}{2} \qquad \text{Solution set: } \left\{ \dfrac{3 \pm i}{2} \right\}$$

27. $-2x^2 + 3x - 6 = 0$

$a = -2, b = 3, c = -6$

$$x = \frac{-3 \pm \sqrt{(3)^2 - 4(-2)(-6)}}{2(-2)} = \frac{-3 \pm \sqrt{-39}}{-4}$$

$$= \frac{-3 \pm i\sqrt{39}}{-4} = \frac{3 \pm i\sqrt{39}}{4}$$

Solution set: $\left\{ \dfrac{3 \pm i\sqrt{39}}{4} \right\}$

29. $\frac{1}{2}x^2 + 13 = 5x, \quad x^2 - 10x + 26 = 0$

$$x = \frac{10 \pm \sqrt{100 - 4(1)(26)}}{2(1)} = \frac{10 \pm \sqrt{-4}}{2}$$

$$= \frac{10 \pm 2i}{2} = 5 \pm i \quad \text{Solution set: } \{5 \pm i\}$$

31. $x^2 - 6x + 2 = 0$

$a = 1, b = -6, c = 2$

$b^2 - 4ac = 36 - 4(1)(2) = 28$

So there are two real solutions to the equation.

33. $-2x^2 + 5x - 6 = 0$

$a = -2, b = 5, c = -6$

$$b^2 - 4ac = 25 - 4(-2)(-6) = -23$$
So there are no real solutions to the equation.

35. $\quad 4m^2 - 20m + 25 = 0$

$\qquad a = 4, b = -20, c = 25$

$\qquad b^2 - 4ac = 400 - 4(4)(25) = 0$

So there is one real solution to the equation.

37. $\quad y^2 - \frac{1}{2}y + \frac{1}{4} = 0$

$\qquad a = 1, b = -\frac{1}{2}, c = \frac{1}{4}$

$\qquad b^2 - 4ac = \frac{1}{4} - 4(1)(\frac{1}{4}) = -\frac{3}{4}$

So there are no real solutions to the equation.

39. $\quad -3t^2 + 5t + 6 = 0$

$\qquad a = -3, \ b = 5, \ c = 6$

$\qquad b^2 - 4ac = 25 - 4(-3)(6) = 97$

So there are two real solutions to the equation.

41. $\quad 16z^2 - 24z + 9 = 0$

$\qquad a = 16, b = -24, c = 9$

$\qquad b^2 - 4ac = (-24)^2 - 4(16)(9) = 0$

So there is one real solution to the equation.

43. $\quad 5x^2 - 7 = 0$

$\qquad a = 5, b = 0, c = -7$

$\qquad b^2 - 4ac = 0 - 4(5)(-7) = 140$

So there are two real solutions to the equation.

45. $\quad x^2 - x = 0$

$\qquad a = 1, b = -1, c = 0$

$\qquad b^2 - 4ac = 1 - 4(1)(0) = 1$

So there are two real solutions to the equation.

47. $\quad \frac{1}{4}y^2 + y = 1$

$\qquad 4\left(\frac{1}{4}y^2 + y\right) = 4 \cdot 1$

$\qquad\qquad y^2 + 4y = 4$

$\qquad\qquad y^2 + 4y - 4 = 0$

$\qquad\qquad a = 1, b = 4, c = -4$

$y = \dfrac{-4 \pm \sqrt{16 - 4(1)(-4)}}{2(1)} = \dfrac{-4 \pm \sqrt{32}}{2}$

$\ = \dfrac{-4 \pm 4\sqrt{2}}{2} = -2 \pm 2\sqrt{2}$

Solution set: $\left\{-2 \pm 2\sqrt{2}\right\}$

49. $\qquad \frac{1}{3}x^2 + \frac{1}{2}x = \frac{1}{3}$

$\qquad 6\left(\frac{1}{3}x^2 + \frac{1}{2}x\right) = 6\left(\frac{1}{3}\right)$

$\qquad\qquad 2x^2 + 3x = 2$

$\qquad\qquad 2x^2 + 3x - 2 = 0$

$\qquad\qquad a = 2, b = 3, c = -2$

$x = \dfrac{-3 \pm \sqrt{9 - 4(2)(-2)}}{2(2)} = \dfrac{-3 \pm 5}{4}$

$\qquad\qquad = \dfrac{2}{4}, \dfrac{-8}{4}$

Solution set: $\left\{-2, \frac{1}{2}\right\}$

51. $\qquad 3y^2 + 2y - 4 = 0$

$\qquad\quad a = 3, b = 2, c = -4$

$y = \dfrac{-2 \pm \sqrt{4 - 4(3)(-4)}}{2(3)} = \dfrac{-2 \pm \sqrt{52}}{6}$

$\ = \dfrac{-2 \pm 2\sqrt{13}}{6} = \dfrac{-1 \pm \sqrt{13}}{3}$

Solution set: $\left\{\dfrac{-1 \pm \sqrt{13}}{3}\right\}$

53. $\qquad \dfrac{w}{w-2} = \dfrac{w}{w-3}$

$\qquad w^2 - 3w = w^2 - 2w$

$\qquad\qquad 0 = w$

Solution set: $\{0\}$

55. $\qquad \dfrac{9(3x-5)^2}{4} = 1$

$\qquad\qquad (3x-5)^2 = \dfrac{4}{9}$

$\qquad\qquad 3x - 5 = \pm\dfrac{2}{3}$

$3x = 5 + \dfrac{2}{3}$ or $\quad 3x = 5 - \dfrac{2}{3}$

$3x = \dfrac{17}{3}$ $\quad$ or $3x = \dfrac{13}{3}$

$x = \dfrac{17}{9}$ $\quad$ or $\quad x = \dfrac{13}{9}$

Solution set: $\left\{\dfrac{13}{9}, \dfrac{17}{9}\right\}$

57. $\ 25 - \frac{1}{3}x^2 = 0$

$\qquad -\frac{1}{3}x^2 = -25$

$\qquad\qquad x^2 = 75$

$\qquad\qquad x = \pm\sqrt{75} = \pm 5\sqrt{3}$

Solution set: $\left\{\pm 5\sqrt{3}\right\}$

59. $\qquad 1 + \dfrac{20}{x^2} = \dfrac{8}{x}$

$\qquad x^2\left(1 + \dfrac{20}{x^2}\right) = x^2 \cdot \dfrac{8}{x}$

$\qquad x^2 - 8x + 20 = 0$

$\qquad x^2 - 8x + 16 = -20 + 16$

$\qquad\qquad (x-4)^2 = -4$

$\qquad\qquad x - 4 = \pm 2i$

$\qquad\qquad x = 4 \pm 2i$

Solution set: $\{4 \pm 2i\}$

61. $\ (x-8)(x+4) = -42$

$\qquad x^2 - 4x + 10 = 0$

$\qquad x^2 - 4x + 4 = -10 + 4$

$\qquad\qquad (x-2)^2 = -6$

$$x - 2 = \pm i\sqrt{6}$$
$$x = 2 \pm i\sqrt{6}$$
Solution set: $\{2 \pm i\sqrt{6}\}$

63.
$$y = \frac{3(2y+5)}{8(y-1)}$$
$$8y^2 - 8y = 6y + 15$$
$$8y^2 - 14y - 15 = 0$$
$$(4y+3)(2y-5) = 0$$
$$4y + 3 = 0 \quad \text{or} \quad 2y - 5 = 0$$
$$y = -\frac{3}{4} \quad \text{or} \quad y = \frac{5}{2}$$
Solution set: $\left\{-\frac{3}{4}, \frac{5}{2}\right\}$

65. $x = \dfrac{-3.2 \pm \sqrt{(3.2)^2 - 4(1)(-5.7)}}{2(1)}$

$= \dfrac{-3.2 \pm \sqrt{33.04}}{2} \qquad \{-4.474, 1.274\}$

67. $x = \dfrac{7.4 \pm \sqrt{(-7.4)^2 - 4(1)(13.69)}}{2(1)}$

$= \dfrac{7.4 \pm \sqrt{0}}{2(1)} = 3.7 \qquad \{3.7\}$

69. $x = \dfrac{-6.72 \pm \sqrt{(6.72)^2 - 4(1.85)(3.6)}}{2(1.85)}$

$= \dfrac{-6.72 \pm \sqrt{18.5184}}{3.7} \qquad \{-2.979, -0.653\}$

71. $x = \dfrac{-14379 \pm \sqrt{14379^2 - 4(3)(243)}}{2(3)}$

$= \dfrac{-14379 \pm 14378.8986}{6}$

Solution set: $\{-4792.983, -0.017\}$

73. $x = \dfrac{-0.00075 \pm \sqrt{0.00075^2 - 4 \cdot 1(-0.0062)}}{2(1)}$

$= \dfrac{-0.00075 \pm 0.1574819}{2} \qquad \{-0.079, 0.078\}$

75. Let $x =$ one number and $x + 1 =$ the other. Since their product is 16, we can write
$$x(x + 1) = 16$$
$$x^2 + x = 16$$
$$x^2 + x - 16 = 0$$
$$x = \frac{-1 \pm \sqrt{1^2 - 4(1)(-16)}}{2(1)} = \frac{-1 \pm \sqrt{65}}{2}$$
Since the numbers are positive,
$$x = \frac{-1 + \sqrt{65}}{2} \text{ and } x + 1 = \frac{-1 + \sqrt{65}}{2} + \frac{2}{2}$$
$$= \frac{1 + \sqrt{65}}{2}. \text{ The numbers are } \frac{1 + \sqrt{65}}{2}$$
and $\dfrac{-1 + \sqrt{65}}{2}$ or approximately 4.5 and 3.5.

77. Let $x =$ one of the numbers. If the numbers are to have a sum of 6, then

$6 - x =$ the other number. Since their product is 4, we can write
$$x(6 - x) = 4$$
$$-x^2 + 6x - 4 = 0$$
$$x^2 - 6x + 4 = 0$$
$$x = \frac{6 \pm \sqrt{36 - 4(1)(4)}}{2(1)} = \frac{6 \pm 2\sqrt{5}}{2}$$
$$= 3 \pm \sqrt{5}$$
If $x = 3 + \sqrt{5}$, then
$6 - x = 6 - (3 + \sqrt{5}) = 3 - \sqrt{5}$.
If $x = 3 - \sqrt{5}$, then
$6 - x = 6 - (3 - \sqrt{5}) = 3 + \sqrt{5}$.
So the numbers are $3 + \sqrt{5}$ and $3 - \sqrt{5}$ or approximately 5.2 and 0.8

79. Let $x =$ the width and $x + 1 =$ the length. Since the diagonal is the hypotenuse of a right triangle, we can write
$$x^2 + (x + 1)^2 = (\sqrt{3})^2$$
$$x^2 + x^2 + 2x + 1 = 3$$
$$2x^2 + 2x - 2 = 0$$
$$x^2 + x - 1 = 0$$
$$x = \frac{-1 \pm \sqrt{1 - 4(1)(-1)}}{2(1)} = \frac{-1 \pm \sqrt{5}}{2}$$
Since the width of a rectangle is a positive number, we have width $= \dfrac{-1 + \sqrt{5}}{2}$ feet and length $= \dfrac{-1 + \sqrt{5}}{2} + \dfrac{2}{2} = \dfrac{1 + \sqrt{5}}{2}$ feet or width approximately 0.6 ft and length approximately 1.6 ft.

81. Let $x =$ the width, and $x + 4 =$ the length. Since the area is 10 square feet, we can write
$$x(x + 4) = 10$$
$$x^2 + 4x - 10 = 0$$
$$x = \frac{-4 \pm \sqrt{16 - 4(1)(-10)}}{2(1)} = \frac{-4 \pm 2\sqrt{14}}{2}$$
$$= -2 \pm \sqrt{14}$$
Since $x > 0$, the width is $-2 + \sqrt{14}$ feet, and the length is $-2 + \sqrt{14} + 4 = 2 + \sqrt{14}$ feet or approximately 1.7 ft and 5.7 ft.

83. The time it takes to reach the ground is found by solving the equation
$$-16t^2 + 16t + 96 = 0$$
$$t^2 - t - 6 = 0$$
$$(t - 3)(t + 2) = 0$$
$$t = 3 \quad \text{or} \quad t = -2$$
The pine cone reaches the earth 3 seconds after it is tossed.

85. The time it takes to reach the river is found by solving the equation

$$-16t^2 - 30t + 1000 = 0$$
$$8t^2 + 15t - 500 = 0$$
$$t = \frac{-15 \pm \sqrt{15^2 - 4(8)(-500)}}{2(8)}$$
$$= \frac{-15 \pm \sqrt{16225}}{16} = \frac{-15 \pm 127.377}{16}$$

Since the time is positive, the time is $\frac{-15 + 127.377}{16}$, or 7.0 seconds.

87. If x is the width of the border, then

$$(30 - 2x)(40 - 2x) = 704$$
$$1200 - 140x + 4x^2 = 704$$
$$4x^2 - 140x + 496 = 0$$
$$x^2 - 35x + 124 = 0$$
$$x = \frac{35 \pm \sqrt{35^2 - 4 \cdot 1 \cdot 124}}{2} = \frac{35 \pm \sqrt{729}}{2}$$
$$= \frac{35 \pm 27}{2} = 31 \text{ or } 4$$

The width of the border is 4 inches.

89. Let $x =$ the number presently sharing the cost. The cost is $\frac{100}{x}$ dollars each. If 6 more join, the cost is $\frac{100}{x} - 15$. We have

$$(x + 6)(\frac{100}{x} - 15) = 100$$
$$100 + \frac{600}{x} - 15x - 90 = 100$$
$$\frac{600}{x} - 15x - 90 = 0$$
$$600 - 15x^2 - 90x = 0$$
$$x^2 + 6x - 40 = 0$$
$$(x - 4)(x + 10) = 0$$
$$x - 4 = 0 \quad \text{or} \quad x + 10 = 0$$
$$x = 4 \quad \text{or} \quad x = -10$$

There are 4 workers in the original group.

91.
$$6x^2 + 5x - 4 = 0$$
$$(3x + 4)(2x - 1) = 0$$
$$x = -\frac{4}{3} \quad \text{or} \quad x = \frac{1}{2}$$
$$-\frac{4}{3} + \frac{1}{2} = -\frac{8}{6} + \frac{3}{6} = -\frac{5}{6}$$
$$-\frac{b}{a} = -\frac{5}{6}$$

The sum of the solutions is equal to $-\frac{b}{a}$.

93. The solutions to $6x^2 + 5x - 4 = 0$ are $-\frac{4}{3}$ and $\frac{1}{2}$. Their product is $-\frac{2}{3}$, which is the value of $\frac{c}{a}$

95. Since $y = x^2 - 6.33x + 3.7$ crosses the x-axis twice there are 2 real solutions.

97. Since $y = 4x^2 - 67.1x + 344$ does not cross the x-axis there are no real solutions.

99. Since the graph does not cross the x-axis there are no real solutions.

8.3 WARM-UPS

1. True, because $-1 = (-2)^2 - 5$ is correct.
2. False, the y-intercept is $(0, 9)$.
3. True, because the solution to $x^2 - 5 = 0$ is $x = \pm \sqrt{5}$.
4. True, because $a > 0$.
5. False, because in $y = x^2 + 4$, $a > 0$.
6. True, because $x = \frac{-2}{2(1)} = -1$ and $(-1)^2 + 2(-1) = -1$.
7. True, because $x^2 + 1 = 0$ has no real solution.
8. True, because if $x = 0$, then $y = c$.
9. True, because the parabola opens downward from the vertex $(0, 9)$
10. False, the minimum value of y occurs when $x = 7/6$.

8.3 EXERCISES

1. A quadratic function is a function of the form $f(x) = ax^2 + bx + c$ with $a \neq 0$.
3. If $a > 0$ then the parabola opens upward. If $a < 0$ then the parabola opens downward.
5. The vertex is the highest point on a parabola that opens downward or the lowest point on a parabola the opens upward.
7. If $x = 3$, then $f(3) = 3^2 - 3 - 12 = -6$. If $y = 0$, then $x^2 - x - 12 = 0$.

$$(x - 4)(x + 3) = 0$$
$$x - 4 = 0 \text{ or } x + 3 = 0$$
$$x = 4 \text{ or } \quad x = -3$$

So the ordered pairs are $(3, -6)$, $(4, 0)$, and $(-3, 0)$.
9. If $t = 4$, then $s = -16 \cdot 4^2 + 32(4) = -128$. It $s = 0$, then $-16t^2 + 32t = 0$.

$$-16t(t - 2) = 0$$
$$-16t = 0 \text{ or } t - 2 = 0$$
$$t = 0 \text{ or } \quad t = 2$$

The ordered pairs are $(4, -128)$, $(0, 0)$ and $(2, 0)$.
11. The ordered pairs $(-2, 6)$, $(-1, 3)$, $(0, 2)$, $(1, 3)$, and $(2, 6)$ satisfy $f(x) = x^2 + 2$. The domain is $(-\infty, \infty)$ and the range is $[2, \infty)$.

13. The ordered pairs $(-4, 4)$, $(-2, -2)$, $(0, -4)$, $(2, -2)$, and $(4, 4)$ satisfy $y = \frac{1}{2}x^2 - 4$. The domain is $(-\infty, \infty)$ and the range is $[-4, \infty)$.

15. The ordered pairs $(-2, -3)$, $(-1, 3)$, $(0, 5)$, $(1, 3)$, and $(2, -3)$ satisfy $f(x) = -2x^2 + 5$. The domain is $(-\infty, \infty)$ and the range is $(-\infty, 5]$.

17. The ordered pairs $(-6, -7)$, $(-3, 2)$, $(0, 5)$, $(3, 2)$, and $(6, -7)$ satisfy $y = -\frac{1}{3}x^2 + 5$. The domain is $(-\infty, \infty)$ and the range is $(-\infty, 5]$.

19. The ordered pairs $(0, 4)$, $(1, 1)$, $(2, 0)$, $(3, 1)$, and $(4, 4)$ satisfy $h(x) = (x - 2)^2$. The domain is $(-\infty, \infty)$ and the range is $[0, \infty)$.

21. $x = \frac{-b}{2a} = \frac{1}{2(1)} = \frac{1}{2}$.

$f\left(\frac{1}{2}\right) = \left(\frac{1}{2}\right)^2 - \frac{1}{2} - 2 = -\frac{9}{4}$

The vertex is $\left(\frac{1}{2}, -\frac{9}{4}\right)$. The y-intercept is $(0, -2)$.

$$x^2 - x - 2 = 0$$
$$(x - 2)(x + 1) = 0$$
$$x - 2 = 0 \quad \text{or} \quad x + 1 = 0$$
$$x = 2 \quad \text{or} \quad x = -1$$

The x-intercepts are $(-1, 0)$ and $(2, 0)$. The domain is $(-\infty, \infty)$ and the range is $\left[-\frac{9}{4}, \infty\right)$.

23. $x = \frac{-b}{2a} = \frac{-2}{2(1)} = -1$

$g(-1) = (-1)^2 + 2(-1) - 8 = -9$

The vertex is $(-1, -9)$. The y-intercept is $(0, -8)$.

$$x^2 + 2x - 8 = 0$$
$$(x - 2)(x + 4) = 0$$
$$x - 2 = 0 \quad \text{or} \quad x + 4 = 0$$
$$x = 2 \quad \text{or} \quad x = -4$$

The x-intercepts are $(-4, 0)$ and $(2, 0)$. The domain is $(-\infty, \infty)$ and the range is $[-9, \infty)$.

25. $y = -x^2 - 4x - 3$

$x = \frac{-b}{2a} = \frac{4}{2(-1)} = -2$

169

$y = -(-2)^2 - 4(-2) - 3 = 1$
Vertex $(-2, 1)$, y-intercept $(0, -3)$.
$$-x^2 - 4x - 3 = 0$$
$$x^2 + 4x + 3 = 0$$
$$(x + 1)(x + 3) = 0$$
$$x + 1 = 0 \quad \text{or} \quad x + 3 = 0$$
$$x = -1 \quad \text{or} \quad x = -3$$
The x-intercepts are $(-1, 0)$ and $(-3, 0)$.
The domain is $(-\infty, \infty)$ and the range is $(-\infty, 1]$.

27. $h(x) = -x^2 + 3x + 4$
$$x = \frac{-b}{2a} = \frac{-3}{2(-1)} = \frac{3}{2}$$
$$h\left(\frac{3}{2}\right) = -\left(\frac{3}{2}\right)^2 + 3\left(\frac{3}{2}\right) + 4 = \frac{25}{4}$$
Vertex $\left(\frac{3}{2}, \frac{25}{4}\right)$, y-intercept $(0, 4)$.
$$-x^2 + 3x + 4 = 0$$
$$x^2 - 3x - 4 = 0$$
$$(x - 4)(x + 1) = 0$$
$$x - 4 = 0 \quad \text{or} \quad x + 1 = 0$$
$$x = 4 \quad \text{or} \quad x = -1$$
The x-intercepts are $(4, 0)$ and $(-1, 0)$.
The domain is $(-\infty, \infty)$ and the range is $\left(-\infty, \frac{25}{4}\right]$.

29. $a = b^2 - 6b - 16$
$$b = \frac{6}{2(1)} = 3$$
$$a = 3^2 - 6(3) - 16 = -25$$
Vertex $(3, -25)$, b-intercept $(0, -16)$.
$$b^2 - 6b - 16 = 0$$
$$(b - 8)(b + 2) = 0$$
$$b - 8 = 0 \quad \text{or} \quad b + 2 = 0$$
$$b = 8 \quad \text{or} \quad b = -2$$
The b-intercepts are $(8, 0)$ and $(-2, 0)$.

The domain is $(-\infty, \infty)$ and the range is $[-25, \infty)$.

31. $y = x^2 - 8$
$$x = \frac{-b}{2a} = \frac{0}{2(1)} = 0$$
$$y = 0^2 - 8 = -8$$
The minimum value of y is -8.

33. $y = -3x^2 + 14$
$$x = \frac{-b}{2a} = \frac{0}{2(-3)} = 0$$
$$y = -3 \cdot 0^2 + 14 = 14$$
The maximum value of y is 14.

35. $y = x^2 + 2x + 3$
$$x = \frac{-b}{2a} = \frac{-2}{2(1)} = -1$$
$$y = (-1)^2 + 2(-1) + 3 = 2$$
The minimum value of y is 2.

37. $y = -2x^2 - 4x$
$$x = \frac{-b}{2a} = \frac{4}{2(-2)} = -1$$
$$y = -2(-1)^2 - 4(-1) = 2$$
The maximum value of y is 2.

39. $s(t) = -16t^2 + 64t$
$$t = \frac{-64}{2(-16)} = 2$$
$$s(2) = -16 \cdot 2^2 + 64(2) = 64$$
Maximum height is 64 feet.

41. The minimum value of C in the formula $C = 0.009x^2 - 1.8x + 100$ is attained when
$$x = \frac{-(-1.8)}{2(0.009)} = 100.$$

The cost per hour will be at a minimum when they produce 100 balls per hour.

43. $A = -w^2 + 50w$

$w = \dfrac{-50}{2(-1)} = 25$

$A = -25^2 + 50(25) = 625$

Maximum area is 625 square meters.

45. $A(t) = -2t^2 + 32t + 12$

$t = \dfrac{-32}{2(-2)} = 8$

The nitrogen dioxide is at its maximum 8 hours after 6 A.M. or at 2 P.M.

47. Since the x-coordinate of the tower on the right is 20, we get $y = 0.0375(20)^2 = 15$. So the height of the towers is 15 meters. Use the Pythagorean theorem to find z:

$$15^2 + 20^2 = z^2$$
$$z^2 = 625$$
$$z = 25$$

So the length of the cable marked z is 25 meters.

49. The graph of $y = kx^2$ gets narrower as k gets larger.

51. The graph of $y = x^2$ has the same shape as $x = y^2$.

53. a) Vertex $(0.84, -0.68)$, domain $(-\infty, \infty)$, range $[-0.68, \infty)$, $(1.30, 0)$, $(0.38, 0)$

b) Vertex $(5.96, 46.26)$, domain $(-\infty, \infty)$, range $(-\infty, 46.26]$, $(12.48, 0)$, $(-0.55, 0)$

8.4 WARM-UPS

1. True, because if $w = x^2$ the equation becomes $w^2 - 5w + 6 = 0$.

2. False, the equation cannot be solved by substitution.

3. False, we can use factoring.

4. True, because $(x^{1/6})^2 = x^{2/6} = x^{1/3}$.

5. False, we should let $w = \sqrt{x}$.

6. False, because $(2^{1/2})^2 = 2^{2/2} = 2^1 = 2$.

7. False, his rate is $1/x$ of the fence per hour.

8. True, because $R = D/T$.

9. False, against a 5 mph current it will go 5 mph.

10. False, the dimensions of the bottom will be $11 - 2x$ by $14 - 2x$.

8.4 EXERCISES

1. If the coefficients are integers and the discriminant is a perfect square, then the quadratic polynomial can be factored.

3. If the solutions are a and b, then the quadratic equation $(x - a)(x - b) = 0$ has those solutions.

5. $b^2 - 4ac = (-1)^2 - 4(2)4 = -31$
Polynomial is prime.

7. $b^2 - 4ac = 6^2 - 4(2)(-5) = 76$
Polynomial is prime.

9. $b^2 - 4ac = 19^2 - 4(6)(-36) = 1225$
Since $\sqrt{1225} = 35$, the polynomial is not prime.
$6x^2 + 19x - 36 = (3x - 4)(2x + 9)$

11. $b^2 - 4ac = 25 - 4(4)(-12) = 217$
Polynomial is prime.

13. $b^2 - 4ac = (-18)^2 - 4 \cdot 8(-45) = 1764$
Since $\sqrt{1764} = 42$, the polynomial is not prime.
$8x^2 - 18x - 45 = (4x - 15)(2x + 3)$

15. $(x - 3)(x + 7) = 0$
$\qquad x^2 + 4x - 21 = 0$

17. $(x - 4)(x - 1) = 0$
$\qquad x^2 - 5x + 4 = 0$

19. $(x - \sqrt{5})(x + \sqrt{5}) = 0$
$\qquad x^2 - 5 = 0$

21. $(x - 4i)(x + 4i) = 0$
$\qquad x^2 + 16 = 0$

23. $(x - i\sqrt{2})(x + i\sqrt{2}) = 0$
$\qquad x^2 + 2 = 0$

25. $(2x - 1)(3x - 1) = 0$
$\qquad 6x^2 - 5x + 1 = 0$

27. Let $w = 2a - 1$ in
$(2a - 1)^2 + 2(2a - 1) - 8 = 0$:
$\qquad w^2 + 2w - 8 = 0$
$\qquad (w + 4)(w - 2) = 0$
$\quad w + 4 = 0$ or $w - 2 = 0$
$\qquad w = -4$ or $\quad w = 2$
$\quad 2a - 1 = -4$ or $2a - 1 = 2$
$\qquad 2a = -3$ or $\quad 2a = 3$
$\qquad a = -\dfrac{3}{2}$ or $\quad a = \dfrac{3}{2}$
Solution set: $\left\{ -\dfrac{3}{2}, \dfrac{3}{2} \right\}$

29. Let $y = w - 1$ in
$(w - 1)^2 + 5(w - 1) + 5 = 0$:
$\qquad y^2 + 5y + 5 = 0$

$$y = \frac{-5 \pm \sqrt{25 - 4(1)(5)}}{2(1)} = \frac{-5 \pm \sqrt{5}}{2}$$

$$w - 1 = \frac{-5 \pm \sqrt{5}}{2}$$

$$w = \frac{-5 \pm \sqrt{5}}{2} + 1 = \frac{-5 \pm \sqrt{5}}{2} + \frac{2}{2} = \frac{-3 \pm \sqrt{5}}{2}$$

Solution set: $\left\{ \frac{-3 \pm \sqrt{5}}{2} \right\}$

31. Let $w = x^2$ and $w^2 = x^4$ in
$x^4 - 14x^2 + 45 = 0$:
$$w^2 - 14w + 45 = 0$$
$$(w - 5)(w - 9) = 0$$
$$w - 5 = 0 \text{ or } w - 9 = 0$$
$$w = 5 \quad \text{ or } \quad w = 9$$
$$x^2 = 5 \quad \text{ or } \quad x^2 = 9$$
$$x = \pm\sqrt{5} \quad \text{ or } x = \pm 3$$
Solution set: $\{ \pm\sqrt{5}, \pm 3 \}$

33. Let $w = x^3$, and $w^2 = x^6$ in
$x^6 + 7x^3 = 8$:
$$w^2 + 7w = 8$$
$$w^2 + 7w - 8 = 0$$
$$(w - 1)(w + 8) = 0$$
$$w - 1 = 0 \quad \text{ or } \quad w + 8 = 0$$
$$w = 1 \quad \text{ or } \quad w = -8$$
$$x^3 = 1 \quad \text{ or } \quad x^3 = -8$$
$$x = 1 \quad \text{ or } \quad x = -2$$
Solution set: $\{-2, 1\}$

35. Let $w = x^2 + 2x$ in
$(x^2 + 2x)^2 - 7(x^2 + 2x) + 12 = 0$:
$$w^2 - 7w + 12 = 0$$
$$(w - 3)(w - 4) = 0$$
$$w - 3 = 0 \quad \text{ or } \quad w - 4 = 0$$
$$w = 3 \quad \text{ or } \quad w = 4$$
$$x^2 + 2x = 3 \quad \text{ or } \quad x^2 + 2x = 4$$
$$x^2 + 2x - 3 = 0 \quad \text{ or } \quad x^2 + 2x - 4 = 0$$
$$(x + 3)(x - 1) = 0 \text{ or }$$
$$x = \frac{-2 \pm \sqrt{4 - 4(1)(-4)}}{2}$$
$$x = -3 \text{ or } x = 1 \quad \text{ or } \quad x = \frac{-2 \pm 2\sqrt{5}}{2}$$
Solution set: $\{-1 \pm \sqrt{5}, -3, 1\}$

37. Let $w = y^2 + y$ in
$(y^2 + y)^2 - 8(y^2 + y) + 12 = 0$:
$$w^2 - 8w + 12 = 0$$
$$(w - 6)(w - 2) = 0$$
$$w = 6 \quad \text{ or } \quad w = 2$$
$$y^2 + y = 6 \quad \text{ or } \quad y^2 + y = 2$$

$$y^2 + y - 6 = 0 \text{ or } \quad y^2 + y - 2 = 0$$
$$(y + 3)(y - 2) = 0 \text{ or } (y + 2)(y - 1) = 0$$
$$y = -3 \text{ or } y = 2 \quad \text{ or } \quad y = -2 \text{ or } y = 1$$
Solution set: $\{-3, -2, 1, 2\}$

39. Let $w = x^{1/4}$ and $w^2 = x^{1/2}$ in
$x^{1/2} - 5x^{1/4} + 6 = 0$:
$$w^2 - 5w + 6 = 0$$
$$(w - 3)(w - 2) = 0$$
$$w - 3 = 0 \quad \text{ or } \quad w - 2 = 0$$
$$w = 3 \quad \text{ or } \quad w = 2$$
$$x^{1/4} = 3 \quad \text{ or } \quad x^{1/4} = 2$$
$$(x^{1/4})^4 = 3^4 \quad \text{ or } \quad (x^{1/4})^4 = 2^4$$
$$x = 81 \quad \text{ or } \quad x = 16$$
Solution set: $\{16, 81\}$

41. Let $w = x^{1/2}$ and $w^2 = x$ in
$2x - 5x^{1/2} - 3 = 0$.
$$2w^2 - 5w - 3 = 0$$
$$(2w + 1)(w - 3) = 0$$
$$2w + 1 = 0 \quad \text{ or } \quad w - 3 = 0$$
$$w = -\frac{1}{2} \text{ or } \quad w = 3$$
$$x^{1/2} = -\frac{1}{2} \text{ or } \quad x^{1/2} = 3$$
$$x = \frac{1}{4} \quad \text{ or } \quad x = 9$$
The solution $1/4$ does not check in the original
equation. So the solution set is $\{9\}$.

43. Let $t = x^{-1}$ in $x^{-2} + x^{-1} - 6 = 0$:
$$t^2 + t - 6 = 0$$
$$(t + 3)(t - 2) = 0$$
$$t = -3 \quad \text{ or } \quad t = 2$$
$$x^{-1} = -3 \quad \text{ or } \quad x^{-1} = 2$$
$$x = -\frac{1}{3} \quad \text{ or } \quad x = \frac{1}{2}$$
Solution set: $\left\{ -\frac{1}{3}, \frac{1}{2} \right\}$

45. Let $w = x^{1/6}$ and $w^2 = x^{1/3}$ in
$x^{1/6} - x^{1/3} + 2 = 0$:
$$w - w^2 + 2 = 0$$
$$w^2 - w - 2 = 0$$
$$(w - 2)(w + 1) = 0$$
$$w = 2 \quad \text{ or } \quad w = -1$$
$$x^{1/6} = 2 \quad \text{ or } \quad x^{1/6} = -1$$
$$(x^{1/6})^6 = 2^6 \quad \text{ or } (x^{1/6})^6 = (-1)^6$$
$$x = 64 \quad \text{ or } \quad x = 1$$

The original equation is not satisfied for $x = 1$.
So the solution set is $\{64\}$.

47. Let $w = \dfrac{1}{y-1}$ in

$$\left(\dfrac{1}{y-1}\right)^2 + \left(\dfrac{1}{y-1}\right) = 6.$$

$$w^2 + w = 6$$
$$w^2 + w - 6 = 0$$
$$(w+3)(w-2) = 0$$
$$w = -3 \quad \text{or} \quad w = 2$$
$$\dfrac{1}{y-1} = -3 \quad \text{or} \quad \dfrac{1}{y-1} = 2$$
$$-3y + 3 = 1 \quad \text{or} \quad 2y - 2 = 1$$
$$-3y = -2 \quad \text{or} \quad 2y = 3$$
$$y = \dfrac{2}{3} \quad \text{or} \quad y = \dfrac{3}{2}$$

Solution set: $\left\{\dfrac{2}{3}, \dfrac{3}{2}\right\}$

49. Let $w = \sqrt{2x^2 - 3}$ and $w^2 = 2x^2 - 3$ in
$2x^2 - 3 - 6\sqrt{2x^2 - 3} + 8 = 0$:

$$w^2 - 6w + 8 = 0$$
$$(w-4)(w-2) = 0$$
$$w = 4 \quad \text{or} \quad w = 2$$
$$\sqrt{2x^2 - 3} = 4 \quad \text{or} \quad \sqrt{2x^2 - 3} = 2$$

Square each side:

$$2x^2 - 3 = 16 \quad \text{or} \quad 2x^2 - 3 = 4$$
$$2x^2 = 19 \quad \text{or} \quad 2x^2 = 7$$
$$x^2 = \dfrac{19}{2} \quad \text{or} \quad x^2 = \dfrac{7}{2}$$
$$x = \pm\sqrt{\dfrac{19}{2}} \quad \text{or} \quad x = \pm\sqrt{\dfrac{7}{2}}$$
$$x = \pm\dfrac{\sqrt{38}}{2} \quad \text{or} \quad x = \pm\dfrac{\sqrt{14}}{2}$$

Solution set: $\left\{\pm\dfrac{\sqrt{14}}{2}, \pm\dfrac{\sqrt{38}}{2}\right\}$

51. Let $t = x^{-1}$ and $t^2 = x^{-2}$ in
$x^{-2} - 2x^{-1} - 1 = 0$:

$$t^2 - 2t - 1 = 0$$

$$t = \dfrac{2 \pm \sqrt{4 - 4(1)(-1)}}{2} = \dfrac{2 \pm 2\sqrt{2}}{2}$$
$$= 1 \pm \sqrt{2}$$
$$x^{-1} = 1 \pm \sqrt{2}$$
$$x = \dfrac{1}{1 + \sqrt{2}} \quad \text{or} \quad x = \dfrac{1}{1 - \sqrt{2}}$$
$$x = \dfrac{1(1 - \sqrt{2})}{(1 + \sqrt{2})(1 - \sqrt{2})} \quad \text{or}$$
$$x = \dfrac{1(1 + \sqrt{2})}{(1 - \sqrt{2})(1 + \sqrt{2})}$$
$$x = \dfrac{1 - \sqrt{2}}{-1} \quad \text{or} \quad x = \dfrac{1 + \sqrt{2}}{-1}$$
$$x = -1 + \sqrt{2} \quad \text{or} \quad x = -1 - \sqrt{2}$$

Solution set: $\{-1 + \sqrt{2}, -1 - \sqrt{2}\}$

53. Let x = Gary's travel time and $x + 1$ = Harry's travel time. Since $R = D/T$, Gary's speed is $300/x$ and Harry's speed is $300/(x+1)$. Since Gary travels 10 mph faster, we can write the following equation.

$$\dfrac{300}{x} = \dfrac{300}{x+1} + 10$$
$$x(x+1)\left(\dfrac{300}{x}\right) = x(x+1)\left(\dfrac{300}{x+1} + 10\right)$$
$$300x + 300 = 300x + 10x^2 + 10x$$
$$0 = 10x^2 + 10x - 300$$
$$0 = x^2 + x - 30$$
$$0 = (x+6)(x-5)$$
$$x + 6 = 0 \quad \text{or} \quad x - 5 = 0$$
$$x = -6 \quad \text{or} \quad x = 5$$

Gary travels 5 hours and they arrive at 2 P.M.

55. Let x = her speed before lunch and $x - 4$ = her speed after lunch. Since $T = D/R$, her time before lunch was $60/x$ and her time after lunch was $46/(x-4)$. Since she put in one hour more after lunch, we can write the following equation.

$$\dfrac{46}{x-4} - 1 = \dfrac{60}{x}$$
$$x(x-4)\left(\dfrac{46}{x-4} - 1\right) = x(x-4)\left(\dfrac{60}{x}\right)$$
$$46x - x^2 + 4x = 60x - 240$$
$$-x^2 - 10x + 240 = 0$$
$$x^2 + 10x - 240 = 0$$

$$x = \dfrac{-10 \pm \sqrt{100 - 4(1)(-240)}}{2(1)}$$
$$= \dfrac{-10 \pm \sqrt{1060}}{2} = \dfrac{-10 \pm 2\sqrt{265}}{2}$$
$$= -5 \pm \sqrt{265}$$

Since $-5 - \sqrt{265}$ is a negative number, we disregard that answer. Her speed before lunch was $-5 + \sqrt{265} \approx 11.3$ mph, and her speed after lunch was 4 mph slower: $-9 + \sqrt{265} \approx 7.3$ mph.

57. Let x = Andrew's time and $x + 3$ = John's time. Andrew's rate is $1/x$ job/hr and John's rate is $1/(x+3)$ job/hr. In 8 hours Andrew does $8/x$ job and John does $8/(x+3)$ job.

$$\frac{8}{x} + \frac{8}{x+3} = 1$$

$$x(x+3)\left(\frac{8}{x} + \frac{8}{x+3}\right) = x(x+3)1$$
$$8x + 24 + 8x = x^2 + 3x$$
$$0 = x^2 - 13x - 24$$

$$x = \frac{13 \pm \sqrt{169 - 4(1)(-24)}}{2(1)} = \frac{13 \pm \sqrt{265}}{2}$$

Since $\frac{13 - \sqrt{265}}{2}$ is a negative number, we disregard that solution. Andrew's time is $\frac{13 + \sqrt{265}}{2} \approx 14.6$ hr and John's time is 3 hours more :

$$3 + \frac{13 + \sqrt{265}}{2} = \frac{19 + \sqrt{265}}{2} \approx 17.6 \text{ hr}$$

59. Let $x =$ the amount of increase. The new length and width will be $30 + x$ and $20 + x$. Since the new area is to be 1000, we can write the following equation.
$$(20 + x)(30 + x) = 1000$$
$$x^2 + 50x - 400 = 0$$
$$x = \frac{-50 \pm \sqrt{2500 - 4(1)(-400)}}{2(1)}$$

$$= \frac{-50 \pm \sqrt{4100}}{2} = \frac{-50 \pm 10\sqrt{41}}{2}$$
$$= -25 \pm 5\sqrt{41}$$
Disregard the negative solution.
Use $x = -25 + 5\sqrt{41}$ to get
$30 + x = 5 + 5\sqrt{41} \approx 37.02$ feet, and
$20 + x = -5 + 5\sqrt{41} \approx 27.02$ feet.
61. Let $x =$ the number of hours for A to empty the pool and $x + 2 =$ the number of hours for B to empty the pool. A's rate is $\frac{1}{x}$ pool/hr and B's rate is $\frac{1}{x+2}$ pool/hr. A works for 9 hrs and does $\frac{9}{x}$ of the pool while B works for 6 hours and does $\frac{6}{x+2}$ of the pool. Since the pool is half full we have the following equation.
$$\frac{9}{x} + \frac{6}{x+2} = \frac{1}{2}$$
$$2x(x+2)\left(\frac{9}{x} + \frac{6}{x+2}\right) = 2x(x+2)\frac{1}{2}$$
$$18x + 36 + 12x = x^2 + 2x$$
$$-x^2 + 28x + 36 = 0$$

$$x^2 - 28x - 36 = 0$$
$$x = \frac{28 \pm \sqrt{28^2 - 4(1)(-36)}}{2} = \frac{28 \pm \sqrt{928}}{2}$$
$$= \frac{28 \pm 4\sqrt{58}}{2} = 14 \pm 2\sqrt{58} = -1.2 \text{ or } 29.2$$
A would take 29.2 hours working alone.
63. $\quad \frac{10}{W} = \frac{W}{10 - W}$
$$100 - 10W = W^2$$
$$W^2 + 10W - 100 = 0$$

$$W = \frac{-10 \pm \sqrt{10^2 - 4(1)(-100)}}{2}$$
$$= \frac{-10 \pm 10\sqrt{5}}{2}$$
$$W = -5 \pm 5\sqrt{5} \approx -16.2 \text{ or } 6.2$$
So the width is $-5 + 5\sqrt{5}$ or 6.2 meters.

65. a) $\left(x - (3 + \sqrt{5})\right)\left(x - (3 - \sqrt{5})\right) = 0$
$x^2 - 6x + 4 = 0$
b) $(x - (4 - 2i))(x - (4 + 2i)) = 0$
$x^2 - 8x + 20 = 0$
c) $\left(x - \frac{1 + i\sqrt{3}}{2}\right)\left(x - \frac{1 - i\sqrt{3}}{2}\right) = 0$
$x^2 - x + 1 = 0$
67. $\{-10, -4, 4, 10\}$
69. $\{4.27\}$

8.5 WARM-UPS

1. False, solution set is $(-\infty, -2) \cup (2, \infty)$.
2. False, we do not multiply each side by a variable.
3. False, that is not how we solve a quadratic inequality.
4. False, we can solve any quadratic inequality.
5. True, that is why we make the sign graph.
6. True, because inequalities change direction when multiplied by a negative number and we do not know if the variable is positive or negative.
7. True, because the solution is based on rules for multiplying or dividing signed numbers.
8. True, multiply each side by 2.
9. True, subtract 1 from each side.
10. False, because 4 causes the denominator to be 0 and so it cannot be in the solution set.

8.5 EXERCISES

1. A quadratic inequality has the form $ax^2 + bx + c > 0$. In place of $>$ we can also use $<$, $\leq$, or $\geq$.

3. A rational inequality is an inequality involving a rational expression.

5. $\qquad x^2 + x - 6 < 0$
$\qquad\qquad (x+3)(x-2) < 0$

$x+3 \ \ - - - -0+ + + + + + + +$
$x-2 \ \ - - - - - - - -0+ + + +$
$\qquad\qquad -3 \qquad\qquad 2$

The product is negative only if the factors have opposite signs. That happens when x is chosen between -3 and 2. Solution set: $(-3, 2)$

$\longleftarrow\!\!+\!\!(\!+\!+\!+\!+\!+\!)\!+\!\longrightarrow$
$\quad -4\,-3\,-2\,-1\ 0\ 1\ 2\ 3$

7. $\qquad y^2 - 4 > 0$
$\qquad\qquad (y-2)(y+2) > 0$

$y+2 \ - - - -0+ + + + + + + +$
$y-2 \ - - - - - - - -0+ + + +$
$\qquad\qquad -2 \qquad\quad 2$

The product is positive if both factors are positive or both are negative.
$(-\infty, -2) \cup (2, \infty)$

$\longleftarrow\!\!+\!\!+\!\!)\!+\!+\!+\!(\!+\!+\!\longrightarrow$
$\quad -4\,-3\,-2\,-1\ 0\ 1\ 2\ 3\ 4$

9. $\qquad 2u^2 + 5u \geq 12$
$\qquad\qquad 2u^2 + 5u - 12 \geq 0$
$\qquad\qquad (2u-3)(u+4) \geq 0$

$2u-3 \ - - - - - - - -0+ + +$
$u+4 \ - - - 0+ + + + + + + +$
$\qquad\qquad -4 \qquad\qquad 3/2$

The product is positive only if the factors have the same sign. That happens if u is chosen to the left of -4 or to the right of $3/2$. The product is 0 if u is either -4 or $3/2$.
$(-\infty, -4] \cup \left[\frac{3}{2}, \infty\right)$

$\qquad\qquad\qquad\qquad\quad \frac{3}{2}$
$\longleftarrow\!\!+\!\![\!+\!+\!+\!+\!+\!+\![\!+\!\longrightarrow$
$\ -6\,-5\,-4\,-3\,-2\,-1\ 0\ 1\ 2\ 3\ 4$

11. $\qquad 4x^2 - 8x \geq 0$
$\qquad\qquad 4x(x-2) \geq 0$

$4x \quad - - -0+ + + + + + + +$
$x-2 - - - - - - - -0+ + + +$
$\qquad\qquad 0 \quad 2$

The product is greater than zero when the signs are the same.
$(-\infty, 0] \cup [2, \infty)$

$\longleftarrow\!\!+\!\!]\!+\!+\![\!+\!+\!+\!\longrightarrow$
$\quad -2\,-1\ 0\ 1\ 2\ 3\ 4$

13. $\qquad 5x - 10x^2 < 0$
$\qquad\qquad 5x(1 - 2x) < 0$

$5x \qquad - - 0+ + + + + + + +$
$1-2x + + + + + + +0- - - -$
$\qquad\qquad 0 \qquad\quad 1/2$

The product is negative only if the factors have opposite signs: $(-\infty, 0) \cup \left(\frac{1}{2}, \infty\right)$

$\qquad\qquad\qquad\qquad\qquad \frac{1}{2}$
$\longleftarrow\!\!+\!\!+\!\!+\!\!)\!(\!+\!+\!+\!\longrightarrow$
$\quad -3\,-2\,-1\ 0\ 1\ 2\ 3$

15. $\quad x^2 + 6x + 9 \geq 0$
$\qquad\qquad (x+3)(x+3) \geq 0$

$x+3 - - - -0+ + + +$
$x+3 - - - -0+ + + +$
$\qquad\qquad -3$

The product is positive only if the factors have the same sign. That happens for every value of x except -3, in which case the product is 0. The solution set is the set of all real numbers, $(-\infty, \infty)$.

17. $\qquad \frac{x}{x-3} > 0$

$x \qquad - - - -0+ + + + + + + +$
$x-3 - - - - - - -0+ + + +$
$\qquad\qquad 0 \qquad\quad 3$

The quotient is positive only if the numerator and denominator have the same sign.
$(-\infty, 0) \cup (3, \infty)$

$\longleftarrow\!\!+\!\!+\!\!)\!+\!+\![\!+\!+\!\longrightarrow$
$\quad -2\,-1\ 0\ 1\ 2\ 3\ 4\ 5$

19. $\qquad \frac{x+2}{x} \leq 0$

$x+2 - - - -0+ + + + + + + +$
$x \qquad - - - - - - -0+ + + +$
$\qquad\qquad -2 \qquad\quad 0$

The quotient is negative only if the numerator and denominator have opposite signs. Since the denominator must not be 0, we do not include 0 in the solution set: $[-2, 0)$

21. $\dfrac{t-3}{t+6} > 0$

$t+6 - - - 0 + + + + + + +$
$t-3 - - - - - - - - 0 + + +$
$\qquad\qquad -6 \qquad\quad 3$

The quotient is positive only if the numerator and denominator have the same sign.
$(-\infty, -6) \cup (3, \infty)$

23. $\dfrac{x}{x+2} > -1$

$\dfrac{x}{x+2} + 1 > 0$

$\dfrac{x}{x+2} + \dfrac{x+2}{x+2} > 0$

$\dfrac{2x+2}{x+2} > 0$

$x+2 - - - - 0 + + + + + + + +$
$2x+2 - - - - - - 0 + + + +$
$\qquad\quad -2 \qquad\quad -1$

The quotient is positive only if the numerator and denominator have the same sign.
$(-\infty, -2) \cup (-1, \infty)$

25. $\dfrac{2}{x-5} > \dfrac{1}{x+4}$

$\dfrac{2}{x-5} - \dfrac{1}{x+4} > 0$

$\dfrac{2(x+4)}{(x-5)(x+4)} - \dfrac{1(x-5)}{(x+4)(x-5)} > 0$

$\dfrac{x+13}{(x-5)(x+4)} > 0$

$x+4 - - - - 0 + + + + + + +$
$x+13 - 0 + + + + + + + + + +$
$x-5 - - - - - - - - 0 + + +$
$\qquad -13 \qquad -4 \qquad\quad 5$

This quotient will be positive only if an even number of the factors have negative values.
$(-13, -4) \cup (5, \infty)$

27. $\dfrac{m}{m-5} + \dfrac{3}{m-1} > 0$

$\dfrac{m(m-1)}{(m-5)(m-1)} + \dfrac{3(m-5)}{(m-1)(m-5)} > 0$

$\dfrac{m^2 + 2m - 15}{(m-1)(m-5)} > 0$

$\dfrac{(m+5)(m-3)}{(m-1)(m-5)} > 0$

$m-3 - - - - - - - 0 + + + + +$
$m-1 - - - - 0 + + + + + + + +$
$m+5 - - 0 + + + + + + + + + +$
$m-5 - - - - - - - - - - - 0 + +$
$\qquad -5 \qquad 1 \qquad 3 \qquad 5$

This quotient is positive only if there is an even number of factors with negative signs.
$(-\infty, -5) \cup (1, 3) \cup (5, \infty)$

29. $\dfrac{x}{x-3} \le \dfrac{-8}{x-6}$

$\dfrac{x}{x-3} + \dfrac{8}{x-6} \le 0$

$\dfrac{x(x-6)}{(x-3)(x-6)} + \dfrac{8(x-3)}{(x-6)(x-3)} \le 0$

$\dfrac{x^2 + 2x - 24}{(x-3)(x-6)} \le 0$

$\dfrac{(x+6)(x-4)}{(x-3)(x-6)} \le 0$

$x-4 - - - - - - - 0 + + + + +$
$x-3 - - - - 0 + + + + + + + +$
$x+6 - - 0 + + + + + + + + + +$
$x-6 - - - - - - - - - - - 0 + +$
$\qquad -6 \qquad 3 \qquad 4 \qquad 6$

The quotient is negative only if an odd number of factors have negative signs. Note that 3 and 6 cause the denominator to have a value of 0, and so they are excluded from the solution set.
$[-6, 3) \cup [4, 6)$

31. To solve $x^2 - 2x - 4 > 0$, first solve $x^2 - 2x - 4 = 0$:

$x = \dfrac{2 \pm \sqrt{4 - 4(1)(-4)}}{2(1)} = \dfrac{2 \pm 2\sqrt{5}}{2}$

$\qquad\qquad = 1 \pm \sqrt{5}$

The two solutions to the equation divide the number line into three regions. Choose one number in each region, say -10, 0, and 10.

Now evaluate $x^2 - 2x - 4$ for the numbers $-10, 0,$ and 10.
$$(-10)^2 - 2(-10) - 4 = 116$$
$$0^2 - 2(0) - 4 = -4$$
$$10^2 - 2(10) - 4 = 76$$
The signs of these answers indicate the solution
to the inequality.
$$(-\infty, 1 - \sqrt{5}) \cup (1 + \sqrt{5}, \infty)$$

33. To solve $2x^2 - 6x + 3 \geq 0$, we first solve $2x^2 - 6x + 3 = 0$.

$$x = \frac{6 \pm \sqrt{36 - 4(2)(3)}}{2(2)} = \frac{3 \pm \sqrt{3}}{2}$$

The two solutions to the equation divide the number line into three regions. Choose a test point in each region, say $-10, 2,$ and 10. Evaluate the polynomial $2x^2 - 6x + 3$ at each of these points.
$$2(-10)^2 - 6(-10) + 3 = 263$$
$$2(2)^2 - 6(2) + 3 = -1$$
$$2(10)^2 - 6(10) + 3 = 143$$
The signs of these answers indicate the regions that satisfy the inequality.
$$\left(-\infty, \frac{3-\sqrt{3}}{2}\right] \cup \left[\frac{3+\sqrt{3}}{2}, \infty\right)$$

35. To solve $y^2 - 3y - 9 \leq 0$ we first solve $y^2 - 3y - 9 = 0$:
$$y = \frac{3 \pm \sqrt{9 - 4(1)(-9)}}{2} = \frac{3 \pm 3\sqrt{5}}{2}$$
Pick test points $-10, 4,$ and 10, and evaluate $y^2 - 3y - 9$ for these numbers.
$$(-10)^2 - 3(-10) - 9 = 121$$
$$4^2 - 3(4) - 9 = -5$$
$$10^2 - 3(10) - 9 = 61$$
The signs of these answers indicate which regions satisfy the original inequality.
$$\left[\frac{3 - 3\sqrt{5}}{2}, \frac{3 + 3\sqrt{5}}{2}\right]$$

37.
$$x^2 \leq 9$$
$$x^2 - 9 \leq 0$$

$$(x - 3)(x + 3) \leq 0$$

The product is negative only if the factors have opposite signs. That happens if x is chosen between -3 and 3. At ± 3 the product is 0. The solution set is $[-3, 3]$.

39.
$$16 - x^2 > 0$$
$$-1(16 - x^2) < -1(0)$$
$$x^2 - 16 < 0$$
$$(x - 4)(x + 4) < 0$$

The product is negative only if the value of x is between -4 and 4. The solution set is $(-4, 4)$.

41.
$$x^2 - 4x \geq 0$$
$$x(x - 4) \geq 0$$

The product is positive only if the factors have the same sign. The solution set is $(-\infty, 0] \cup [4, \infty)$.

43.
$$3(2w^2 - 5) < w$$
$$6w^2 - 15 < w$$
$$6w^2 - w - 15 < 0$$
$$(2w + 3)(3w - 5) < 0$$

The product is negative only if the factors have opposite signs. The solution set is $\left(-\frac{3}{2}, \frac{5}{3}\right)$.

45.
$$z^2 \geq 4(z + 3)$$
$$z^2 \geq 4z + 12$$
$$z^2 - 4z - 12 \geq 0$$
$$(z - 6)(z + 2) \geq 0$$

The product is positive only if the factors have the same sign. The solution set is $(-\infty, -2] \cup [6, \infty)$.

47.
$$(q + 4)^2 > 10q + 31$$
$$q^2 + 8q + 16 > 10q + 31$$
$$q^2 - 2q - 15 > 0$$
$$(q + 3)(q - 5) > 0$$

$$
\begin{array}{l}
q + 3 \; - \; - \; - \; -\;0+ + + + + + + + \\
q - 5 \; - \; - \; - \; - \; - \; - \; - \; - \;0+ + + + \\
\hline
\qquad\qquad -3 \qquad\quad 5
\end{array}
$$

The product is positive only if the factors have the same sign. The solution set is $(-\infty, -3) \cup (5, \infty)$.

49.
$$\tfrac{1}{2}x^2 \geq 4 - x$$
$$x^2 \geq 8 - 2x$$
$$x^2 + 2x - 8 \geq 0$$
$$(x + 4)(x - 2) \geq 0$$

$$
\begin{array}{l}
x + 4 \; - \; - \; -\;0+ + + + + + + + + \\
x - 2 \; - \; - \; - \; - \; - \; - \; - \; - \;0+ + + + \\
\hline
\qquad\quad -4 \qquad\qquad 2
\end{array}
$$

The product is positive only if the factors have the same sign. The solution set is $(-\infty, -4] \cup [2, \infty)$.

51. $\dfrac{x - 4}{x + 3} \leq 0$

$$
\begin{array}{l}
x + 3 \; - \; - \; - \; -\;0+ + + + + + + + \\
x - 4 \; - \; - \; - \; - \; - \; - \; - \; -\;0+ + + + \\
\hline
\qquad\qquad -3 \qquad\qquad 4
\end{array}
$$

The quotient is negative if the factors have opposite signs. The solution set is $(-3, 4]$.

53. $(x - 2)(x + 1)(x - 5) \geq 0$

$$
\begin{array}{l}
x - 2 \; - \; - \; - \; - \;0+ + + + + + + + \\
x + 1 \; - \; -\;0+ + + + + + + + + + + \\
x - 5 \; - \; - \; - \; - \; - \; - \; - \;0+ + + + \\
\hline
\qquad -1 \qquad 2 \qquad\qquad 5
\end{array}
$$

The product of these three factors is positive only if an even number of the factors have negative signs. This happens between -1 and 2, and also above 5 where no factors have negative signs. The solution set is $[-1, 2] \cup [5, \infty)$.

55.
$$x^3 + 3x^2 - x - 3 < 0$$
$$x^2(x + 3) - 1(x + 3) < 0$$

$$(x^2 - 1)(x + 3) < 0$$
$$(x - 1)(x + 1)(x + 3) < 0$$

$$
\begin{array}{l}
x + 1 \; - \; - \; - \; - \; -\;0+ + + + + + + \\
x + 3 \; - \; -\;0+ + + + + + + + + + + \\
x - 1 \; - \; - \; - \; - \; - \; - \; - \; - \; - \; -\;0+ + + \\
\hline
\qquad -3 \qquad -1 \qquad\qquad 1
\end{array}
$$

The product of these three factors is negative only if an odd number of them have negative signs. This happens to the left of -3 and also between -1 and 1. The solution set is $(-\infty, -3) \cup (-1, 1)$.

57. To solve $0.23x^2 + 6.5x + 4.3 < 0$, we first solve $0.23x^2 + 6.5x + 4.3 = 0$.
$$x = \frac{-6.5 \pm \sqrt{(6.5)^2 - 4(0.23)(4.3)}}{2(0.23)}$$

$x = -27.58$ or $x = -0.68$
Test the numbers -30, -1, and 0.
$$0.23(-30)^2 + 6.5(-30) + 4.3 = 16.3$$
$$0.23(-1)^2 + 6.5(-1) + 4.3 = -1.97$$
$$0.23(0)^2 + 6.5(0) + 4.3 = 4.3$$

According to the signs of the values of the polynomial at the test points, the value of the polynomial is negative between the two solutions to the equation. The solution set is $(-27.58, -0.68)$.

59.
$$\frac{x}{x - 2} > \frac{-1}{x + 3}$$
$$\frac{x}{x - 2} + \frac{1}{x + 3} > 0$$
$$\frac{x(x + 3)}{(x - 2)(x + 3)} + \frac{1(x - 2)}{(x + 3)(x - 2)} > 0$$
$$\frac{x^2 + 4x - 2}{(x + 3)(x - 2)} > 0$$

Solve $x^2 + 4x - 2 = 0$:
$$x = \frac{-4 \pm \sqrt{16 - 4(1)(-2)}}{2} = \frac{-4 \pm 2\sqrt{6}}{2}$$
$$= -2 \pm \sqrt{6} = -4.4, 0.4$$

The numbers -4.4, -3, 0.4, and 2 divide the number line into 5 regions. Pick a number in each region and test it in the original inequality. We get the solution set $(-\infty, -2 - \sqrt{6}) \cup (-3, -2 + \sqrt{6}) \cup (2, \infty)$.

61. To solve $x^2 + 5x - 50 > 0$, we first solve $x^2 + 5x - 50 = 0$:
$$(x + 10)(x - 5) = 0$$
$$x = -10 \text{ or } x = 5$$

178

Test a point in each region of the number line, to find that the profit is positive if $x < -10$ or if $x > 5$. He cannot sell negative mobile homes so he must sell more than 5 to have a positive profit. He should sell 6, 7, 8, etc.

63. We must solve the inequality
$$-16t^2 + 96t + 6 > 86$$
$$-16t^2 + 96t - 80 > 0$$
$$t^2 - 6t + 5 < 0$$
Solve the equation $t^2 - 6t + 5 = 0$.
$$(t - 5)(t - 1) = 0$$
$$t = 5 \text{ or } t = 1$$
Using a test point we find that the inequality is satisfied for t between 1 and 5 seconds. So the arrow is more than 86 feet high for 4 seconds.

65. a) From the graph it appears that the maximum height is 900 feet.

b) The projectile was above 864 feet for approximately 3 seconds.

c) $S = -16t^2 + \dfrac{240\sqrt{2}}{\sqrt{2}}t + 0$

or $S = -16t^2 + 240t$

Solve
$$-16t^2 + 240t > 864$$
$$-16t^2 + 240t - 864 > 0$$
$$t^2 - 15t + 54 < 0$$
Solve the equation
$$(t - 6)(t - 9) = 0$$
$$t = 6 \text{ or } t = 9$$
Evaluate $t^2 - 15t + 54$ at the test points 0, 7, and 10.
$$0^2 - 15(0) + 54 = 54$$
$$7^2 - 15(7) + 54 = -2$$
$$10^2 - 15(10) + 54 = 4$$
So the inequality is satisfied for t in the interval $(6, 9)$. The projectile was above 864 ft for 3 sec.

67. a) (h, k)

b) $(-\infty, h) \cup (k, \infty)$ **c)** $(-k, -h)$

d) $(-\infty, -k] \cup [-h, \infty)$

e) $(-\infty, h] \cup (k, \infty)$ **f)** $(-k, -h]$

69. c   **71.** b

Enriching Your Mathematical Word Power
1. b   **2.** a   **3.** d   **4.** c   **5.** b   **6.** c
7. c   **8.** a   **9.** c   **10.** a   **11.** c

CHAPTER 8   REVIEW

1. $x^2 - 2x - 15 = 0$
$$(x - 5)(x + 3) = 0$$
$$x - 5 = 0 \text{ or } x + 3 = 0$$
$$x = 5 \text{ or } \quad x = -3$$
Solution set: $\{-3, 5\}$

3. $2x^2 + x = 15$
$$2x^2 + x - 15 = 0$$
$$(2x - 5)(x + 3) = 0$$
$$2x - 5 = 0 \quad \text{or} \quad x + 3 = 0$$
$$x = \frac{5}{2} \quad \text{or} \quad x = -3$$
Solution set: $\left\{-3, \frac{5}{2}\right\}$

5. $w^2 - 25 = 0$
$$(w - 5)(w + 5) = 0$$
$$w - 5 = 0 \quad \text{or} \quad w + 5 = 0$$
$$w = 5 \quad \text{or} \quad w = -5$$
Solution set: $\{-5, 5\}$

7. $4x^2 - 12x + 9 = 0$
$$(2x - 3)^2 = 0$$
$$2x - 3 = 0$$
$$x = \frac{3}{2}$$
Solution set: $\left\{\frac{3}{2}\right\}$

9. $x^2 = 12$
$$x = \pm\sqrt{12} = \pm 2\sqrt{3}$$
Solution set: $\left\{\pm 2\sqrt{3}\right\}$

11. $(x - 1)^2 = 9$
$$x - 1 = \pm 3$$
$$x = 1 \pm 3$$
Solution set: $\{-2, 4\}$

13. $(x - 2)^2 = \frac{3}{4}$
$$x - 2 = \pm\sqrt{\frac{3}{4}}$$
$$x = 2 \pm \frac{\sqrt{3}}{2} = \frac{4}{2} \pm \frac{\sqrt{3}}{2}$$
Solution set: $\left\{\dfrac{4 \pm \sqrt{3}}{2}\right\}$

15. $4x^2 = 9$
$$x^2 = \frac{9}{4}$$
$$x = \pm\sqrt{\frac{9}{4}} = \pm\frac{3}{2}$$
Solution set: $\left\{\pm\frac{3}{2}\right\}$

17. $x^2 - 6x + 8 = 0$
$$x^2 - 6x = -8$$
$$x^2 - 6x + 9 = -8 + 9$$
$$(x - 3)^2 = 1$$
$$x - 3 = \pm 1$$
$$x = 3 \pm 1$$
Solution set: $\{2, 4\}$

19. $x^2 - 5x + 6 = 0$
$$x^2 - 5x = -6$$
$$x^2 - 5x + \frac{25}{4} = -\frac{24}{4} + \frac{25}{4}$$
$$\left(x - \frac{5}{2}\right)^2 = \frac{1}{4}$$
$$x - \frac{5}{2} = \pm \frac{1}{2}$$
$$x = \frac{5}{2} \pm \frac{1}{2}$$
Solution set: $\{2, 3\}$

21. $2x^2 - 7x + 3 = 0$
$$2x^2 - 7x = -3$$
$$x^2 - \frac{7}{2}x = -\frac{3}{2}$$
$$x^2 - \frac{7}{2}x + \frac{49}{16} = -\frac{24}{16} + \frac{49}{16}$$
$$\left(x - \frac{7}{4}\right)^2 = \frac{25}{16}$$
$$x - \frac{7}{4} = \pm \frac{5}{4}$$
$$x = \frac{7}{4} \pm \frac{5}{4} \quad \left(\frac{12}{4} \text{ or } \frac{2}{4}\right)$$
Solution set: $\left\{\frac{1}{2}, 3\right\}$

23. $x^2 + 4x + 1 = 0$
$$x^2 + 4x = -1$$
$$x^2 + 4x + 4 = -1 + 4$$
$$(x + 2)^2 = 3$$
$$x + 2 = \pm \sqrt{3}$$
$$x = -2 \pm \sqrt{3}$$
Solution set: $\{-2 \pm \sqrt{3}\}$

25. $x^2 - 3x - 10 = 0$

$$x = \frac{3 \pm \sqrt{9 - 4(1)(-10)}}{2(1)} = \frac{3 \pm \sqrt{49}}{2}$$
$$= \frac{3 \pm 7}{2}$$
Solution set: $\{-2, 5\}$

27. $6x^2 - 7x = 3$
$$6x^2 - 7x - 3 = 0$$
$$x = \frac{7 \pm \sqrt{49 - 4(6)(-3)}}{2(6)} = \frac{7 \pm \sqrt{121}}{12}$$
$$= \frac{7 \pm 11}{12}$$
Solution set: $\left\{-\frac{1}{3}, \frac{3}{2}\right\}$

29. $x^2 + 4x + 2 = 0$

$$x = \frac{-4 \pm \sqrt{16 - 4(1)(2)}}{2(1)} = \frac{-4 \pm 2\sqrt{2}}{2}$$
$$= -2 \pm \sqrt{2}$$
Solution set: $\left\{-2 \pm \sqrt{2}\right\}$

31. $3x^2 + 1 = 5x$
$$3x^2 - 5x + 1 = 0$$
$$x = \frac{5 \pm \sqrt{25 - 4(3)(1)}}{2(3)} = \frac{5 \pm \sqrt{13}}{6}$$
Solution set: $\left\{\frac{5 \pm \sqrt{13}}{6}\right\}$

33. $25x^2 - 20x + 4 = 0$
$b^2 - 4ac = (-20)^2 - 4(25)(4) = 0$
One real solution.

35. $x^2 - 3x + 7 = 0$
$b^2 - 4ac = (-3)^2 - 4(1)(7) = -19$
No real solutions

37. $2x^2 - 5x + 1 = 0$
$b^2 - 4ac = (-5)^2 - 4(2)(1) = 17$
Two real solutions

39. $2x^2 - 4x + 3 = 0$
$$x = \frac{4 \pm \sqrt{16 - 4(2)(3)}}{2(2)} = \frac{4 \pm \sqrt{-8}}{4}$$
$$= \frac{4 \pm 2i\sqrt{2}}{4} = \frac{2 \pm i\sqrt{2}}{2}$$
Solution set: $\left\{\frac{2 \pm i\sqrt{2}}{2}\right\}$

41. $2x^2 - 3x + 3 = 0$
$$x = \frac{3 \pm \sqrt{9 - 4(2)(3)}}{2(2)} = \frac{3 \pm \sqrt{-15}}{4}$$
$$= \frac{3 \pm i\sqrt{15}}{4}$$
Solution set: $\left\{\frac{3 \pm i\sqrt{15}}{4}\right\}$

43. $3x^2 + 2x + 2 = 0$
$$x = \frac{-2 \pm \sqrt{4 - 4(3)(2)}}{2(3)} = \frac{-2 \pm \sqrt{-20}}{6}$$
$$= \frac{-2 \pm 2i\sqrt{5}}{6} = \frac{-1 \pm i\sqrt{5}}{3}$$
Solution set: $\left\{\frac{-1 \pm i\sqrt{5}}{3}\right\}$

45. $\frac{1}{2}x^2 + 3x + 8 = 0$
$$x^2 + 6x + 16 = 0$$
$$x = \frac{-6 \pm \sqrt{36 - 4(1)(16)}}{2(1)} = \frac{-6 \pm \sqrt{-28}}{2}$$
$$= \frac{-6 \pm 2i\sqrt{7}}{2} = -3 \pm i\sqrt{7}$$
Solution set: $\{-3 \pm i\sqrt{7}\}$

47. $f(x) = x^2 - 6x$
$x = \frac{-b}{2a} = \frac{6}{2(1)} = 3$, $f(3) = 3^2 - 6(3) = -9$
The vertex is $(3, -9)$. The y-intercept is $(0, 0)$.
$$x^2 - 6x = 0$$
$$x(x - 6) = 0$$
$$x = 0 \text{ or } x - 6 = 0$$
$$x = 0 \text{ or } \qquad x = 6$$
The x-intercepts are $(0, 0)$ and $(6, 0)$.

49. $g(x) = x^2 - 4x - 12$
$x = \frac{-b}{2a} = \frac{4}{2(1)} = 2$,
$g(2) = 2^2 - 4(2) - 12 = -16$
The vertex is $(2, -16)$. The y-intercept is
$(0, -12)$. The solutions to $x^2 - 4x - 12 = 0$
are 6 and -2. So the x-intercepts are $(-2, 0)$
and $(6, 0)$.

51. $h(x) = -2x^2 + 8x$
$x = \frac{-b}{2a} = \frac{-8}{2(-2)} = 2$,
$h(2) = -2 \cdot 2^2 + 8(2) = 8$
The vertex is $(2, 8)$. The y-intercept is $(0, 0)$.
The solutions to $-2x^2 + 8x = 0$ are 0 and 4.
So the x-intercepts are $(0, 0)$ and $(4, 0)$.

53. $y = -x^2 + 2x + 3$

$x = \frac{-b}{2a} = \frac{-2}{2(-1)} = 1$,
$y = -1^2 + 2(1) + 3 = 4$
The vertex is $(1, 4)$. The y-intercept is $(0, 3)$.
The solutions to $-x^2 + 2x + 3 = 0$ are -1 and
3. So the x-intercepts are $(-1, 0)$ and $(3, 0)$.

55. $f(x) = x^2 + 4x + 1$
$x = \frac{-b}{2a} = \frac{-4}{2(1)} = -2$,
$f(-2) = (-2)^2 + 4(-2) + 1 = -3$
The domain is $(-\infty, \infty)$ and the range is
$[-3, \infty)$.
57. $y = -2x^2 - x + 4$
$x = \frac{-b}{2a} = \frac{1}{2(-2)} = -\frac{1}{4}$,
$y = -2(-\frac{1}{4})^2 - (-\frac{1}{4}) + 4 = 4.125$
The domain is $(-\infty, \infty)$ and the range is
$(-\infty, 4.125]$.
59. $b^2 - 4ac = (-10)^2 - 4(8)(-3) = 196$
Since 196 is a perfect square, the polynomial is
not prime.
$8x^2 - 10x - 3 = (4x + 1)(2x - 3)$

61. $b^2 - 4ac = (-5)^2 - 4(4)(2) = -7$
Since -7 is a not a perfect square, the
polynomial is prime.
63. $b^2 - 4ac = (10)^2 - 4(8)(-25) = 900$
Since 900 is a perfect square, the polynomial is
not prime.
$8y^2 + 10y - 25 = (4y - 5)(2y + 5)$
65. $(x + 3)(x + 6) = 0$
$$x^2 + 9x + 18 = 0$$
67. $(x + 5\sqrt{2})(x - 5\sqrt{2}) = 0$
$$x^2 - 50 = 0$$

69. $\qquad x^6 + 7x^3 - 8 = 0$
$$(x^3 + 8)(x^3 - 1) = 0$$
$$x^3 + 8 = 0 \text{ or } x^3 - 1 = 0$$
$$x^3 = -8 \text{ or } \quad x^3 = 1$$
$$x = -2 \text{ or } \quad x = 1$$
Solution set: $\{-2, 1\}$

181

71.
$$x^4 - 13x^2 + 36 = 0$$
$$(x^2 - 4)(x^2 - 9) = 0$$
$$x^2 - 4 = 0 \quad \text{or} \quad x^2 - 9 = 0$$
$$x^2 = 4 \quad \text{or} \qquad x^2 = 9$$
$$x = \pm 2 \quad \text{or} \quad x = \pm 3$$
Solution set: $\{\pm 2, \pm 3\}$

73. Let $w = x^2 + 3x$ in
$$(x^2 + 3x)^2 - 28(x^2 + 3x) + 180 = 0.$$
$$w^2 - 28w + 180 = 0$$
$$(w - 10)(w - 18) = 0$$
$$w = 10 \quad \text{or} \qquad w = 18$$
$$x^2 + 3x = 10 \quad \text{or} \quad x^2 + 3x = 18$$
$$x^2 + 3x - 10 = 0 \text{ or } x^2 + 3x - 18 = 0$$
$$(x + 5)(x - 2) = 0 \text{ or } (x + 6)(x - 3) = 0$$
$$x = -5 \text{ or } x = 2 \text{ or } x = -6 \text{ or } x = 3$$
Solution set: $\{-6, -5, 2, 3\}$

75. Let $w = \sqrt{x^2 - 6x}$ and $w^2 = x^2 - 6x$
in $x^2 - 6x + 6\sqrt{x^2 - 6x} - 40 = 0.$
$$w^2 + 6w - 40 = 0$$
$$(w + 10)(w - 4) = 0$$
$$w = -10 \quad \text{or} \quad w = 4$$
$$\sqrt{x^2 - 6x} = -10 \quad \text{or} \quad \sqrt{x^2 - 6x} = 4$$
No real solution here. $\qquad x^2 - 6x = 16$
$$x^2 - 6x - 16 = 0$$
$$(x - 8)(x + 2) = 0$$
$$x = 8 \quad \text{or} \quad x = -2$$

Solution set: $\{-2, 8\}$

77. Let $w = t^{-1}$ and $w^2 = t^{-2}$ in
$t^{-2} + 5t^{-1} - 36 = 0.$
$$w^2 + 5w - 36 = 0$$
$$(w + 9)(w - 4) = 0$$
$$w = -9 \quad \text{or} \qquad w = 4$$
$$t^{-1} = -9 \quad \text{or} \quad t^{-1} = 4$$
$$t = -\frac{1}{9} \quad \text{or} \qquad t = \frac{1}{4}$$
Solution set: $\left\{-\frac{1}{9}, \frac{1}{4}\right\}$

79. Let $y = \sqrt{w}$ and $y^2 = w$ in
$w - 13\sqrt{w} + 36 = 0.$
$$y^2 - 13y + 36 = 0$$
$$(y - 9)(y - 4) = 0$$
$$y = 9 \quad \text{or} \qquad y = 4$$
$$\sqrt{w} = 9 \quad \text{or} \quad \sqrt{w} = 4$$
$$w = 81 \quad \text{or} \qquad w = 16$$
Solution set: $\{16, 81\}$

81.
$$a^2 + a > 6$$
$$a^2 + a - 6 > 0$$
$$(a + 3)(a - 2) > 0$$
$$a + 3 \ - \ - \ -0+ + + + + + + + + +$$
$$a - 2 \ - \ - \ - \ - \ - \ - \ - \ - -0+ + + +$$
$$\qquad\qquad -3 \qquad\qquad 2$$
The inequality is satisfied if both factors are
the same sign. $(-\infty, -3) \cup (2, \infty)$

83. $x^2 - x - 20 \le 0$
$$(x - 5)(x + 4) \le 0$$

$$x + 4 \ - \ - \ -0+ + + + + + + + + +$$
$$x - 5 \ - \ - \ - \ - \ - \ - \ - \ - -0+ + + +$$
$$\qquad\quad -4 \qquad\qquad 5$$
The inequality is satisfied if the factors have
opposite signs, or if one of the factors is zero.
$[-4, 5]$

85.
$$w^2 - w < 0$$
$$w(w - 1) < 0$$
$$w \ - \ - \ -0+ + + + + + + + + +$$
$$w - 1 \ - \ - \ - \ - \ - \ - \ - \ - -0+ + +$$
$$\qquad\quad 0 \qquad\qquad 1$$
The inequality is satisfied if the factors have
opposite signs. $\qquad (0, 1)$

87. $\dfrac{x - 4}{x + 2} \ge 0$

$$x + 2 \ - \ - \ -0+ + + + + + + + + +$$
$$x - 4 \ - \ - \ - \ - \ - \ - \ - \ - -0+ + + +$$
$$\qquad\quad -2 \qquad\qquad 4$$

The inequality is satisfied if the factors have
the same sign, or if the numerator is equal to
zero. $\qquad (-\infty, -2) \cup [4, \infty)$

89. $\dfrac{x - 2}{x + 3} < 1$
$$\frac{x - 2}{x + 3} - 1 < 0$$
$$\frac{x - 2}{x + 3} - \frac{x + 3}{x + 3} < 0$$
$$\frac{-5}{x + 3} < 0$$

182

Since the numerator is definitely negative, the inequality is satisfied only if the denominator is positive:

$$x + 3 > 0$$
$$x > -3$$

$(-3, \infty)$

91.
$$\frac{3}{x+2} > \frac{1}{x+1}$$
$$\frac{3}{x+2} - \frac{1}{x+1} > 0$$
$$\frac{3(x+1)}{(x+2)(x+1)} - \frac{1(x+2)}{(x+1)(x+2)} > 0$$
$$\frac{2x+1}{(x+2)(x+1)} > 0$$

$$
\begin{array}{l}
x+1 \quad - \; - \; - \; - \; 0 + + + + + + + + + \\
x+2 \quad - \; - \; 0 + + + + + + + + + + + \\
2x+1 \quad - \; - \; - \; - \; - \; - \; - \; - \; - \; 0 + + + \\
\hline
\qquad -2 \qquad -1 \qquad \quad -1/2
\end{array}
$$

The inequality is satisfied when an even number of the factors have negative signs.

$(-2, -1) \cup \left(-\frac{1}{2}, \infty\right)$

93. $144x^2 - 120x + 25 = 0$
$$(12x - 5)^2 = 0$$
$$12x - 5 = 0$$
$$x = \frac{5}{12}$$
Solution set: $\left\{\frac{5}{12}\right\}$

95. $(2x+3)^2 + 7 = 12$
$$(2x+3)^2 = 5$$
$$2x + 3 = \pm\sqrt{5}$$
$$x = \frac{-3 \pm \sqrt{5}}{2}$$
Solution set: $\left\{\frac{-3 \pm \sqrt{5}}{2}\right\}$

97. $1 + \frac{20}{9x^2} = \frac{8}{3x}$
$$9x^2\left(1 + \frac{20}{9x^2}\right) = 9x^2 \cdot \frac{8}{3x}$$
$$9x^2 + 20 = 24x$$
$$9x^2 - 24x + 20 = 0$$
$$x = \frac{24 \pm \sqrt{(-24)^2 - 4(9)(20)}}{2(9)} = \frac{24 \pm 12i}{18}$$

$$= \frac{4 \pm 2i}{3}$$
Solution set: $\left\{\frac{4 \pm 2i}{3}\right\}$

99. $\sqrt{3x^2 + 7x - 30} = x$
$$3x^2 + 7x - 30 = x^2$$
$$2x^2 + 7x - 30 = 0$$
$$(2x - 5)(x + 6) = 0$$
$$2x - 5 = 0 \quad \text{or} \quad x + 6 = 0$$
$$x = \frac{5}{2} \quad \text{or} \qquad x = -6$$
Since -6 does not check, solution set is $\left\{\frac{5}{2}\right\}$.

101. $2(2x+1)^2 + 5(2x+1) = 3$
Let $y = 2x + 1$:
$$2y^2 + 5y - 3 = 0$$
$$(2y - 1)(y + 3) = 0$$
$$2y - 1 = 0 \quad \text{or} \quad y + 3 = 0$$
$$y = \frac{1}{2} \quad \text{or} \qquad y = -3$$
$$2x + 1 = \frac{1}{2} \text{ or } 2x + 1 = -3$$
$$x = -\frac{1}{4} \quad \text{or} \qquad x = -2$$
Solution set: $\left\{-2, -\frac{1}{4}\right\}$

103. $x^{1/2} - 15x^{1/4} + 50 = 0 \quad$ Let $y = x^{1/4}$
$$y^2 - 15y + 50 = 0$$
$$(y - 5)(y - 10) = 0$$
$$y = 5 \quad \text{or} \quad y = 10$$
$$x^{1/4} = 5 \text{ or } x^{1/4} = 10$$
$$x = 5^4 \text{ or } \quad x = 10^4$$
$$x = 625 \text{ or } \quad x = 10,000$$
Solution set: $\{625, 10,000\}$

105. If $x =$ one of the numbers, then $x + 4 =$ the other number. Since their product is 4, we can write the equation
$$x(x + 4) = 4$$
$$x^2 + 4x = 4$$
$$x^2 + 4x + 4 = 4 + 4$$
$$(x + 2)^2 = 8$$
$$x + 2 = \pm 2\sqrt{2}$$
$$x = -2 \pm 2\sqrt{2}$$
Disregard $-2 - 2\sqrt{2}$ since it is not positive. If $x = -2 + 2\sqrt{2}$ then $x + 4 = 2 + 2\sqrt{2}$. The two numbers are $-2 + 2\sqrt{2} \approx 0.83$ and $2 + 2\sqrt{2} \approx 4.83$.

107. Let $x =$ the width and $x - 4 =$ the height. The diagonal 19 is the hypotenuse of a right triangle with legs x and $x - 4$. By the Pythagorean theorem we can write
$$x^2 + (x - 4)^2 = 19^2$$

$$x^2 + x^2 - 8x + 16 = 361$$
$$2x^2 - 8x - 345 = 0$$
$$x = \frac{8 \pm \sqrt{64 - 4(2)(-345)}}{2(2)}$$
$$= \frac{8 \pm \sqrt{2824}}{4} = \frac{8 \pm 2\sqrt{706}}{4}$$
$$= \frac{4 \pm \sqrt{706}}{2}$$

Since the width must be positive, we have
$$x = \frac{4 + \sqrt{706}}{2} \text{ and}$$
$$x - 4 = \frac{4 + \sqrt{706}}{2} - \frac{8}{2} = \frac{-4 + \sqrt{706}}{2}.$$

So the width is $\dfrac{4 + \sqrt{706}}{2} \approx 15.3$ inches and

the height is $\dfrac{-4 + \sqrt{706}}{2} \approx 11.3$ inches.

109. Let $x =$ the width of the border. The dimensions of the printed area will be $8 - 2x$ by $10 - 2x$. Since the printed area is to be 24 square inches, we can write the equation
$$(8 - 2x)(10 - 2x) = 24$$
$$80 - 36x + 4x^2 = 24$$
$$4x^2 - 36x + 56 = 0$$
$$x^2 - 9x + 14 = 0$$
$$(x - 7)(x - 2) = 0$$
$$x = 7 \quad \text{or} \quad x = 2$$
Since 7 inches is too wide for a border on an 8 by 10 piece of paper, the border must be 2 inches wide.

111. Let $x =$ width and $x + 4 =$ the length.
$$x(x + 4) = 45$$
$$x^2 + 4x - 45 = 0$$
$$(x + 9)(x - 5) = 0$$
$$x + 9 = 0 \text{ or } x - 5 = 0$$
$$x = -9 \text{ or } \quad x = 5$$
$$x + 4 = 9$$
The table is 5 ft wide and 9 feet long.

113. $C(n) = 0.004n^2 - 3.2n + 660$
If $n = 390$, then
$$C(390) = 0.004(390)^2 - 3.2(390) + 660$$
$$= \$20.40$$
$$n = \frac{3.2}{2(0.004)} = 400$$
The unit cost is at a minimum for 400 starters.

115. $\qquad 12 = -16t^2 + 32t$

$$16t^2 - 32t + 12 = 0$$
$$4t^2 - 8t + 3 = 0$$
$$t = \frac{8 \pm \sqrt{64 - 4(4)(3)}}{2(4)} = \frac{8 \pm \sqrt{16}}{8}$$
$$= 1 \pm \frac{1}{2} = 1.5 \text{ or } 0.5$$
His height was 12 ft for $t = 0.5$ sec and $t = 1.5$ sec.

CHAPTER 8   TEST

1. $2x^2 - 3x + 2 = 0$
$(-3)^2 - 4(2)(2) = 9 - 16 = -7$
The equation has no real solutions.

2. $-3x^2 + 5x - 1 = 0$
$(5)^2 - 4(-3)(-1) = 25 - 12 = 13$
The equation has 2 real solutions.

3. $4x^2 - 4x + 1 = 0$
$(-4)^2 - 4(4)(1) = 16 - 16 = 0$
The equation has 1 real solution.

4. $2x^2 + 5x - 3 = 0$
$$x = \frac{-5 \pm \sqrt{25 - 4(2)(-3)}}{2(2)} = \frac{-5 \pm \sqrt{49}}{4}$$
$$= \frac{-5 \pm 7}{4} \quad \left(\frac{2}{4} \text{ or } \frac{-12}{4}\right)$$
Solution set: $\left\{-3, \frac{1}{2}\right\}$

5. $x^2 + 6x + 6 = 0$
$$x = \frac{-6 \pm \sqrt{36 - 4(1)(6)}}{2} = \frac{-6 \pm \sqrt{12}}{2}$$
$$= \frac{-6 \pm 2\sqrt{3}}{2} = -3 \pm \sqrt{3}$$
Solution set: $\{-3 \pm \sqrt{3}\}$

6. $x^2 + 10x + 25 = 0$
$$(x + 5)^2 = 0$$
$$x + 5 = 0$$
$$x = -5$$
Solution set: $\{-5\}$

7. $2x^2 + x - 6 = 0$
$$2x^2 + x = 6$$
$$x^2 + \frac{1}{2}x = 3$$
$$x^2 + \frac{1}{2}x + \frac{1}{16} = 3 + \frac{1}{16}$$

$$\left(x + \frac{1}{4}\right)^2 = \frac{49}{16}$$
$$x + \frac{1}{4} = \pm\frac{7}{4}$$
$$x = -\frac{1}{4} \pm \frac{7}{4} \quad \left(\frac{6}{4} \text{ or } \frac{-8}{4}\right)$$
Solution set: $\left\{-2, \frac{3}{2}\right\}$

184

8.
$$x(x + 1) = 12$$
$$x^2 + x - 12 = 0$$
$$(x + 4)(x - 3) = 0$$
$$x = -4 \quad \text{or} \quad x = 3$$
Solution set: $\{-4, 3\}$

9.
$$a^4 - 5a^2 + 4 = 0$$
$$(a^2 - 4)(a^2 - 1) = 0$$
$$a^2 - 4 = 0 \quad \text{or} \quad a^2 - 1 = 0$$
$$a^2 = 4 \quad \text{or} \quad a^2 = 1$$
$$a = \pm 2 \quad \text{or} \quad a = \pm 1$$
Solution set: $\{\pm 1, \pm 2\}$

10. Let $w = \sqrt{x - 2}$ and $w^2 = x - 2$ in
$$x - 2 - 8\sqrt{x - 2} + 15 = 0$$
$$w^2 - 8w + 15 = 0$$
$$(w - 3)(w - 5) = 0$$
$$w = 3 \quad \text{or} \quad w = 5$$
$$\sqrt{x - 2} = 3 \quad \text{or} \quad \sqrt{x - 2} = 5$$
$$x - 2 = 9 \quad \text{or} \quad x - 2 = 25$$
$$x = 11 \quad \text{or} \quad x = 27$$
Solution set: $\{11, 27\}$

11.
$$x^2 + 36 = 0$$
$$x^2 = -36$$
$$x = \pm\sqrt{-36} = \pm 6i$$
Solution set: $\{\pm 6i\}$

12. $x^2 + 6x + 10 = 0$
$$x = \frac{-6 \pm \sqrt{36 - 4(1)(10)}}{2(1)} = \frac{-6 \pm \sqrt{-4}}{2}$$
$$= \frac{-6 \pm 2i}{2} = -3 \pm i$$
Solution set: $\{-3 \pm i\}$

13. $3x^2 - x + 1 = 0$
$$x = \frac{1 \pm \sqrt{1 - 4(3)(1)}}{2(3)} = \frac{1 \pm \sqrt{-11}}{6}$$
Solution set: $\left\{\dfrac{1 \pm i\sqrt{11}}{6}\right\}$

14. The graph of $f(x) = 16 - x^2$ goes through $(-2, 12)$, $(-1, 15)$, $(0, 16)$, $(1, 15)$, and $(2, 12)$. The vertex is $(0, 16)$. The domain is $(-\infty, \infty)$ and the range is $(-\infty, 16]$.

15. The graph of $g(x) = x^2 - 3x$ goes through $(0, 0)$, $(1, -2)$, $(2, -2)$, and $(3, 0)$. The vertex is $\left(\dfrac{3}{2}, -\dfrac{9}{4}\right)$. The domain is $(-\infty, \infty)$ and the range is $\left[-\dfrac{9}{4}, \infty\right)$.

16. $(x + 4)(x - 6) = 0$
$$x^2 - 2x - 24 = 0$$

17. $(x - 5i)(x + 5i) = 0$
$$x^2 + 25 = 0$$

18.
$$w^2 + 3w < 18$$
$$w^2 + 3w - 18 < 0$$
$$(w + 6)(w - 3) < 0$$

$$w + 6 \quad - - - - 0 + + + + + + + + +$$
$$w - 3 \quad - - - - - - - - - 0 + + + +$$
$$\overline{\qquad\qquad -6 \qquad\qquad 3 \qquad\qquad}$$

The inequality is satisfied when the factors have opposite signs. $(-6, 3)$

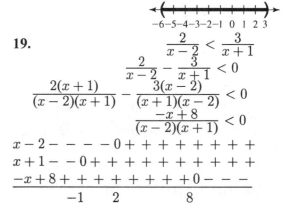

19.
$$\frac{2}{x - 2} < \frac{3}{x + 1}$$
$$\frac{2}{x - 2} - \frac{3}{x + 1} < 0$$
$$\frac{2(x + 1)}{(x - 2)(x + 1)} - \frac{3(x - 2)}{(x + 1)(x - 2)} < 0$$
$$\frac{-x + 8}{(x - 2)(x + 1)} < 0$$

$$x - 2 \quad - - - - 0 + + + + + + + + +$$
$$x + 1 \quad - - 0 + + + + + + + + + + +$$
$$-x + 8 \quad + + + + + + + + + 0 - - -$$
$$\overline{\qquad -1 \qquad 2 \qquad\qquad 8 \qquad\qquad}$$

This quotient will be negative only if an odd number of factors are negative.
$(-1, 2) \cup (8, \infty)$

20. Let $x =$ the width and $x + 2 =$ the length. Since the area is 16 square feet, we can write the equation
$$x(x + 2) = 16$$
$$x^2 + 2x - 16 = 0$$
$$x = \frac{-2 \pm \sqrt{4 - 4(1)(-16)}}{2} = \frac{-2 \pm \sqrt{68}}{2}$$

$$= \frac{-2 \pm 2\sqrt{17}}{2} = -1 \pm \sqrt{17}$$

Since the width must be positive, we have $x = -1 + \sqrt{17}$ and $x + 2 = 1 + \sqrt{17}$. The width is $-1 + \sqrt{17}$ feet and the length is $1 + \sqrt{17}$ feet.

21. Let $x =$ time for the new computer and $x + 1 =$ time for the old computer. New computer's rate is $\frac{1}{x}$ payroll/hr and old computer's rate is $\frac{1}{x+1}$ payroll/hr. In 3 hrs new computer does $\frac{3}{x}$ payroll and old computer does $\frac{3}{x+1}$ payroll.

$$\frac{3}{x} + \frac{3}{x+1} = 1$$
$$x(x+1)\left(\frac{3}{x} + \frac{3}{x+1}\right) = x(x+1)1$$
$$3x + 3 + 3x = x^2 + x$$
$$0 = x^2 - 5x - 3$$
$$x = \frac{5 \pm \sqrt{5^2 - 4(1)(-3)}}{2} = \frac{5 \pm \sqrt{37}}{2}$$
$$\approx 5.5 \text{ or } -0.5$$

It takes the new computer $\frac{5 + \sqrt{37}}{2}$ or 5.5 hrs to do the payroll by itself.

22.
$$x^2 - 6x + 5 = 0$$
$$(x - 5)(x - 1) = 0$$
$$x - 5 = 0 \text{ or } x - 1 = 0$$
$$x = 5 \text{ or } \quad x = 1$$

The x-intercepts are $(1, 0)$ and $(5, 0)$.

23. $s(t) = -16t^2 + 48t$
$$t = \frac{-48}{2(-16)} = \frac{3}{2}, \; s = -16(\tfrac{3}{2})^2 + 48(\tfrac{3}{2}) = 36$$

The maximum height reached by the ball is 36 feet.

Making Connections

Chapters 1 - 8

1.
$$2x - 15 = 0$$
$$2x = 15$$
$$x = \frac{15}{2}$$
Solution set: $\left\{ \frac{15}{2} \right\}$

2.
$$2x^2 - 15 = 0$$
$$2x^2 = 15$$
$$x^2 = \frac{15}{2}$$
$$x = \pm\sqrt{\frac{15}{2}} = \pm\frac{\sqrt{15}\sqrt{2}}{\sqrt{2}\sqrt{2}} = \pm\frac{\sqrt{30}}{2}$$

Solution set: $\left\{ \pm\frac{\sqrt{30}}{2} \right\}$

3.
$$2x^2 + x - 15 = 0$$
$$(2x - 5)(x + 3) = 0$$
$$2x - 5 = 0 \text{ or } x + 3 = 0$$
$$x = \frac{5}{2} \text{ or } \quad x = -3$$
Solution set: $\left\{ -3, \frac{5}{2} \right\}$

4.
$$2x^2 + 4x - 15 = 0$$
$$x = \frac{-4 \pm \sqrt{16 - 4(2)(-15)}}{2(2)} = \frac{-4 \pm \sqrt{136}}{4}$$
$$= \frac{-4 \pm 2\sqrt{34}}{4} = \frac{-2 \pm \sqrt{34}}{2}$$
Solution set: $\left\{ \frac{-2 \pm \sqrt{34}}{2} \right\}$

5.
$$|4x + 11| = 3$$
$$4x + 11 = 3 \text{ or } 4x + 11 = -3$$
$$4x = -8 \text{ or } \quad 4x = -14$$
$$x = -2 \text{ or } \quad x = -\frac{7}{2}$$
Solution set: $\left\{ -\frac{7}{2}, -2 \right\}$

6.
$$|4x^2 + 11x| = 3$$
$$4x^2 + 11x = 3 \text{ or } \quad 4x^2 + 11x = -3$$
$$4x^2 + 11x - 3 = 0 \text{ or } 4x^2 + 11x + 3 = 0$$
$$(4x - 1)(x + 3) = 0 \text{ or } \quad x = \frac{-11 \pm \sqrt{73}}{8}$$
$$4x - 1 = 0 \text{ or } x + 3 = 0$$
$$x = \frac{1}{4} \text{ or } \quad x = -3$$
Solution set: $\left\{ -3, \frac{1}{4}, \frac{-11 \pm \sqrt{73}}{8} \right\}$

7.
$$\sqrt{x} = x - 6$$
$$(\sqrt{x})^2 = (x - 6)^2$$
$$x = x^2 - 12x + 36$$
$$0 = x^2 - 13x + 36$$
$$0 = (x - 4)(x - 9)$$
$$x - 4 = 0 \text{ or } x - 9 = 0$$
$$x = 4 \text{ or } \quad x = 9$$
Since $\sqrt{4} = 4 - 6$ is incorrect and $\sqrt{9} = 9 - 6$ is correct, the solution set is $\{9\}$.

8.
$$(2x - 5)^{2/3} = 4$$
$$\left((2x - 5)^{2/3} \right)^3 = 4^3$$
$$(2x - 5)^2 = 64$$
$$2x - 5 = \pm 8$$
$$2x = 5 \pm 8$$
$$x = \frac{5 \pm 8}{2}$$
Solution set: $\left\{ -\frac{3}{2}, \frac{13}{2} \right\}$

9. $1 - 2x < 5 - x$
$\qquad -x < 4$
$\qquad\quad x > -4$
$(-4, \infty)$

10. $(1 - 2x)(5 - x) \leq 0$

$$
\begin{array}{l}
1 - 2x + \;+\;+\; 0 - - - - - - - - - - \\
5 - x + + + + + + + + 0 - - - - \\
\hline
\qquad\quad 1/2 \qquad\qquad 5
\end{array}
$$

The inequality is satisfied when the factors have opposite signs. The solution set is $\left[\frac{1}{2}, 5\right]$.

11. $\dfrac{1 - 2x}{5 - x} \leq 0$ Same as last exercise, but 5 is excluded from the solution set because of the denominator $5 - x$. The solution set is $\left[\frac{1}{2}, 5\right)$.

12. $\quad |\,5 - x\,| < 3$
$\qquad -3 < 5 - x < 3$
$\qquad -8 < -x < -2$
$\qquad\quad 8 > x > 2$
The solution set is $(2, 8)$.

13. $3x - 1 < 5$ and $-3 \leq x$
$\qquad x < 2$ and $\quad x \geq -3$
The solution set is $[-3, 2)$.

14. $x - 3 < 1$ or $2x \geq 8$
$\qquad x < 4 \quad$ or $\quad x \geq 4$
The solution set is $(-\infty, \infty)$.

15. $\quad 2x - 3y = 9$
$\qquad\quad -3y = -2x + 9$
$\qquad\qquad y = \dfrac{-2x + 9}{-3}$
$\qquad\qquad y = \dfrac{2}{3}x - 3$

16. $\quad \dfrac{y - 3}{x + 2} = -\dfrac{1}{2}$
$\qquad y - 3 = -\dfrac{1}{2}(x + 2)$
$\qquad y - 3 = -\dfrac{1}{2}x - 1$
$\qquad\quad y = -\dfrac{1}{2}x + 2$

17. $\quad 3y^2 + cy + d = 0$
$\qquad y = \dfrac{-c \pm \sqrt{c^2 - 4(3)(d)}}{2(3)}$
$\qquad y = \dfrac{-c \pm \sqrt{c^2 - 12d}}{6}$

18. $\quad my^2 - ny - w = 0$
$\qquad y = \dfrac{-(-n) \pm \sqrt{(-n)^2 - 4(m)(-w)}}{2m}$
$\qquad y = \dfrac{n \pm \sqrt{n^2 + 4mw}}{2m}$

19. $\qquad \dfrac{1}{3}x - \dfrac{2}{5}y = \dfrac{5}{6}$
$\qquad 30\left(\dfrac{1}{3}x - \dfrac{2}{5}y\right) = 30\left(\dfrac{5}{6}\right)$
$\qquad\quad 10x - 12y = 25$
$\qquad\qquad -12y = -10x + 25$
$\qquad\qquad\quad y = \dfrac{-10}{-12}x + \dfrac{25}{-12}$
$\qquad\qquad\quad y = \dfrac{5}{6}x - \dfrac{25}{12}$

20. $\quad y - 3 = -\dfrac{2}{3}(x - 4)$
$\qquad y - 3 = -\dfrac{2}{3}x + \dfrac{8}{3}$
$\qquad\quad y = -\dfrac{2}{3}x + \dfrac{8}{3} + \dfrac{9}{3}$
$\qquad\quad y = -\dfrac{2}{3}x + \dfrac{17}{3}$

21. $m = \dfrac{7 - 3}{5 - 2} = \dfrac{4}{3}$

22. $m = \dfrac{-6 - 5}{4 - (-3)} = -\dfrac{11}{7}$

23. $m = \dfrac{0.4 - 0.8}{0.5 - 0.3} = \dfrac{-0.4}{0.2} = -2$

24. $m = \dfrac{-\frac{4}{3} - \frac{3}{5}}{\frac{1}{3} - \frac{1}{2}} = \dfrac{-\frac{29}{15}}{-\frac{1}{6}} = \dfrac{29}{15} \cdot \dfrac{6}{1} = \dfrac{58}{5}$

25. At \$20 per ticket,
$\qquad n = 48{,}000 - 400(20) = 40{,}000.$
At \$25 per ticket
$\qquad n = 48{,}000 - 400(25) = 38{,}000$
If 35,000 tickets are sold, then the price is \$32.50 per ticket.

26. If $p = \$20$, then
$R = 20(48{,}000 - 400 \cdot 20)$
$\quad = \$800{,}000$
If $p = \$25$, then $R = 25(48{,}000 - 400 \cdot 25)$
$\qquad\qquad\qquad = \$950{,}000$
$\qquad 1{,}280{,}000 = p(48{,}000 - 400p)$
$\qquad 1{,}280{,}000 = 48{,}000p - 400p^2$
$400p^2 - 48{,}000p + 1{,}280{,}000 = 0$
$\qquad p^2 - 120p + 3{,}200 = 0$
$\qquad (p - 80)(p - 40) = 0$
$\qquad p = 80 \quad$ or $\quad p = 40$
A revenue of \$1.28 million occurs at a price of \$40 and at a price of \$80.
The price that determines the maximum revenue is \$60 per ticket.

9.1 WARM-UPS

1. True, because the graph of a function consists of all ordered pairs that are in the function.

2. True, that is why they are called linear functions.

3. True, because absolute value functions are generally v-shaped.

4. False, because any real number can be used in place of x in $f(x) = 3$.

5. True, because the graph of $y = ax^2 + bx + c$ for $a \neq 0$ is a parabola.

6. False, the domain of a quadratic function is $(-\infty, \infty)$ or R.

7. True, because $f(x)$ is just another name for the dependent variable y.

8. True, because $y^2 \geq 0$ and x must also be greater than or equal to 0.

9. False, because 1 is in the domain of the function and $(1, \infty)$ does not include 1.

10. True, because any real number can be used for x in $f(x) = ax^2 + bx + c$.

9.1 EXERCISES

1. A linear function is a function of the form $f(x) = mx + b$ where m and b are real numbers with $m \neq 0$.

3. The graph of a constant function is a horizontal line.

5. The graph of a quadratic function is a parabola.

7. The graph of $h(x) = -2$ is the same as the graph of the horizontal line $y = -2$.

The domain of the function is R or $(-\infty, \infty)$. The only y-coordinate used is -2, and so the range is $\{-2\}$.

9. The graph of $f(x) = 2x - 1$ is the same as the graph of the linear equation $y = 2x - 1$. To draw the graph, start at the y-intercept

$(0, -1)$, and use the slope of $2 = 2/1$. Rise 2 and run 1 to locate a second point on the line.

From the graph, we can see that the domain is $(-\infty, \infty)$ and the range is also $(-\infty, \infty)$.

11. Graph the line $y = (1/2)x + 2$ by locating the y-intercept $(0, 2)$ and using a slope of $1/2$. The domain is $(-\infty, \infty)$ and the range is $(-\infty, \infty)$.

13. The graph of $y = -\frac{2}{3}x + 3$ is a straight line with y-intercept $(0, 3)$ and slope $-2/3$.

The domain is $(-\infty, \infty)$ and range is $(-\infty, \infty)$.

15. The graph of $y = -0.3x + 6.5$ is a straight line with y-intercept $(0, 6.5)$ and slope $-3/10$.

The domain is $(-\infty, \infty)$ and range is $(-\infty, \infty)$.

17. The graph of $f(x) = |x| + 1$ contains the points $(0, 1)$, $(3, 4)$, and $(-3, 4)$. Plot these points and draw the v-shaped graph.

The domain is $(-\infty, \infty)$, and from the graph we can see that the range is $[1, \infty)$.

19. The graph of $h(x) = |x + 1|$ includes the points $(0, 1)$, $(1, 2)$, $(-1, 0)$, and $(-2, 1)$.

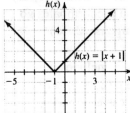

From the graph we can see that the domain is $(-\infty, \infty)$. In the vertical direction the graph is on or above the x-axis. Since all y-coordinates on the graph are greater than or equal to 0, the range is $[0, \infty)$.

21. The graph of $g(x) = |3x|$ includes the points $(0, 0)$, $(1, 3)$, and $(-1, 3)$. Plot these points and draw the graph.

From the graph we can see that the domain is $(-\infty, \infty)$. Since all of the y-coordinates on the graph are greater than or equal to zero, the range is $[0, \infty)$.

23. The graph of $f(x) = |2x - 1|$ includes the points $(0, 1)$, $(1/2, 0)$, $(1, 1)$, and $(2, 3)$. Plot these points and draw a v-shaped graph.

From the graph we can see that the domain is $(-\infty, \infty)$, and the range is $[0, \infty)$.

25. The graph of $f(x) = |x - 2| + 1$ includes the points $(2, 1)$, $(3, 2)$, $(4, 3)$, $(1, 2)$, and $(0, 3)$. Plot these points and draw a v-shaped graph. Form the graph we can see that the domain is $(-\infty, \infty)$. Since all y-coordinates are greater than or equal to 1, the range is $[1, \infty)$.

27. The graph of $g(x) = x^2 + 2$ includes the points $(-2, 6)$, $(-1, 3)$, $(0, 2)$, $(1, 3)$, and $(2, 6)$. Plot these points and draw the parabola through them.

The domain is $(-\infty, \infty)$. Since no y-coordinate on the parabola is below 2, the range is $[2, \infty)$.

29. The graph of $f(x) = 2x^2$ includes the points $(0, 0)$, $(1, 2)$, $(-1, 2)$, $(2, 8)$, and $(-2, 8)$. Plot these points and draw a parabola through them.

The domain is $(-\infty, \infty)$ and the range is $[0, \infty)$.

31. The graph of $y = 6 - x^2$ includes the points $(-3, -3)$, $(0, 6)$, $(3, -3)$, $(-2, 2)$, and $(2, 2)$. Plot these points and draw a parabola through them.

The domain is $(-\infty, \infty)$ and range is $(-\infty, 6]$.

33. The graph of $g(x) = 2\sqrt{x}$ includes the points $(0, 0)$, $(1, 2)$, and $(4, 4)$. Plot these points and draw a curve through them.

Since we must have $x \geq 0$ in $\sqrt{x}$, the domain is $[0, \infty)$. The range is $[0, \infty)$

35. The graph of $f(x) = \sqrt{x-1}$ includes the points $(1, 0)$, $(2, 1)$, and $(5, 2)$. Plot these points and draw a smooth curve through them. The curve is half of a parabola, positioned on its side.

Since we must have $x - 1 \geq 0$, or $x \geq 1$ the domain is $[1, \infty)$. The range is $[0, \infty)$.

37. The graph of $h(x) = -\sqrt{x}$ includes the points $(0, 0)$, $(1, -1)$, and $(4, -2)$. Plot these points and draw a curve through them.

Since we must have $x \geq 0$ in $\sqrt{x}$, the domain is $[0, \infty)$. The range is $(-\infty, 0]$.

39. The graph of $y = \sqrt{x} + 2$ includes the points $(0, 2)$, $(1, 3)$, and $(4, 4)$. Graph these points and draw a curve through them.

Because $\sqrt{x}$ is a real number only for nonnegative values of x, the domain is $[0, \infty)$. From the graph we see that y-coordinates go no lower than 2, so the range is $[2, \infty)$.

41. The graph of $x = |y|$ includes the points $(0, 0)$, $(1, -1)$, $(1, 1)$, $(2, -2)$, and $(2, 2)$. Note that in this case it is easier to pick the y-coordinate and then find the appropriate x-coordinate using $x = |y|$. Draw the v-shaped graph through these points.

Since the x-coordinates are nonnegative, the domain is $[0, \infty)$. Since we are allowed to select any number for y, the range is $(-\infty, \infty)$.

43. To find pairs that satisfy $x = -y^2$, pick the y-coordinate first and then find the x-coordinate. The points $(0, 0)$, $(-1, 1)$, $(1, 1)$, $(-4, 2)$, and $(-4, -2)$ are on the graph.

From the graph we see that only nonpositive x-coordinates are used, so the domain is $(-\infty, 0]$. Since any real number is allowable for y, the range is $(-\infty, \infty)$.

45. The equation $x = 5$ is the equation of a vertical line with an x-intercept of $(5, 0)$.

All points on this line have x-coordinate 5, so the domain is $\{5\}$. Every real number occurs as a y-coordinate, so the range is $(-\infty, \infty)$.

47. Rewrite $x + 9 = y^2$ as $x = y^2 - 9$. Select some y-coordinates and calculate the appropriate x-coordinates. The points $(-9, 0)$, $(0, 3)$, and $(0, -3)$ are on the parabola.

Since no x-coordinate is below -9, the domain is $[-9, \infty)$. Since we could select any value for y, the range is $(-\infty, \infty)$.

49. For the equation $x = \sqrt{y}$, we must select nonnegative values for y and find the corresponding value for x. The points $(0, 0)$, $(1, 1)$, and $(2, 4)$ are on the graph.

Since the x-coordinates are also nonnegative, the domain is $[0, \infty)$. Since we could only use nonnegative values for y, the range is $[0, \infty)$.

51. The graph of $x = (y - 1)^2$ includes the points $(0, 1)$, $(1, 0)$, $(1, 2)$, $(4, 3)$, and $(4, -1)$. Plot the points and draw a curve through them. From the graph we see that the domain is $[0, \infty)$ and the range is $(-\infty, \infty)$.

53. To graph $f(x) = 1 - |x|$, we arbitrarily select a value for x and find the corresponding value for y. The points $(-2, -1)$, $(-1, 0)$, $(0, 1)$, $(1, 0)$, and $(2, -1)$ are on the graph.

Since any real number could be used for x, the domain is $(-\infty, \infty)$. From the graph we see that the y-coordinates are not higher than 1, so the range is $(-\infty, 1]$.

55. The graph of $y = (x - 3)^2 - 1$ includes the points $(1, 3)$, $(2, 0)$, $(3, -1)$, $(4, 0)$, and $(5, 3)$.

The domain is $(-\infty, \infty)$. Since the y-coordinates are greater than or equal to -1, the range is $[-1, \infty)$.

57. Graph of $y = |x + 3| + 1$ goes through $(-4, 2)$, $(-3, 1)$, $(-2, 2)$, $(-1, 3)$, $(0, 4)$.

The domain is $(-\infty, \infty)$ and range is $[1, \infty)$.

59. Graph of $y = \sqrt{x} - 3$ goes through $(0, -3)$, $(1, -2)$, and $(4, -1)$. Domain is $[0, \infty)$ and range is $[-3, \infty)$.

61. The graph of $y = 3x - 5$ is a straight line with y-intercept $(0, -5)$, and a slope of 3.

The domain is $(-\infty, \infty)$. From the graph we see that any real number could occur as a y-coordinate, so the range is $(-\infty, \infty)$.

63. The graph of $y = -x^2 + 4x - 4$ goes through $(0, -4)$, $(1, -1)$, $(2, 0)$, $(3, -1)$, and $(4, -4)$. From the graph we can see that the domain is $(-\infty, \infty)$ and range is $(-\infty, 0]$.

65. Graph of $f(x) = \sqrt{x^2}$ is same as graph of $y = |x|$. This graph shows us that $\sqrt{x^2} \neq x$ when $x < 0$.

67. If $k > 0$, then for large values of k the graph gets narrower and for smaller values of k the graph gets broader.

69. The graph of $y = (x - k)^2$ is a shift of the graph of $y = x^2$ to the right for $k > 0$ and to the left for $k < 0$.

71. The graph of $y = f(x - k)$ lies to the right of the graph of $y = f(x)$ when $k > 0$.

9.2 WARM-UPS

1. False, because $(-x)^2 = x^2$ and the graphs of $f(x) = x^2$ and $g(x) = (-x)^2$ are identical.

2. True, because $-2 = -(2)$.

3. True. The graph of $y = x + 3$ is 3 units above $y = x$ or 3 units to the left of $y = x$.

4. False, because $y = |x - 3|$ is 3 units to the right of $y = |x|$.

5. True, because subtracting 3 after the absolute value moves each point on $y = |x|$ down 3 units.

6. True, because the negative sign causes the reflection and the multiplication by 2 causes the stretching.

7. False, because $f(-x) = (-x - 2)^2$ $= (-1)^2(x + 2)^2 = (x + 2)^2 \neq (x - 2)^2$.

8. True, because $y = \sqrt{\frac{x}{9}} = \frac{1}{3}\sqrt{x}$. The y-coordinates on $y = \sqrt{x}$ are 3 times as large as the y-coordinates on $y = \frac{1}{3}\sqrt{x}$.

9. True, because $y = \sqrt{x - 3} + 5$ is the same as $y = \sqrt{x}$ moved 3 units to the right and 5 units upward.

10. False, because the graph of $y = x^2$ must be moved 2 units to the left, reflected in the x-axis, and then moved 7 units downward to obtain the graph of $y = -(x + 2)^2 - 7$.

9.2 EXERCISES

1. The graph of $y = -f(x)$ is a reflection in the x-axis of the graph of $y = f(x)$.

3. The graph of $y = f(x) + k$ for $k < 0$ is a downward translation of the graph of $y = f(x)$.

5. The graph of $y = f(x + k)$ for $k > 0$ is a translation to the left of the graph of $y = f(x)$.

7. The graph of $f(x) = \sqrt{2x}$ goes through $(0, 0)$, $(2, 2)$, and $(8, 4)$. The graph of $g(x) = -\sqrt{2x}$ is a reflection in the x-axis of the graph of $f(x) = \sqrt{2x}$.

9. The graph of $f(x) = x^2 + 1$ is a parabola through $(0,0)$, $(\pm 1, 2)$, and $(\pm 2, 5)$. The graph of $g(x) = -(x^2 + 1)$ is a reflection in the x-axis of the graph of $f(x)$.

11. The graph of $y = \sqrt{x - 2}$ is half of a parabola. The graph contains the points $(2, 0)$, $(3, 1)$, and $(6, 2)$. The graph of $y = -\sqrt{x - 2}$ is a reflection in the x-axis of the first graph.

13. The graph of $f(x) = x - 3$ is a straight line through $(0, -3)$ with slope 1. The graph of $g(x) = 3 - x = -(x - 3)$ is a reflection in the x-axis of the graph of $f(x)$.

15. The graph of $g(x) = x^2$ is a parabola opening upward with vertex $(0,0)$. The graph of $f(x) = x^2 - 4$ is a downward translation of

4 units from the graph of g. It includes the points $(0, -4)$, $(\pm 1, -3)$, and $(\pm 2, 0)$.

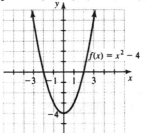

17. The graph of $y = x$ is a straight line with slope 1 and y-intercept $(0, 0)$. The graph of $y = x + 3$ is a translation 3 units upward of the graph of $y = x$.

19. The graph of $f(x) = (x - 3)^2$ is a translation 3 units to the right of the graph of the parabola $y = x^2$. The graph of $f(x) = (x - 3)^2$ goes through the points $(2, 1)$, $(3, 0)$, and $(4, 1)$.

21. The graph of $y = \sqrt{x} + 1$ is a translation 1 unit upward of the graph of $y = \sqrt{x}$. The graph of $y = \sqrt{x} + 1$ goes through $(0, 1)$, $(1, 2)$, and $(4, 3)$.

23. The graph of $f(x) = |x + 2|$ is a translation 2 units to the left of the graph of $y = |x|$. The graph of $f(x) = |x + 2|$ includes the points $(-3, 1)$, $(-2, 0)$, and $(-1, 1)$.

25. The graph of $y = |x| + 2$ is a translation 2 units upward of the graph of $y = |x|$. The graph of $y = |x| + 2$ goes through the points $(-1, 3)$, $(0, 2)$, and $(1, 3)$.

27. The graph of $f(x) = \sqrt{x-1}$ is a translation 1 unit to the right of the graph of $y = \sqrt{x}$. The graph of $f(x) = \sqrt{x-1}$ goes through the points $(1, 0)$, $(2, 1)$, and $(5, 2)$.

29. The graph of $f(x) = 3x^2$ is obtained by stretching the graph of $y = x^2$. The graph of $f(x) = 3x^2$ is a parabola through $(-1, 3)$, $(0, 0)$, and $(1, 3)$.

31. The graph of $y = \frac{1}{5}x$ is obtained by shrinking the graph of $y = x$. The graph of $y = \frac{1}{5}x$ is a straight line through the points $(-5, -1)$, $(0, 0)$, and $(5, 1)$.

33. The graph of $f(x) = 3\sqrt{x}$ is obtained by stretching the graph of $y = \sqrt{x}$. The graph of $f(x) = 3\sqrt{x}$ goes through $(0, 0)$, $(1, 3)$, and $(4, 6)$.

35. The graph of $y = \frac{1}{4}|x|$ is obtained by shrinking the graph of $y = |x|$. The graph of $y = \frac{1}{4}|x|$ goes through the points $(-4, 1)$, $(0, 0)$, and $(4, 1)$.

37. The graph of $y = \sqrt{x-2} + 1$ is obtained by translating the graph of $y = \sqrt{x}$ two units to the right and one unit upward. The graph of $y = \sqrt{x-2} + 1$ goes through the points $(2, 1)$, $(3, 2)$, and $(6, 3)$.

39. The graph of $y = -|x+3|$ is obtained from the graph of $y = |x|$ by translating it 3 units to the left and then reflecting in the x-axis. The graph of $y = -|x+3|$ contains the points $(-4, -1)$, $(-3, 0)$, and $(-2, -1)$.

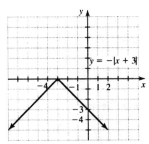

41. The graph of $f(x) = (x+3)^2 - 5$ is a translation of $y = x^2$ three units to the left and five units downward. The graph goes through the points $(-4, -4)$, $(-3, -5)$, $(-2, -4)$, and $(0, 4)$.

43. The graph of $y = -\sqrt{x+1} - 2$ is obtained from the graph of $y = \sqrt{x}$ by translating it 1 unit to the left, then reflecting in the x-axis, then translating 2 units downward. The graph of $y = -\sqrt{x+1} - 2$ includes the points $(-1, -2)$, $(0, -3)$, and $(3, -4)$.

45. The graph of $f(x) = -2\,|\,x - 3\,| + 4$ is obtained from the graph of $y = \,|\,x\,|$ by translating it 3 units to the right, stretching by a factor of 2, reflecting in the x-axis, and then translating 4 units upward. The graph of $f(x) = -2\,|\,x - 3\,| + 4$ includes the points $(2, 2)$, $(3, 4)$, and $(4, 2)$.

47. The graph of $y = -2x + 3$ is obtained from the graph of $y = x$ by stretching by a factor of 2, reflecting in the x-axis, and then translating it 3 units upward. The graph of $y = -2x + 3$ includes the points $(0, 3)$, $(1, 1)$, and $(2, -1)$.

49. The graph of $y = 2(x+3)^2 + 1$ is a parabola with vertex at $(-3, 1)$ and it opens upward. The graph includes the points $(-4, 3)$ and $(-2, 3)$.

51. The graph of $y = -2(x-4)^2 + 2$ is a parabola opening downward with vertex at $(4, 2)$. The graph includes the points $(5, 0)$ and $(3, 0)$.

53. The graph of $y = -3(x-1)^2 + 6$ is a parabola opening downward with vertex at $(1, 6)$. The graph includes the points $(0, 3)$ and $(2, 3)$.

55. The graph of $y = 2 + \sqrt{x}$ is obtained from $y = \sqrt{x}$ by translating it 2 units upward. The graph of $y = 2 + \sqrt{x}$ includes the points $(0, 2)$, $(1, 3)$, and $(4, 4)$. The graph is (d).

57. The graph of $y = 2\sqrt{x}$ is obtained by stretching the graph of $y = \sqrt{x}$. The graph of $y = 2\sqrt{x}$ goes through $(0, 0)$, $(1, 2)$, and $(4, 4)$. The graph is (e).

59. The graph of $y = \frac{1}{2}\sqrt{x}$ is obtained by shrinking the graph of $y = \sqrt{x}$. The graph of $y = \frac{1}{2}\sqrt{x}$ goes through the points $(0, 0)$, $(1, 0.5)$, and $(4, 1)$. The graph is (h).

61. The graph of $y = -2\sqrt{x}$ is obtained by stretching the graph of $y = \sqrt{x}$ and then reflecting in the x-axis. The graph of $y = -2\sqrt{x}$ contains the points $(0, 0)$, $(1, -2)$, and $(4, -4)$. The graph is (c).

63. If we translate the graph of $y = x^2$ upward 8 units, then the equation is $y = x^2 + 8$.

65. If we translate the graph of $y = \sqrt{x}$ to the left 5 units, then the equation is $y = \sqrt{x+5}$.

67. If we translate the graph of $y = |x|$ upward 5 units and the left 3 units, then the equation is $y = |x+3| + 5$.

69. Translate the graph of $f(x) = |x|$ to the right 20 units and upward 30 units to get the graph of $g(x) = |x - 20| + 30$.

9.3 WARM-UPS

1. True, because $(f - g)(x) = f(x) - g(x)$
$= x - 2 - (x + 3) = -5$.

2. True, because $(f/g)(2) = f(2)/g(2)$
$= (2 + 4)/(3 \cdot 2) = 6/6 = 1$.

3. False, because if $f(x) = 3x$ and $g(x) = x + 1$, then $f(g(x)) = 3(x + 1)$
$= 3x + 3$ and $g(f(x)) = 3x + 1$.

4. False, because $(f \circ g)(x) = f(x + 2)$
$= (x + 2)^2 = x^2 + 4x + 4$.

5. False, because if $f(x) = 3x$ and $g(x) = x + 1$, then
$(f \circ g)(x) = f(x + 1) = 3x + 3$
and $(f \cdot g)(x) = 3x(x + 1) = 3x^2 + 3x$.

6. False, because $g[f(x)] = \sqrt{x} - 9$ and $f[g(x)] = \sqrt{x - 9}$ which are not equal for all values of x.

7. True, because
$(f \circ g)(x) = f(\frac{x}{3}) = 3 \cdot \frac{x}{3} = x$.

8. True, because b determines a and a determines c so b determines the value of c.

9. True, $F(x) = \sqrt{x - 5}$ is a composition of $g(x) = x - 5$ and $h(x) = \sqrt{x}$.

10. True, because $(g \circ h)(x) = g[h(x)]$
$= g[x - 1] = (x - 1)^2 = F(x)$.

9.3 EXERCISES

1. The basic operations of functions are addition, subtraction, multiplication, and division.

3. In the composition of functions the second function is evaluated on the result of the first function.

5. $(f + g)(x) = f(x) + g(x)$
$= 4x - 3 + x^2 - 2x$
$= x^2 + 2x - 3$

7. $(f \cdot g)(x) = f(x) \cdot g(x)$
$= (4x - 3)(x^2 - 2x)$
$= 4x^3 - 11x^2 + 6x$

9. Use $x = 3$ in the formula of Exercise 5:
$(f + g)(3) = 3^2 + 2(3) - 3 = 12$

11. $(f - g)(-3) = f(-3) - g(-3)$.
$= 4(-3) - 3 - [(-3)^2 - 2(-3)]$
$= -12 - 3 - [9 + 6]$
$= -30$

13. Use $x = -1$ in the formula of Exercise 7:
$(f \cdot g)(-1) = 4(-1)^3 - 11(-1)^2 + 6(-1)$
$= -4 - 11 - 6$
$= -21$

15. $(f/g)(4) = \dfrac{f(4)}{g(4)} = \dfrac{4(4) - 3}{4^2 - 2(4)} = \dfrac{13}{8}$

17. Since $a = 2x - 6$, and $y = 3a - 2$ we have

$$y = 3(2x - 6) - 2$$
$$y = 6x - 18 - 2$$
$$y = 6x - 20$$

19. Let $d = (x + 1)/2$ in $y = 2d + 1$.

$$y = 2\left(\frac{x+1}{2}\right) + 1$$
$$y = x + 1 + 1$$
$$y = x + 2$$

21. Let $m = x + 1$ in $y = m^2 - 1$.

$$y = (x + 1)^2 - 1$$
$$y = x^2 + 2x + 1 - 1$$
$$y = x^2 + 2x$$

23.
$$y = \frac{a - 3}{a + 2}$$
$$y = \frac{\frac{2x+3}{1-x} - 3}{\frac{2x+3}{1-x} + 2}$$
$$y = \frac{\left(\frac{2x+3}{1-x} - 3\right)(1-x)}{\left(\frac{2x+3}{1-x} + 2\right)(1-x)}$$
$$y = \frac{2x + 3 - 3(1 - x)}{2x + 3 + 2(1 - x)}$$
$$y = \frac{2x + 3 - 3 + 3x}{2x + 3 + 2 - 2x} = \frac{5x}{5}$$
$$y = x$$

25. $(g \circ f)(1) = g[f(1)] = g(-1)$
$$= (-1)^2 + 3(-1) = -2$$

27. $(f \circ g)(1) = f[g(1)] = f[4]$
$$= 2(4) - 3 = 5$$

29. $(f \circ f)(4) = f[f(4)] = f[5]$
$$= 2(5) - 3 = 7$$

31. $(h \circ f)(5) = h[f(5)] = h[7] = \dfrac{7 + 3}{2} = 5$

33. $(f \circ h)(5) = f[h(5)] = f[4] = 2(4) - 3$
$$= 5$$

35. $(g \circ h)(-1) = g[h(-1)] = g[1] = 4$

37. $(f \circ g)(2.36) = f[g(2.36)] = f[12.6496]$
$$= 22.2992$$

39. $(g \circ f)(x) = g[f(x)] = g[2x - 3]$
$$= (2x - 3)^2 + 3(2x - 3)$$
$$= 4x^2 - 12x + 9 + 6x - 9$$
$$= 4x^2 - 6x$$

41. $(f \circ g)(x) = f[g(x)] = f[x^2 + 3x]$
$$= 2(x^2 + 3x) - 3 = 2x^2 + 6x - 3$$

43. $(h \circ f)(x) = h[f(x)] = h[2x - 3]$
$$= \frac{2x - 3 + 3}{2} = x$$

45. $(f \circ f)(x) = f[f(x)] = f[2x - 3]$
$$= 2(2x - 3) - 3 = 4x - 9$$

47. $(h \circ h)(x) = h[h(x)] = h[\frac{x+3}{2}]$
$$= \frac{\frac{x+3}{2} + 3}{2} = \frac{\left(\frac{x+3}{2} + 3\right)2}{2 \cdot 2}$$
$$= \frac{x + 3 + 6}{4} = \frac{x + 9}{4}$$

49. $F(x) = \sqrt{x - 3} = \sqrt{h(x)} = f[h(x)]$
Therefore, $F = f \circ h$.

51. $G(x) = x^2 - 6x + 9 = (x - 3)^2$
$$= [h(x)]^2 = g[h(x)]$$
Therefore $G = g \circ h$.

53. $H(x) = x^2 - 3 = g(x) - 3 = h[g(x)]$
Therefore $H = h \circ g$.

55. $J(x) = x - 6 = x - 3 - 3$
$$= h(x) - 3 = h[h(x)]$$
Therefore, $J = h \circ h$.

57. $K(x) = x^4 = (x^2)^2 = [g(x)]^2 = g[g(x)]$
Therefore, $K = g \circ g$.

59. $f[g(x)] = f\left(\dfrac{x - 5}{3}\right) = 3\left(\dfrac{x - 5}{3}\right) + 5$
$$= x$$
$g[f(x)] = g(3x + 5) = \dfrac{3x + 5 - 5}{3} = x$

61. $f[g(x)] = f\left(\sqrt[3]{x + 9}\right)$
$$= \left(\sqrt[3]{x + 9}\right)^3 - 9 = x + 9 - 9 = x$$
$g[f(x)] = g[x^3 - 9] = \sqrt[3]{x^3 - 9 + 9}$
$$= \sqrt[3]{x^3} = x$$

63. $f[g(x)] = f\left(\dfrac{x + 1}{1 - x}\right) = \dfrac{\dfrac{x+1}{1-x} - 1}{\dfrac{x+1}{1-x} + 1}$
$$= \frac{\left(\frac{x+1}{1-x} - 1\right)(1 - x)}{\left(\frac{x+1}{1-x} + 1\right)(1 - x)} = \frac{x + 1 - (1 - x)}{x + 1 + 1 - x}$$
$$= \frac{2x}{2} = x$$

$g[f(x)] = g\left(\dfrac{x - 1}{x + 1}\right) = \dfrac{\dfrac{x-1}{x+1} + 1}{1 - \dfrac{x-1}{x+1}}$
$$= \frac{\left(\frac{x-1}{x+1} + 1\right)(x + 1)}{\left(1 - \frac{x-1}{x+1}\right)(x + 1)} = \frac{x - 1 + x + 1}{x + 1 - (x - 1)}$$
$$= \frac{2x}{2} = x$$

65. $f[g(x)] = f\left(\frac{1}{x}\right) = \frac{1}{\frac{1}{x}} = x$

$g[f(x)] = g\left(\frac{1}{x}\right) = \frac{1}{\frac{1}{x}} = x$

67. a) Let s = the length of a side of the square.

$$10^2 = s^2 + s^2$$
$$2s^2 = 100$$
$$s^2 = 50$$
$$s = \sqrt{50}$$

Since $A = s^2$, we have $A = 50$ ft^2

b) In general if d is the diagonal of a square and s is the length of a side, we have $d^2 = s^2 + s^2$ or $s^2 = \frac{d^2}{2}$. Since $A = s^2$, we have $A = \frac{d^2}{2}$.

69. $P(x) = R(x) - C(x)$
$= x^2 - 10x + 30 - (2x^2 - 30x + 200)$
$= -x^2 + 20x - 170$

71. Substitute $F = 0.25I$ into $J = 0.10F$.

$$J = 0.10(0.25I)$$
$$J = 0.025I$$

73. a) $x = (30.25/100)^3 \approx 0.0277$
$D = (24665/2240)/0.0277 \approx 397.8$

b) $x = (L/100)^3$
$D = (25000/2240)/x$
$D = (25000/2240)/(L/100)^3$
$D = \frac{25000}{2240} \cdot \frac{100^3}{L^3}$
$D = \frac{1.116 \times 10^7}{L^3}$

c) From the graph it appears that the displacement-length ratio decreases as the length increases.

75. The domain of $f(x) = \sqrt{x} - 4$ is $[0, \infty)$. The domain of $g(x) = \sqrt{x}$ is $[0, \infty)$.
$(g \circ f)(x) = g(\sqrt{x} - 4) = \sqrt{\sqrt{x} - 4}$
So that $\sqrt{x} - 4 \geq 0$ we must have $x \geq 16$. So the domain of $g \circ f$ is $[16, \infty)$.

77. The graph of $y = x + \sqrt{x}$ appears only for $x \geq 0$. So the domain is $[0, \infty)$. The graph starts at $(0, 0)$ and goes up from there. So the range is $[0, \infty)$.

9.4 WARM-UPS

1. False, because the inverse function is obtained by interchanging the coordinates in every ordered pair.
2. False, because the points (0, 3) and (1, 3) both satisfy $f(x) = 3$.
3. False, because the inverse of multiplication by 2 is division by 2. If $g(x) = 2x$, then $g^{-1}(x) = x/2$.
4. True, because if we interchange the coordinates in a function that is not one-to-one, we do not obtain a function.
5. True, because in an inverse function the domain and range of the function are reversed.
6. False, because it is not one-to-one. Both $(2, 16)$ and $(-2, 16)$ satisfy $f(x) = x^4$.
7. True, because taking the opposite of a number twice gives back the original number.
8. True, because the inverse function interchanges the coordinates in all of the ordered pairs of the function.
9. True, the inverse of $k(x) = 3x - 6$ is $k^{-1}(x) = (x + 6)/3 = (1/3)x + 2$.
10. False, because $f^{-1}(x) = (x + 4)/3$.

9.4 EXERCISES

1. The inverse of a function is a function with the same ordered pairs except that the coordinates are reversed.
3. The range of f^{-1} is the domain of f.
5. A function is one-to-one if no two ordered pairs have the same second coordinate with different first coordinates.
7. The switch and solve strategy is used to find a formula for an inverse function.
9. The function is not one-to-one because of the pairs $(-2, 2)$ and $(2, 2)$. Therefore, the function is not invertible.
11. The function is one-to-one. The inverse is $\{(4, 16), (3, 9), (0, 0)\}$.
It is obtained by interchanging the coordinates in each ordered pair.
13. The function is not one-to-one because of the pairs $(5, 0)$ and $(6, 0)$. Therefore, it is not invertible.

15. This function is one-to-one. Its inverse is obtained by interchanging the coordinates in each ordered pair: $\{(0, 0), (2, 2), (9, 9)\}$.

17. This function is not one-to-one because we can draw a horizontal line that crosses the graph twice.

19. This function is one-to-one because we cannot draw a horizontal line that crosses the graph more than once.

21. Since $(f \circ g)(x) = f(g(x)) = f(0.5x)$
$= 2(0.5x) = x$ and $(g \circ f)(x) = g(f(x))$
$= g(2x) = 0.5(2x) = x$, the functions are inverses of each other.

23. $(f \circ g)(x) = f(g(x)) = f(\frac{1}{2}x + 5)$
$= 2(\frac{1}{2}x + 5) - 10 = x + 10 - 10 = x$
$(g \circ f)(x) = g(f(x)) = g(2x - 10)$
$= \frac{1}{2}(2x - 10) + 5 = x - 5 + 5 = x$
Therefore, the functions are inverses of each other.

25. $(f \circ g)(x) = f(g(x)) = f(-x) = -(-x)$
$= x$
$(g \circ f)(x) = g(f(x)) = g(-x) = -(-x) = x$
Therefore, the functions are inverses of each other.

27. $(g \circ f)(x) = g(f(x)) = g(x^4) = (x^4)^{1/4}$
$= |x|$
For example, $g(f(-2)) = g(16) = 2$. So the functions are not inverses of each other.

29. $y = 5x$, $x = 5y$, $y = \frac{1}{5}x$
So $f^{-1}(x) = \frac{x}{5}$.

31. $y = x - 9$, $x = y - 9$, $y = x + 9$
So $g^{-1}(x) = x + 9$.

33. $y = 5x - 9$, $x = 5y - 9$, $5y = x + 9$,
$y = \frac{x + 9}{5}$
So $k^{-1}(x) = \frac{x + 9}{5}$.

35. $y = \frac{2}{x}$, $x = \frac{2}{y}$, $xy = 2$, $y = \frac{2}{x}$
So $m^{-1}(x) = \frac{2}{x}$.

37. $y = \sqrt[3]{x - 4}$, $x = \sqrt[3]{y - 4}$, $x^3 = y - 4$,
$y = x^3 + 4$ So $f^{-1}(x) = x^3 + 4$.

39. $y = \frac{3}{x - 4}$, $x = \frac{3}{y - 4}$, $x(y - 4) = 3$,
$y - 4 = \frac{3}{x}$, $y = \frac{3}{x} + 4$
So $f^{-1}(x) = \frac{3}{x} + 4$.

41. $f(x) = \sqrt[3]{3x + 7}$
           $y = \sqrt[3]{3x + 7}$

$$
\begin{aligned}
x &= \sqrt[3]{3y + 7} \\
x^3 &= 3y + 7 \\
x^3 - 7 &= 3y \\
\frac{x^3 - 7}{3} &= y \\
f^{-1}(x) &= \frac{x^3 - 7}{3}
\end{aligned}
$$

43. $f(x) = \frac{x + 1}{x - 2}$
$$
\begin{aligned}
y &= \frac{x + 1}{x - 2} \\
x &= \frac{y + 1}{y - 2} \\
x(y - 2) &= y + 1 \\
xy - 2x &= y + 1 \\
xy - y &= 2x + 1 \\
y(x - 1) &= 2x + 1 \\
y &= \frac{2x + 1}{x - 1} \\[4pt]
f^{-1}(x) &= \frac{2x + 1}{x - 1}
\end{aligned}
$$

45. $f(x) = \frac{x + 1}{3x - 4}$
$$
\begin{aligned}
y &= \frac{x + 1}{3x - 4} \\
x &= \frac{y + 1}{3y - 4} \\
x(3y - 4) &= y + 1 \\
3xy - 4x &= y + 1 \\
3xy - y &= 1 + 4x \\
y(3x - 1) &= 1 + 4x \\
y &= \frac{1 + 4x}{3x - 1} \\
f^{-1}(x) &= \frac{1 + 4x}{3x - 1}
\end{aligned}
$$

47. The function $p(x) = \sqrt[4]{x}$ finds the fourth root. The inverse must be the fourth power. Since the domain and range of p are both the set of nonnegative real numbers, $p^{-1}(x) = x^4$ for $x \geq 0$.

49. $y = (x - 2)^2$, $x = (y - 2)^2$,
$y - 2 = \pm\sqrt{x}$, $y = 2 \pm \sqrt{x}$
Since $x \geq 2$ in the function, we must have $y \geq 2$ in the inverse function. So $y = 2 - \sqrt{x}$ is not the inverse and $f^{-1}(x) = 2 + \sqrt{x}$.

51. $y = x^2 + 3$, $x = y^2 + 3$, $y^2 = x - 3$,
$y = \pm\sqrt{x - 3}$. Since $x \geq 0$ in the function, $y \geq 0$ in the inverse function. So $f^{-1}(x) = \sqrt{x - 3}$.

53. $f(x) = \sqrt{x + 2}$
$$
\begin{aligned}
y &= \sqrt{x + 2} \\
x &= \sqrt{y + 2}
\end{aligned}
$$

$$x^2 = y + 2$$
$$x^2 - 2 = y$$

Since $y \geq 0$ in the function, $x \geq 0$ in the inverse. So $f^{-1}(x) = x^2 - 2$ for $x \geq 0$.

55. The inverse of $f(x) = 2x + 3$ is a function that subtracts 3 and then divides by 2, $f^{-1}(x) = \dfrac{x-3}{2} = \dfrac{1}{2}x - \dfrac{3}{2}$. Use y-intercepts and slopes to graph each straight line.

57. The graph of $f(x) = x^2 - 1$ for $x \geq 0$ contains the points $(0, -1)$, $(1, 0)$, and $(2, 3)$. For the inverse function we add 1 and then take the square root, $f^{-1}(x) = \sqrt{x + 1}$. The function f^{-1} contains $(-1, 0)$, $(0, 1)$, and $(3, 2)$.

59. The graph of f is a line through $(0, 0)$ with slope 5, and the graph of $f^{-1}(x) = x/5$ is a straight line through $(0, 0)$ with slope $1/5$.

61. The inverse of cubing is the cube root. So $f^{-1}(x) = \sqrt[3]{x}$. The graph of f contains the points $(0, 0)$, $(1, 1)$, $(-1, -1)$, $(2, 8)$, and $(-2, -8)$. The graph of f^{-1} contains the points $(0, 0)$, $(1, 1)$, $(-1, -1)$, $(8, 2)$, and $(-8, -2)$.

63. The inverse of subtracting 2 and then taking a square root is squaring and then adding 2. So $f^{-1}(x) = x^2 + 2$ for $x \geq 0$. The graph of f contains $(2, 0)$, $(3, 1)$, and $(6, 2)$. The graph of f^{-1} contains $(0, 2)$, $(1, 3)$, and $(2, 6)$.

65. $(f^{-1} \circ f)(x) = f^{-1}(f(x)) = f^{-1}(x^3 - 1)$
$$= \sqrt[3]{x^3 - 1 + 1} = x$$

67. $(f^{-1} \circ f)(x) = f^{-1}(f(x)) = f^{-1}(\tfrac{1}{2}x - 3)$
$$= 2(\tfrac{1}{2}x - 3) + 6 = x - 6 + 6 = x$$

69. $(f^{-1} \circ f)(x) = f^{-1}(f(x)) = f^{-1}(\tfrac{1}{x} + 2)$
$$= \frac{1}{\frac{1}{x} + 2 - 2} = \frac{1}{\frac{1}{x}} = x$$

71. $(f^{-1} \circ f)(x) = f^{-1}(f(x)) = f^{-1}(\tfrac{x+1}{x-2})$
$$= \frac{2(\frac{x+1}{x-2}) + 1}{\frac{x+1}{x-2} - 1} = \frac{\left(\frac{2x+2}{x-2} + 1\right)(x-2)}{\left(\frac{x+1}{x-2} - 1\right)(x-2)}$$
$$= \frac{2x + 2 + x - 2}{x + 1 - x + 2} = \frac{3x}{3} = x$$

73. a) $S = \sqrt{30 \cdot 50 \cdot 0.75} \approx 33.5$ mph

b) From the graph you can see that the drag factor decreases when a road gets wet because the skid marks for a given speed will be longer.

c)
$$S = \sqrt{30L \cdot 1}$$
$$S^2 = 30L$$
$$L = \frac{S^2}{30}$$

75. If x is the selling price, then the total cost is given by $T(x) = x + 0.09x + 125$ or $T(x) = 1.09x + 125$. If x is the total cost, then the function $T^{-1}(x) = \dfrac{x - 125}{1.09}$ gives the selling price as a function of the total cost.

77. Only the odd powers of x are one-to-one functions. So $f(x) = x^n$ is invertible if n is an odd positive integer. If n is an even positive integer, the function $f(x) = x^n$ is not one-to-one and not invertible.

79. The functions f and g are not inverses of each other because the function $y_2 = \sqrt{y_1} = \sqrt{x^2}$ is the absolute value function and not the identity function.

9.5 WARM-UPS

1. True, because direct variation means that a is a constant multiple of b.

2. False, because a inversely proportional to b means $a = k/b$.

3. False, because jointly proportional means $a = kbc$ for some constant k.

4. True, a is a constant multiple of the square root of c.

5. False, b is directly proportional to a means that $b = ka$.

6. True, because a is a multiple of b and b is divided by c.

7. False, a is jointly proportional to c and the square of b means that $a = kcb^2$.

8. False, $a = kc/\sqrt{b}$ is correct.

9. False, b is directly proportional to a and inversely proportional to the square of c means that $b = ka/c^2$.

10. True, b is equal to a constant divided by the square of c.

9.5 EXERCISES

1. If y varies directly as x, then $y = kx$ for some constant k.

3. If y is inversely proportional to x, then $y = k/x$.

5. If y is jointly proportional to x and z, then $y = kxz$ for some constant k.

7. If a varies directly as m, then a is equal to a constant multiple of m, $a = km$.

9. If d varies inversely as e, then d is equal to a constant divided by e, $d = k/e$.

11. If I varies jointly as r and t, then I is a constant multiple of the product of r and t, $I = krt$.

13. If m is directly proportional to the square of p, then m is a constant multiple of the square of p, $m = kp^2$.

15. If B is directly proportional to the cube root of w, then B is a constant multiple of the cube root of w, $B = k\sqrt[3]{w}$.

17. If t is inversely proportional to the square of x, then t is equal to a constant divided by the square of x, $t = \dfrac{k}{x^2}$.

19. If v varies directly as m and inversely as n, then v is equal to a constant multiple of m divided by n, $v = \dfrac{km}{n}$.

21. If y varies directly as x, then $y = kx$. If $y = 6$ when $x = 4$, then $6 = k \cdot 4$, or $k = \dfrac{6}{4} = \dfrac{3}{2}$. Therefore, $y = kx$ can be written as $y = \dfrac{3}{2}x$.

23. If A varies inversely as B, then $A = \dfrac{k}{B}$. If $A = 10$ when $B = 3$, then $10 = \dfrac{k}{3}$ or $k = 30$. Therefore, $A = \dfrac{k}{B}$ can be written as $A = \dfrac{30}{B}$.

25. If m varies inversely as the square root of p, then $m = \dfrac{k}{\sqrt{p}}$. If $m = 12$ when $p = 9$, then $12 = \dfrac{k}{\sqrt{9}}$ or $k = 12\sqrt{9} = 36$. Therefore, $m = \dfrac{k}{\sqrt{p}}$ can be written as $m = \dfrac{36}{\sqrt{p}}$.

27. If A varies jointly as t and u, then $A = ktu$. If $A = 6$ when $t = 5$ and $u = 3$, then $6 = k(5)(3)$ or $k = \dfrac{2}{5}$. Therefore, $A = \dfrac{2}{5}tu$.

29. If y varies directly as x and inversely as z, then $y = \dfrac{kx}{z}$. If $y = 2.37$ when $x = \pi$ and $z = \sqrt{2}$, then $2.37 = \dfrac{k\pi}{\sqrt{2}}$ or $k = 2.37\sqrt{2}/\pi \approx 1.067$. So, $y = \dfrac{1.067x}{z}$.

31. If y varies directly as x, then $y = kx$. Since $y = 7$ when $x = 5$, we have $7 = k(5)$ or $k = \frac{7}{5}$. So, $y = kx$ can be written as $y = \frac{7}{5}x$. To find y when $x = -3$, we get $y = \frac{7}{5}(-3) = -\frac{21}{5}$.

33. If w varies inversely as z, then $w = \frac{k}{z}$. Since $w = 6$ when $z = 2$, we have $6 = \frac{k}{2}$ or $k = 12$. Therefore, $w = \frac{12}{z}$. Now when $z = -8$, we get $w = \frac{12}{-8} = -\frac{3}{2}$.

35. If A varies jointly as F and T, then $A = kFT$. Since $A = 6$ when $F = 3\sqrt{2}$ and $T = 4$, we have $6 = k3\sqrt{2}(4)$ or $k = \frac{\sqrt{2}}{4}$. Therefore, $A = \frac{\sqrt{2}}{4}FT$. When $F = 2\sqrt{2}$ and $T = \frac{1}{2}$, we get $A = \frac{\sqrt{2}}{4}(2\sqrt{2})(\frac{1}{2}) = \frac{1}{2}$.

37. If D varies directly with t and inversely with the square of s, then $D = \frac{kt}{s^2}$. Since $D = 12.35$ when $t = 2.8$ and $s = 2.48$, we get
$$12.35 = \frac{k2.8}{(2.48)^2}$$ or
$$k = \frac{12.35(2.48)^2}{2.8} \approx 27.128.$$
When $t = 5.63$ and $s = 6.81$, we get
$$D = \frac{(27.128)(5.63)}{(6.81)^2} \approx 3.293.$$

39. Inverse variation, y varies inversely as x.

41. Direct variation, y varies directly as x.

43. Combined variation, y varies directly as x and inversely as w.

45. Joint variation, y varies jointly as x and z.

47. The cost c varies directly with the size of the lawn s means that $c = ks$. Since $c = \$280$ when $s = 4000$, we get $280 = k(4000)$ or $k = 0.07$. The formula is now written as $c = 0.07s$. For a 6000 square foot lawn, the cost is
$$c = 0.07(6000) = \$420.$$

49. The volume v is inversely proportional to the weight w means that $v = k/w$. Since the volume is 6 when the weight is 30, we have $6 = k/30$ or $k = 180$. Therefore, $v = 180/w$. If the weight is 20 kg, then
$$v = 180/20 = 9 \, \text{cm}^3.$$

51. The price is jointly proportional to the radius and length means that $P = kRL$. Since

a 12 foot culvert with a 6 inch radius costs $324, we have $324 = k(6)(12)$ or $k = 4.5$. Therefore, the formula is $P = 4.5RL$. The price of a 10 foot culvert with an 8 inch radius is $P = 4.5(8)(10) = \$360$.

53. The price varies jointly as the length and the square of the diameter means that $P = kLD^2$. Since an 18 foot rod with a diameter of 2 inches has a price of $12.60, we get $12.60 = k(18)(2)^2$ or $k = 0.175$. Now the formula is written $P = 0.175LD^2$. The cost of a 12 foot rod with a 3 inch diameter is $P = 0.175(12)(3)^2 = \$18.90$.

55. a) The distance varies directly as the square of the time means that $d = kt^2$. If an object falls 0.16 feet in 0.1 seconds, then $0.16 = k(0.1)^2$ or $k = 16$. So the formula is $d = 16t^2$.

b) In 0.5 seconds the object falls $d = 16(0.5)^2 = 4$ feet.

c)
$$16t^2 = 100$$
$$t^2 = 6.25$$
$$t = \pm 2.5$$
It takes 2.5 seconds to reach the ground.

57. The force is inversely proportional to the length means that $F = k/L$. If a force of 2000 pounds is needed at 2 feet, then $2000 = k/2$ or $k = 4000$. The formula is now written as $F = 4000/L$. At 10 feet the force would be $F = 4000/10 = 400$ pounds.

59. Since $t = \frac{kc}{n}$ and $t = 8$ when $n = 3$ and $c = 6$, we have $8 = \frac{k6}{3}$, or $k = 4$. So $t = \frac{4c}{n}$. Now if $c = 9$ and $n = 5$, $t = \frac{4 \cdot 9}{5} = 7.2$ days.

61. a) $G = \frac{kNd}{c}$
$$54 = \frac{k \cdot 52 \cdot 27}{26}$$
$$54 = k \cdot 54$$
$$1 = k$$
So $G = \frac{Nd}{c}$.

b) $G = \frac{42 \cdot 26}{13} = 84$

c)
$$G = \frac{Nd}{c}$$
$$cG = Nd$$
$$c = \frac{Nd}{G}$$

$$c = \frac{44 \cdot 27}{52} \approx 23 \qquad c = \frac{44 \cdot 27}{59} \approx 20$$

$$c = \frac{44 \cdot 27}{70} \approx 17 \qquad c = \frac{44 \cdot 27}{79} \approx 15$$

$$c = \frac{44 \cdot 27}{91} \approx 13$$

d) The gear ratio decreases as the number of teeth on the cog increases.

63. The curves cross at $(1, 1)$. The function y_1 is increasing and y_2 is decreasing. The function y_1 represents direct variation and y_2 represents inverse variation.

Enriching Your Mathematical Word Power

1. a **2.** b **3.** d **4.** d **5.** a
6. b **7.** d **8.** c **9.** d **10.** b
11. a **12.** d

CHAPTER 9 REVIEW

1. The graph of $f(x) = 3x - 4$ is a straight line with y-intercept $(0, -4)$ and slope 3. Since any number can be used for x, the domain is $(-\infty, \infty)$. From the graph we see that any real number can occur as a y-coordinate, so the range is $(-\infty, \infty)$.

3. The graph of $h(x) = |x| - 2$ includes the points $(0, -2)$, $(1, -1)$, $(2, 0)$, $(-1, -1)$, and $(-2, 0)$. Draw a v-shaped graph through these points.

Since any real number can be used for x, the domain is $(-\infty, \infty)$. From the graph we see

that all y-coordinates are greater than or equal to -2, and so the range is $[-2, \infty)$.

5. The graph of $y = x^2 - 2x + 1$ includes the points $(0, 1)$, $(1, 0)$, $(2, 1)$, $(3, 4)$, and $(-1, 4)$. Draw a parabola through these points.

Since any number can be used for x, the domain is $(-\infty, \infty)$. From the graph we see that the y-coordinates are nonnegative, and so the range is $[0, \infty)$.

7. The graph of $k(x) = \sqrt{x} + 2$ includes the points $(0, 2)$, $(1, 3)$, and $(4, 4)$.

Since we can use only nonnegative numbers for x, the domain is $[0, \infty)$. From the graph we see that no y-coordinate is less than 2, and so the range is $[2, \infty)$.

9. The graph of $y = 30 - x^2$ includes the points $(-5, 5)$, $(0, 30)$, and $(5, 5)$.

Since x can be any real number, the domain is $(-\infty, \infty)$. From the graph we see that y is at most 30. So the range is $(-\infty, 30]$.

11. The graph of $x = 2$ is the vertical line with x-intercept $(2, 0)$.

Since the only x-coordinate used on this graph is 2, the domain is $\{2\}$. Since all real numbers occur as y-coordinates, the range is $(-\infty, \infty)$.

13. The graph of $x = |y| + 1$ includes the points $(1, 0)$, $(2, 1)$, $(2, -1)$, $(3, 2)$, and $(3, -2)$. Draw a v-shaped graph through these points.

From the graph we see that the x-coordinates are not less than 1, and so the domain is $[1, \infty)$. Any number can be used for y, and so the range is $(-\infty, \infty)$.

15. The graph of $y = \sqrt{x}$ includes the points $(0, 0)$, $(1, 1)$, and $(4, 2)$.

17. The graph of $y = -2\sqrt{x}$ includes the points $(0, 0)$, $(1, -2)$, and $(4, -4)$. It can be obtained from the graph of $y = \sqrt{x}$ by stretching and reflecting in the x-axis.

19. The graph of $y = \sqrt{x - 2}$ can be obtained by shifting the graph of $y = \sqrt{x}$ two spaces to the right. The graph of $y = \sqrt{x - 2}$ includes the points $(2, 0)$, $(3, 1)$, and $(6, 2)$.

21. The graph of $y = \frac{1}{2}\sqrt{x}$ is obtained from the graph of $y = \sqrt{x}$ by shrinking. The graph of $y = \frac{1}{2}\sqrt{x}$ includes the points $(0, 0)$, $(1, 0.5)$, and $(4, 1)$.

23. The graph of $y = -\sqrt{x + 1} + 3$ is obtained from the graph of $y = \sqrt{x}$ by shifting 1 unit to the left, reflecting in the x-axis, and then shifting 3 units upward. The graph of $y = -\sqrt{x + 1} + 3$ includes the points $(-1, 3)$, $(0, 2)$, and $(3, 1)$.

25. $f(-3) = 3(-3) + 5 = -4$

27. $(h \circ f)(\sqrt{2}) = h(f(\sqrt{2})) = h(3\sqrt{2} + 5)$
$$= \frac{3\sqrt{2} + 5 - 5}{3} = \frac{3\sqrt{2}}{3} = \sqrt{2}$$

29. $(g \circ f)(2) = g(f(2))$
$$= g(11) = 11^2 - 2(11) = 99$$

31. $(f + g)(3) = f(3) + g(3)$
$$= 3 \cdot 3 + 5 + 3^2 - 2 \cdot 3 = 17$$

33. $(f \cdot g)(x) = f(x) \cdot g(x)$
$= (3x + 5)(x^2 - 2x)$
$= 3x^3 + 5x^2 - 6x^2 - 10x = 3x^3 - x^2 - 10x$

35. $(f \circ f)(0) = f(f(0)) = f(5) = 3 \cdot 5 + 5$
$$= 20$$

37. $F(x) = |x+2| = |g(x)| = f(g(x))$
$$= (f \circ g)(x)$$
Therefore, F is the same function as f composite g, and we write $F = f \circ g$.

39. $H(x) = x^2 + 2 = h(x) + 2 = g(h(x))$
$$= (g \circ h)(x)$$
Therefore, H is the same function as g composite h, and we write $H = g \circ h$.

41. $I(x) = x + 4 = x + 2 + 2 = g(x) + 2$
$$= g(g(x)) = (g \circ g)(x)$$
Therefore, I is the same function as g composite g, and we write $I = g \circ g$.

43. This function is not invertible, because it is not one-to-one. The ordered pairs have different first coordinates and the same second coordinate.

45. The function $f(x) = 8x$ is invertible. The inverse of multiplication by 8 is division by 8 and $f^{-1}(x) = x/8$.

47. The function $g(x) = 13x - 6$ is one-to-one and so it is invertible. The inverse of multiplying by 13 and subtracting 6 is adding 6 and then dividing by 13, $g^{-1}(x) = \dfrac{x+6}{13}$.

49. $j(x) = \dfrac{x+1}{x-1}$
$$y = \dfrac{x+1}{x-1}$$
$$x = \dfrac{y+1}{y-1}$$
$$x(y-1) = y+1$$
$$xy - x = y + 1$$
$$xy - y = x + 1$$
$$y(x-1) = x + 1$$
$$y = \dfrac{x+1}{x-1}$$
$$j^{-1}(x) = \dfrac{x+1}{x-1}$$

51. The function $m(x) = (x-1)^2$ is not invertible because it is not one-to-one. The ordered pairs $(2, 1)$ and $(0, 1)$ are both in the function.

53. The function $f(x) = 3x - 1$ is inverted by adding 1 and then dividing by 3:

$$f^{-1}(x) = \frac{x+1}{3} = \tfrac{1}{3}x + \tfrac{1}{3}$$

Use slope and y-intercept to graph both functions on the same coordinate system.

55. The function $f(x) = x^3/2$ is inverted by multiplying by 2 and then taking the cube root:
$$f^{-1}(x) = \sqrt[3]{2x}$$
The graph of f includes the points $(0, 0)$, $(1, 1/2)$, $(-1, -1/2)$, $(2, 4)$, and $(-2, -4)$. The graph of f^{-1} includes the points $(0, 0)$, $(1/2, 1)$, $(-1/2, -1)$, $(4, 2)$, and $(-4, -2)$. Plot these points and graph both functions on the same coordinate system.

57. If y varies directly as m, then $y = km$. If $y = -3$ when $m = 1/4$, then
$$-3 = k \cdot \tfrac{1}{4}$$
or
$$-12 = k.$$
Since $y = -12m$, when $m = -2$ we get
$$y = -12(-2) = 24.$$

59. If c varies directly as m and inversely as n, then $c = km/n$. If $c = 20$ when $m = 10$ and $n = 4$, then
$$20 = \frac{k(10)}{4}$$
or
$$8 = k.$$
Since $c = 8m/n$, when $m = 6$ and $n = -3$, we get
$$c = \frac{8(6)}{-3} = -16.$$

205

61. If the distance varies directly with the square of the time, then $d = kt^2$. If the ball travels 144 feet in 3 seconds, then $144 = k3^2$, or $k = 16$ and $d = 16t^2$. So in 4 seconds the ball falls $d = 16 \cdot 4^2 = 256$ feet.

63. The area of the circle is $A = \pi r^2$. Let $s = $ the length of the side of the square. Since the square is inscribed in the circle, the length of the diagonal of the square is the same as the diameter of the circle $2r$. Since the diagonal of the square is the hypotenuse of a right triangle whose sides are s and s, we can write the following equation.

$$s^2 + s^2 = (2r)^2$$
$$2s^2 = 4r^2$$
$$s^2 = 2r^2$$

Since the area of the square is s^2, we have $B = s^2$, or $B = 2r^2$. Since $r^2 = \frac{A}{\pi}$, we can write $B = 2 \cdot \frac{A}{\pi}$, or $B = \frac{2A}{\pi}$.

65. Substitute $k = 5w - 6$ into the equation $a = 3k + 2$:

$$a = 3(5w - 6) + 2$$
$$a = 15w - 18 + 2$$
$$a = 15w - 16$$

67. The area of a square as a function of the side is $A = s^2$. Solve for s to get the side as a function of the area: $s = \sqrt{A}$.

CHAPTER 9   TEST

1. The line $f(x) = -\frac{2}{3}x + 1$ has a y-intercept of $(0, 1)$ and a slope of $-2/3$. Start at $(0, 1)$ and rise -2 and go 3 units to the right to locate a second point on the line.

2. The graph of $y = |x| - 4$ contains the points $(-2, -2)$, $(-1, -3)$, $(0, -4)$, $(1, -3)$, and $(2, -2)$. Draw a v-shaped graph through these points.

3. The graph of $g(x) = x^2 + 2x - 8$ contains the points $(-2, -8)$, $(-1, -9)$, $(0, -8)$, $(1, -5)$, and $(2, 0)$. Draw a parabola through these points.

4. The graph of $x = y^2$ contains the points $(0, 0)$, $(1, 1)$, $(1, -1)$, $(4, -2)$, and $(4, 2)$. Draw a parabola through these points.

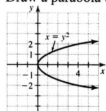

5. The graph of $y = -\,|\,x - 2\,|$ is a translation and reflection of the graph of $y = \,|\,x\,|$. The graph contains $(2, 0)$, $(0, -2)$, and $(4, -2)$.

6. The graph of $y = \sqrt{x + 5} - 2$ is obtained from the graph of $y = \sqrt{x}$ by translating 5 units to the left and 2 units downward. The graph of $y = \sqrt{x+5} - 2$ goes through $(-5, -2)$, $(-4, -1)$, and $(-1, 0)$.

7. $f(-3) = -2(-3) + 5 = 6 + 5 = 11$

8. $(g \circ f)(-3) = g(f(-3)) = g(11) = 11^2 + 4$
$$= 125$$

9. Because $f(-3) = 11$, $f^{-1}(11) = -3$.

10. Because f consists of multiplying by -2 and adding 5, f^{-1} consists of subtracting 5 and dividing by -2:
$$f^{-1}(x) = \frac{x-5}{-2} \text{ or } f^{-1}(x) = -\frac{1}{2}x + \frac{5}{2}$$

11. $(g + f)(x) = g(x) + f(x)$
$$= x^2 + 4 + (-2x + 5)$$
$$= x^2 - 2x + 9.$$

12. $(f \cdot g)(1) = f(1) \cdot g(1) = 3 \cdot 5 = 15$

13. Because f^{-1} is the inverse of f we have $(f^{-1} \circ f)(1776) = 1776$

14. $(f/g)(2) = f(2)/g(2) = 1/8$

15. $(f \circ g)(x) = f(g(x)) = f(x^2 + 4)$
$$= -2(x^2 + 4) + 5 = -2x^2 - 8 + 5$$
$$= -2x^2 - 3$$

16. $(g \circ f)(x) = g(f(x)) = g(-2x + 5)$
$$= (-2x + 5)^2 + 4 = 4x^2 - 20x + 29$$

17. $f(g(x)) = f(x^2) = x^2 - 7 = H(x)$
So $H = f \circ g$.

18. $g(f(x)) = g(x - 7) = (x - 7)^2$
$$= x^2 - 14x + 49$$
So $W = g \circ f$.

19. The function is not invertible because it is not one-to-one.

20. The function is invertible and its inverse is $f^{-1}(x) = (x - 9)^3$.

21. Since we can use any real number for x in the expression $|x|$, the domain is $(-\infty, \infty)$. Since $|x| \geq 0$ for any real number x, the range is $[0, \infty)$.

22.
$$f(x) = \frac{2x + 1}{x - 1}$$
$$y = \frac{2x + 1}{x - 1}$$
$$x = \frac{2y + 1}{y - 1}$$
$$x(y - 1) = 2y + 1$$

$$xy - x = 2y + 1$$
$$xy - 2y = x + 1$$
$$y(x - 2) = x + 1$$
$$y = \frac{x + 1}{x - 2}$$
$$f^{-1}(x) = \frac{x + 1}{x - 2}$$

23. Since the volume varies directly as the cube of the radius, we have $V = kr^3$. If $V = 36\pi$ when $r = 3$, we have $36\pi = k3^3$ or $k = 4\pi/3$. To find the volume of a sphere with a radius of 2, use $r = 2$ and $k = 4\pi/3$ in the formula $V = kr^3$:
$$V = \frac{4\pi}{3}(2)^3 = \frac{32\pi}{3} \text{ cubic feet}$$

24. If y varies directly as x and inversely as the square root of z, then $y = kx/\sqrt{z}$. If $y = 12$ when $x = 7$ and $z = 9$, we have
$$12 = \frac{k(7)}{\sqrt{9}} \text{ or } k = \frac{36}{7}.$$

25. If the cost varies jointly as the length and width, then $C = kLW$.
$$2256 = k(6)(8)$$
$$k = 47$$
So $C = 47LW$. So the cost of a 9 by 12 rug is
$$C = 47(9)(12) = \$5076.$$

Making Connections
Chapters 1 - 9

1. $125^{-2/3} = 5^{-2} = \frac{1}{25}$

2. $\left(\frac{8}{27}\right)^{-1/3} = \left(\frac{27}{8}\right)^{1/3} = \frac{3}{2}$

3. $\sqrt{18} - \sqrt{8} = 3\sqrt{2} - 2\sqrt{2} = \sqrt{2}$

4. $x^5 \cdot x^3 = x^{5+3} = x^8$

5. $16^{1/4} = 2$, because $2^4 = 16$.

6. $\frac{x^{12}}{x^3} = x^{12-3} = x^9$

7. $x^2 = 9$
$$x = \pm\sqrt{9} = \pm 3$$
The solution set is $\{\pm 3\}$.

8. $x^2 = 8$, $x = \pm\sqrt{8} = \pm 2\sqrt{2}$
The solution set is $\{\pm 2\sqrt{2}\}$.

9. $x^2 - x = 0$
$$x(x - 1) = 0$$
$$x = 0 \text{ or } x - 1 = 0$$
The solution set is $\{0, 1\}$.

10. $x^2 - 4x - 6 = 0$

$$x = \frac{4 \pm \sqrt{(-4)^2 - 4(1)(-6)}}{2(1)}$$

$$= \frac{4 \pm \sqrt{40}}{2} = \frac{4 \pm 2\sqrt{10}}{2} = 2 \pm \sqrt{10}$$

The solution set is $\{2 \pm \sqrt{10}\}$.

11. $\quad x^{1/4} = 3$

$$(x^{1/4})^4 = 3^4$$

$$x = 81$$

Since we raised each side to an even power we must check. Since the fourth root of 81 is 3, the solution set is $\{81\}$.

12. If we raise each side to the sixth power, we will get $x = 64$. However, the sixth root of 64 is 2 and not -2. So 64 does not check. The solution set is $\emptyset$.

13. $\quad |x| = 8$

$$x = 8 \text{ or } x = -8$$

The solution set is $\{\pm 8\}$.

14. $\quad |5x - 4| = 21$

$$5x - 4 = 21 \text{ or } 5x - 4 = -21$$

$$5x = 25 \quad \text{or} \quad 5x = -17$$

$$x = 5 \quad \text{or} \quad x = -\frac{17}{5}$$

The solution set is $\left\{-\frac{17}{5}, 5\right\}$.

15. $\quad x^3 = 8$

$$(x^3)^{1/3} = 8^{1/3}$$

$$x = 2$$

The solution set is $\{2\}$.

16. $(3x - 2)^3 = 27$

$$3x - 2 = 3$$

$$3x = 5$$

$$x = \frac{5}{3}$$

The solution set is $\left\{\frac{5}{3}\right\}$.

17. $\quad \sqrt{2x - 3} = 9$

$$(\sqrt{2x - 3})^2 = 9^2$$

$$2x - 3 = 81$$

$$2x = 84$$

$$x = 42$$

The solution set is $\{42\}$.

18. $\quad \sqrt{x - 2} = x - 8$

$$x - 2 = (x - 8)^2$$

$$x - 2 = x^2 - 16x + 64$$

$$x^2 - 17x + 66 = 0$$

$$(x - 6)(x - 11) = 0$$

$$x = 6 \text{ or } x = 11$$

Only 11 satisfies the original equation. The solution set is $\{11\}$.

19. The graph of the set of points that satisfy the equation $y = 5$ is a horizontal straight line with y-intercept $(0, 5)$.

20. The graph of the set of points that satisfy $y = 2x - 5$ is a straight line with y-intercept $(0, -5)$ and slope 2.

21. The graph of the set of points that satisfy $x = 5$ is a vertical line with x-intercept $(5, 0)$.

22. The graph of $3y = x$ is the same as the graph of $y = (1/3)x$, a straight line with y-intercept $(0, 0)$ and slope $1/3$.

208

23. The graph of $y = 5x^2$ is a parabola going through $(0, 0)$, $(1, 5)$, and $(-1, 5)$.

24. The graph of $y = -2x^2$ is a parabola going through $(0, 0)$, $(1, -2)$, and $(-1, -2)$.

25. Because $2^2 = 4$, and $2^3 = 8$, we have $(2, 4)$ and $(3, 8)$. Because $2^1 = 2$ and $2^4 = 16$, we have $(1, 2)$ and $(4, 16)$.

26. Because $4^{1/2} = 2$ and $4^{-1} = 1/4$, we have $(1/2, 2)$ and $(-1, 1/4)$. Because $4^2 = 16$ and $4^0 = 1$, we have $(2, 16)$ and $(0, 1)$.

27. In $\sqrt{x}$ we must have $x \geq 0$. So the domain is $[0, \infty)$.

28. In $\sqrt{6 - 2x}$ we must have $6 - 2x \geq 0$ or $6 \geq 2x$, or $3 \geq x$, or $x \leq 3$. So the domain is $(-\infty, 3]$.

29. Because there are no real solutions to $x^2 + 1 = 0$, we can use any real number for x in this expression. The domain is $(-\infty, \infty)$

30. Solve $x^2 - 10x + 9 = 0$ by factoring.

$$(x - 9)(x - 1) = 0$$
$$x - 9 = 0 \text{ or } x - 1 = 0$$
$$x = 9 \text{ or } \quad x = 1$$

So the domain is all real numbers except 1 and 9, which is written in interval notation as $(-\infty, 1) \cup (1, 9) \cup (9, \infty)$

31. a) $C = 0.12x + 3000$
b) Find the equation of the line through $(0, 0.15)$ and $(100,000, 0.25)$

$$m = \frac{0.25 - 0.15}{100,000 - 0} = 0.000001$$
$$P\text{-intercept is } (0, 0.15)$$

$$P = 0.000001x + 0.15$$
$$P = 1 \times 10^{-6}x + 0.15$$

32. a) $T = \frac{C}{x} + P$

$$T = \frac{3000 + 0.12x}{x} + 0.000001x + 0.15$$
$$T = \frac{3000}{x} + 0.000001x + 0.27$$

If $x = 20,000$, then $T = \$0.44$
If $x = 30,000$, then $T = \$0.40$
If $x = 90,000$, then $T = \$0.39$
b)

c) $\frac{3000}{x} + 0.000001x + 0.27 = 0.38$
$$\frac{3000}{x} + 0.000001x = 0.11$$
$$3000 + 0.000001x^2 = 0.11x$$
$$0.000001x^2 - 0.11x + 3000 = 0$$

$$x = \frac{0.11 \pm \sqrt{0.11^2 - 4(0.000001)(3000)}}{2(0.000001)}$$
$$= 50,000 \text{ or } 60,000$$

The car will be replaced at 60,000 miles.

d) The total cost is less than or equal to \$0.38 per mile for mileage in the interval $[50,000, 60,000]$

209

10.1 WARM-UPS

1. True, when dividing by $x - 5$ we use 5 for synthetic division.
2. True, when dividing by $x + 7$, or $x - (-7)$, -7 is used in synthetic division.
3. False, because the remainder of synthetic division is 2 and not 0.
4. True, because 2 is a solution to $x^3 - 8 = 0$ implies that $x - 2$ is a factor of $x^3 - 8$ and so the remainder is 0.
5. True, because of the factor theorem.
6. True, because of the factor theorem. If c is a solution to $P(x) = 0$, then $x - c$ is a factor of $P(x)$.
7. True, because $1^{35} - 3(1)^{24} + 2(1)^{18} = 0$ and so by the factor theorem $x - 1$ is a factor of the polynomial.
8. True, because $(-1)^3 - 3(-1)^2 + (-1) + 5 = 0$ and so by the factor theorem $x + 1$ is a factor of the polynomial.
9. True, because $1^3 - 5(1) + 4 = 0$ implies that $x - 1$ is a factor and so the remainder is 0.
10. True, because of the factor theorem.

10.1 EXERCISES

1. A zero of the function f is a number a such that $f(a) = 0$.
3. Two statements are equivalent means that they are either both true or both false.
5. If the remainder is zero when $P(x)$ is divided by $x - c$, then $P(c) = 0$.

7.

$$\begin{array}{r|rrrr} 1 & 1 & -1 & 1 & -1 \\ & & 1 & 0 & 1 \\ \hline & 1 & 0 & 1 & 0 \end{array}$$

Since the remainder is 0, 1 is a zero of the polynomial function.

9.

$$\begin{array}{r|rrrrr} -3 & -1 & -3 & -2 & 0 & 18 \\ & & 3 & 0 & 6 & -18 \\ \hline & -1 & 0 & -2 & 6 & 0 \end{array}$$

Since the remainder is 0, -3 is a zero of the function.

11.

$$\begin{array}{r|rrrr} 2 & 2 & -4 & -5 & 9 \\ & & 4 & 0 & -10 \\ \hline & 2 & 0 & -5 & -1 \end{array}$$

Since the remainder is -1 and not 0, 2 is not a zero of the function.

13.

$$\begin{array}{r|rrrr} -3 & 1 & 5 & 2 & -12 \\ & & -3 & -6 & 12 \\ \hline & 1 & 2 & -4 & 0 \end{array}$$

Since the remainder is 0, -3 is a solution to the equation.

15.

$$\begin{array}{r|rrrrr} -2 & 1 & 3 & -5 & -10 & 5 \\ & & -2 & -2 & 14 & -8 \\ \hline & 1 & 1 & -7 & 4 & -3 \end{array}$$

Since the remainder is -3 and not 0, -2 is not a solution to the equation.

17.

$$\begin{array}{r|rrrrr} 4 & -2 & 0 & 30 & 5 & 12 \\ & & -8 & -32 & -8 & -12 \\ \hline & -2 & -8 & -2 & -3 & 0 \end{array}$$

Since the remainder is 0, 4 is a solution to the equation.

19.

$$\begin{array}{r|rrr} 3 & 0.8 & -0.3 & -6.3 \\ & & 2.4 & 6.3 \\ \hline & 0.8 & 2.1 & 0 \end{array}$$

Since the remainder is 0, 3 is a solution to the equation.

21.

$$\begin{array}{r|rrrr} 3 & 1 & 0 & -6 & -9 \\ & & 3 & 9 & 9 \\ \hline & 1 & 3 & 3 & 0 \end{array}$$

Since the remainder is 0, $x - 3$ is a factor. The other factor is the quotient.
$$x^3 - 6x - 9 = (x - 3)(x^2 + 3x + 3)$$

23.

$$\begin{array}{r|rrrr} -5 & 1 & 9 & 23 & 15 \\ & & -5 & -20 & -15 \\ \hline & 1 & 4 & 3 & 0 \end{array}$$

Since the remainder is 0, $x + 5$ is a factor of the polynomial. The quotient is the other factor.
$$x^3 + 9x^2 + 23x + 15 = (x + 5)(x^2 + 4x + 3)$$
$$= (x + 5)(x + 3)(x + 1)$$

25.

$$\begin{array}{r|rrrr} 2 & 1 & -8 & 4 & -6 \\ & & 2 & -12 & -16 \\ \hline & 1 & -6 & -8 & -22 \end{array}$$

Since the remainder is -22 and not 0, $x - 2$ is not a factor of the polynomial.

27.

$$\begin{array}{r|rrrrr} -1 & 1 & 1 & 0 & -8 & -8 \\ & & -1 & 0 & 0 & 8 \\ \hline & 1 & 0 & 0 & -8 & 0 \end{array}$$

Since the remainder is 0, $x + 1$ is a factor. The other factor is the quotient.

$$x^4 + x^3 - 8x - 8 = (x+1)(x^3 - 8)$$
$$= (x+1)(x-2)(x^2 + 2x + 4)$$

29.
$$
\begin{array}{r|rrrr}
0.5 & 2 & -3 & -11 & 6 \\
    &   & 1 & -1 & -6 \\
\hline
    & 2 & -2 & -12 & 0 \\
\end{array}
$$
Since the remainder is 0, $x - 0.5$ is a factor.
$$2x^3 - 3x^2 - 11x + 6$$
$$= (x - 0.5)(2x^2 - 2x - 12)$$
$$= (x - 0.5)(2)(x^2 - x - 6)$$
$$= (x - 0.5)(2)(x - 3)(x + 2)$$
$$= (2x - 1)(x - 3)(x + 2)$$

31. Try integers between -5 and 5 with synthetic division until we find a solution. To save space we will not show any failures. Try 1 first.
$$
\begin{array}{r|rrrr}
1 & 1 & 0 & -13 & 12 \\
  &   & 1 & 1 & -12 \\
\hline
  & 1 & 1 & -12 & 0 \\
\end{array}
$$
$$x^3 - 13x + 12 = 0$$
$$(x - 1)(x^2 + x - 12) = 0$$
$$(x - 1)(x + 4)(x - 3) = 0$$
$$x - 1 = 0 \text{ or } x + 4 = 0 \text{ or } x - 3 = 0$$
$$x = 1 \text{ or } x = -4 \text{ or } x = 3$$
The solution set is $\{-4, 1, 3\}$.

33. Try integers between -5 and 5 with synthetic division to find a solution. Try 2 first.
$$
\begin{array}{r|rrrr}
2 & 2 & -9 & 7 & 6 \\
  &   & 4 & -10 & -6 \\
\hline
  & 2 & -5 & -3 & 0 \\
\end{array}
$$
$$2x^3 - 9x^2 + 7x + 6 = 0$$
$$(x - 2)(2x^2 - 5x - 3) = 0$$
$$(x - 2)(2x + 1)(x - 3) = 0$$
$$x - 2 = 0 \text{ or } 2x + 1 = 0 \text{ or } x - 3 = 0$$
$$x = 2 \text{ or } x = -1/2 \text{ or } x = 3$$
The solution set is $\left\{-\frac{1}{2}, 2, 3\right\}$.

35. The only integral solution between -5 and 5 is -4:
$$
\begin{array}{r|rrrr}
-4 & 2 & -3 & -50 & -24 \\
   &   & -8 & 44 & 24 \\
\hline
   & 2 & -11 & -6 & 0 \\
\end{array}
$$
$$2x^3 - 3x^2 - 50x - 24 = 0$$
$$(x + 4)(2x^2 - 11x - 6) = 0$$
$$(x + 4)(2x + 1)(x - 6) = 0$$
$$x + 4 = 0 \text{ or } 2x + 1 = 0 \text{ or } x - 6 = 0$$
$$x = -4 \text{ or } x = -1/2 \text{ or } x = 6$$

The solution set is $\left\{-4, -\frac{1}{2}, 6\right\}$.

37. Check to see if 1 is a solution.
$$
\begin{array}{r|rrrr}
1 & 1 & 5 & 3 & -9 \\
  &   & 1 & 6 & 9 \\
\hline
  & 1 & 6 & 9 & 0 \\
\end{array}
$$
$$x^3 + 5x^2 + 3x - 9 = 0$$
$$(x - 1)(x^2 + 6x + 9) = 0$$
$$(x - 1)(x + 3)^2 = 0$$
$$x - 1 = 0 \text{ or } x + 3 = 0$$
$$x = 1 \text{ or } x = -3$$
The solution set is $\{-3, 1\}$.

39. Check to see if 2 is a solution.
$$
\begin{array}{r|rrrrr}
2 & 1 & -4 & 3 & 4 & -4 \\
  &   & 2 & -4 & -2 & 4 \\
\hline
  & 1 & -2 & -1 & 2 & 0 \\
\end{array}
$$
$$x^4 - 4x^3 + 3x^2 + 4x - 4 = 0$$
$$(x - 2)(x^3 - 2x^2 - x + 2) = 0$$
$$(x - 2)(x^2[x - 2] - 1[x - 2]) = 0$$
$$(x - 2)(x^2 - 1)(x - 2) = 0$$
$$(x - 2)(x - 1)(x + 1)(x - 2) = 0$$
$$x - 2 = 0 \text{ or } x - 1 = 0 \text{ or } x + 1 = 0$$
$$x = 2 \text{ or } x = 1 \text{ or } x = -1$$
The solution set is $\{-1, 1, 2\}$.

41. a) $f(x) = x + 2$
b) $f(x) = x^2 - 25$
c) $f(x) = (x - 1)(x + 3)(x - 4)$
d) Yes, the degree is the same as the number of zeros.

43. a) $\{-2, 1, 5\}$ **b)** $\left\{-\frac{3}{2}, \frac{3}{2}, \frac{5}{2}\right\}$

10.2 WARM-UPS

1. False, because in a polynomial function the exponents on x are whole numbers.
2. False, because to find $f(3)$ we use 3 in the synthetic division.
3. True, because of the remainder theorem.
4. True, because the remainder is equal to the value of $x^4 - 1$ when $x = 2$: $2^4 - 1 = 15$.
5. False, because $P(x) = 5$ has no zeros.
6. True, because $P(c) = c^3 - 7c + 3$.
7. False, because any rational root for this polynomial must be a factor of 14, and 3 is not a factor of 14.
8. True, because any rational root to this polynomial must be a factor of 9 over a factor of 5, and 1/3 does not have that form.

9. False, because $1^4 - 1^3 + 1^2 - 1 + 1 \neq 0$.
10. False, because if $x + 3$ is a factor, then -3 is a zero of the polynomial, but any whole number zero to this polynomial must be a factor of 2.

10.2 EXERCISES

1. A polynomial function is a function for which the value of the dependent variable is determined by a polynomial.

3. A zero of a function $f(x)$ is a number c such that $f(c) = 0$.

5. The rational root theorem says that if a rational number p/q (in lowest terms) is a zero of a polynomial, then p is a factor of the constant term and q is a factor of the leading coefficient.

7. $f(1) = 1^4 - 1 = 0$

$$
\begin{array}{r|rrrrr}
1 & 1 & 0 & 0 & 0 & -1 \\
  &   & 1 & 1 & 1 & 1 \\
\hline
  & 1 & 1 & 1 & 1 & 0
\end{array}
$$

Since the remainder is 0, $f(1) = 0$.

9. $f(-2) = (-2)^4 - 1 = 16 - 1 = 15$

$$
\begin{array}{r|rrrrr}
-2 & 1 & 0 & 0 & 0 & -1 \\
   &   & -2 & 4 & -8 & 16 \\
\hline
   & 1 & -2 & 4 & -8 & 15
\end{array}
$$

Since the remainder of the synthetic division is 15, $f(-2) = 15$.

11. $g(1) = 1^3 - 3(1)^2 + 5 = 1 - 3 + 5 = 3$

$$
\begin{array}{r|rrrr}
1 & 1 & -3 & 0 & 5 \\
  &   & 1 & -2 & -2 \\
\hline
  & 1 & -2 & -2 & 3
\end{array}
$$

Since the remainder of the synthetic division is 3, $g(1) = 3$.

13. $g(-2) = (-2)^3 - 3(-2)^2 + 5$
$= -8 - 12 + 5 = -15$

$$
\begin{array}{r|rrrr}
-2 & 1 & -3 & 0 & 5 \\
   &   & -2 & 10 & -20 \\
\hline
   & 1 & -5 & 10 & -15
\end{array}
$$

Since the remainder of the synthetic division is -15, $g(-2) = -15$.

15. $h\left(\frac{1}{2}\right) = 4\left(\frac{1}{2}\right)^4 - 3\left(\frac{1}{2}\right)^2 + 3\left(\frac{1}{2}\right) - 1$
$= \frac{1}{4} - \frac{3}{4} + \frac{3}{2} - \frac{2}{2} = 0$

$$
\begin{array}{r|rrrrr}
\frac{1}{2} & 4 & 0 & -3 & 3 & -1 \\
            &   & 2 & 1 & -1 & 1 \\
\hline
            & 4 & 2 & -2 & 2 & 0
\end{array}
$$

So $h\left(\frac{1}{2}\right) = 0$

17. $h\left(\frac{3}{2}\right) = 4\left(\frac{3}{2}\right)^4 - 3\left(\frac{3}{2}\right)^2 + 3\left(\frac{3}{2}\right) - 1$
$= \frac{81}{4} - \frac{27}{4} + \frac{9}{2} - \frac{2}{2} = 17$

$$
\begin{array}{r|rrrrr}
\frac{3}{2} & 4 & 0 & -3 & 3 & -1 \\
            &   & 6 & 9 & 9 & 18 \\
\hline
            & 4 & 6 & 6 & 12 & 17
\end{array}
$$

So $h\left(\frac{3}{2}\right) = 17$.

19.
$$
\begin{array}{r|rrrr}
3 & 3 & -6 & -4 & -15 \\
  &   & 9 & 9 & 15 \\
\hline
  & 3 & 3 & 5 & 0
\end{array}
$$

Since the remainder is 0, $f(3) = 0$ and 3 is a zero of the polynomial function.

21.
$$
\begin{array}{r|rrrr}
-2 & 1 & 2 & 3 & 2 \\
   &   & -2 & 0 & -6 \\
\hline
   & 1 & 0 & 3 & -4
\end{array}
$$

Since the remainder is -4, $P(-2) = -4$ and -2 is not a zero of the polynomial.

23.
$$
\begin{array}{r|rrrrr}
-1 & 1 & 3 & 5 & 5 & 2 \\
   &   & -1 & -2 & -3 & -2 \\
\hline
   & 1 & 2 & 3 & 2 & 0
\end{array}
$$

Since the remainder is 0, $g(-1) = 0$ and -1 is a zero of the polynomial.

25.
$$
\begin{array}{r|rrrr}
\frac{1}{2} & 1 & 3 & -4 & 1 \\
            &   & \frac{1}{2} & \frac{7}{4} & -\frac{9}{8} \\
\hline
            & 1 & \frac{7}{2} & -\frac{9}{4} & -\frac{1}{8}
\end{array}
$$

Since the remainder is $-1/8$, $h(1/2) = -1/8$ and $1/2$ is not a zero of the polynomial.

27. $5x - 9 = 0$
$5x = 9$
$x = \frac{9}{5}$

So 9/5 is the only zero of the polynomial function f.

29. $x^2 - 5x + 6 = 0$
$(x - 2)(x - 3) = 0$
$x - 2 = 0$ or $x - 3 = 0$
$x = 2$ or $x = 3$

Both 2 and 3 are zeros of the polynomial function.

31. $t^2 + 9 = 0$
$t^2 = -9$
$t = \pm\sqrt{-9} = \pm 3i$

Both $3i$ and $-3i$ are zeros of the polynomial function $P(t) = t^2 + 9$.

33. $x^3 - 3x^2 + 4x - 12 = 0$
$x^2(x - 3) + 4(x - 3) = 0$
$(x^2 + 4)(x - 3) = 0$
$x^2 + 4 = 0$ or $x - 3 = 0$
$x^2 = -4$ or $x = 3$
$x = \pm 2i$ or $x = 3$

The zeros of the polynomial function are $-2i$, $2i$, and 3.

35. $2s^3 - 5s^2 - 3s = 0$
$s(2s^2 - 5s - 3) = 0$
$s(2s + 1)(s - 3) = 0$
$s = 0$ or $2s + 1 = 0$ or $s - 3 = 0$
$s = 0$ or $s = -\frac{1}{2}$ or $s = 3$

The zeros of the polynomial function are $-\frac{1}{2}$, 0, and 3.

37. $-x^3 - x^2 + 7x + 7 = 0$
$-x^2(x + 1) + 7(x + 1) = 0$
$(-x^2 + 7)(x + 1) = 0$
$-x^2 + 7 = 0$ or $x + 1 = 0$
$x^2 = 7$ or $x = -1$
$x = \pm \sqrt{7}$ or $x = -1$

The zeros of the polynomial function are $\pm \sqrt{7}$ and -1.

39. $x^3 - 1 = 0$
$(x - 1)(x^2 + x + 1) = 0$
$x - 1 = 0$ or $x = \dfrac{-1 \pm \sqrt{1^2 - 4(1)(1)}}{2(1)}$
$x = 1$ or $x = \dfrac{-1 \pm i\sqrt{3}}{2}$

The zeros of the polynomial function are $\dfrac{-1 \pm i\sqrt{3}}{2}$ and 1.

41. Each possible rational zero is a factor of 2 divided by a factor of 2. The factors of 2 are 1 and 2. Taking each possible combination we get ± 1, ± 2, and $\pm \frac{1}{2}$ as the possible rational zeros.

43. The possible rational zeros come from a factor of 24 divided by a factor of 1. The factors of 24 are 1, 2, 3, 4, 6, 8, 12, and 24. So the possible rational zeros are ± 1, ± 2, ± 3, ± 4, ± 6, ± 8, ± 12, and ± 24.

45. The possible rational zeros come from a factor of 6 divided by a factor of 1. The factors of 6 are 1, 2, and 3. So the possible rational zeros are ± 1, ± 2, ± 3, and ± 6.

47. The possible rational zeros for $P(x)$ were found for Exercise 41: ± 1, ± 2, and $\pm \frac{1}{2}$. Start testing these possible zeros with synthetic division until we find an actual zero. To save space we will not show any failures.

$$
\begin{array}{r|rrrr}
1 & 2 & -5 & 1 & 2 \\
  &   & 2 & -3 & -2 \\
\hline
  & 2 & -3 & -2 & 0
\end{array}
$$

Since 1 is a zero, $x - 1$ is a factor and the quotient is the other factor.
$2x^3 - 5x^2 + x + 2 = 0$
$(x - 1)(2x^2 - 3x - 2) = 0$
$(x - 1)(2x + 1)(x - 2) = 0$
$x - 1 = 0$ or $2x + 1 = 0$ or $x - 2 = 0$
$x = 1$ or $x = -\frac{1}{2}$ or $x = 2$

The zeros for the polynomial function are $-\frac{1}{2}$, 1, and 2.

49. The possible rational zeros were listed for Exercise 43: ± 1, ± 2, ± 3, ± 4, ± 6, ± 8, ± 12, and ± 24. Use synthetic division to determine which of these possible zeros are actually zeros.

$$
\begin{array}{r|rrrr}
-2 & 1 & 9 & 26 & 24 \\
   &   & -2 & -14 & -24 \\
\hline
   & 1 & 7 & 12 & 0
\end{array}
$$

Since -2 is a zero of the polynomial function, $x + 2$ is a factor of the polynomial and the other factor is the quotient from the division.
$x^3 + 9x^2 + 26x + 24 = 0$
$(x + 2)(x^2 + 7x + 12) = 0$
$(x + 2)(x + 3)(x + 4) = 0$
$x + 2 = 0$ or $x + 3 = 0$ or $x + 4 = 0$
$x = -2$ or $x = -3$ or $x = -4$

The zeros of the polynomial function are -2, -3, and -4.

51. The possible rational zeros were listed for Exercise 45: ± 1, ± 2, ± 3, ± 6. Check each of these possibilities with synthetic division until we find an actual zero.

$$
\begin{array}{r|rrrr}
1 & 1 & 5 & 0 & -6 \\
  &   & 1 & 6 & 6 \\
\hline
  & 1 & 6 & 6 & 0
\end{array}
$$

Since 1 is a zero of the polynomial function, $x - 1$ is a factor of the polynomial. The other factor is the quotient from the synthetic division.
$x^3 + 5x^2 - 6 = 0$

$$(x - 1)(x^2 + 6x + 6) = 0$$
$$x - 1 = 0 \text{ or } x = \frac{-6 \pm \sqrt{6^2 - 4(1)(6)}}{2(1)}$$
$$x = 1 \text{ or } x = \frac{-6 \pm \sqrt{12}}{2} = -3 \pm \sqrt{3}$$
The zeros of the polynomial function are $-3 - \sqrt{3}$, $-3 + \sqrt{3}$, and 1.

53. The possible rational zeros for the function $P(x) = x^3 + 2x^2 - 16$ are ± 1, ± 2, ± 4, ± 8, and ± 16. Check these possibilities with synthetic division.

$$\begin{array}{r|rrr} 2 & 1 & 2 & 0 & -16 \\ & & 2 & 8 & 16 \\ \hline & 1 & 4 & 8 & 0 \end{array}$$

Since 2 is a zero, $x - 2$ is a factor of the polynomial. The other factor is the quotient from the synthetic division.
$$x^3 + 2x^2 - 16 = 0$$
$$(x - 2)(x^2 + 4x + 8) = 0$$
$$x - 2 = 0 \text{ or } x^2 + 4x + 8 = 0$$
$$x = 2 \text{ or } x = \frac{-4 \pm \sqrt{4^2 - 4(1)(8)}}{2(1)}$$
$$x = 2 \text{ or } x = \frac{-4 \pm \sqrt{-16}}{2} = -2 \pm 2i$$
The zeros of the function are $-2 \pm 2i$ and 2.

55. The possible rational zeros of the function $P(x) = 24x^3 - 26x^2 + 9x - 1$ are ± 1, $\pm \frac{1}{2}$, $\pm \frac{1}{3}$, $\pm \frac{1}{4}$, $\pm \frac{1}{6}$, $\pm \frac{1}{8}$, $\pm \frac{1}{12}$, and $\pm \frac{1}{24}$. Use synthetic division to find an actual zero.

$$\begin{array}{r|rrrr} \frac{1}{3} & 24 & -26 & 9 & -1 \\ & & 8 & -6 & 1 \\ \hline & 24 & -18 & 3 & 0 \end{array}$$

Since $1/3$ is a zero of the function, $x - 1/3$ is a factor of the polynomial. The other factor is the quotient from the synthetic division.
$$\left(x - \frac{1}{3}\right)(24x^2 - 18x + 3) = 0$$
$$(3x - 1)(8x^2 - 6x + 1) = 0$$
$$(3x - 1)(2x - 1)(4x - 1) = 0$$
$$3x - 1 = 0 \text{ or } 2x - 1 = 0 \text{ or } 4x - 1 = 0$$
$$x = \frac{1}{3} \text{ or } \quad x = \frac{1}{2} \text{ or } \quad x = \frac{1}{4}$$
The zeros of the function are $1/2$, $1/3$, and $1/4$.

57. The profit when 10 are sold is $P(10)$.
$$P(10) = (10)^3 - 40(10)^2 + 400(10) = 1000$$
The profit when 10 are sold is \$1000. To find how many must be sold to get a profit of 0

dollars we find the zeros to the polynomial function $P(x) = x^3 - 40x^2 + 400x$.
$$x^3 - 40x^2 + 400x = 0$$
$$x(x^2 - 40x + 400) = 0$$
$$x(x - 20)^2 = 0$$
$$x = 0 \text{ or } x - 20 = 0$$
$$x = 0 \text{ or } \quad\quad x = 20$$
So there is no profit when 0 or 20 are sold.

59. -0.2, 0.1, 0.3

61. 10, 20, 30

10.3 WARM-UPS

1. False, because $x^2 - 9 = (x - 3)(x + 3)$ and the multiplicity of 3 is 1.

2. True, the number of solutions counting multiplicity is equal to the degree of the polynomial.

3. True, because the factor $3x - 2$ occurs 4 times in the polynomial.

4. False, because $(x - 2)^3(x^2 + x - 6)$
$= (x - 2)^3(x - 2)(x + 3)$ and 2 is a root with multiplicity 4.

5. False, because the complex conjugate of $2 - 3i$ is $2 + 3i$.

6. True, because $5 + 4i$ and $3 - 6i$ must also be solutions to $P(x) = 0$.

7. False, because if $-1 - i\sqrt{2}$ and $1 - i\sqrt{2}$ are solutions, then their conjugates $-1 + i\sqrt{2}$ and $1 + i\sqrt{2}$ must also be solutions and the degree of the polynomial would be at least 4.

8. False, because
$$P(-x) = (-x)^3 - 6(-x)^2 + 3(-x) - 2$$
$$= -x^3 - 6x^2 - 3x - 2.$$

9. True, because there are no sign changes in the polynomial.

10. True, because it has one positive solution and no negative solutions by Descartes' rule of signs, and a third degree equation must have 3 solutions.

10.3 EXERCISES

1. Multiplicity of a root c is the number of times that $x - c$ occurs as a factor.

3. The conjugate pairs theorem says that for a polynomial equation with real coefficients if

$a + bi$ $(b \neq 0)$ is a root , then $a - bi$ is also a root.

5. If there are no roots greater than c, then c is an upper bound for the roots.

7.
$$x^5 - 4x^3 = 0$$
$$x^3(x^2 - 4) = 0$$
$$x^3(x - 2)(x + 2) = 0$$
$$x^3 = 0 \text{ or } x - 2 = 0 \text{ or } x + 2 = 0$$
$$x = 0 \text{ or } \quad x = 2 \text{ or } \quad x = -2$$
The degree of the equation is 5. The roots are $-2, 2,$ and 0. The root 0 has multiplicity 3.

9.
$$x^4 + 2x^3 + x^2 = 0$$
$$x^2(x^2 + 2x + 1) = 0$$
$$x^2(x + 1)^2 = 0$$
$$x^2 = 0 \text{ or } \quad (x + 1)^2 = 0$$
$$x = 0 \text{ or } \quad x = -1$$
The roots to this fourth degree equation are 0 with multiplicity 2 and -1 with multiplicity 2.

11.
$$x^4 - 6x^2 + 9 = 0$$
$$(x^2 - 3)^2 = 0$$
$$x^2 - 3 = 0 \text{ or } x^2 - 3 = 0$$
$$x^2 = 3 \text{ or } \quad x^2 = 3$$
$$x = \pm\sqrt{3} \text{ or } x = \pm\sqrt{3}$$
The roots to this fourth degree equation are $-\sqrt{3}$ with multiplicity 2, and $\sqrt{3}$ with multiplicity 2.

13.
$$(x - 1)^2(x + 2)^2 = 0$$
$$x - 1 = 0 \text{ or } x + 2 = 0$$
$$x = 1 \text{ or } \quad x = -2$$
The roots to this fourth degree equation are 1 with multiplicity 2, and -2 with multiplicity 2.

15.
$$x^4 - 2x^2 + 1 = 0$$
$$(x^2 - 1)^2 = 0$$
$$x^2 - 1 = 0$$
$$x^2 = 1$$
$$x = \pm 1$$
The roots to this fourth degree equation are -1 with multiplicity 2, and 1 with multiplicity 2.

17. If 3 is a root, then $x - 3$ is a factor. If $2 - i$ is a root, then $2 + i$ must be a root also. If $2 - i$ is a root, the $x - (2 - i)$ is a factor. If $2 + i$ is a root, then $x - (2 + i)$ is a factor.
$$(x - 3)(x - (2 - i))(x - (2 + i)) = 0$$
$$(x - 3)(x^2 - 4x + 4 - i^2) = 0$$
$$(x - 3)(x^2 - 4x + 5) = 0$$
$$(x - 3)x^2 - (x - 3)(4x) + (x - 3)(5) = 0$$
$$x^3 - 3x^2 - 4x^2 + 12x + 5x - 15 = 0$$
$$x^3 - 7x^2 + 17x - 15 = 0$$

19. If -2 is a root, then $x + 2$ is a factor. If i is a root, then $-i$ is also a root.
$$(x + 2)(x - i)(x + i) = 0$$
$$(x + 2)(x^2 - i^2) = 0$$
$$(x + 2)(x^2 + 1) = 0$$
$$x^3 + 2x^2 + x + 2 = 0$$

21. If 0 is a root, then $x - 0$ or x is a factor. If $i\sqrt{2}$ is a root, then $-i\sqrt{2}$ is also a root.
$$x(x - i\sqrt{2})(x + i\sqrt{2}) = 0$$
$$x(x^2 - 2i^2) = 0$$
$$x(x^2 + 2) = 0$$
$$x^3 + 2x = 0$$

23. If i is a root, then $-i$ is also a root. If $1 - i$ is a root, then $1 + i$ is also a root.
$$(x - i)(x + i)(x - (1 - i))(x - (1 + i)) = 0$$
$$(x^2 - i^2)(x^2 - 2x + 1 - i^2) = 0$$
$$(x^2 + 1)(x^2 - 2x + 2) = 0$$
$$x^2(x^2 - 2x + 2) + 1(x^2 - 2x + 2) = 0$$
$$x^4 - 2x^3 + 2x^2 + x^2 - 2x + 2 = 0$$
$$x^4 - 2x^3 + 3x^2 - 2x + 2 = 0$$

25. If 1 is a root, then $x - 1$ is a factor. If 2 is a root, then $x - 2$ is a factor.
$$(x - 1)(x - 2) = 0$$
$$x^2 - 3x + 2 = 0$$

27. If -1 is a root, then $x + 1$ is a factor. If 2 is a root, then $x - 2$ is a factor. If 3 is a root, then $x - 3$ is a factor.
$$(x + 1)(x - 2)(x - 3) = 0$$
$$(x + 1)(x^2 - 5x + 6) = 0$$
$$x(x^2 - 5x + 6) + 1(x^2 - 5x + 6) = 0$$
$$x^3 - 5x^2 + 6x + x^2 - 5x + 6 = 0$$
$$x^3 - 4x^2 + x + 6 = 0$$

29. Because there are no changes of sign in $P(x) = x^3 + 3x^2 + 5x + 7$, there are no positive roots to the equation. Because there are 3 changes of sign in $P(-x) = -x^3 + 3x^2 - 5x + 7$, there are either 3 or 1 negative roots to the equation. There are two possibilities for the 3 roots to the equation: 3 negative roots, or 1 negative root and 2 imaginary roots.

31. Because there is 1 change of sign in $P(x) = -2x^3 - x^2 + 3x + 2$, there is 1 positive root to the equation. Because there are 2 changes of sign in $P(-x) = 2x^3 - x^2 - 3x + 2$, there are either 2 or 0 negative roots to the equation. There are two possibilities for the 3 roots to the equation:

1 positive root and 2 negative roots, or 1 positive root and 2 imaginary roots.

33. Because there are 4 changes of sign in $P(x) = x^4 - x^3 + x^2 - x + 1 = 0$, there are either 4, 2, or 0 positive roots to the equation. Because there are no changes of sign in $P(-x) = x^4 + x^3 + x^2 + x + 1$, there are no negative roots to the equation. So there are 3 possibilities for the 4 roots to the equation: 4 positive roots, or 2 positive and 2 imaginary roots, or 4 imaginary roots.

35. Because there are no changes of sign in $P(x) = x^4 + x^2 + 1$, there are no positive roots. Because there are no changes of sign in $P(-x) = x^4 + x^2 + 1$, there are no negative roots. So there is only one possibility for the 4 roots to the equation: 4 imaginary roots.

37. Because there is 1 change of sign in $P(x) = x^3 + x - 1$, there is 1 positive root to the equation. Because there are no changes of sign in $P(-x) = -x^3 - x - 1$, there are no negative roots to the equation. So there is only one possibility for the 3 roots to the equation: 1 positive root and 2 imaginary roots.

39. Because there are no changes of sign in $P(x) = x^5 + x^3 + 3x$, there are no positive roots. Because there are no changes of sign in $P(-x) = -x^5 - x^3 - 3x$, there are no negative roots. We can see that 0 is a root because x is a factor of the polynomial. So there is only one possibility for the 5 roots to the equation: 0 is a root and there are 4 imaginary roots.

41. Try synthetic division with positive integers:

```
1 |  1  0  -5   0   7
  |     1   1  -4  -4
  +---------------------
     1  1  -4  -4   3

2 |  1  0  -5   0   7
  |     2   4  -2  -4
  +---------------------
     1  2  -1  -2   3

3 |  1  0  -5   0   7
  |     3   9  12  36
  +---------------------
     1  3   4  12  43
```

Since all terms in the bottom row are nonnegative, no number larger than 3 is a root to the equation.

```
-1 |  1  0  -5   0   7
   |    -1   1   4  -4
   +---------------------
      1 -1  -4   4   3

-2 |  1  0  -5   0   7
   |    -2   4   2  -4
   +---------------------
      1 -2  -1   2   3

-3 |  1  0  -5   0   7
   |    -3   9 -12  36
   +---------------------
      1 -3   4 -12  43
```

Since the terms in the last row alternate in sign, no number smaller than -3 can be a root to the equation. So all real roots to the equation are between -3 and 3.

43. Try synthetic division with positive integers:

```
1 |  2  -5   9  -18
  |      2  -3    6
  +------------------
     2  -3   6  -12

2 |  2  -5   9  -18
  |      4  -2   14
  +------------------
     2  -1   7   -4

3 |  2  -5   9  -18
  |      6   3   36
  +------------------
     2   1  12   18
```

Since all terms in the bottom row are nonnegative, no number larger than 3 is a root to the equation.

```
-1 |  2  -5   9  -18
   |     -2   7  -16
   +------------------
      2  -7  16  -34
```

Since the terms in the last row alternate in sign, no number smaller than -1 can be a root to the equation. So all real roots to the equation are between -1 and 3.

45. Try synthetic division with positive integers:

```
3 |  1   1  -13
  |      3   12
  +-------------
     1   4   -1
```

4 | 1 1 −13
| 4 20
1 5 7

Since all terms in the bottom row are nonnegative, no number larger than 4 is a root to the equation.

−4 | 1 1 −13
| −4 12
1 −3 −1

−5 | 1 1 −13
| −5 20
1 −4 7

Since the terms in the last row alternate in sign, no number smaller than −5 can be a root to the equation. So all real roots to the equation are between −5 and 4.

47. Try synthetic division with positive integers:

6 | 2 −13 25 −14
| 12 −6 114
2 −1 19 100

7 | 2 −13 25 −14
| 14 7 224
2 1 32 210

Since all terms in the bottom row are nonnegative, no number larger than 7 is a root to the equation.

−1 | 2 −13 25 −14
| −2 15 −40
2 −15 40 −54

Since the terms in the last row alternate in sign, no number smaller than −1 can be a root to the equation. So all real roots to the equation are between −1 and 7.

49. Possible rational roots are ± 1, ± 2, ± 5, and ± 10. By Descartes' rule there are no positive roots.

−2 | 1 0 1 10
| −2 4 −10
1 −2 5 0

Since −2 is a root, $x + 2$ is a factor of the polynomial $x^3 + x + 10$.
$$x^3 + x + 10 = 0$$
$$(x + 2)(x^2 - 2x + 5) = 0$$
$$x + 2 = 0 \quad \text{or} \quad x^2 - 2x + 5 = 0$$

$$x = -2 \quad \text{or} \quad x = \frac{2 \pm \sqrt{4 - 4(1)(5)}}{2(1)}$$
$$x = -2 \quad \text{or} \quad x = 1 \pm 2i$$
The solutions are -2, $1 + 2i$, and $1 - 2i$.

51. The possible rational roots are ± 1, ± 2, ± 4, and $\pm \frac{1}{2}$. By Descartes' rule there are either 0 or 2 positive roots, and 1 negative root.

$\frac{1}{2}$ | 2 −5 −6 4
| 1 −2 −4
2 −4 −8 0

Since $1/2$ is a root, $x - 1/2$ is a factor.
$$2x^3 - 5x^2 - 6x + 4 = 0$$
$$(x - \tfrac{1}{2})(2x^2 - 4x - 8) = 0$$
$$(x - \tfrac{1}{2})(2)(x^2 - 2x - 4) = 0$$
$$(2x - 1)(x^2 - 2x - 4) = 0$$
$$2x - 1 = 0 \quad \text{or} \quad x^2 - 2x - 4 = 0$$
$$2x = 1 \quad \text{or} \quad x = \frac{2 \pm \sqrt{4 - 4(1)(-4)}}{2}$$
$$x = \tfrac{1}{2} \quad \text{or} \quad x = 1 \pm \sqrt{5}$$
The solutions are $\frac{1}{2}$, $1 + \sqrt{5}$, and $1 - \sqrt{5}$.

53. The possible rational roots are ± 1, $\pm \frac{1}{2}$, and $\pm \frac{1}{4}$. Since there are 2 variations in sign for $P(x)$, there are 0 or 2 positive roots. Since there is 1 variation in sign for $P(-x)$, there is 1 negative root. By the Theorem on Bounds, the real roots are between −1 and 2.

$-\frac{1}{2}$ | 4 −6 −2 1
| −2 4 −1
4 −8 2 0

Since $-1/2$ is a root, $x + 1/2$ is a factor.
$$4x^3 - 6x^2 - 2x + 1 = 0$$
$$(x + \tfrac{1}{2})(4x^2 - 8x + 2) = 0$$
$$(x + \tfrac{1}{2})(2)(2x^2 - 4x + 1) = 0$$
$$(2x + 1)(2x^2 - 4x + 1) = 0$$
$$2x + 1 = 0 \quad \text{or} \quad 2x^2 - 4x + 1 = 0$$
$$2x = -1 \quad \text{or} \quad x = \frac{4 \pm \sqrt{16 - 4(2)(1)}}{2(2)}$$
$$x = -\tfrac{1}{2} \quad \text{or} \quad x = \frac{2 \pm \sqrt{2}}{2}$$
The solutions are $-\frac{1}{2}$, $\frac{2 + \sqrt{2}}{2}$, and $\frac{2 - \sqrt{2}}{2}$.

55. The possible rational roots are ± 1, ± 2, ± 3, and ± 6. Since there are 3 variations of sign for $P(x)$, there are 1 or 3 positive roots. Since there is 1 variation of sign for $P(-x)$, there is 1 negative root. By the Theorem on Bounds, the real roots are between -1 and 5.

$$\begin{array}{r|rrrrr} -1 & 1 & -5 & 5 & 5 & -6 \\ & & -1 & 6 & -11 & 6 \\ \hline & 1 & -6 & 11 & -6 & 0 \end{array}$$

$$\begin{array}{r|rrrr} 1 & 1 & -6 & 11 & -6 \\ & & 1 & -5 & 6 \\ \hline & 1 & -5 & 6 & 0 \end{array}$$

Since -1 and 1 are roots, $x+1$ and $x-1$ are factors.
$$x^4 - 5x^3 + 5x^2 + 5x - 6 = 0$$
$$(x+1)(x-1)(x^2 - 5x + 6) = 0$$
$$(x+1)(x-1)(x-2)(x-3) = 0$$
$x = -1$ or $x = 1$ or $x = 2$ or $x = 3$
The solutions are -1, 1, 2, and 3.

57. The rational roots are ± 1, ± 2, ± 3, and ± 6. Since there are 4 variations of sign for $P(x)$, there are 0, 2, or 4 positive roots. Since there are 0 variations of sign for $P(-x)$, there are 0 negative roots.

$$\begin{array}{r|rrrrr} 1 & 1 & -7 & 17 & -17 & 6 \\ & & 1 & -6 & 11 & -6 \\ \hline & 1 & -6 & 11 & -6 & 0 \end{array}$$

$$\begin{array}{r|rrrr} 1 & 1 & -6 & 11 & -6 \\ & & 1 & -5 & 6 \\ \hline & 1 & -5 & 6 & 0 \end{array}$$

Since 1 is a root with multiplicity 2, $x - 1$ is a double factor.
$$x^4 - 7x^3 + 17x^2 - 17x + 6 = 0$$
$$(x-1)(x-1)(x^2 - 5x + 6) = 0$$
$$(x-1)(x-1)(x-2)(x-3) = 0$$
$x = 1$ or $x = 1$ or $x = 2$ or $x = 3$
The roots are 1 (with multiplicity 2), 2, and 3.

59. First factor out the common factor x from each term.
$$x^6 - x^5 + 2x^4 - 2x^3 - 15x^2 + 15x = 0$$
$$x(x^5 - x^4 + 2x^3 - 2x^2 - 15x + 15) = 0$$
The possible rational roots to $x^5 - x^4 + 2x^3 - 2x^2 - 15x + 15 = 0$ are ± 1, ± 3, ± 5, and ± 15. Since there are 4 variations in sign for $P(x)$, there are 0, 2, or 4 positive roots. Since there is 1 variation in sign for $P(-x)$, there is 1 negative root. By the Theorem on Bounds, the real roots are between -2 and 3.

$$\begin{array}{r|rrrrrr} 1 & 1 & -1 & 2 & -2 & -15 & 15 \\ & & 1 & 0 & 2 & 0 & -15 \\ \hline & 1 & 0 & 2 & 0 & -15 & 0 \end{array}$$

Since 1 is a root, $x - 1$ is a factor.
$$x(x^5 - x^4 + 2x^3 - 2x^2 - 15x + 15) = 0$$
$$x(x-1)(x^4 + 2x^2 - 15) = 0$$
$$x(x-1)(x^2 + 5)(x^2 - 3) = 0$$
$x = 0$ or $x - 1 = 0$ or $x^2 + 5 = 0$ or $x^2 - 3 = 0$
$x = 0$ or $x = 1$ or $x^2 = -5$ or $x^2 = 3$
$x = 0$ or $x = 1$ or $x = \pm i\sqrt{5}$ or $x = \pm\sqrt{3}$
The roots are 0, 1, $\pm i\sqrt{5}$, and $\pm\sqrt{3}$.

61. The volume of a cone with height 2 and radius x is $\frac{2}{3}\pi x^2$. The volume of a cylinder with radius x and height $x + 20$ is $\pi x^2(x + 20)$. The tank consists of two cones and one cylinder with a total volume of 984π. So we can write the following equation.
$$2\left(\frac{2}{3}\pi x^2\right) + \pi x^2(x + 20) = 984\pi$$
$$\frac{4}{3}\pi x^2 + \pi x^3 + 20\pi x^2 = 984\pi$$
$$4x^2 + 3x^3 + 60x^2 = 2952$$
$$3x^3 + 64x^2 - 2952 = 0$$
There is only one positive solution to this equation, because there is one change of sign.

$$\begin{array}{r|rrrr} 6 & 3 & 64 & 0 & -2952 \\ & & 18 & 492 & 2952 \\ \hline & 3 & 82 & 492 & 0 \end{array}$$

The radius of the cylinder is 6 feet.

63. If the thickness of the insulating material is x inches, then the dimensions of the box are $4 + 2x$ by $5 + 2x$ by $3 + 2x$. Since the volume is 120 cubic inches, we can write the following equation.
$$(4 + 2x)(5 + 2x)(3 + 2x) = 120$$
$$(20 + 18x + 4x^2)(3 + 2x) = 120$$
$$60 + 94x + 48x^2 + 8x^3 = 120$$
$$8x^3 + 48x^2 + 94x - 60 = 0$$
$$4x^3 + 24x^2 + 47x - 30 = 0$$
Since there is only one change of sign, the equation has only one positive solution. Try possible rational solutions.

$$\begin{array}{r|rrrr} \frac{1}{2} & 4 & 24 & 47 & -30 \\ & & 2 & 13 & 30 \\ \hline & 4 & 26 & 60 & 0 \end{array}$$

The thickness of the insulation is $\frac{1}{2}$ in.

65. 3.3315, 0.9492, −0.9492, −3.3315

67. −2.4495, −1.4142, −1, 1, 1.4142, 2.4495

10.4 WARM-UPS

1. False, because $f(-x) = -x^3 + x$.

2. False, because $y = 2x - 1$ has neither symmetry. If $f(x) = 2x - 1$, then $f(-x) = -2x - 1$.

3. False, because $P(x) = P(-x)$ for graphs that are symmetric about the y-axis.

4. False, because $f(-x) = -3x$ for the function $f(x) = 3x$.

5. True, because $f(-x)$
$= 3(-x)^4 - 5(-x)^3 + 2(-x)^2 - 6(-x) + 7$
$= 3x^4 + 5x^3 + 2x^2 + 6x + 7$.

6. True because $x^2 - 4x + 4 = 0$ is equivalent to $(x - 2)^2 = 0$, which has only one solution.

7. False, because the x-intercepts are $(2, 0)$ and $(-3, 0)$.

8. False, because $P(0) = -9$. The y-intercept is $(0, -9)$.

9. True, because each x-intercept comes from a factor with an even exponent.

10. False, because $x^3 - 8 = 0$ has only one real solution. The only x-intercept is $(2, 0)$.

10.4 EXERCISES

1. The graph of $y = f(x)$ is symmetric about the y-axis if $f(x) = f(-x)$ for all x in the domain of the function.

3. An x-intercept is a point at which a graph intersects the x-axis.

5. Since $f(-x) = (-x)^4 = x^4$, the graph is symmetric about the y-axis.

7. Since $f(-x) = (-x)^3 - 5(-x) + 1$
$= -x^3 + 5x + 1$, the graph has neither symmetry.

9. Since $f(-x) = 6(-x)^6 - 3(-x)^2 - (-x)$
$= 6x^6 - 3x^2 + x$,
the graph has neither symmetry.

11. Since $f(-x) = (-x - 3)^2 = x^2 + 6x + 9$ and $f(x) = x^2 - 6x + 9$, the graph has neither symmetry.

13. Since $f(-x) = ((-x)^2 - 5)^3 = (x^2 - 5)^3$, the graph is symmetric about the y-axis.

15. Since $f(-x) = -x$ and $f(x) = x$, the graph is symmetric about the origin.

17. To find x-intercepts we solve $f(x) = 0$.
$$(x - 5)^2 = 0$$
$$x - 5 = 0$$
$$x = 5$$
The x-intercept is $(5, 0)$. Since $(x - 5)$ occurs with an even exponent in $f(x) = (x - 5)^2$, the graph does not cross the x-axis at the x-intercept.

19. To find x-intercepts we solve $f(x) = 0$.
$$x^2 - 3x - 4 = 0$$
$$(x - 4)(x + 1) = 0$$
$$x - 4 = 0 \text{ or } x + 1 = 0$$
$$x = 4 \text{ or } \quad x = -1$$
The x-intercepts are $(4, 0)$ and $(-1, 0)$. Since $(x - 4)$ and $(x + 1)$ each occur with an odd exponent in $f(x) = (x - 4)(x + 1)$, the graph crosses the x-axis at each x-intercept.

21. To find x-intercepts we solve $f(x) = 0$.
$$x^4 = 0$$
$$x = 0$$
The x-intercept is $(0, 0)$. Since x occurs with an even exponent in $f(x) = x^4$, the graph does not cross the x-axis at the x-intercept.

23. To find x-intercepts we solve $f(x) = 0$.
$$(x - 3)^2(x + 5) = 0$$
$$x - 3 = 0 \text{ or } x + 5 = 0$$
$$x = 3 \text{ or } \quad x = -5$$
The x-intercepts are $(3, 0)$ and $(-5, 0)$. Since $(x - 3)$ occurs with an even exponent and $(x + 5)$ occurs with an odd exponent in $f(x) = (x - 3)^2(x + 5)$, the graph does not cross the x-axis at $(3, 0)$ but does cross at $(-5, 0)$.

25. To find x-intercepts we solve $f(x) = 0$.
$$x^3 - x^2 = 0$$
$$x^2(x - 1) = 0$$
$$x = 0 \text{ or } x = 1$$

The x-intercepts are $(0, 0)$ and $(1, 0)$. Since x occurs with an even exponent and $(x - 1)$ occurs with an odd exponent in $f(x) = x^2(x - 1)$, the graph does not cross the x-axis at $(0, 0)$ but crosses the x-axis at $(1, 0)$.

27. To find x-intercepts we solve $f(x) = 0$.
$$2x^3 + 3x^2 - 1 = 0$$

$$\begin{array}{r|rrrr} -1 & 2 & 3 & 0 & -1 \\ & & -2 & -1 & 1 \\ \hline & 2 & 1 & -1 & 0 \end{array}$$

$$(x+1)(2x^2 + x - 1) = 0$$
$$(x+1)(2x - 1)(x + 1) = 0$$
$$x = -1 \text{ or } x = \tfrac{1}{2} \text{ or } x = -1$$

The x-intercepts are $(-1, 0)$ and $(1/2, 0)$. Since $(x + 1)$ occurs with an even exponent and $(2x - 1)$ occurs with an odd exponent in $f(x) = (x + 1)^2(2x - 1)$, the graph does not cross the x-axis at $(-1, 0)$ but does cross at $(1/2, 0)$.

29. The graph of $f(x) = -2x + 3$ is a straight line through $(0, 3)$ and $(1, 1)$. The graph is (d).

31. The graph of $f(x) = -2x^3 + 3$ crosses the y-axis at $(0, 3)$ and contains the points $(1, 1)$ and $(-1, 5)$. The graph is (a).

33. The graph of $f(x) = -x^4 + 3$ contains the points $(0, 3)$, $(1, 2)$, and $(-1, 2)$. The graph is (c).

35. The function $f(x) = x^3 + 3x^2 - x - 3$ can be written as
$f(x) = (x - 1)(x + 1)(x + 3)$. The graph crosses the x-axis at $(1, 0)$, $(-1, 0)$, and $(-3, 0)$. The graph is (e).

37. The graph of $f(x) = 2x - 6$ is a straight line with a y-intercept $(0, -6)$ and a slope of 2.

39. The graph of $f(x) = -x^2$ is a parabola opening downward. The graph includes the points $(0, 0)$, $(1, -1)$, $(-1, -1)$, $(2, -4)$, and $(-2, -4)$.

41. The function $f(x) = x^3 - 2x^2$ can be written as $f(x) = x^2(x - 2)$. The x-intercepts are $(0, 0)$ and $(2, 0)$. The graph crosses the x-axis at $(2, 0)$ but does not cross at $(0, 0)$. The graph also goes through the points $(1, -1)$, $(3, 9)$, and $(-1, -3)$.

43. The graph of $f(x) = (x - 1)^2(x + 1)^2$ does not cross the x-axis at its intercepts $(1, 0)$ and $(-1, 0)$. The graph also includes the points $(0, 1)$, $(2, 9)$, $(-2, 9)$.

45. Use synthetic division to find that $f(3) = 0$. So $f(x) = (x - 3)(x^2 - 2x + 1)$ $= (x - 3)(x - 1)^2$. The graph crosses the x-axis at the x-intercept $(3, 0)$, but does not cross at the intercept $(1, 0)$. The graph also includes the points $(0, -3)$, $(2, -1)$, and $(4, 9)$.

47. Write $f(x) = x^4 - 4x^3 + 4x^2$ as $f(x) = x^2(x^2 - 4x + 4) = x^2(x - 2)^2$. The graph does not cross the x-axis at its x-intercepts $(0, 0)$ and $(2, 0)$. The graph also includes the points $(1, 1)$, $(3, 9)$, and $(-3, 225)$.

49. Get an image of the function on a graphing calculator, then graph it on paper. Adjust the window to show the x-intercepts.

51. Get an image of the function on a graphing calculator, then graph it on paper. Adjust the window to show the x-intercepts.

53. Get an image of the function on a graphing calculator, then graph it on paper. Adjust the window to show the x-intercepts.

10.5 WARM-UPS

1. False, because the domain is $\{x \mid x \neq 9\}$.
2. False, because the domain is $\{x \mid x \neq -2\}$ or $(-\infty, -2) \cup (-2, \infty)$.

3. False, because the equation $x^2 + 1 = 0$ has no real solutions. The domain is the set of all real numbers, $(-\infty, \infty)$.
4. False, because the vertical asymptotes are $x = 2$ and $x = -2$.
5. True, because the degree of the denominator is larger than the degree of the numerator.
6. False, because $\dfrac{3x - 5}{x - 2} = 3 + \dfrac{1}{x - 2}$ and $y = 3$ is the horizontal asymptote.
7. True, because when x is very large the value of $2x - 5 + \dfrac{1}{x}$ is approximately $2x - 5$.
8. False, because the graph of $f(x) = 2x - 5 + x^2$ is a parabola and it has no asymptotes.
9. True, because
$$f(-x) = \frac{(-x)^2}{(-x)^2 - 9} = \frac{x^2}{x^2 - 9}.$$
10. True, because
$$f(-x) = \frac{3(-x)}{(-x)^2 - 25} = -\frac{3x}{x^2 - 25} = -f(x).$$

10.5 EXERCISES

1. A rational function is of the form $f(x) = P(x)/Q(x)$ where $P(x)$ and $Q(x)$ are polynomials with no common factor and $Q(x) \neq 0$.
3. A vertical asymptote is a vertical line that is approached by the graph of a rational function.
5. An oblique asymptote is a nonhorizontal nonvertical line that is approached by the graph of a rational function.
7. First find any values of x that would make the denominator have a value of 0.
$$x - 1 = 0$$
$$x = 1$$
The domain is $\{x \mid x \neq 1\}$ or $(-\infty, 1) \cup (1, \infty)$.
9. First find any values of x that would make the denominator have a value of 0.
$$x = 0$$
The domain is $\{x \mid x \neq 0\}$ or $(-\infty, 0) \cup (0, \infty)$.
11. First find any values of x that would make the denominator have a value of 0.
$$x^2 - 16 = 0$$
$$x^2 = 16$$

$x = \pm 4$

The domain is $\{x \mid x \neq -4 \text{ and } x \neq 4\}$ or $(-\infty, -4) \cup (-4, 4) \cup (4, \infty)$.

13. If $x + 4 = 0$, then $x = -4$. So the vertical line $x = -4$ is an asymptote for the graph of f. The x-axis is a horizontal asymptote.

15. If $x^2 - 16 = 0$, then $x = \pm 4$. So the vertical lines $x = 4$ and $x = -4$ are asymptotes for the graph of f. The x-axis is a horizontal asymptote.

17. If $x - 7 = 0$, then $x = 7$. So the vertical line $x = 7$ is an asymptote for the graph of f. The ratio of the leading coefficients is 5. So $y = 5$ is the horizontal asymptote

19. The vertical line $x = 3$ is the vertical asymptote. Since $\dfrac{2x^2}{x-3} = 2x + 6 + \dfrac{18}{x-3}$, the line $y = 2x + 6$ is an oblique asymptote for the graph.

21. The graph of $f(x) = -\dfrac{2}{x}$ has the x-axis and the y-axis as asymptotes. The graph is (c).

23. The function $f(x) = \dfrac{x}{x-2}$ can be written as $f(x) = 1 + \dfrac{2}{x-2}$. The asymptotes are the lines $y = 1$ and $x = 2$. The graph is (b).

25. Write the function as $f(x) = \dfrac{1}{x(x-2)}$. The lines $x = 0$ and $x = 2$ are vertical asymptotes. The graph is (g).

27. Write the function as $f(x) = -\dfrac{1}{2}x - 2$. The graph is a straight line through $(0, -2)$ with slope $-1/2$. The graph is (f).

29. First find any values of x that would make the denominator have a value of 0.
$$x + 4 = 0$$
$$x = -4$$
The asymptotes are $x = -4$ and the x-axis. Draw the asymptotes that we found as dashed lines. The graph goes through the points $(-3, 2)$, $(-2, 1)$, $(-5, -2)$, and $(-6, -1)$.

31. First find any values of x that would make the denominator have a value of 0.
$$x^2 - 9 = 0$$
$$x^2 = 9$$
$$x = \pm 3$$
The asymptotes are $x = \pm 3$ and the x-axis. Draw the asymptotes that we found as dashed lines. The graph goes through the points $(0, 0)$, $(-2, \frac{2}{5})$, $(2, -\frac{2}{5})$, $(4, \frac{4}{7})$, and $(-4, -\frac{4}{7})$.

33. First find any values of x that would make the denominator have a value of 0.
$$x + 3 = 0$$
$$x = -3$$
The asymptotes are $x = -3$ and $y = 2$, from the ratio of the leading coefficients. Draw the asymptotes that we found as dashed lines. The graph goes through the points $(0, -\frac{1}{3})$, $(\frac{1}{2}, 0)$, $(-2, -5)$, and $(-4, 9)$.

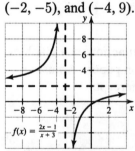

35. First find any values of x that would make the denominator have a value of 0.
$$x = 0$$
The asymptotes are $x = 0$, and $y = x - 3$, obtained from dividing the denominator into the numerator. Draw the asymptotes that we found as dashed lines. The graph goes through the points $(1, -1)$, $(-2, -\frac{11}{2})$, $(-1, -5)$, and $(3, \frac{1}{3})$.

37. First find any values of x that would make the denominator have a value of 0.

$$x - 1 = 0$$
$$x = 1$$

The asymptotes are $x = 1$ and $y = 3x + 1$, obtained from dividing the denominator into the numerator. Draw the asymptotes that we found as dashed lines. The graph goes through the points $(0, 0)$, $(2, 8)$, $(\frac{2}{3}, 0)$, and $(-1, -2.5)$.

39. Since $x^2 = 0$ is equivalent to $x = 0$, the y-axis is a vertical asymptote. The x-axis is a horizontal asymptote. There are no x-intercepts or y-intercepts. The graph goes through $(1, 1)$, and $(-1, 1)$.

41. Write the function as
$$f(x) = \frac{2x - 3}{(x + 3)(x - 2)}.$$ The vertical asymptotes are $x = -3$ and $x = 2$. The x-axis is a horizontal asymptote. The graph goes through $(0, 0.5)$, $(1.5, 0)$, $(3, 0.5)$, and $(-4, -\frac{11}{6})$.

43. The graph of $f(x) = \dfrac{x + 1}{x^2}$ has a vertical asymptote at $x = 0$ and the x-axis is a horizontal asymptote. The graph goes through $(1, 2)$, $(2, \frac{3}{4})$, $(-1, 0)$, and $(-2, -\frac{1}{4})$.

45. $f(x) = \dfrac{2x - 1}{x(x - 3)(x + 3)}.$
The vertical asymptotes are $x = 0$, $x = 3$, and $x = -3$. The x-axis is a horizontal asymptote. The graph goes through $(\frac{1}{2}, 0)$, $(-1, -\frac{3}{8})$, $(-4, \frac{9}{28})$, and $(4, \frac{7}{28})$.

47. Write the function as
$$f(x) = \frac{x}{(x - 1)(x + 1)}.$$ The vertical asymptotes are $x = 1$ and $x = -1$. The x-axis is a horizontal asymptote. The graph goes through $(0, 0)$, $(2, \frac{2}{3})$, and $(-2, -\frac{2}{3})$.

49. The function $f(x) = \dfrac{2}{x^2 + 1}$ has no vertical asymptotes, because the equation $x^2 + 1 = 0$ has no real solutions. The x-axis is a horizontal asymptote. The graph goes through the points $(0, 2)$, $(1, 1)$, and $(-1, 1)$.

51. Use long division to write the function as $f(x) = x - 1 + \dfrac{1}{x + 1}$. The line $y = x - 1$ is an oblique asymptote and the line $x = -1$ is a vertical asymptote. The graph goes through the points $(0, 0)$, $(1, 0.5)$, $(-2, -4)$, and $(-3, -4.5)$.

53. Write the function as $f(t) = \dfrac{36t + 500}{t(t + 6)}$.
The vertical asymptotes are at $t = 0$ and $t = -6$, but neither 0 nor -6 is in the domain of the function. The t-axis is a horizontal asymptote. The average cost at time $t = 20$ is
$$f(20) = \frac{36(20) + 500}{20(20 + 6)} \approx \$2.35.$$
The average cost at time $t = 30$ is
$$f(30) = \frac{36(30) + 500}{30(30 + 6)} \approx \$1.46.$$
As t gets very large the average cost per modulator approaches 0, because the horizontal asymptote is the t-axis.

55. a) The horizontal asymptote is $y = 25,000$ because the ratio of the leading coefficients is 25,000.
b) $A(50,000)$

$$= \frac{25,000(50,000) + 700,000,000}{50,000}$$
$$= \$39,000$$

c) $\dfrac{25,000x + 700,000,000}{x} = 30,000$

$$25,000x + 700,000,000 = 30,000x$$
$$700,000,000 = 5,000x$$
$$x = 140,000$$

d)

57. The graph of $f(x)$ is an asymptote for the graph of $g(x)$.
59. The graph of $f(x)$ is an asymptote for the graph of $g(x)$.
61. The graph of $f(x)$ is an asymptote for the graph of $g(x)$.

Enriching Your Mathematical Word Power
1. c   **2.** b   **3.** a   **4.** d   **5.** a
6. a   **7.** d   **8.** d   **9.** a   **10.** b
11. a

CHAPTER 10 REVIEW

1. Use synthetic division to check the possible solutions. Try 2 first.

```
2 |  1  -4  -11   30
   |      2   -4  -30
   ------------------
      1  -2  -15    0
```

$$x^3 - 4x^2 - 11x + 30 = (x - 2)(x^2 - 2x - 15)$$
$$= (x - 2)(x - 5)(x + 3)$$

There is one solution for each factor. The solution set is $\{-3, 2, 5\}$.

3. Use synthetic division to check the possible solutions. Try -2 first.

```
-2 |  1  -5  -2   24
   |     -2  14  -24
   ------------------
      1  -7  12    0
```

$$x^3 - 5x^2 - 2x + 24 = (x + 2)(x^2 - 7x + 12)$$
$$= (x + 2)(x - 3)(x - 4)$$

There is one solution for each factor of the polynomial. The solution set is $\{-2, 3, 4\}$.

5. Use synthetic division to check for possible factors.

$$3 \quad \begin{array}{ccccc} 1 & 4 & -11 & -30 \\ & 3 & 21 & 30 \\ \hline 1 & 7 & 10 & 0 \end{array}$$

$$x^3 + 4x^2 - 11x - 30 = (x-3)(x^2 + 7x + 10)$$
$$= (x-3)(x+2)(x+5)$$

7. Use synthetic division to check for possible factors.

$$-2 \quad \begin{array}{ccccc} 1 & 0 & -5 & -6 \\ & -2 & 4 & 2 \\ \hline 1 & -2 & -1 & -4 \end{array}$$

Since the remainder is not 0, $x + 2$ is not a factor of the polynomial.

9. $P(-2) = 4(-2)^3 - 2(-2)^2 + 3(-2) + 1$
$$= -32 - 8 - 6 + 1 = -45$$

Use synthetic division.

$$-2 \quad \begin{array}{cccc} 4 & -2 & 3 & 1 \\ & -8 & 20 & -46 \\ \hline 4 & -10 & 23 & -45 \end{array}$$

So $P(-2) = -45$

$$P\left(\tfrac{1}{2}\right) = 4(\tfrac{1}{2})^3 - 2(\tfrac{1}{2})^2 + 3(\tfrac{1}{2}) + 1$$
$$= \tfrac{1}{2} - \tfrac{1}{2} + \tfrac{3}{2} + 1 = \tfrac{5}{2}$$

Use synthetic division.

$$\tfrac{1}{2} \quad \begin{array}{cccc} 4 & -2 & 3 & 1 \\ & 2 & 0 & 1.5 \\ \hline 4 & 0 & 3 & 2.5 \end{array}$$

So $P(\tfrac{1}{2}) = 2.5$

11. The factors of 4 are 1, 2, and 4. The factors of 3 are 1 and 3. We form all possible rational numbers using a factor of 4 over a factor of 3. The possible rational roots are $\pm 1, \pm 2, \pm 4, \pm \tfrac{1}{3}, \pm \tfrac{2}{3}$, and $\pm \tfrac{4}{3}$.

13. The factors of 3 are 1 and 3. The factors of 4 are 1, 2, and 4. We form all possible rational numbers using a factor of 3 over a factor of 4. The possible rational roots are $\pm 1, \pm 3, \pm \tfrac{1}{2}, \pm \tfrac{3}{2}, \pm \tfrac{1}{4}$, and $\pm \tfrac{3}{4}$.

15. To find the zeros of the polynomial function, we solve the following equation.
$$2x - 3 = 0$$
$$2x = 3$$
$$x = \tfrac{3}{2}$$
The polynomial function has one zero, $\tfrac{3}{2}$.

17. To find the zeros of the polynomial function, we solve the following equation.
$$x^2 - 100 = 0$$
$$x^2 = 100$$
$$x = \pm 10$$
The zeros to the function are 10 and -10.

19. To find the zeros solve the following equation.
$$x^3 - 64 = 0$$
$$(x - 4)(x^2 + 4x + 16) = 0$$
$$x - 4 = 0 \text{ or } x^2 + 4x + 16 = 0$$
$$x = 4 \text{ or } x = \frac{-4 \pm \sqrt{16 - 4(1)(16)}}{2(1)}$$
$$x = 4 \text{ or } x = \frac{-4 \pm \sqrt{-48}}{2}$$
$$x = 4 \text{ or } x = -2 \pm 2i\sqrt{3}$$
The zeros of the function are 4 and $-2 \pm 2i\sqrt{3}$.

21. To find the zeros solve the following equation.
$$x^4 - 64 = 0$$
$$(x^2 - 8)(x^2 + 8) = 0$$
$$x^2 - 8 = 0 \quad \text{or} \quad x^2 + 8 = 0$$
$$x^2 = 8 \quad \text{or} \quad x^2 = -8$$
$$x = \pm 2\sqrt{2} \quad \text{or} \quad x = \pm 2i\sqrt{2}$$
The zeros are $\pm 2\sqrt{2}$ and $\pm 2i\sqrt{2}$.

23. Solve the following equation.
$$12x^3 - 4x^2 - 3x + 1 = 0$$
$$4x^2(3x - 1) - 1(3x - 1) = 0$$
$$(4x^2 - 1)(3x - 1) = 0$$
$$(2x - 1)(2x + 1)(3x - 1) = 0$$
$$2x - 1 = 0 \text{ or } 2x + 1 = 0 \text{ or } 3x - 1 = 0$$
$$x = \tfrac{1}{2} \text{ or } \quad x = -\tfrac{1}{2} \text{ or } \quad x = \tfrac{1}{3}$$
The zeros of the function are $\tfrac{1}{3}$ and $\pm \tfrac{1}{2}$.

25. If $1/2$ is a root, then $x - 1/2$ or $2x - 1$ is a factor of the polynomial. If 2 is a root, the $x - 2$ is a factor of the polynomial.
$$(2x - 1)(x - 2) = 0$$
$$2x^2 - 5x + 2 = 0$$
The last equation has the roots 2 and $\tfrac{1}{2}$.

27. If $5 - 2i$ is a root and the polynomial is to have real coefficients, then $5 + 2i$ must be a root also. If $5 - 2i$ and $5 + 2i$ are roots, then $x - (5 - 2i)$ and $x - (5 + 2i)$ are factors.
$$(x - (5 - 2i))(x - (5 + 2i)) = 0$$
$$x^2 - (5 - 2i)x - (5 + 2i)x + (5 - 2i)(5 + 2i)$$
$$= 0$$

$$x^2 - 5x + 2ix - 5x - 2ix + 25 - 4i^2 = 0$$
$$x^2 - 10x + 29 = 0$$

The last equation has real coefficients and the given root.

29. If $1 + 3i$ is a root and the polynomial is to have real coefficients, then $1 - 3i$ must be a root also. If $1 - 3i, 1 + 3i$, and 3 are roots, then $x - (1 - 3i)$, $x - (1 + 3i)$, and $x - 3$ are factors.
$$(x - (1 - 3i))(x - (1 + 3i))(x - 3) = 0$$
$$(x^2 - 2x + 10)(x - 3) = 0$$
$$x(x^2 - 2x + 10) - 3(x^2 - 2x + 10) = 0$$
$$x^3 - 2x^2 + 10x - 3x^2 + 6x - 30 = 0$$
$$x^3 - 5x^2 + 16x - 30 = 0$$

The last equation has real coefficients and the given roots.

31. Since there are no sign changes in the polynomial $P(x) = x^6 + x^4 + x^2$, there are no positive roots to the equation.
$$P(-x) = (-x)^6 + (-x)^4 + (-x)^2$$
$$= x^6 + x^4 + x^2$$

Since there are no sign changes in $P(-x)$, there are no negative roots to the equation. Since x^2 is a common factor of $P(x)$, 0 is a root with multiplicity 2. Since the equation must have 6 roots and there are no positive or negative roots, the other 4 roots are imaginary numbers. So the equation has 4 imaginary roots and 0 is a root with multiplicity 2.

33. Since there are 3 sign changes in $P(x) = x^3 - 5x^2 + 3x - 9$, there are either 3 or 1 positive roots to the equation.
$$P(-x) = (-x)^3 - 5(-x)^2 + 3(-x) - 9$$
$$= -x^3 - 5x^2 - 3x - 9$$

Since there are no sign changes in $P(-x)$, there are no negative roots to the equation. Since there must be a total of 3 roots to the equation, there are either 3 positive roots, or 1 positive and 2 imaginary roots.

35. To establish the best upper bound for the roots we try synthetic division with positive integers until all entries in the bottom line are nonnegative.

```
4 |  1  -2  -4
   |      4   8
   |_____
      1   2   4
```

To establish the best lower bound for the roots we try synthetic division with negative integers

until the entries in the bottom line alternate in sign.

```
-2 |  1  -2  -4
   |     -2   8
   |_____
      1  -4   4
```

According to the theorem on bounds, the roots of the equation are between -2 and 4.

37. To establish the best upper bound for the roots we try synthetic division with positive integers until all entries in the bottom line are nonnegative.

```
5 |  1  -3  -5  -6
  |      5  10  25
  |_____
     1   2   5  19
```

To establish the best lower bound for the roots we try synthetic division with negative integers until the entries in the bottom line alternate in sign.

```
-2 |  1  -3  -5   -6
   |     -2  10  -10
   |_____
      1  -5   5  -16
```

According to the theorem on bounds, the roots of the equation are between -2 and 5.

39. Try possible rational roots using synthetic division.

```
4 |  2  -19   49  -20
  |       8  -44   20
  |_____
     2  -11    5    0
```

Since 4 is a root to the equation, $x - 4$ is a factor of the polynomial.
$$2x^3 - 19x^2 + 49x - 20 = 0$$
$$(x - 4)(2x^2 - 11x + 5) = 0$$
$$(x - 4)(2x - 1)(x - 5) = 0$$
$$x - 4 = 0 \text{ or } 2x - 1 = 0 \text{ or } x - 5 = 0$$
$$x = 4 \text{ or } \quad x = \frac{1}{2} \quad \text{ or } \quad x = 5$$

The solutions to the equation are $\frac{1}{2}$, 4, and 5.

41. Try possible rational roots using synthetic division.

```
5/2 |  8  -12  -26   15
    |      20   20  -15
    |_____
       8    8   -6    0
```

Since $5/2$ is a root to the equation, $x - 5/2$ is a factor of the polynomial.
$$8x^3 - 12x^2 - 26x + 15 = 0$$
$$(x - \frac{5}{2})(8x^2 + 8x - 6) = 0$$
$$(2x - 5)(4x^2 + 4x - 3) = 0$$

$(2x - 5)(2x - 1)(2x + 3) = 0$

$x = \frac{5}{2}$ or $x = \frac{1}{2}$ or $x = -\frac{3}{2}$

The solutions are $-\frac{3}{2}$, $\frac{1}{2}$, and $\frac{5}{2}$.

43. Try possible rational roots using synthetic division.

$$
\begin{array}{r|rrrr}
5 & 4 & -24 & 25 & -25 \\
  &   & 20 & -20 & 25 \\
\hline
  & 4 & -4 & 5 & 0
\end{array}
$$

Since 5 is a root to the equation, $x - 5$ is a factor of the polynomial.

$4x^3 - 24x^2 + 25x - 25 = 0$

$(x - 5)(4x^2 - 4x + 5) = 0$

$x - 5 = 0$ or $4x^2 - 4x + 5 = 0$

$x = 5$ or $x = \dfrac{4 \pm \sqrt{16 - 4(4)(5)}}{2(4)}$

$x = 5$ or $x = \dfrac{4 \pm 8i}{8} = \dfrac{1}{2} \pm i$

The solutions to the equation are 5 and $\frac{1}{2} \pm i$.

45. Try possible rational roots using synthetic division.

$$
\begin{array}{r|rrrrr}
3 & 4 & -28 & 59 & -30 & -9 \\
  &   & 12 & -48 & 33 & 9 \\
\hline
  & 4 & -16 & 11 & 3 & 0
\end{array}
$$

Look for another root with synthetic division.

$$
\begin{array}{r|rrrr}
3 & 4 & -16 & 11 & 3 \\
  &   & 12 & -12 & -3 \\
\hline
  & 4 & -4 & -1 & 0
\end{array}
$$

Since 3 is a root of multiplicity 2, $(x - 3)^2$ is a factor of the polynomial.

$4x^4 - 28x^3 + 59x^2 - 30x - 9 = 0$

$(x - 3)^2(4x^2 - 4x - 1) = 0$

$x - 3 = 0$ or $4x^2 - 4x - 1 = 0$

$x = 3$ or $x = \dfrac{4 \pm \sqrt{16 - 4(4)(-1)}}{2(4)}$

$x = 3$ or $x = \dfrac{4 \pm 4\sqrt{2}}{8} = \dfrac{1 \pm \sqrt{2}}{2}$

The solutions to the equation are 3 (with multiplicity 2) and $\dfrac{1 \pm \sqrt{2}}{2}$.

47. Try possible rational roots using synthetic division.

$$
\begin{array}{r|rrrr}
1 & 1 & -3 & 3 & -1 \\
  &   & 1 & -2 & 1 \\
\hline
  & 1 & -2 & 1 & 0
\end{array}
$$

Look for another root with synthetic division.

$$
\begin{array}{r|rrr}
1 & 1 & -2 & 1 \\
  &   & 1 & -1 \\
\hline
  & 1 & -1 & 0
\end{array}
$$

Look for another root with synthetic division.

$$
\begin{array}{r|rr}
1 & 1 & -1 \\
  &   & 1 \\
\hline
  & 1 & 0
\end{array}
$$

The only solution to the equation is 1 (with multiplicity 3).

49. Since $f(-x) = 3(-x)^4 - 2(-x)^2$
$= 3x^4 - 2x^2$, the graph is symmetric about the y-axis.

51. Since $f(-x) = -(-x)^3 + 3(-x)$
$= x^3 - 3x = -(-x^3 + 3x)$, the graph is symmetric about the origin.

53. To find the x-intercepts solve the equation $x^3 - 3x + 2 = 0$. Try possible rational roots using synthetic division.

$$
\begin{array}{r|rrrr}
1 & 1 & 0 & -3 & 2 \\
  &   & 1 & 1 & -2 \\
\hline
  & 1 & 1 & -2 & 0
\end{array}
$$

Since 1 is a root, $x - 1$ is a factor.

$(x - 1)(x^2 + x - 2) = 0$

$(x - 1)(x + 2)(x - 1) = 0$

$x = 1$ or $x = -2$ or $x = 1$

The x-intercepts are $(1, 0)$ and $(-2, 0)$. To find the y-intercept find $f(0)$:

$f(0) = 0^3 - 3(0) + 2 = 2$

The y-intercept is $(0, 2)$. The graph goes through the intercepts and the points $(-1, 4)$ and $(2, 4)$.

$f(x) = x^3 - 3x + 2$

55. First find the x-intercepts.

$-x^4 + 2x^2 - 1 = 0$

$x^4 - 2x^2 + 1 = 0$

$(x^2 - 1)(x^2 - 1) = 0$

$(x - 1)(x + 1)(x - 1)(x + 1) = 0$

$x = 1$ or $x = -1$ or $x = 1$ or $x = -1$

The x-intercepts are $(1, 0)$ and $(-1, 0)$. To find the y-intercept find $f(0)$:

$f(0) = -0^4 + 2(0)^2 - 1 = -1$
The y-intercept is $(0, -1)$. The graph goes
through the intercepts and the points $(-2, -9)$
and $(2, -9)$.

57. First find the x-intercepts.
$$\tfrac{1}{2}x^3 - 2x = 0$$
$$x^3 - 4x = 0$$
$$x(x^2 - 4) = 0$$
$$x(x - 2)(x + 2) = 0$$
$$x = 0 \text{ or } x = 2 \text{ or } x = -2$$
The x-intercepts are $(0, 0)$, $(2, 0)$, and $(-2, 0)$.
To find the y-intercept find $f(0)$:
$$f(0) = \tfrac{1}{2}(0)^3 - 2(0) = 0$$
The y-intercept is $(0, 0)$. The graph goes
through the intercepts and the points $(-1, 1.5)$
and $(1, -1.5)$.

59. If $2x + 3 = 0$, the $x = -\tfrac{3}{2}$. So the
domain of the rational function is
$\left\{x \mid x \neq -\tfrac{3}{2}\right\}$ or $(-\infty, -3/2) \cup (3/2, \infty)$.

61. Since there is no real solution to
$x^2 + 9 = 0$, the domain is the set of all real
numbers or $(-\infty, \infty)$.

63. If $x - 3 = 0$, then $x = 3$. So there is a
vertical asymptote at $x = 3$. If x is very large,
the value of $f(x)$ is approximately 0. So the x-
axis is a horizontal asymptote. The graph
includes the points $(0, -2/3)$ and $(4, 2)$.

65. If $x^2 - 4 = 0$, then $x = \pm 2$. So the
vertical asymptotes are $x = 2$ and $x = -2$. If
x is very large, then $f(x)$ is approximately 0.
So the x-axis is a horizontal asymptote. The
graph goes through $(0, 0)$, $(3, \tfrac{3}{5})$, and
$\left(-3, -\tfrac{3}{5}\right)$.

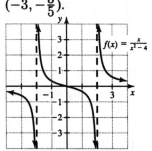

67. Use long division to rewrite the function
as $f(x) = 2 + \dfrac{1}{x - 1}$. If $x - 1 = 0$, then
$x = 1$. So the graph has a vertical asymptote
at $x = 1$. If x is very large, then $f(x)$ is
approximately 2. So the line $y = 2$ is a
horizontal asymptote. The ratio of the leading
coefficients is 2. The graph includes the points
$(0, 1)$ and $(2, 3)$.

69. Use long division to rewrite the function
as $f(x) = x + \dfrac{1}{x - 2}$. If $x - 2 = 0$, then
$x = 2$. So the line $x = 2$ is a vertical
asymptote for the graph. If x is very large,
then the value of $\dfrac{1}{x - 2}$ is approximately 0 and
the value of $f(x)$ is approximately x. So the
line $y = x$ is an oblique asymptote for the
graph. The graph includes the points $(0, -0.5)$,
$(1, 0)$, $(3, 4)$, and $(4, 4.5)$.

71. The graph of $f(x) = 3$ is a horizontal line with a y-intercept $(0, 3)$.

73. The graph of $f(x) = x^2 - 3$ is a parabola with vertex at $(0, -3)$. The graph also goes through $(1, -2)$ and $(-1, -2)$.

75. The graph of $f(x) = \dfrac{1}{x^2 - 3}$ has vertical asymptotes at $x = \sqrt{3}$ and $x = -\sqrt{3}$. The graph includes the points $(0, -\frac{1}{3})$, $(2, 1)$, and $(-2, 1)$.

77. The graph of $f(x) = x(x - 1)(x + 2)$ has x-intercepts $(0, 0)$, $(1, 0)$, and $(-2, 0)$. The graph also includes the points $(-1, 2)$ and $(2, 8)$.

79. The volume of the sphere is $\frac{4}{3}\pi x^3$ and the volume of the base is $(2x)(2x)(\pi)$ or $4\pi x^2$. Since the total volume is 72π cubic inches, we can write the following equation.

$$4\pi x^2 + \frac{4}{3}\pi x^3 = 72\pi$$
$$12\pi x^2 + 4\pi x^3 = 216\pi$$
$$12x^2 + 4x^3 = 216$$
$$4x^3 + 12x^2 - 216 = 0$$
$$x^3 + 3x^2 - 54 = 0$$

Try possible rational roots using synthetic division.

$$
\begin{array}{r|rrrr}
3 & 1 & 3 & 0 & -54 \\
  &   & 3 & 18 & 54 \\
\hline
  & 1 & 6 & 18 & 0
\end{array}
$$

Since there is only one sign change, there is only one positive root, and 3 is that root. So the radius of the sphere is 3 inches.

CHAPTER 10 TEST

1. $f(2) = 3(2)^4 + 5(2)^3 - 10(2)^2 + 3(2) - 2$
$= 48 + 40 - 40 + 6 - 2$
$= 52$
Find $f(2)$ using synthetic division.

$$
\begin{array}{r|rrrrr}
2 & 3 & 5 & -10 & 3 & -2 \\
  &   & 6 & 22 & 24 & 54 \\
\hline
  & 3 & 11 & 12 & 27 & 52
\end{array}
$$

So $f(2) = 52$

2. $f(-3) =$
$3(-3)^4 + 5(-3)^3 - 10(-3)^2 + 3(-3) - 2$
$= 243 - 135 - 90 - 9 - 2 = 7$
Find $f(-3)$ using synthetic division.

$$
\begin{array}{r|rrrrr}
-3 & 3 & 5 & -10 & 3 & -2 \\
   &   & -9 & 12 & -6 & 9 \\
\hline
   & 3 & -4 & 2 & -3 & 7
\end{array}
$$

So $f(-3) = 7$

3. The factors of 6 are 1, 2, 3, and 6. The only factor of 1 (the coefficient of x^4) is 1. Form all possible rational number using a factor of 6 over a factor of 1. The possible rational zeros are ± 1, ± 2, ± 3, and ± 6.

4. The factors of 3 are 1 and 3. The factors of 2 are 1 and 2. Form all possible rational numbers using a factor of 3 over a factor of 2. The possible rational zeros are ± 1, ± 3, $\pm \frac{1}{2}$, and $\pm \frac{3}{2}$.

5. Check the possible rational zeros using synthetic division.

$$\begin{array}{r|rrrr} 1 & 1 & 0 & -13 & 12 \\ & & 1 & 1 & -12 \\ \hline & 1 & 1 & -12 & 0 \end{array}$$

Since 1 is a zero, $x - 1$ is a factor.

$$x^3 - 13x + 12 = 0$$
$$(x - 1)(x^2 + x - 12) = 0$$
$$(x - 1)(x + 4)(x - 3) = 0$$
$$x = 1 \text{ or } x = -4 \text{ or } x = 3$$

The zeros of the function are -4, 1, and 3.

6. Check the possible rational zeros using synthetic division.

$$\begin{array}{r|rrrr} \frac{1}{2} & 2 & -9 & 14 & -5 \\ & & 1 & -4 & 5 \\ \hline & 2 & -8 & 10 & 0 \end{array}$$

$$2x^2 - 8x + 10 = 0$$
$$x^2 - 4x + 5 = 0$$
$$x = \frac{4 \pm \sqrt{16 - 4(1)(5)}}{2}$$
$$= \frac{4 \pm 2i}{2} = 2 \pm i$$

The zeros of the function are $\frac{1}{2}$ and $2 \pm i$.

7. If -1, 1, and 2 are roots of the polynomial equation, then $x + 1$, $x - 1$, and $x - 2$ are factors of the polynomial.

$$(x + 1)(x - 1)(x - 2) = 0$$
$$(x^2 - 1)(x - 2) = 0$$
$$x^3 - 2x^2 - x + 2 = 0$$

The last equation is a polynomial equation with the given roots.

8. If $3 - i$ is a root and the polynomial equation has real coefficients, then $3 + i$ is also a root. If 2, $3 - i$, and $3 + i$ are roots, then $x - 2$, $x - (3 - i)$, and $x - (3 + i)$ are factors of the polynomial.

$$(x - 2)(x - (3 - i))(x - (3 + i)) = 0$$
$$(x - 2)(x^2 - 6x + 10) = 0$$
$$x(x^2 - 6x + 10) - 2(x^2 - 6x + 10) = 0$$
$$x^3 - 8x^2 + 22x - 20 = 0$$

9. If $x + 1$ is a factor, then $f(-1) = 0$.

Find $f(-1)$ for
$$f(x) = 2x^4 - 5x^3 + 3x^2 + 6x - 4$$
using synthetic division.

$$\begin{array}{r|rrrrr} -1 & 2 & -5 & 3 & 6 & -4 \\ & & -2 & 7 & -10 & 4 \\ \hline & 2 & -7 & 10 & -4 & 0 \end{array}$$

Since $f(-1) = 0$, $x + 1$ is a factor of $2x^4 - 5x^3 + 3x^2 + 6x - 4$.

10. Use synthetic division to find a zero of the function $f(x) = x^3 - 12x^2 + 47x - 60$. The possible rational zeros are divisors of 60: ± 1, ± 2, ± 3, ± 4, ± 5, ± 6, ± 10, ± 12, ± 15, ± 20, ± 30, ± 60

$$\begin{array}{r|rrrr} 3 & 1 & -12 & 47 & -60 \\ & & 3 & -27 & 60 \\ \hline & 1 & -9 & 20 & 0 \end{array}$$

$$x^3 - 12x^2 + 47x - 60$$
$$= (x - 3)(x^2 - 9x + 20)$$
$$= (x - 3)(x - 4)(x - 5)$$

11. Since $P(x) = 2x^3 - 6x^2 + 5x - 9$ has 3 sign changes, there are either 3 or 1 positive roots to the equation.

$$P(-x) = 2(-x)^3 - 6(-x)^2 + 5(-x) - 9$$
$$= -2x^3 - 6x^2 - 5x - 9$$

Since there are no sign changes in $P(-x)$, there are no negative roots to the polynomial. Since there are 3 roots to the polynomial equation, there are either 3 positive roots, or 1 positive root and 2 imaginary roots.

12. Since $P(x) = x^5 + 5x^3 + x$ has no changes of sign, there are no positive roots to the polynomial equation.

$$P(-x) = (-x)^5 + 5(-x)^3 + (-x)$$
$$= -x^5 - 5x^3 - x$$

Since there are no changes of sign in $P(-x)$, there are no negative roots to the equation. Since 0 is a root and there must be 5 roots, the other 4 roots are imaginary numbers.

13.

$$\begin{array}{r|rrr} 3 & 4 & -4 & -15 \\ & & 12 & 24 \\ \hline & 4 & 8 & 9 \end{array}$$

$$\begin{array}{r|rrr} -2 & 4 & -4 & -15 \\ & & -8 & 24 \\ \hline & 4 & -12 & 9 \end{array}$$

The real roots are between -2 and 3.

230

14.

$$\begin{array}{c|rrrr} 3 & 2 & -3 & -7 & 4 \\ & & 6 & 9 & 6 \\ \hline & 2 & 3 & 2 & 10 \end{array}$$

$$\begin{array}{c|rrrr} -2 & 2 & -3 & -7 & 4 \\ & & -4 & 14 & -14 \\ \hline & 2 & -7 & 7 & -10 \end{array}$$

The real roots are between -2 and 3.

15. Since $f(-x) = (-x)^3 - 16(-x)$
$$= -x^3 + 16x = -(x^3 - 16x),$$
the graph of the function is symmetric about the origin.

16. Since $f(-x) = (-x)^4 - 25(-x)^2 + 1$
$$= x^4 - 25x^2 + 1,$$
the graph of the function is symmetric about the y-axis.

17. The graph of $f(x) = (x+2)(x-2)^2$ has x-intercepts $(-2, 0)$ and $(2, 0)$. The graph crosses the x-axis at $(-2, 0)$, but does not cross at $(2, 0)$. The graph includes the points $(0, 8)$, $(1, 3)$, and $(3, 5)$.

18. Write the function as $f(x) = \dfrac{1}{(x-2)^2}$.

The graph has $x = 2$ as a vertical asymptote and the x-axis as a horizontal asymptote. The graph goes through the points $(0, \frac{1}{4})$, $(1, 1)$, and $(3, 1)$.

19. Use long division to write the function as $f(x) = 2 + \dfrac{1}{x-2}$. The graph has a vertical asymptote at $x = 2$ and a horizontal asymptote

at $y = 2$. The graph includes the points $(0, 1.5)$, $(1, 1)$, and $(3, 3)$.

20. Write the function as $f(x) = (x-2)(x+2)(x-1)$. The x-intercepts are $(2, 0)$, $(-2, 0)$, and $(1, 0)$. The y-intercept is $(0, 4)$. The graph also goes through $(-1, 6)$ and $(3, 10)$.

Making Connections

Chapters 1 - 10

1. $2x - 3 = x - 5$
$$2x = x - 2$$
$$x = -2$$
The only solution to the equation is -2.

2. $(2x - 3)(x - 5) = 0$
$$2x - 3 = 0 \text{ or } x - 5 = 0$$
$$x = \tfrac{3}{2} \text{ or } \quad x = 5$$
The solutions to the equation are $\frac{3}{2}$ and 5.

3. $\dfrac{2x - 3}{x - 5} = 0$
$$2x - 3 = 0$$
$$x = \tfrac{3}{2}$$
The solution to the equation is $\frac{3}{2}$.

4. $2x^2 - 3 = 0$
$$x^2 = \tfrac{3}{2}$$
$$x = \pm \sqrt{\tfrac{3}{2}} = \pm \dfrac{\sqrt{6}}{2}$$
The solutions are $\pm \dfrac{\sqrt{6}}{2}$.

5.
$$\frac{3}{2x-3} = \frac{2}{x-5}$$
$$2(2x-3) = 3(x-5)$$
$$4x - 6 = 3x - 15$$
$$x - 6 = -15$$
$$x = -9$$

The solution to the equation is -9.

6.
$$2x^4 + 7x^2 - 15 = 0$$
$$(2x^2 - 3)(x^2 + 5) = 0$$
$$2x^2 - 3 = 0 \quad \text{or} \quad x^2 + 5 = 0$$
$$x^2 = \frac{3}{2} \quad \text{or} \quad x^2 = -5$$
$$x = \pm\sqrt{\frac{3}{2}} \quad \text{or} \quad x = \pm\sqrt{-5}$$
$$x = \pm\frac{\sqrt{6}}{2} \quad \text{or} \quad x = \pm i\sqrt{5}$$

The solutions to the equation are $\pm\dfrac{\sqrt{6}}{2}$ and $\pm i\sqrt{5}$.

7.
$$\frac{1}{x} + \frac{3}{x-5} = -\frac{1}{2}$$
$$2x(x-5)\frac{1}{x} + 2x(x-5)\frac{3}{x-5}$$
$$= -\frac{1}{2} \cdot 2x(x-5)$$
$$2x - 10 + 6x = -x^2 + 5x$$
$$x^2 + 3x - 10 = 0$$
$$(x+5)(x-2) = 0$$
$$x + 5 = 0 \text{ or } x - 2 = 0$$
$$x = -5 \text{ or } \quad x = 2$$

The solutions to the equation are -5 and 2.

8.

$$-\frac{1}{3}$$

30	−29	−22	−3
	−10	13	3
30	−39	−9	0

$$-\frac{1}{5}$$

30	−39	−9
	−6	9
30	−45	0

$$\frac{3}{2}$$

30	−45
	45
30	0

The solutions are $-\dfrac{1}{3}$, $-\dfrac{1}{5}$, and $\dfrac{3}{2}$.

9. The graph of $y = 2x$ is a straight line with slope 2 and y-intercept $(0, 0)$.

10. The graph of $y = 2 - x$ is a line with slope -1 and y-intercept $(0, 2)$.

11. The graph of $y = x - 2$ is a line with slope 1 and y-intercept $(0, -2)$.

12. The graph of $y = \dfrac{1}{x-2}$ has a vertical asymptote at $x = 2$. The x-axis is a horizontal asymptote. The graph goes through $(1, -1)$, $(0, -0.5)$, and $(3, 1)$.

13. The graph of $y = x^2 - 2$ is a parabola opening upward with vertex $(0, -2)$. The graph also goes through $(1, -1)$ and $(-1, -1)$.

14. The graph of $y = (x - 2)^2$ is a parabola obtained by shifting the parabola $y = x^2$ two spaces to the right. The graph of $y = (x - 2)^2$ goes through $(2, 0)$, $(1, 1)$, and $(3, 1)$.

15. The graph of $y = \dfrac{1}{(x - 2)^2}$ is obtained from the graph of $y = \dfrac{1}{x^2}$ by shifting two spaces to the right. The graph of $y = \dfrac{1}{(x - 2)^2}$ has a vertical asymptote at $x = 2$ and the x-axis is a horizontal asymptote. The graph goes through $(1, 1)$ and $(3, 1)$.

16. The graph of $y = x^3 - 2$ is obtained by shifting the graph of $y = x^3$ two units downward. The graph of $y = x^3 - 2$ goes through $(0, -2)$, $(1, -1)$, and $(2, 6)$.

17. The function $y = x^3 - x$ can be written as $y = x(x - 1)(x + 1)$. The x-intercepts are $(0, 0)$, $(1, 0)$, and $(-1, 0)$. The graph goes through $(2, 6)$ and $(-2, -6)$.

18. Write the function as $y = x^2(x - 1)(x + 1)$. The x-intercepts are $(0, 0)$, $(1, 0)$, and $(-1, 0)$. The graph crosses the x-axis at $(-1, 0)$ and $(1, 0)$, but does not cross at $(0, 0)$. The graph goes through $(2, 12)$ and $(-2, 12)$.

19. The function $y = \sqrt{x + 5} - 2$ can be obtained from the graph of $y = \sqrt{x}$ by translating 5 units to the left and down two units. The graph goes through $(-5, -2)$, $(-4, -1)$, $(-1, 0)$, and $(4, 1)$.

20. The function $y = -|x - 2|$ can be obtained from the graph of $y = |x|$ by translating 2 units to the right and then reflecting in the x-axis. The graph goes through $(2, 0)$, $(3, -1)$, and $(1, -1)$.

11.1 WARM-UPS

1. False, because $f(-1/2) = 4^{-1/2} = 1/2$.
2. True, because $(1/3)^{-1} = 3$.
3. False, because the variable is not in the exponent.
4. True, because $(1/2)^x = (2^{-1})^x = 2^{-x}$.
5. True, because it is a one-to-one function.
6. False, because the equation $(1/3)^x = 0$ has no solution.
7. True, because $e^0 = 1$.
8. False, because the domain of the exponential function $f(x) = 2^x$ is the set of all real numbers.
9. True, because $2^{-x} = 1/(2^x)$.
10. False, at the end of 3 years the investment is worth $500(1.005)^{36}$ dollars

11.1 EXERCISES

1. An exponential function has the form $f(x) = a^x$ where $a > 0$ and $a \neq 1$.
3. The two most popular bases are e and 10.
5. The compound interest formula is $A = P(1 + i)^n$.
7. $f(2) = 4^2 = 16$
9. $f(1/2) = 4^{1/2} = 2$

11. $g(-2) = (1/3)^{-2+1} = (1/3)^{-1} = 3$

13. $g(0) = (1/3)^{0+1} = (1/3)^1 = \frac{1}{3}$
15. $h(0) = -2^0 = -1$
17. $h(-2) = -2^{-2} = -1/4$
19. $h(0) = 10^0 = 1$
21. $h(2) = 10^2 = 100$
23. $j(1) = e^1 = e \approx 2.718$
25. $j(-2) = e^{-2} \approx 0.135$

27. The graph of $f(x) = 4^x$ includes the points $(0, 1)$, $(1, 4)$, $(2, 16)$, and $(-1, 1/4)$.

29. The graph of $h(x) = \left(\frac{1}{3}\right)^x$ includes the points $(0, 1)$, $(1, 1/3)$, $(2, 1/9)$, $(-1, 3)$, and $(-2, 9)$.

31. The graph of $y = 10^x$ includes the points $(0, 1)$, $(1, 10)$, and $(-1, 1/10)$.

33. The graph of $y = 10^{x+2}$ includes the points $(0, 100)$, $(-1, 10)$, $(-2, 1)$, and $(-3, 1/10)$.

35. The graph of $f(x) = -2^x$ includes the points $(0, -1)$, $(1, -2)$, $(2, -4)$, $(-1, -1/2)$, and $(-2, -1/4)$.

37. The graph of $g(x) = 2^{-x}$ includes the points $(0, 1)$, $(-1, 2)$, $(-2, 4)$, and $(1, 1/2)$.

39. The graph of $f(x) = -e^x$ includes the points $(0, -1)$, $(1, -2.7)$, $(2, -7.4)$, $(-1, -0.4)$.

41. The graph of $H(x) = 10^{|x|}$ includes the points $(0, 1)$, $(1, 10)$, and $(-1, 10)$.

43. The graph of $P = 5000(1.05)^t$ includes the points $(0, 5000)$, $(20, 13{,}266)$, and $(-10, 3070)$.

45. $\quad 2^x = 64$
$\qquad 2^x = 2^6$
$\qquad\quad x = 6 \quad$ By the one-to-one property
The solution set is $\{6\}$.

47. $\quad 10^x = 0.001$
$\qquad 10^x = 10^{-3}$
$\qquad\quad x = -3$
The solution set is $\{-3\}$.

49. $\quad 2^x = \frac{1}{4}$
$\qquad 2^x = 2^{-2}$
$\qquad\quad x = -2$
The solution set is $\{-2\}$.

51. $\quad \left(\frac{2}{3}\right)^{x-1} = \frac{9}{4}$
$\qquad \left(\frac{2}{3}\right)^{x-1} = \left(\frac{2}{3}\right)^{-2}$
$\qquad\quad x - 1 = -2$
$\qquad\qquad x = -1$
The solution set is $\{-1\}$.

53. $\quad 5^{-x} = 25$
$\qquad 5^{-x} = 5^2$
$\qquad\quad -x = 2 \quad$ By the one-to-one property

$\qquad\quad x = -2$
The solution set is $\{-2\}$.

55. $\quad -2^{1-x} = -8$
$\qquad -2^{1-x} = -2^3$
$\qquad\quad 2^{1-x} = 2^3$
$\qquad\quad 1 - x = 3$
$\qquad\qquad x = -2$
The solution set is $\{-2\}$.

57. $\quad 10^{|x|} = 1000$
$\qquad 10^{|x|} = 10^3$
$\qquad\quad |x| = 3$
$\qquad\quad x = 3 \text{ or } x = -3$
The solution set is $\{-3, 3\}$.

59. If $f(x) = 2^x$ and $f(x) = 4 = 2^2$, then we must have $x = 2$ by the one-to-one property of exponential functions.

61. Note that $4^{2/3} = (2^2)^{2/3} = 2^{4/3}$. So if $2^x = 2^{4/3}$, then $x = \frac{4}{3}$.

63. Note that $9 = 3^2 = (1/3)^{-2}$. So if $(1/3)^x = (1/3)^{-2}$, then $x = -2$.

65. Note that $1 = (1/3)^0$. So if $(1/3)^x = (1/3)^0$, then $x = 0$.

67. Note that $16 = 4^2$. So if $h(x) = 4^{2x-1} = 4^2$, then
$\qquad 2x - 1 = 2$
$\qquad\quad 2x = 3$
$\qquad\quad\; x = \frac{3}{2}$.

69. Since $1 = 4^0$, $h(x) = 4^{2x-1} = 4^0$ implies that $2x - 1 = 0$ or $x = \frac{1}{2}$.

71. Compounded quarterly for 10 years means that interest will be paid 40 times at 1.25%.
$$S = 6000\left(1 + \frac{0.05}{4}\right)^{40} = 6000(1.0125)^{40}$$
$$= \$9861.72$$

73. a) $\$10{,}000(1 + 0.1821)^{10} \approx \$53{,}277.30$
b) From the graph it appears that the $\$10{,}000$ will be worth $\$200{,}000$ approximately 18 years after 1992 or around 2010.

75. When the book is new, $t = 0$, and the value is $V = 45 \cdot 2^{-0.9(0)} = \$45$. When $t = 2$, the value is $V = 45 \cdot 2^{-0.9(2)} \approx \$12.92$.

77. A deposit of $\$500$ for 3 years at 7% compounded continuously amounts to $S = 500 \cdot e^{0.07(3)} \approx \$616.84$.

79. A deposit of $80,000 at 7.5% compounded continuously for 1 year amounts to $S = 80{,}000 \cdot e^{0.075(1)} \approx \$86{,}230.73$. The interest earned is
$\$86{,}230.73 - \$80{,}000 = \$6230.73$.

81. The amount at time $t = 0$ is
$A = 300 \cdot e^{-0.06(0)} = 300$ grams. The amount present after 20 years, $t = 20$, is
$A = 300 \cdot e^{-0.06(20)} \approx 90.4$ grams.
One-half of the substance decays in about 12 years. The substance will never completely disappear because the t-axis is the horizontal asymptote.

83. $1 + 1 + \frac{1}{2} + \frac{1}{6} \approx 2.66666667$
$e - 2.66666667 \approx 0.0516$
$e - \left(1 + 1 + \frac{1}{2} + \frac{1}{6} + \frac{1}{24} + \frac{1}{120} + \frac{1}{720} + \frac{1}{5040}\right)$
$\approx 2.8 \times 10^{-5}$

85. The graph of $y = 3^{x-k}$ lies k units to the right of $y = 3^x$ when $k > 0$ and $|k|$ units to the left of $y = 3^x$ when $k < 0$.

11.2 WARM-UPS

1. True, because 3 is the exponent of a that produces 2.

2. False, because $b = 8^a$ is equivalent to $\log_8(b) = a$.

3. True, because the inverse of the base a exponential function is the base a logarithm function.

4. True, because the inverse of the base e exponential function is the base e logarithm function.

5. False, the domain is $(0, \infty)$.

6. False, $\log_{25}(5) = 1/2$.

7. False, $\log(-10)$ is undefined because -10 is not in the domain of the base 10 logarithm function.

8. False, $\log(0)$ is undefined.

9. True, because $\log_5(125) = 3$ and $5^3 = 125$.

10. True, because $(1/2)^{-5} = 2^5 = 32$.

11.2 EXERCISES

1. If $f(x) = 2^x$ then $f^{-1}(x) = \log_2(x)$.

3. The common logarithm uses the base 10 and the natural logarithm uses base e.

5. The one-to-one property for logarithmic functions says that if $\log_a(m) = \log_a(n)$ then $m = n$.

7. $\log_2(8) = 3$ is equivalent to $2^3 = 8$.

9. $10^2 = 100$ is equivalent to $\log(100) = 2$.

11. $y = \log_5(x)$ is equivalent to $5^y = x$.

13. $2^a = b$ is equivalent to $\log_2(b) = a$.

15. $\log_3(x) = 10$ is equivalent to $3^{10} = x$.

17. $e^3 = x$ is equivalent to $\ln(x) = 3$.

19. Because $2^2 = 4$, $\log_2(4) = 2$.

21. Because $2^4 = 16$, $\log_2(16) = 4$.

23. Because $2^6 = 64$, $\log_2(64) = 6$.

25. Because $4^3 = 64$, $\log_4(64) = 3$.

27. Because $2^{-2} = \frac{1}{4}$, $\log_2(1/4) = -2$.

29. Because $10^2 = 100$, $\log(100) = 2$.

31. Because $10^{-2} = 0.01$, $\log(0.01) = -2$.

33. Because $(1/3)^1 = 1/3$, $\log_{1/3}(1/3) = 1$.

35. Because $(1/3)^{-3} = 27$, $\log_{1/3}(27) = -3$.

37. Because $25^{1/2} = 5$, $\log_{25}(5) = 1/2$.

39. Because $e^2 = e^2$, $\ln(e^2) = 2$.

41. Use a calculator with a base 10 logarithm key to find $\log(5) \approx 0.6990$.

43. Use a calculator with a natural logarithm key to find $\ln(6.238) \approx 1.8307$.

45. The graph of $f(x) = \log_3(x)$ includes the points $(3, 1)$, $(1, 0)$, and $(1/3, -1)$. All graphs of logarithm functions have similar shapes.

47. The graph of $y = \log_4(x)$ includes the points $(4, 1)$, $(1, 0)$, and $(1/4, -1)$.

49. The graph of $y = \log_{1/4}(x)$ includes the points $(1, 0)$, $(4, -1)$, and $(1/4, 1)$.

51. The graph of $h(x) = \log_{1/5}(x)$ includes the points $(1, 0)$, $(5, -1)$, and $(1/5, 1)$.

53. The inverse of $f(x) = 6^x$ is $f^{-1}(x) = \log_6(x)$.

55. The inverse of $f(x) = \ln(x)$ is $f^{-1}(x) = e^x$.

57. If $f(x) = \log_{1/2}(x)$ then $f^{-1}(x) = \left(\frac{1}{2}\right)^x$.

59. $x = (1/2)^{-2} = 2^2 = 4$
The solution set is $\{4\}$.

61. $5 = 25^x$
    $x = \log_{25}(5) = 1/2$
The solution set is $\left\{\frac{1}{2}\right\}$.

63. $\log(x) = -3$
        $x = 10^{-3} = 0.001$
The solution set is $\{0.001\}$.

65. $\log_x(36) = 2$
            $x^2 = 36$
            $x = \pm 6$
Omit -6 because the base of any logarithm function is positive. The solution set is $\{6\}$.

67. $\log_x(5) = -1$
        $x^{-1} = 5$
        $(x^{-1})^{-1} = 5^{-1}$
        $x = \frac{1}{5}$
The solution set is $\left\{\frac{1}{5}\right\}$.

69. $\log(x^2) = \log(9)$
            $x^2 = 9$
            $x = \pm 3$

Both 3 and -3 check in the original equation. The solution set is $\{\pm 3\}$.

71. $3 = 10^x$
        $x = \log(3) \approx 0.4771$
The solution set is $\{0.4771\}$.

73. $10^x = \frac{1}{2}$
        $x = \log(1/2) = \log(0.5) \approx -0.3010$
The solution set is $\{-0.3010\}$.

75. $e^x = 7.2$
        $x = \ln(7.2) \approx 1.9741$
The solution set is $\{1.9741\}$.

77. Use the continuous compounding formula.
    $10,000 = 5000 \cdot e^{0.12t}$
        $2 = e^{0.12t}$
    $0.12t = \ln(2)$
        $t = \dfrac{\ln(2)}{0.12} \approx 5.776$ years

79. To earn \$1000 in interest, the principal must increase from \$6000 to \$7000 in t years.
    $7000 = 6000 \cdot e^{0.08t}$
    $7/6 = e^{0.08t}$
    $0.08t = \ln(7/6)$
        $t = \dfrac{\ln(7/6)}{0.08} \approx 1.9269$ years

81. a) $20,733 = 10,000 \cdot e^{r5}$
        $2.0733 = e^{5r}$
            $5r = \ln(2.0733)$
        $r = \dfrac{\ln(2.0733)}{5} \approx 0.1458 = 14.58\%$

b) In 2010 the investment grows for 13 years:
$A = 10,000e^{0.1458(13)} \approx \$66,576.60$

83. pH $= -\log_{10}(10^{-4.1}) = -(-4.1) = 4.1$

85. pH $= -\log(1.58 \times 10^{-7}) \approx 6.8$

87. $L = 10 \cdot \log(0.001 \times 10^{12}) = 90$ db

89. $f(x) = 5 + \log_2(x - 3)$
        $y = 5 + \log_2(x - 3)$
        $x = 5 + \log_2(y - 3)$
    $x - 5 = \log_2(y - 3)$
    $y - 3 = 2^{x-5}$
        $y = 2^{x-5} + 3$
    $f^{-1}(x) = 2^{x-5} + 3$
Domain of f^{-1} is $(-\infty, \infty)$ and range is $(3, \infty)$.

91. $y = \ln(e^x) = x$ for $-\infty < x < \infty$
$y = e^{\ln(x)} = x$ for $0 < x < \infty$

11.3 WARM-UPS

1. True, because
$$\log_2(x^2/8) = \log_2(x^2) - \log_2(8)$$
$$= \log_2(x^2) - 3.$$
2. False, because
$\log(100) = 2$ and $\log(10) = 1$, and
$2 \div 1 \neq 2 - 1$.
3. True, because
$$\ln(\sqrt{2}) = \ln(2^{1/2}) = \frac{1}{2} \cdot \ln(2) = \frac{\ln(2)}{2}.$$
4. True, because $\log_3(17)$ is the exponent that we place on 3 to obtain 17.
5. False, because $\log_2(1/8) = -3$ and $\log_2(8) = 3$.
6. True, because $\ln(8) = \ln(2^3) = 3 \cdot \ln(2)$.
7. False, because $\ln(1) = 0$ and $e \neq 0$.
8. False, because $\log(100) = 2$, $\log(10) = 1$, and $2 \div 10 \neq 1$.
9. False, because $\log_2(8) = 3$, $\log_2(2) = 1$, $\log_2(4) = 2$, and $3 \div 1 \neq 2$.
10. True, because $\ln(2) + \ln(3) - \ln(7) = \ln(2 \cdot 3) - \ln(7) = \ln(6/7)$.

11.3 EXERCISES

1. The product rule for logarithms says that
$\log_a(MN) = \log_a(M) + \log_a(N)$.
3. The power rule for logarithms says that
$\log_a(M^N) = N \cdot \log_a(M)$.
5. Since $\log_a(M)$ is the exponent that you would use on a to obtain M, using $\log_a(M)$ as the exponent produces M: $a^{\log_a(M)} = M$.
7. $\log(3) + \log(7) = \log(3 \cdot 7) = \log(21)$
9. $\log_3(\sqrt{5}) + \log_3(\sqrt{x}) = \log_3(\sqrt{5x})$
11. $\log(x^2) + \log(x^3) = \log(x^2 \cdot x^3) = \log(x^5)$
13. $\ln(2) + \ln(3) + \ln(5) = \ln(2 \cdot 3 \cdot 5)$
$$= \ln(30)$$
15. $\log(x) + \log(x + 3) = \log(x^2 + 3x)$
17. $\log_2(x - 3) + \log_2(x + 2)$
$$= \log_2(x^2 - x - 6)$$
19. $\log(8) - \log(2) = \log(8/2) = \log(4)$
21. $\log_2(x^6) - \log_2(x^2) = \log_2\left(\frac{x^6}{x^2}\right)$
$$= \log_2(x^4)$$
23. $\log(\sqrt{10}) - \log(\sqrt{2}) = \log\left(\frac{\sqrt{10}}{\sqrt{2}}\right)$
$$= \log\left(\sqrt{5}\right)$$

25. $\ln(4h - 8) - \ln(4) = \ln\left(\frac{4h - 8}{4}\right)$
$$= \ln(h - 2)$$
27. $\log_2\left(w^2 - 4\right) - \log_2(w + 2)$
$$= \log_2\left(\frac{w^2 - 4}{w + 2}\right) = \log_2\left(\frac{(w - 2)(w + 2)}{w + 2}\right)$$
$$= \log_2(w - 2)$$
29. $\ln\left(x^2 + x - 6\right) - \ln(x + 3)$
$$= \ln\left(\frac{x^2 + x - 6}{x + 3}\right) = \ln\left(\frac{(x + 3)(x - 2)}{x + 3}\right)$$
$$= \ln(x - 2)$$
31. $\log(27) = \log(3^3) = 3\log(3)$
33. $\log(\sqrt{3}) = \log(3^{1/2}) = \frac{1}{2}\log(3)$
35. $\log(3^x) = x\log(3)$
37. $\log_2(2^{10}) = 10$
39. $5^{\log_5(19)} = 19$
41. $\log(10^8) = 8$
43. $e^{\ln(4.3)} = 4.3$
45. $\log(15) = \log(3 \cdot 5) = \log(3) + \log(5)$
47. $\log(5/3) = \log(5) - \log(3)$
49. $\log(25) = \log(5^2) = 2\log(5)$
51. $\log(75) = \log(5^2 \cdot 3) = 2\log(5) + \log(3)$
53. $\log\left(\frac{1}{3}\right) = \log(1) - \log(3)$
$$= 0 - \log(3) = -\log(3)$$
55. $\log(0.2) = \log(1/5) = \log(1) - \log(5)$
$$= 0 - \log(5) = -\log(5)$$
57. $\log(xyz) = \log(x) + \log(y) + \log(z)$
59. $\log_2(8x) = \log_2(8) + \log_2(x)$
$$= 3 + \log_2(x)$$
61. $\ln(x/y) = \ln(x) - \ln(y)$
63. $\log(10x^2) = \log(10) + \log(x^2)$
$$= 1 + 2\log(x)$$
65. $\log_5\left[\frac{(x - 3)^2}{\sqrt{w}}\right]$
$$= \log_5\left[(x - 3)^2\right] - \log_5(\sqrt{w})$$
$$= 2\log_5(x - 3) - \frac{1}{2}\log_5(w)$$
67. $\ln\left[\frac{yz\sqrt{x}}{w}\right] = \ln[yz\sqrt{x}] - \ln(w)$
$$= \ln(y) + \ln(z) + \ln(\sqrt{x}) - \ln(w)$$
$$= \ln(y) + \ln(z) + \frac{1}{2}\ln(x) - \ln(w)$$
69. $\log(x) + \log(x - 1) = \log(x^2 - x)$
71. $\ln(3x - 6) - \ln(x - 2) = \ln\left(\frac{3x - 6}{x - 2}\right)$
$$= \ln(3)$$
73. $\ln(x) - \ln(w) + \ln(z) = \ln(xz) - \ln(w)$
$$= \ln\left(\frac{xz}{w}\right)$$

75. $3 \cdot \ln(y) + 2 \cdot \ln(x) - \ln(w)$
$$= \ln(y^3) + \ln(x^2) - \ln(w)$$
$$= \ln(x^2 y^3) - \ln(w) = \ln\left(\frac{x^2 y^3}{w}\right)$$

77. $\frac{1}{2} \cdot \log(x - 3) - \frac{2}{3} \cdot \log(x + 1)$
$$= \log((x - 3)^{1/2}) - \log((x + 1)^{2/3})$$
$$= \log\left(\frac{(x - 3)^{1/2}}{(x + 1)^{2/3}}\right)$$

79. $\frac{2}{3} \cdot \log_2(x - 1) - \frac{1}{4} \cdot \log_2(x + 2)$
$$= \log_2\left((x - 1)^{2/3}\right) - \log_2\left((x + 2)^{1/4}\right)$$
$$= \log_2\left(\frac{(x - 1)^{2/3}}{(x + 2)^{1/4}}\right)$$

81. False, because
$$\log(56) = \log(7 \cdot 8) = \log(7) + \log(8).$$

83. True, because $\log_2(4^2) = \log_2(16) = 4$
and $(\log_2(4))^2 = (2)^2 = 4.$

85. True, because $\ln(25) = \ln(5^2) = 2 \cdot \ln(5).$

87. False, because $\log_2(64) = 6$ and
$\log_2(8) = 3$ and $6 \div 3 \neq 3.$

89. True, because $\log(1/3) = \log(1) - \log(3)$
$$= 0 - \log(3) = -\log(3).$$

91. True, because
$$\log_2(16^5) = 5 \cdot \log_2(16) = 5 \cdot 4 = 20.$$

93. True, $\log(10^3) = 3 \cdot \log(10) = 3 \cdot 1 = 3.$

95. False, because $\log(100 + 3) = \log(103)$
and
$2 + \log(3) = \log(100) + \log(3) = \log(300).$

97. $r = \log(I) - \log(I_0)$
$$r = \log\left(\frac{I}{I_0}\right)$$
If $I = 100 \cdot I_0$, then
$$r = \log\left(\frac{100 \cdot I_0}{I_0}\right) = \log(100) = 2.$$

99. Only (b) is an identity, because it is the only one that is a correct application of a property of logarithms.

101. The graphs are the same because
$\ln(\sqrt{x}) = \ln(x^{1/2}) = \frac{1}{2}\ln(x) = 0.5 \cdot \ln(x).$

103. The graph is a straight line because
$\log(e^x) = x \log(e) \approx 0.434x.$ The slope is
$\log(e)$ or approximately $0.434.$

11.4 WARM-UPS

1. True, because $\log(x - 2) + \log(x + 2) =$
$\log[(x - 2)(x + 2)] = \log(x^2 - 4).$

2. True, because of the one-to-one property of logarithms.

3. True, because of the one-to-one property of exponential functions.

4. False, because the bases are different and the one-to-one property does not apply.

5. True, because
$\log_2(x^2 - 3x + 5) = 3$ is equivalent to
$x^2 - 3x + 5 = 2^3 = 8.$

6. True, $a^x = y$ is equivalent to $\log_a(y) = x.$

7. True, if $5^x = 23$, then $\ln(5^x) = \ln(23)$, or
$x \cdot \ln(5) = \ln(23).$

8. False, because $\log_3(5) = \dfrac{\ln(5)}{\ln(3)}.$

9. True, because $\log_6(2) = \dfrac{\ln(2)}{\ln(6)}$ and
$\log_6(2) = \dfrac{\log(2)}{\log(6)}.$

10. False, $\log(5) \approx 0.699$ and $\ln(5) \approx 1.609.$

11.4 EXERCISES

1. Equivalent to $\log_a(x) = y$ is $a^y = x.$

3. $\log_2(x + 1) = 3$
$$x + 1 = 2^3$$
$$x + 1 = 8$$
$$x = 7$$
The solution set is $\{7\}.$

5. $3\log_2(x + 1) - 2 = 13$
$$3\log_2(x + 1) = 15$$
$$\log_2(x + 1) = 5$$
$$x + 1 = 2^5$$
$$x = 31$$
The solution set is $\{31\}.$

7. $12 + 2\ln(x) = 14$
$$2\ln(x) = 2$$
$$\ln(x) = 1$$
$$x = e^1$$
The solution set is $\{e\}.$

9. $\log(x) + \log(5) = 1$
$$\log(5x) = 1$$
$$5x = 10^1$$
$$x = 2$$
The solution set is $\{2\}.$

11. $\log_2(x-1) + \log_2(x+1) = 3$
$$\log_2[(x-1)(x+1)] = 3$$
$$\log_2(x^2 - 1) = 3$$
$$x^2 - 1 = 2^3$$
$$x^2 = 9$$
$$x = \pm 3$$

If $x = -3$ in the original equation, $\log_2(-3-1)$ is undefined. The solution set is $\{3\}$.

13. $\log_2(x-1) - \log_2(x+2) = 2$
$$\log_2\left(\frac{x-1}{x+2}\right) = 2$$
$$\frac{x-1}{x+2} = 2^2$$
$$x - 1 = 4(x+2)$$
$$x - 1 = 4x + 8$$
$$-9 = 3x$$
$$-3 = x$$

If $x = -3$ in the original equation, we get the logarithm of a negative number, which is undefined. The solution set is $\emptyset$.

15. $\log_2(x-4) + \log_2(x+2) = 4$
$$\log_2(x^2 - 2x - 8) = 4$$
$$x^2 - 2x - 8 = 2^4$$
$$x^2 - 2x - 8 = 16$$
$$x^2 - 2x - 24 = 0$$
$$(x-6)(x+4) = 0$$
$$x - 6 = 0 \text{ or } x + 4 = 0$$
$$x = 6 \text{ or } x = -4$$

If $x = -4$ in the original equation, we get $\log_2(-8)$, which is undefined. The solution set is $\{6\}$.

17. $\ln(x) + \ln(x+5)$
$$= \ln(x+1) + \ln(x+3)$$
$$\ln(x^2 + 5x) = \ln(x^2 + 4x + 3)$$
$$x^2 + 5x = x^2 + 4x + 3$$
$$x = 3$$

The solution set is $\{3\}$.

19. $\log(x+3) + \log(x+4)$
$$= \log(x^3 + 13x^2) - \log(x)$$
$$\log(x^2 + 7x + 12) = \log(x^2 + 13x)$$
$$x^2 + 7x + 12 = x^2 + 13x$$
$$12 = 6x$$
$$2 = x$$

The solution set is $\{2\}$.

21. $2 \cdot \log(x) = \log(20 - x)$
$$\log(x^2) = \log(20 - x)$$
$$x^2 = 20 - x$$
$$x^2 + x - 20 = 0$$

$$(x-4)(x+5) = 0$$
$$x - 4 = 0 \text{ or } x + 5 = 0$$
$$x = 4 \text{ or } x = -5$$

If $x = -5$ in the original equation we get a logarithm of a negative number, which is undefined. The solution set is $\{4\}$.

23. $3^x = 7$
$$x = \log_3(7)$$
The solution set is $\{\log_3(7)\}$.

25. $e^{2x} = 7$
$$2x = \ln(7)$$
$$x = \frac{\ln(7)}{2}$$
The solution set is $\left\{\frac{\ln(7)}{2}\right\}$.

27. $2^{3x+4} = 4^{x-1}$
$$2^{3x+4} = (2^2)^{x-1}$$
$$2^{3x+4} = (2)^{2x-2}$$
$$3x + 4 = 2x - 2$$
$$x = -6$$
The solution set is $\{-6\}$.

29. $(1/3)^x = 3^{1+x}$
$$(3^{-1})^x = 3^{1+x}$$
$$3^{-x} = 3^{1+x}$$
$$-x = 1 + x$$
$$-2x = 1$$
$$x = -1/2$$
The solution set is $\left\{-\frac{1}{2}\right\}$.

31. $2^x = 3^{x+5}$
$$\ln(2^x) = \ln(3^{x+5})$$
$$x \cdot \ln(2) = (x+5)\ln(3)$$
$$x \cdot \ln(2) = x \cdot \ln(3) + 5 \cdot \ln(3)$$
$$x \cdot \ln(2) - x \cdot \ln(3) = 5 \cdot \ln(3)$$
$$x(\ln(2) - \ln(3)) = 5 \cdot \ln(3)$$
$$x = \frac{5 \cdot \ln(3)}{\ln(2) - \ln(3)}$$

This is the exact solution. Use a calculator to find an approximate solution: $\frac{5\ln(3)}{\ln(2) - \ln(3)} \approx -13.548$.

33. $5^{x+2} = 10^{x-4}$
$$\log(5^{x+2}) = \log(10^{x-4})$$
$$(x+2)\log(5) = x - 4$$
$$x \cdot \log(5) + 2 \cdot \log(5) = x - 4$$
$$x \cdot \log(5) - x = -4 - 2 \cdot \log(5)$$
$$x[\log(5) - 1] = -4 - 2 \cdot \log(5)$$
$$x = \frac{-4 - 2 \cdot \log(5)}{\log(5) - 1}$$

$$= \frac{4 + 2\log(5)}{1 - \log(5)}$$
$$\approx 17.932$$

35. $8^x = 9^{x-1}$
$$\ln(8^x) = \ln(9^{x-1})$$
$$x \cdot \ln(8) = (x-1)\ln(9)$$
$$x \cdot \ln(8) = x \cdot \ln(9) - \ln(9)$$
$$x \cdot \ln(8) - x \cdot \ln(9) = -\ln(9)$$
$$x[\ln(8) - \ln(9)] = -\ln(9)$$
$$x = \frac{-\ln(9)}{\ln(8) - \ln(9)} = \frac{\ln(9)}{\ln(9) - \ln(8)}$$
$$\approx 18.655$$

37. $\log_2(3) = \dfrac{\ln(3)}{\ln(2)} \approx 1.5850$

39. $\log_3(1/2) = \dfrac{\ln(0.5)}{\ln(3)} \approx -0.6309$

41. $\log_{1/2}(4.6) = \dfrac{\ln(4.6)}{\ln(0.5)} \approx -2.2016$

43. $\log_{0.1}(0.03) = \dfrac{\ln(0.03)}{\ln(0.1)} \approx 1.5229$

45. $x \cdot \ln(2) = \ln(7)$
$$x = \frac{\ln(7)}{\ln(2)}$$
This is the exact solution.
Use a calculator to find the approximate value
of x, $x \approx 2.807$.

47. $3x - x \cdot \ln(2) = 1$
$$x(3 - \ln(2)) = 1$$
$$x = \frac{1}{3 - \ln(2)}$$
Exact solution.
Use a calculator to find an approximate value
for x: $\dfrac{1}{3 - \ln(2)} \approx 0.433$.

49. $3^x = 5$
$$x = \log_3(5) = \frac{\ln(5)}{\ln(3)}$$
This is the exact solution.
Use a calculator to find an approximate value
for x that satisfies the equation,
$\dfrac{\ln(5)}{\ln(3)} \approx 1.465$.

51. $2^{x-1} = 9$
$$\ln(2^{x-1}) = \ln(9)$$
$$(x-1)\ln(2) = \ln(9)$$
$$x - 1 = \frac{\ln(9)}{\ln(2)}$$
$$x = 1 + \frac{\ln(9)}{\ln(2)} \quad \text{Exact solution}$$
Use a calculator to find an approximate value
for x: $1 + \dfrac{\ln(9)}{\ln(2)} \approx 4.170$.

53. $3^x = 20$
$$x = \log_3(20) = \frac{\ln(20)}{\ln(3)} \approx 2.727$$

55. $\log_3(x) + \log_3(5) = 1$
$$\log_3(5x) = 1$$
$$5x = 3^1$$
$$x = \frac{3}{5}$$

57. $8^x = 2^{x+1}$
$$(2^3)^x = 2^{x+1}$$
$$2^{3x} = 2^{x+1}$$
$$3x = x + 1$$
$$2x = 1$$
$$x = \frac{1}{2}$$

59. Use the formula $S = P(1 + i)^n$.
$$1500 = 1000(1 + 0.01)^n$$
$$1.5 = (1.01)^n$$
$$n = \log_{1.01}(1.5) = \frac{\ln(1.5)}{\ln(1.01)}$$
$$\approx 40.749$$
It takes approximately 41 months.

61. Use the formula $S = P(1 + i)^n$.
$$105 = 100(1 + 0.03/365)^n$$
$$1.05 = (1.00008)^n$$
$$n = \log_{1.00008}(1.05) = \frac{\ln(1.05)}{\ln(1.00008)} \approx 593.6$$
It takes approximately 594 days.

63. If $t = 0$, then
$$A = 10e^{-0.0001(0)} = 10 \text{ grams.}$$
If $A = 4$, then
$$4 = 10e^{-0.0001t}$$
$$0.4 = e^{-0.0001t}$$
$$-0.0001t = \ln(0.4)$$
$$t = \frac{\ln(0.4)}{-0.0001} \approx 9162.9$$
The cloth was made about 9163 years ago.

65. $y = 114.308e^{(0.265 \cdot 15.8)} \approx 7524 \text{ ft}^3/\text{sec}$

67. $40 = 28e^{0.05t}$
$$\frac{10}{7} = e^{0.05t}$$
$$0.05t = \ln(10/7)$$
$$t = \frac{\ln(10/7)}{0.05} \approx 7.133$$
There will be 40 million people above the
poverty level in approximately 7.1 years.

69. $28e^{0.05t} = 20e^{0.07t}$
$$\frac{e^{0.05t}}{e^{0.07t}} = \frac{20}{28}$$
$$e^{-0.02t} = 5/7$$
$$-0.02t = \ln(5/7)$$

$$t = \frac{\ln(5/7)}{-0.02} \approx 16.824$$

The number of people above the poverty level will equal the number below the poverty level in approximately 16.8 years.

71.
$$\text{pH} = -\log(\text{H}^+)$$
$$3.7 = -\log(\text{H}^+)$$
$$-3.7 = \log(\text{H}^+)$$
$$\text{H}^+ = 10^{-3.7} \approx 2.0 \times 10^{-4}$$

73. $d = \log_2\left(\dfrac{3\sqrt[3]{2}}{2}\right) = \dfrac{\ln\left(\dfrac{3\sqrt[3]{2}}{2}\right)}{\ln(2)}$
$$\approx 0.9183$$

75. Using logarithms:
$$x^3 = 12$$
$$3\ln(x) = \ln(12)$$
$$\ln(x) = \ln(12)/3$$
$$x = e^{\ln(12)/3} \approx 2.2894$$

Using roots:
$$x^3 = 12$$
$$x = \sqrt[3]{12} = 12^{1/3} \approx 2.2894$$

77. $(2.71, 6.54)$

79. $(1.03, 0.04), (4.73, 2.24)$

Enriching Your Mathematical Word Power

1. a   **2.** d   **3.** b   **4.** d   **5.** d

6. b   **7.** a   **8.** b   **9.** b   **10.** c

CHAPTER 11 REVIEW

1. $f(-2) = 5^{-2} = \dfrac{1}{5^2} = \dfrac{1}{25}$

3. $f(3) = 5^3 = 125$

5. $g(1) = 10^{1-1} = 10^0 = 1$

7. $g(0) = 10^{0-1} = 10^{-1} = \dfrac{1}{10}$

9. $h(-1) = (1/4)^{-1} = 4$

11. $h(1/2) = (1/4)^{1/2} = \sqrt{\dfrac{1}{4}} = \dfrac{1}{2}$

13. If $f(x) = 25$, then $5^x = 25$, $x = 2$.

15. If $g(x) = 1000$, then $10^{x-1} = 1000$.
$$10^{x-1} = 10^3$$
$$x - 1 = 3$$
$$x = 4$$

17. If $h(x) = 32$, then $(1/4)^x = 32$.
$$(2^{-2})^x = 2^5$$
$$2^{-2x} = 2^5$$
$$-2x = 5$$
$$x = -\dfrac{5}{2}$$

19. If $h(x) = 1/16$, then $(1/4)^x = 1/16$, or $(1/4)^x = (1/4)^2$, $x = 2$.

21. $f(1.34) = 5^{1.34} \approx 8.6421$

23. $g(3.25) = 10^{3.25-1} = 10^{2.25} \approx 177.828$

25. $h(2.82) = (1/4)^{2.82} = (0.25)^{2.82}$
$$\approx 0.02005$$

27. $h(\sqrt{2}) = (1/4)^{\sqrt{2}} = \approx 0.1408$

29. The graph of $f(x) = 5^x$ includes the points $(0, 1)$, $(1, 5)$, and $(-1, 1/5)$.

31. The graph of $y = (1/5)^x$ includes the points $(1, 1/5)$, $(0, 1)$ and $(-1, 5)$.

33. The graph of $y = 3^{-x}$ includes the points $(0, 1)$, $(1, 1/3)$, and $(-1, 3)$.

35. The graph of $y = 1 + 2^x$ includes the points $(0, 2)$, $(1, 3)$, $(2, 5)$, and $(-1, 1.5)$.

37. $\log(n) = m$

39. $k^h = t$

41. $f(1/8) = \log_2(1/8) = -3$, because $2^{-3} = 1/8$.

43. $g(0.1) = \log(0.1) = -1$, because $10^{-1} = 0.1$.

45. $g(100) = \log(100) = 2$, because $10^2 = 100$.

47. $h(1) = \log_{1/2}(1) = 0$, because $(1/2)^0 = 1$.

49. If $f(x) = 8$, then $\log_2(x) = 8$,
$x = 2^8 = 256$.

51. $f(77) = \log_2(77) = \dfrac{\ln(77)}{\ln(2)} \approx 6.267$

53. $h(33.9) = \log_{1/2}(33.9) = \dfrac{\ln(33.9)}{\ln(0.5)}$
≈ -5.083

55. If $f(x) = 2.475$, then $\log_2(x) \approx 2.475$.
$x = 2^{2.475} \approx 5.560$

57. The inverse of the function $f(x) = 10^x$ is the base 10 logarithm function,
$f^{-1}(x) = \log(x)$.
The graph of $f(x)$ includes the points $(1, 10)$, $(0, 1)$, and $(-1, 0.1)$. The graph of $f^{-1}(x)$ includes the points $(10, 1)$, $(1, 0)$, and $(0.1, -1)$.

59. The inverse of the function $f(x) = e^x$ is $f^{-1}(x) = \ln(x)$. The graph of $f(x)$ includes the points $(1, e)$, $(0, 1)$, and $(-1, 1/e)$. The graph of $f^{-1}(x)$ includes the points $(e, 1)$, $(1, 0)$, and $(1/e, -1)$.

61. $\log(x^2 y) = \log(x^2) + \log(y)$
$= 2\log(x) + \log(y)$

63. $\ln(16) = \ln(2^4) = 4\ln(2)$

65. $\log_5(1/x) = \log_5(1) - \log_5(x)$
$= -\log_5(x)$

67. $\frac{1}{2}\log(x + 2) - 2\log(x - 1)$
$= \log\big((x + 2)^{1/2}\big) - \log\big((x - 1)^2\big)$
$= \log\left(\dfrac{\sqrt{x + 2}}{(x - 1)^2}\right)$

69. $\log_2(x) = 8$
$x = 2^8 = 256$

The solution set is $\{256\}$.

71. $\log_2(8) = x$
$3 = x$
The solution set is $\{3\}$.

73. $x^3 = 8$
$x = \sqrt[3]{8} = 2$
The solution set is $\{2\}$.

75. $\log_x(27) = 3$
$x^3 = 27$
$x = \sqrt[3]{27} = 3$
The solution set is $\{3\}$.

77. $x \cdot \ln(3) - x = \ln(7)$
$x[\ln(3) - 1] = \ln(7)$
$x = \dfrac{\ln(7)}{\ln(3) - 1}$
The solution set is $\left\{ \dfrac{\ln(7)}{\ln(3) - 1} \right\}$.

79. $3^x = 5^{x-1}$
$\ln(3^x) = \ln(5^{x-1})$
$x \cdot \ln(3) = (x - 1)\ln(5)$
$x \cdot \ln(3) = x \cdot \ln(5) - \ln(5)$
$x \cdot \ln(3) - x \cdot \ln(5) = -\ln(5)$
$x[\ln(3) - \ln(5)] = -\ln(5)$
$x = \dfrac{-\ln(5)}{\ln(3) - \ln(5)} = \dfrac{\ln(5)}{\ln(5) - \ln(3)}$
The solution set is $\left\{ \dfrac{\ln(5)}{\ln(5) - \ln(3)} \right\}$.

81. $4^{2x} = 2^{x+1}$
$(2^2)^{2x} = 2^{x+1}$
$2^{4x} = 2^{x+1}$
$4x = x + 1$
$3x = 1$
$x = \dfrac{1}{3}$
The solution set is $\left\{ \dfrac{1}{3} \right\}$.

83. $\ln(x + 2) - \ln(x - 10) = \ln(2)$
$\ln\left(\dfrac{x + 2}{x - 10} \right) = \ln(2)$
$\dfrac{x + 2}{x - 10} = 2$
$x + 2 = 2(x - 10)$
$x + 2 = 2x - 20$
$22 = x$
The solution set is $\{22\}$.

85. $\log(x) - \log(x-2) = 2$

$$\log\left(\frac{x}{x-2}\right) = 2$$

$$\frac{x}{x-2} = 10^2$$

$$x = 100(x-2)$$

$$x = 100x - 200$$

$$-99x = -200$$

$$x = \frac{-200}{-99} = \frac{200}{99}$$

The solution set is $\left\{\frac{200}{99}\right\}$.

87. $6^x = 12$

$$x = \log_6(12) = \frac{\ln(12)}{\ln(6)} \approx 1.3869$$

The solution set is $\{1.3869\}$.

89. $3^{x+1} = 5$

$$x + 1 = \log_3(5)$$

$$x = -1 + \log_3(5) = -1 + \frac{\ln(5)}{\ln(3)}$$

$$\approx 0.4650$$

The solution set is $\{0.4650\}$.

91. Use the formula $S = P(1+i)^n$ with $i = 11.5\% = 0.115$, $n = 15$, and $P = \$10,000$.

$S = 10,000(1.115)^{15} \approx \$51,182.68$

93. Use the formula $A = A_0 e^{-0.0003t}$ with $A_0 = 218$ and $t = 1000$.

$$A = 218e^{-0.0003(1000)} = 218e^{-0.3}$$

$$\approx 161.5 \text{ grams}$$

95. The amount in Melissa's account is given by the formula $S = 1000(1.05)^t$ for any number of years t.   The amount in Frank's account is given by the formula $S = 900e^{0.07t}$ for any number of years t. To find when they have the same amount, we set the two expressions equal and solve for t.

$$1000(1.05)^t = 900e^{0.07t}$$

$$\ln\left(1000(1.05)^t\right) = \ln\left(900e^{0.07t}\right)$$

$$\ln(1000) + t \cdot \ln(1.05) = \ln(900) + \ln(e^{0.07t})$$

$$\ln(1000) + t \cdot \ln(1.05) = \ln(900) + 0.07t$$

$$t \cdot \ln(1.05) - 0.07t = \ln(900) - \ln(1000)$$

$$t[\ln(1.05) - 0.07] = \ln(900) - \ln(1000)$$

$$t = \frac{\ln(900) - \ln(1000)}{\ln(1.05) - 0.07} \approx 4.9675$$

The amounts will be equal in approximately 5 years.

97. $114.308e^{0.265(20.6-6.87)} \approx 4347.5 \text{ ft}^3/\text{sec}$

CHAPTER 11 TEST

1. $f(2) = 5^2 = 25$

2. $f(-1) = 5^{-1} = \frac{1}{5}$

3. $f(0) = 5^0 = 1$

4. $g(125) = \log_5(125) = 3$, because $5^3 = 125$.

5. $g(1) = \log_5(1) = 0$, because $5^0 = 1$.

6. $g(1/5) = \log_5(1/5) = -1$, because $5^{-1} = 1/5$.

7. The graph of $y = 2^x$ includes the points $(0, 1)$, $(1, 2)$, $(2, 4)$, and $(-1, 1/2)$.

8. The graph of $f(x) = \log_2(x)$ includes the points $(1, 0)$, $(2, 1)$, $(4, 2)$, and $(1/2, -1)$.

9. The graph of $y = \left(\frac{1}{3}\right)^x$ includes the points $(1, 1/3)$, $(0, 1)$, and $(-1, 3)$.

10. The graph of $g(x) = \log_{1/3}(x)$ includes the points $(1, 0)$, $(3, -1)$, and $(1/3, 1)$.

11. $\log_a(MN) = \log_a(M) + \log_a(N) = 6 + 4$
$$= 10$$

12. $\log_a(M^2/N) = 2 \cdot \log_a(M) - \log_a(N)$
$$= 2 \cdot 6 - 4 = 8$$

13. $\dfrac{\log_a(M)}{\log_a(N)} = \dfrac{6}{4} = \dfrac{3}{2}$

14. $\log_a(a^3 M^2) = 3 \cdot \log_a(a) + 2 \cdot \log_a(M)$
$$= 3 \cdot 1 + 2 \cdot 6 = 15$$

15. $\log_a(1/N) = \log_a(1) - \log_a(N)$
$$= 0 - 4 = -4$$

16. $\quad 3^x = 12$
$$x = \log_3(12)$$
The solution set is $\{\log_3(12)\}$.

17. $\quad \log_3(x) = 1/2$
$$x = 3^{1/2} = \sqrt{3}$$
The solution set is $\{\sqrt{3}\}$.

18. $\qquad 5^x = 8^{x-1}$
$$\ln(5^x) = \ln(8^{x-1})$$
$$x \cdot \ln(5) = (x-1)\ln(8)$$
$$x \cdot \ln(5) = x \cdot \ln(8) - \ln(8)$$
$$x \cdot \ln(5) - x \cdot \ln(8) = -\ln(8)$$
$$x[\ln(5) - \ln(8)] = -\ln(8)$$
$$x = \dfrac{-\ln(8)}{\ln(5) - \ln(8)} = \dfrac{\ln(8)}{-\ln(5) + \ln(8)}$$
The solution set is $\left\{ \dfrac{\ln(8)}{\ln(8) - \ln(5)} \right\}$.

19. $\quad \log(x) + \log(x+15) = 2$
$$\log(x^2 + 15x) = 2$$
$$x^2 + 15x = 10^2$$
$$x^2 + 15x - 100 = 0$$
$$(x+20)(x-5) = 0$$
$$x + 20 = 0 \quad \text{or} \quad x - 5 = 0$$
$$x = -20 \quad \text{or} \quad x = 5$$
If $x = -20$ in the original equation, we get a logarithm of a negative number, which is undefined. So the solution set is $\{5\}$.

20. $\qquad 2 \cdot \ln(x) = \ln(3) + \ln(6 - x)$
$$\ln(x^2) = \ln(18 - 3x)$$
$$x^2 = 18 - 3x$$
$$x^2 + 3x - 18 = 0$$
$$(x+6)(x-3) = 0$$
$$x + 6 = 0 \quad \text{or} \quad x - 3 = 0$$
$$x = -6 \quad \text{or} \quad x = 3$$
If $x = -6$ in the original equation, we get a logarithm of a negative number, which is undefined. So the solution set is $\{3\}$.

21. $\quad 20^x = 5$
$$x = \log_{20}(5) = \dfrac{\ln(5)}{\ln(20)} \approx 0.5372$$
The solution set is $\{0.5372\}$.

22. $\quad \log_3(x) = 2.75$
$$x = 3^{2.75} \approx 20.5156$$
The solution set is $\{20.5156\}$

23. To find the number present initially, let $t = 0$ in the formula:
$$N = 10e^{0.4(0)} = 10e^0 = 10 \cdot 1 = 10$$
To find the number present after 24 hours, let $t = 24$ in the formula:
$$N = 10e^{0.4(24)} = 10e^{9.6} \approx 147{,}648$$

24. To find how long it takes for the population to double, we must find the value of t for which $e^{0.4t} = 2$. Solve for t.
$$0.4t = \ln(2)$$
$$t = \dfrac{\ln(2)}{0.4} \approx 1.733$$
The bacteria population doubles in 1.733 hours.

Making Connections

Chapters 1 - 11

1. $\quad (x-3)^2 = 8$
$$x - 3 = \pm\sqrt{8}$$
$$x = 3 \pm 2\sqrt{2}$$
The solution set is $\left\{ 3 \pm 2\sqrt{2} \right\}$.

2. $\log_2(x-3) = 8$
$$x - 3 = 2^8$$
$$x = 256 + 3 = 259$$
The solution set is $\{259\}$.

3. $\qquad 2^{x-3} = 8$
$$2^{x-3} = 2^3$$
$$x - 3 = 3$$
$$x = 6$$
The solution set is $\{6\}$.

4. $2x - 3 = 8$
$$2x = 11$$
$$x = \dfrac{11}{2}$$
The solution set is $\left\{ \dfrac{11}{2} \right\}$.

5. $\quad |x - 3| = 8$
$$x - 3 = 8 \quad \text{or} \quad x - 3 = -8$$
$$x = 11 \quad \text{or} \quad x = -5$$

The solution set is $\{-5, 11\}$

6. $\qquad \sqrt{x-3} = 8$
$$(\sqrt{x-3})^2 = 8^2$$
$$x - 3 = 64$$
$$x = 67$$
The solution set is $\{67\}$.

7. $\log_2(x-3) + \log_2(x) = \log_2(18)$

$$\log_2(x^2 - 3x) = \log_2(18)$$
$$x^2 - 3x = 18$$
$$x^2 - 3x - 18 = 0$$
$$(x-6)(x+3) = 0$$
$$x - 6 = 0 \quad \text{or} \quad x + 3 = 0$$
$$x = 6 \quad \text{or} \quad x = -3$$

If $x = -3$ in the original equation, then we get a logarithm of a negative number, which is undefined. The solution set is $\{6\}$.

8. $2 \cdot \log_2(x-3) = \log_2(5-x)$

$$\log_2[(x-3)^2] = \log_2(5-x)$$
$$(x-3)^2 = 5 - x$$
$$x^2 - 6x + 9 = 5 - x$$
$$x^2 - 5x + 4 = 0$$
$$(x-4)(x-1) = 0$$
$$x - 4 = 0 \quad \text{or} \quad x - 1 = 0$$
$$x = 4 \quad \text{or} \quad x = 1$$

If $x = 1$ in the original equation, we get a logarithm of a negative number. The solution set is $\{4\}$.

9. $\quad \frac{1}{2}x - \frac{2}{3} = \frac{3}{4}x + \frac{1}{5}$

$$60\left(\frac{1}{2}x - \frac{2}{3}\right) = 60\left(\frac{3}{4}x + \frac{1}{5}\right)$$
$$30x - 40 = 45x + 12$$
$$-52 = 15x$$
$$-\frac{52}{15} = x$$

The solution set is $\left\{-\frac{52}{15}\right\}$.

10. To solve $3x^2 - 6x + 2 = 0$ use the quadratic formula.

$$x = \frac{6 \pm \sqrt{(-6)^2 - 4(3)(2)}}{2(3)} = \frac{6 \pm \sqrt{12}}{6}$$
$$= \frac{6 \pm 2\sqrt{3}}{6} = \frac{3 \pm \sqrt{3}}{3}$$

The solution set is $\left\{\frac{3 \pm \sqrt{3}}{3}\right\}$.

11. The inverse of dividing by 3 is multiplying by 3. So $f^{-1}(x) = 3x$.

12. The inverse of the base 3 logarithm function is the base 3 exponential function, $g^{-1}(x) = 3^x$.

13. The inverse of multiplying by 2 and then subtracting 4 is adding 4 and then dividing by 2, $f^{-1}(x) = \frac{x+4}{2}$.

14. The inverse of the square root function is the squaring function. To keep the domain of one function equal to the range of the inverse function we must restrict the squaring function to the nonnegative numbers: $h^{-1}(x) = x^2$ for $x \geq 0$

15. The reciprocal function is its own inverse, $j^{-1}(x) = \frac{1}{x}$.

16. The inverse of the base 5 exponential function is the base 5 logarithm function, $k^{-1}(x) = \log_5(x)$

17. We will find the inverse for m by using the technique of interchanging x and y and then solving for y.

$$m(x) = e^{x-1}$$
$$y = e^{x-1}$$
$$x = e^{y-1}$$
$$y - 1 = \ln(x)$$
$$y = 1 + \ln(x)$$
$$m^{-1}(x) = 1 + \ln(x)$$

18. The inverse of the natural logarithm function is the base e exponential function, $n^{-1}(x) = e^x$.

19. The graph of $y = 2x$ is a straight line with slope 2 and y-intercept $(0, 0)$. Start at the origin and rise 2 and go 1 to the right to find a second point on the line.

20. The graph of $y = 2^x$ includes the points $(0, 1)$, $(1, 2)$, $(2, 2)$, and $(-1, 1/2)$.

21. The graph of $y = x^2$ is a parabola through $(0, 0)$, $(1, 1)$, $(2, 4)$, $(-1, 1)$, and $(-2, 4)$.

22. The graph of $y = \log_2(x)$ includes the points $(1, 0)$, $(2, 1)$, $(1/2, -1)$ and $(4, 2)$.

23. The graph of $y = \frac{1}{2}x - 4$ is a straight line with slope $1/2$ and y-intercept $(0, -4)$. Start at $(0, -4)$, rise 1 and go 2 to the right to locate a second point on the line.

24. The graph of $y = |\, 2 - x\,|$ is a v-shaped graph through $(-2, 4)$, $(-1, 3)$, $(0, 2)$, $(1, 1)$, $(2, 0)$, $(3, 1)$, $(4, 2)$, and $(5, 3)$.

25. The graph of $y = 2 - x^2$ is a parabola with a vertex at $(0, 2)$. It also includes the points $(1, 1)$, $(-1, 1)$, $(2, -2)$, and $(-2, -2)$.

26. Note that e^2 is not a variable. The value of e^2 is approximately 7.389. The graph of $y = e^2$ is a straight line with 0 slope and y-intercept $(0, 7.389)$.

27. a) The graph of $n = 1.51t + 125.5$ is a straight line through $(0, 125.5)$ and $(10, 140.6)$. The graph of $n = 125.6e^{0.011t}$ is an exponential curve through $(0, 125.6)$, $(5, 132.7)$, $(10, 140.2)$, and $(15, 148.1)$.

b) In 2010, $t = 20$:
$$n = 1.51(20) + 125.5 \approx 155.7 \text{ million}$$
or
$$n = 125.6e^{0.011(20)} \approx 156.5 \text{ million}$$
c) Look up current data at www.bls.gov.

28. a) Use $D = RT$ to get $2d_1 = v(0.270)$ or $d_1 = 0.135v$.
b) Using $D = RT$ we get $2d_2 = v(0.432)$ or $d_2 = 0.216v$.
c) Using the Pythagorean theorem we get $(d_2)^2 = (d_1)^2 + 250^2$. Use $d_1 = 0.135v$ and $d_2 = 0.216v$ to get
$$(0.216v)^2 = (0.135v)^2 + 250^2$$
$$0.028431v^2 = 250^2$$
$$v^2 = 2198304.667$$
$$v = 1482.668091$$
$$d_1 = 0.135v \approx 200.2 \text{ meters}$$

12.1 WARM-UPS

1. True, because any equation of the form $y = ax^2 + bx + c$ has a graph that is a parabola.

2. False, because absolute value has a v-shaped graph.

3. False, because $-4 \neq \sqrt{5(3) + 1}$.

4. True, because $y = \sqrt{x}$ lies entirely in the first quadrant, and $y = -x - 2$ has no points in the first quadrant.

5. False, because we can also use addition.

6. True.

7. True, because a 30-60-90 triangle is half of an equilateral triangle.

8. True, because for any rectangular solid we have $V = LWH$.

9. True, because the surface area consists of 6 rectangles, of which two have area LW, two have area WH, and two have area LH.

10. True, because the area of a triangle is $(bh)/2$.

12.1 EXERCISES

1. If the graph of an equation is not a straight line then it is called nonlinear.

3. Graphing is not an accurate method for solving a system and the graphs might be difficult to draw.

5. The graph of $y = x^2$ is a parabola and the graph of $x + y = 6$ is a straight line.

To solve the system, substitute $y = x^2$ into $x + y = 6$.
$$x + x^2 = 6$$
$$x^2 + x - 6 = 0$$
$$(x + 3)(x - 2) = 0$$
$$x = -3 \ \text{ or } \ x = 2$$
If $x = -3$, $y = (-3)^2 = 9$, and if $x = 2$, $y = 2^2 = 4$. The solution set to the system is $\{(2, 4), (-3, 9)\}$.

7. The graph of $y = |x|$ is v-shaped, and the graph of $2y - x = 6$ is a straight line. To solve the system, substitute $x = 2y - 6$ into $y = |x|$.
$$y = |2y - 6|$$
$$y = 2y - 6 \ \text{or} \ \ y = -(2y - 6)$$
$$6 = y \qquad \text{or} \quad y = -2y + 6$$
$$3y = 6$$
$$y = 2$$
Use $y = 6$ in $x = 2y - 6$ to get $x = 6$. Use $y = 2$ in $x = 2y - 6$ to get $x = -2$. The graphs intersect at the points $(-2, 2)$ and $(6, 6)$.

The solution set is $\{(-2, 2), (6, 6)\}$.

9. The graph of $y = \sqrt{2x}$ includes the points $(0, 0)$, $(2, 2)$, and $(4.5, 3)$. The graph of $x - y = 4$ is a straight line with slope 1 and y-intercept $(0, -4)$.

Substitute $y = \sqrt{2x}$ into $y = x - 4$.
$$\sqrt{2x} = x - 4$$
$$(\sqrt{2x})^2 = (x - 4)^2$$
$$2x = x^2 - 8x + 16$$
$$0 = x^2 - 10x + 16$$
$$0 = (x - 8)(x - 2)$$
$$x = 8 \qquad \text{or} \quad x = 2$$
$$y = \sqrt{2(8)} \ \text{or} \ y = \sqrt{2(2)}$$
$$= 4 \qquad \text{or} \quad = 2$$

Since we squared both sides, we must check. The pair $(8, 4)$ satisfies both equations, but $(2, 2)$ does not. The solution set to the system is $\{(8, 4)\}$.

11. The graph of $4x - 9y = 9$ is a straight line. The equation $xy = 1$ can be written as $y = 1/x$. Substitute $y = 1/x$ into $4x - 9y = 9$.

$$4x - 9(\tfrac{1}{x}) = 9$$
$$4x^2 - 9 = 9x$$
$$4x^2 - 9x - 9 = 0$$
$$(4x + 3)(x - 3) = 0$$

$$x = -\tfrac{3}{4} \quad \text{or} \quad x = 3$$
$$y = \frac{1}{-3/4} = -\tfrac{4}{3} \qquad y = \tfrac{1}{3}$$

The solution set is $\left\{ \left(-\tfrac{3}{4}, -\tfrac{4}{3}\right), \left(3, \tfrac{1}{3}\right) \right\}$.

13. The graph of $y = x^2$ is a parabola opening upward, and the graph of $y = -x^2 + 1$ is a parabola opening downward.

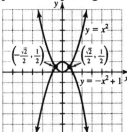

Substitute $y = x^2$ into $y = -x^2 + 1$.

$$x^2 = -x^2 + 1$$
$$2x^2 = 1$$
$$x^2 = \tfrac{1}{2}$$
$$x = \pm\sqrt{\tfrac{1}{2}} = \pm\frac{\sqrt{2}}{2}$$

Since $y = x^2$, $y = \tfrac{1}{2}$ for either value of x. The solution set is $\left\{ \left(\frac{\sqrt{2}}{2}, \tfrac{1}{2}\right), \left(-\frac{\sqrt{2}}{2}, \tfrac{1}{2}\right) \right\}$.

15. Write $y = x^2 - 5$ as $x^2 = y + 5$ and substitute into $x^2 + y^2 = 25$

$$y + 5 + y^2 = 25$$
$$y^2 + y - 20 = 0$$
$$(y + 5)(y - 4) = 0$$
$$y = -5 \quad \text{or} \quad y = 4$$

Use $y = -5$ in $x^2 = y + 5$ to get $x^2 = 0$ or $x = 0$. Use $y = 4$ in $x^2 = y + 5$ to get $x^2 = 9$ or $x = \pm 3$. The solution set to the system is $\{(0, -5), (3, 4), (-3, 4)\}$.

17. Substitute $y = x + 1$ into $xy - 3x = 8$.

$$x(x + 1) - 3x = 8$$
$$x^2 + x - 3x = 8$$
$$x^2 - 2x - 8 = 0$$
$$(x - 4)(x + 2) = 0$$

$$x = 4 \quad \text{or} \quad x = -2$$
$$y = 5 \qquad\qquad y = -1$$

The solution set is $\{(4, 5), (-2, -1)\}$.

19. Write $xy - x = 8$ as $xy = x + 8$, and substitute for xy in $xy + 3x = -4$.

$$x + 8 + 3x = -4$$
$$4x = -12$$
$$x = -3$$

Use $x = -3$ in $xy = x + 8$.

$$-3y = -3 + 8$$
$$-3y = 5$$
$$y = -5/3$$

The solution set is $\left\{ \left(-3, -\tfrac{5}{3}\right) \right\}$.

21. If we add $x^2 + y^2 = 8$ and $x^2 - y^2 = 2$ we get $2x^2 = 10$.

$$x^2 = 5$$
$$x = \pm\sqrt{5}$$

Use $x = \sqrt{5}$ in $y^2 = 8 - x^2$, to get $y = \pm\sqrt{3}$. Use $x = -\sqrt{5}$ in $y^2 = 8 - x^2$, to get $y = \pm\sqrt{3}$.

The solution set contains four points,
$$\{(\sqrt{5}, \sqrt{3}), (\sqrt{5}, -\sqrt{3}), (-\sqrt{5}, \sqrt{3}), (-\sqrt{5}, -\sqrt{3})\}$$

23. Write $x^2 + 2y^2 = 8$ as $x^2 = 8 - 2y^2$ and substitute this equation into $2x^2 - y^2 = 1$.

$$2(8 - 2y^2) - y^2 = 1$$
$$16 - 4y^2 - y^2 = 1$$
$$-5y^2 = -15$$
$$y^2 = 3$$
$$y = \pm\sqrt{3}$$

If $y = \sqrt{3}$, then $x^2 = 8 - 2(\sqrt{3})^2 = 2$, or $x = \pm\sqrt{2}$.

If $y = -\sqrt{3}$, then $x^2 = 8 - 2(-\sqrt{3})^2 = 2$, or $x = \pm\sqrt{2}$. The solution set is
$$\{(\sqrt{2}, \sqrt{3}), (\sqrt{2}, -\sqrt{3}), (-\sqrt{2}, \sqrt{3}), (-\sqrt{2}, -\sqrt{3})\}$$

25. If we just add the equations as they are given, then y is eliminated.

$$\frac{1}{x} - \frac{1}{y} = 5$$
$$\frac{2}{x} + \frac{1}{y} = -3$$
$$\overline{\qquad\qquad}$$
$$\frac{3}{x} \qquad = 2$$
$$3 = 2x$$
$$3/2 = x$$

Use $x = 3/2$ in $\frac{1}{x} - \frac{1}{y} = 5$.

$$\frac{1}{3/2} - \frac{1}{y} = 5$$

$$\frac{2}{3} - \frac{1}{y} = 5$$
$$2y - 3 = 15y$$
$$-3 = 13y$$
$$-3/13 = y$$

The solution set is $\left\{\left(\frac{3}{2}, -\frac{3}{13}\right)\right\}$.

27. Multiply the first equation by -3 and add the result to the second equation.

$$\frac{-6}{x} + \frac{3}{y} = \frac{-15}{12}$$
$$\frac{1}{x} - \frac{3}{y} = -\frac{5}{12}$$
$$\overline{\qquad\qquad}$$
$$\frac{-5}{x} \qquad = \frac{-20}{12}$$
$$-20x = -60$$
$$x = 3$$

Use $x = 3$ in the first equation to find y.

$$\frac{2}{3} - \frac{1}{y} = \frac{5}{12}$$
$$12y\left(\frac{2}{3} - \frac{1}{y}\right) = 12y \cdot \frac{5}{12}$$
$$8y - 12 = 5y$$
$$3y = 12$$
$$y = 4$$

The solution set is $\{(3, 4)\}$.

29. Substitute $y = 20/x^2$ into $xy + 2 = 6x$.

$$x \cdot \frac{20}{x^2} + 2 = 6x$$
$$\frac{20}{x} + 2 = 6x$$
$$20 + 2x = 6x^2$$
$$0 = 6x^2 - 2x - 20$$
$$3x^2 - x - 10 = 0$$
$$(3x + 5)(x - 2) = 0$$
$$x = -5/3 \quad\text{or}\quad x = 2$$

Use $y = 20/x^2$ to get

$$y = 36/5 \qquad y = 5.$$

The solution set is $\left\{\left(-\frac{5}{3}, \frac{36}{5}\right), (2, 5)\right\}$.

31. Substitute $y = 7 - x$ in $x^2 + xy - y^2 = -11$.

$$x^2 + x(7 - x) - (7 - x)^2 = -11$$
$$x^2 + 7x - x^2 - 49 + 14x - x^2 = -11$$
$$-x^2 + 21x - 38 = 0$$
$$x^2 - 21x + 38 = 0$$
$$(x - 2)(x - 19) = 0$$
$$x = 2 \quad\text{or}\qquad x = 19$$

Since $y = 7 - x$

$$y = 5 \qquad\qquad y = -12$$

The solution set is $\{(2, 5), (19, -12)\}$.

33. If $x^2 = y$, then $x^4 = y^2$. Substitute for x^4 in the equation $3y - 2 = x^4$.

$$3y - 2 = y^2$$
$$0 = y^2 - 3y + 2$$
$$0 = (y - 1)(y - 2)$$
$$y = 1 \quad\text{or}\qquad y = 2$$

Use $y = 1$ in $x^2 = y$ to get $x^2 = 1$, or $x = \pm 1$. Use $y = 2$ in $x^2 = y$ to get $x^2 = 2$, or $x = \pm\sqrt{2}$. The solution set is
$$\left\{(\sqrt{2}, 2), (-\sqrt{2}, 2), (1, 1), (-1, 1)\right\}.$$

35. Eliminate y by substitution.

$$\log_2(x - 1) = 3 - \log_2(x + 1)$$
$$\log_2(x - 1) + \log_2(x + 1) = 3$$
$$\log_2[x^2 - 1] = 3$$
$$x^2 - 1 = 8$$
$$x^2 = 9$$
$$x = \pm 3$$

If $x = -3$, we get a logarithm of a negative number. If $x = 3$, then $y = \log_2(3 - 1) = 1$. The solution set is $\{(3, 1)\}$.

37. Use substitution to eliminate y.

$$\log_2(x - 1) = 2 + \log_2(x + 2)$$
$$\log_2(x - 1) - \log_2(x + 2) = 2$$
$$\log_2\left(\frac{x - 1}{x + 2}\right) = 2$$
$$\frac{x - 1}{x + 2} = 4$$
$$x - 1 = 4x + 8$$
$$-9 = 3x$$
$$-3 = x$$

If $x = -3$ in either of the original equations, we get a logarithm of a negative number. So the solution set is the empty set, $\emptyset$.

39. Use substitution to eliminate y.
$$2^{3x+4} = 4^{x-1}$$
$$2^{3x+4} = (2^2)^{x-1}$$
$$2^{3x+4} = 2^{2x-2}$$
$$3x + 4 = 2x - 2$$
$$x = -6$$

If $x = -6$, then $y = 4^{-6-1} = 4^{-7}$. So the solution set is $\left\{ \left(-6, 4^{-7}\right) \right\}$.

41. Let x = the length of one leg and y = the length of the other. We write one equation for the area: $3 = \frac{1}{2}xy$. We write the other equation from the Pythagorean theorem: $x^2 + y^2 = 15$.

Substitute $y = 6/x$ into $x^2 + y^2 = 15$.
$$x^2 + \left(\frac{6}{x}\right)^2 = 15$$
$$x^2 + \frac{36}{x^2} = 15$$
$$x^4 + 36 = 15x^2$$
$$x^4 - 15x^2 + 36 = 0$$
$$(x^2 - 3)(x^2 - 12) = 0$$
$$x = \sqrt{3} \quad \text{or} \quad x = \sqrt{12} = 2\sqrt{3}$$

If $x = \sqrt{3}$, then $y = 6/\sqrt{3} = 2\sqrt{3}$. If $x = 2\sqrt{3}$, then $y = 6/(2\sqrt{3}) = \sqrt{3}$. So the lengths of the legs are $\sqrt{3}$ feet and $2\sqrt{3}$ feet.

43. Let h = the height of each triangle, and b = the length of the base of each triangle. Since the 7 triangles are to have a total area of 3,500, the area of each must be 500: $\frac{1}{2}hb = 500$. The ratio of the height to base must be 1 to 4 is expressed as $\frac{h}{b} = \frac{1}{4}$. These two equations can be written as $hb = 1000$ and $b = 4h$. Use substitution.
$$h(4h) = 1000$$
$$h^2 = 250$$
$$h = \sqrt{250} = 5\sqrt{10}$$

Since $b = 4h$, $b = 20\sqrt{10}$. So the height is $5\sqrt{10}$ inches and the base is $20\sqrt{10}$ inches.

45. Let x = the number of hours for pump A to fill the tank alone, and y = the number of hours for pump B to fill the tank alone.
$$\frac{1}{x} + \frac{1}{y} = \frac{1}{6}$$
$$\frac{1}{y} - \frac{1}{x} = \frac{1}{12}$$

Adding these two equations eliminates x.
$$\frac{2}{y} = \frac{1}{6} + \frac{1}{12}$$
$$\frac{2}{y} = \frac{1}{4}$$

$$y = 8$$

Use $y = 8$ in the first equation.
$$\frac{1}{x} + \frac{1}{8} = \frac{1}{6}$$
$$\frac{1}{x} = \frac{1}{24}$$
$$x = 24$$

It would take pump A 24 hours to fill the tank alone and pump B 8 hours to fill the tank alone.

47. Let x = the time for Jan to do the job alone. Let y = the time for Beth to do the job alone. Since they do the job together in 24 minutes we have the equation
$$\frac{1}{x} + \frac{1}{y} = \frac{1}{24}.$$

In 50 minutes the job was completed, but not by working together. This implies that the total of their times alone is 100, $x + y = 100$. Substitute $y = 100 - x$ into the first equation.

$$\frac{1}{x} + \frac{1}{100 - x} = \frac{1}{24}$$
$$24(100 - x) + 24x = x(100 - x)$$
$$2400 - 24x + 24x = 100x - x^2$$
$$x^2 - 100x + 2400 = 0$$
$$(x - 40)(x - 60) = 0$$
$$x = 40 \quad \text{or} \quad x = 60$$

Since $y = 100 - x$
$$y = 60 \qquad\qquad y = 40$$

Since Jan is the faster worker, it would take her 40 minutes to complete the catfish by herself.

49. Let x = the length and y = the width. The area is 72 is expressed as $xy = 72$. The perimeter is 34 is expressed as $2x + 2y = 34$ or $x + y = 17$. Substitute $y = 17 - x$ into $xy = 72$.

$$x(17 - x) = 72$$
$$-x^2 + 17x - 72 = 0$$
$$x^2 - 17x + 72 = 0$$
$$(x - 9)(x - 8) = 0$$
$$x = 9 \quad \text{or} \quad x = 8$$
$$y = 8 \qquad\qquad y = 9$$

The rectangular area is 8 feet by 9 feet.

51. Let x and y represent the numbers.

$$x + y = 8$$
$$xy = 20$$

Substitute $y = 8 - x$ into $xy = 20$.

$$x(8 - x) = 20$$
$$-x^2 + 8x - 20 = 0$$
$$x^2 - 8x + 20 = 0$$
$$x = \frac{8 \pm \sqrt{64 - 4(1)(20)}}{2} = \frac{8 \pm \sqrt{-16}}{2}$$
$$= 4 \pm 2i$$

If $x = 4 + 2i$, then $y = 8 - (4 + 2i) = 4 - 2i$.
If $x = 4 - 2i$, then $y = 8 - (4 - 2i) = 4 + 2i$.
So the numbers are $4 - 2i$ and $4 + 2i$.

53. Let $x = $ the length of the side of the square, and $y = $ the height of the triangle. The total height of 10 feet means that $x + y = 10$. The total area is 72 means that $x^2 + \frac{1}{2}xy = 72$. Substitute $y = 10 - x$ into the second equation.

$$x^2 + \frac{1}{2}x(10 - x) = 72$$
$$x^2 + 5x - \frac{1}{2}x^2 = 72$$
$$\frac{1}{2}x^2 + 5x - 72 = 0$$
$$x^2 + 10x - 144 = 0$$
$$(x - 8)(x + 18) = 0$$
$$x = 8 \quad \text{or} \quad x = -18$$
$$y = 2$$

The side of the square is 8 feet and the height of the triangle is 2 feet.

55. a) $(1.71, 1.55), (-2.98, -3.95)$
b) $(1, 1), (0.40, 0.16)$
c) $(1.17, 1.62), (-1.17, -1.62)$

12.2 WARM-UPS

1. False, because if the focus is $(2, 3)$ and the directrix is $y = 1$, then the vertex is $(2, 2)$.
2. True, because the focus is 1 unit above the vertex which is $(0, 1)$.
3. True, because in $y = a(x - h)^2 + k$, the vertex is (h, k).
4. False, because $y = 6x + 3x + 2$ is equivalent to $y = 9x + 2$, which is a straight line.
5. False, because $y = -x^2 + 2x + 9$ opens downward.
6. True, the vertex is $(0, 0)$ and so is the y-intercept.

7. True, because it opens upward from $(2, 3)$ and it cannot intersect the x-axis.
8. True, because the focus is above the directrix in a parabola the opens upward.
9. True, because the vertex is $(2, k)$ and the axis of symmetry is a vertical line through the vertex.
10. True, because $1/(4 \cdot \frac{1}{4}) = 1$.

12.2 EXERCISES

1. A parabola is the set of all points in a plane that are equidistant from a given line and a fixed point not on the line.
3. A parabola can be written in the forms $y = ax^2 + bx + c$ or $y = a(x - h)^2 + k$.
5. We use completing the square to convert $y = ax^2 + bx + c$ into $y = a(x - h)^2 + k$.
7. $\sqrt{(4 - 6)^2 + (2 - 5)^2} = \sqrt{4 + 9} = \sqrt{13}$
9. $\sqrt{(1 - 3)^2 + (-3 - 5)^2} = \sqrt{4 + 64}$
$$= 2\sqrt{17}$$
11. $\sqrt{(-3 - 4)^2 + (-6 - (-2))^2}$
$$= \sqrt{49 + 16} = \sqrt{65}$$
13. The vertex is $(0, 0)$. Since $2 = 1/(4p)$, $p = 1/8$. So the focus is $(0, 1/8)$ and the directrix is $y = -1/8$.
15. The vertex is $(0, 0)$. Since $-1/4 = 1/(4p)$, $4p = -4$, or $p = -1$. So the focus is $(0, -1)$ and the directrix is $y = 1$.

17. The vertex is $(3, 2)$. Since $1/2 = 1/(4p)$, $p = 1/2$. So the focus is $(3, 2.5)$ and the directrix is $y = 1.5$.

19. The vertex is $(-1, 6)$. Since $-1 = 1/(4p)$, $p = -1/4$. So the focus is $(-1, 5.75)$ and the directrix is $y = 6.25$.

21. Since the distance between the focus and directrix is 4, $p = 2$ and $a = \frac{1}{4p} = \frac{1}{4(2)} = \frac{1}{8}$.
Since the vertex is half way between the focus and directrix, the vertex is $(0, 0)$. Use the form $y = a(x - h)^2 + k$ to get the equation.
$$y = \frac{1}{8}(x - 0)^2 + 0$$
$$y = \frac{1}{8}x^2$$

23. Since the distance between the focus and directrix is 1 and the focus is below the directrix,

$p = -\frac{1}{2}$ and $a = \frac{1}{4p} = \frac{1}{4(-\frac{1}{2})} = -\frac{1}{2}$. Since

the vertex is half way between the focus and directrix, the vertex is $(0, 0)$. Use the form $y = a(x - h)^2 + k$ to get the equation.

$$y = -\frac{1}{2}(x - 0)^2 + 0$$
$$y = -\frac{1}{2}x^2$$

25. Since the distance between the focus and directrix is 1 and the parabola opens upward,

$p = \frac{1}{2}$ and $a = \frac{1}{4p} = \frac{1}{4(\frac{1}{2})} = \frac{1}{2}$. Since the

vertex is half way between the focus and directrix, the vertex is $\left(3, \frac{3}{2}\right)$. Use the form $y = a(x - h)^2 + k$ to get the equation.

$$y = \frac{1}{2}(x - 3)^2 + \frac{3}{2}$$
$$y = \frac{1}{2}x^2 - 3x + \frac{9}{2} + \frac{3}{2}$$
$$y = \frac{1}{2}x^2 - 3x + 6$$

27. Since the distance between the focus and directrix is 4 and the parabola opens downward, $p = -2$ and

$a = \frac{1}{4p} = \frac{1}{4(-2)} = -\frac{1}{8}$. Since the

vertex is half way between the focus and directrix, the vertex is $(1, 0)$. Use the form $y = a(x - h)^2 + k$ to get the equation.

$$y = -\frac{1}{8}(x - 1)^2 + 0$$
$$y = -\frac{1}{8}x^2 + \frac{1}{4}x - \frac{1}{8}$$

29. Since the distance between the focus and directrix is $\frac{1}{2}$ and the parabola opens upward,

$p = \frac{1}{4}$ and $a = \frac{1}{4p} = \frac{1}{4(0.25)} = 1$. Since the

vertex is half way between the focus and directrix, the vertex is $(-3, 1)$. Use the form $y = a(x - h)^2 + k$ to get the equation.

$$y = 1(x + 3)^2 + 1$$
$$y = x^2 + 6x + 10$$

31. $\quad y = x^2 - 6x + 1$
$$y = x^2 - 6x + 9 - 9 + 1$$
$$y = (x - 3)^2 - 8$$

The vertex is $(3, -8)$. Because $a = 1$, the parabola opens upward, and $p = 1/4$. The focus is $(3, -7.75)$ and the directrix is

$y = -8.25$. The axis of symmetry is the vertical line through the vertex, $x = 3$.

33. $\quad y = 2x^2 + 12x + 5$
$$y = 2(x^2 + 6x) + 5$$
$$y = 2(x^2 + 6x + 9) + 5 - 18$$
$$y = 2(x + 3)^2 - 13$$

The vertex is $(-3, -13)$ and the parabola opens upward. Because $a = 2$, $p = 1/8 = 0.125$ The focus is $(-3, -12.875)$ and the directrix is $y = -13.125$. The axis of symmetry is $x = -3$.

35. $\quad y = -2x^2 + 16x + 1$
$$y = -2(x^2 - 8x) + 1$$
$$y = -2(x^2 - 8x + 16) + 1 + 32$$
$$y = -2(x - 4)^2 + 33$$

The vertex is $(4, 33)$ and the parabola opens downward. Because $a = -2$, $p = -1/8 = -0.125$. The focus is $(4, 32\frac{7}{8})$ and the directrix is $y = 33\frac{1}{8}$. The axis of symmetry is $x = 4$.

37. $\quad y = 5x^2 + 40x$
$$y = 5(x^2 + 8x)$$
$$y = 5(x^2 + 8x + 16) - 80$$
$$y = 5(x + 4)^2 - 80$$

The vertex is $(-4, -80)$ and the parabola opens upward. Because $a = 5$, $p = 1/20$. The focus is $(-4, -79\frac{19}{20})$ and the directrix is $y = -80\frac{1}{20}$. The axis of symmetry is $x = -4$.

39. The x-coordinate of the vertex is $x = \frac{-b}{2a}$.

$$x = \frac{-(-4)}{2(1)} = 2$$
$$y = 2^2 - 4(2) + 1 = -3$$

The vertex is $(2, -3)$ and the parabola opens upward. Because $a = 1$, $p = 1/4$. The focus is $(2, -2\frac{3}{4})$ and the directrix is $y = -3\frac{1}{4}$. Axis of symmetry is $x = 2$.

41. The x-coordinate of the vertex is $x = \frac{-b}{2a}$.

$$x = \frac{-(2)}{2(-1)} = 1$$
$$y = -1^2 + 2(1) - 3 = -2$$

The vertex is $(1, -2)$ and the parabola opens downward. Because $a = -1$, $p = -1/4$. The focus is $(1, -2\frac{1}{4})$ and the directrix is $y = -1\frac{3}{4}$. The axis of symmetry is $x = 1$.

43. The x-coordinate of the vertex is $x = \frac{-b}{2a}$.
$$x = \frac{-(-6)}{2(3)} = 1$$
$$y = 3(1^2) - 6(1) + 1 = -2$$
The vertex is $(1, -2)$ and the parabola opens upward. Because $a = 3$, $p = 1/12$. The focus is $(1, -1\frac{11}{12})$ and the directrix is $y = -2\frac{1}{12}$. The axis of symmetry is $x = 1$.

45. The x-coordinate of the vertex is $x = \frac{-b}{2a}$.
$$x = \frac{-(-3)}{2(-1)} = -\frac{3}{2}$$
$$y = -(-\tfrac{3}{2})^2 - 3(-\tfrac{3}{2}) + 2 = \frac{17}{4}$$
The vertex is $(-\frac{3}{2}, \frac{17}{4})$ and the parabola opens downward. Because $a = -1$, $p = -1/4$. The focus is $(-\frac{3}{2}, 4)$ and the directrix is $y = \frac{9}{2}$. The axis of symmetry is $x = -\frac{3}{2}$.

47. The x-coordinate of the vertex is $x = \frac{-b}{2a}$.
$$x = \frac{-(0)}{2(3)} = 0$$
$$y = 3(0)^2 + 5 = 5$$
The vertex is $(0, 5)$ and the parabola opens upward. Because $a = 3$, $p = 1/12$. The focus is $(0, 5\frac{1}{12})$ and the directrix is $y = 4\frac{11}{12}$. The axis of symmetry is $x = 0$.

49. For $x = (y - 2)^2 + 3$ the vertex is $(3, 2)$. Since $a = 1$ we have $p = 1/4$ and the parabola opens to the right. The focus is $\left(\frac{13}{4}, 2\right)$ and the directrix is $x = \frac{11}{4}$.

51. For $x = \frac{1}{4}(y - 1)^2 - 2$ the vertex is $(-2, 1)$. Since $a = 1/4$ we have $p = 1$ and the parabola opens to the right. The focus is $(-1, 1)$ and the directrix is $x = -3$.

53. For $x = -\frac{1}{2}(y - 2)^2 + 4$ the vertex is $(4, 2)$. Since $a = -1/2$ we have $p = -1/2$ and the parabola opens to the left. The focus is $\left(\frac{7}{2}, 2\right)$ and the directrix is $x = \frac{9}{2}$.

55. For $y = (x - 2)^2 + 3$ the vertex is $(2, 3)$ and the parabola opens upward. The graph also goes through $(0, 7)$, $(1, 4)$, $(3, 4)$, and $(4, 7)$.

57. For $y = -2(x - 1)^2 + 3$ the vertex is $(1, 3)$ and the parabola opens downward. The graph also goes through $(-1, -5)$, $(0, 1)$, $(2, 1)$, and $(3, -5)$.

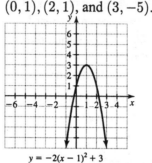

59. For $x = (y - 2)^2 + 3$ the vertex is $(3, 2)$ and the parabola opens to the right. The graph also goes through $(4, 3)$, $(4, 1)$, $(7, 4)$, and $(7, 0)$.

61. For $x = -2(y - 1)^2 + 3$ the vertex is $(3, 1)$ and the parabola opens to the left. The graph also goes through $(1, 2)$, $(1, 0)$, $(-5, 3)$, and $(-5, -1)$.

63. The distance from the vertex to the focus is 15. So $p = 15$ and $a = 1/(4 \cdot 15) = 1/60$. the equation of the parabola is $y = \frac{1}{60}x^2$.

65. The graph of $y = -x^2 + 3$ is a parabola opening downward, and the graph of $y = x^2 + 1$ is a parabola opening upward. To find the points of intersection, eliminate y by substitution.

$$x^2 + 1 = -x^2 + 3$$
$$2x^2 = 2$$
$$x^2 = 1$$
$$x = \pm 1$$

If $x = \pm 1$, then $y = 2$ (since $y = x^2 + 1$). The solution set for the system is $\{(-1, 2), (1, 2)\}$.

67. The graph of $y = x^2 - 2$ is a parabola opening upward, and the graph of $y = 2x - 3$ is a straight line.

We eliminate y by substitution.

$$x^2 - 2 = 2x - 3$$
$$x^2 - 2x + 1 = 0$$
$$(x - 1)^2 = 0$$
$$x = 1$$

If $x = 1$, then $y = -1$ (because $y = 2x - 3$). The solution set for the system is $\{(1, -1)\}$.

69. The graph of $y = x^2 + 3x - 4$ is a parabola opening upward, and the graph of $y = -x^2 - 2x + 8$ is a parabola opening downward.

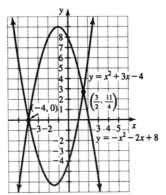

Eliminate y by substitution.

$$x^2 + 3x - 4 = -x^2 - 2x + 8$$
$$2x^2 + 5x - 12 = 0$$
$$(2x - 3)(x + 4) = 0$$
$$x = 3/2 \quad \text{or} \quad x = -4$$

If $x = 3/2$, then

$$y = (3/2)^2 + 3(3/2) - 4 = 11/4.$$

If $x = -4$, then

$$y = (-4)^2 + 3(-4) - 4 = 0.$$

The solution set for the system is

$$\left\{ \left(\frac{3}{2}, \frac{11}{4} \right), (-4, 0) \right\}.$$

71. The graph of $y = x^2 + 3x - 4$ is a parabola opening upward. The graph of $y = 2x + 2$ is a straight line with slope 2 and y-intercept $(0, 2)$.

Use substitution to eliminate y.

$$x^2 + 3x - 4 = 2x + 2$$
$$x^2 + x - 6 = 0$$
$$(x + 3)(x - 2) = 0$$
$$x = -3 \quad \text{or} \quad x = 2$$

Since $y = 2x + 2$

$$y = -4 \qquad y = 6$$

The solution set for the system is $\{(-3, -4), (2, 6)\}$.

73. On the x-axis, the y-coordinate is zero. So we solve $x^2 - 2x - 3 = 0$.
$$(x - 3)(x + 1) = 0$$
$$x - 3 = 0 \quad \text{or} \quad x + 1 = 0$$
$$x = 3 \quad \text{or} \quad x = -1$$
The points of intersection are $(3, 0)$ and $(-1, 0)$.

75. Substitute $y = 4$ into $y = 0.01x^2$:
$$4 = 0.01x^2$$
$$400 = x^2$$
$$\pm 20 = x$$
The points of intersection are $(-20, 4)$ and $(20, 4)$

77. Substitute to get $x = (x^2)^2$ or $x = x^4$.
$$x = x^4$$
$$x^4 - x = 0$$
$$x(x^3 - 1) = 0$$
$$x(x - 1)(x^2 + x + 1) = 0$$
$$x = 0 \text{ or } \quad x - 1 = 0 \quad \text{ or } \quad x^2 + x + 1 = 0$$
$$x = 0 \text{ or } \quad\quad\quad x = 1$$
There are no real solutions to $x^2 + x + 1 = 0$. If $x = 0$, then $y = 0^2 = 0$. If $x = 1$, then $y = 1^2 = 1$. So the points of intersection are $(0, 0)$ and $(1, 1)$.

79. The distance from (x, y) to the focus $(p, 0)$ is $\sqrt{(x - p)^2 + (y - 0)^2}$. The distance from (x, y) to the vertical line $x = -p$ is the distance between the points (x, y) and $(-p, y)$, which is $\sqrt{(x - (-p))^2 + (y - y)^2}$. Set these equal, square each side, then simplify to get
$$x^2 - 2xp + p^2 + y^2 = x^2 + 2xp + p^2.$$
Solve for x to get $x = \frac{1}{4p}y^2$ or $x = ay^2$ where $a = 1/(4p)$.

81. The graphs have identical shapes.

12.3 WARM-UPS

1. False, the radius of a circle is a positive real number.

2. False, the coordinates of the center do not satisfy the equation of the circle, only points on the circle satisfy the equation.

3. True, the center is $(0, 0)$.

4. False, the equation of the circle centered at the origin with radius 9 is $x^2 + y^2 = 81$.

5. False, because $(x - 2)^2 + (y - 3)^2 = 4$ has radius 2.

6. False, because $(x - 3)^2 + (y + 5)^2 = 9$ is a circle of radius 3 centered at $(3, -5)$

7. True, because the distance from $(-3, -1)$ to $(0, 0)$ is the radius.

8. False, the center is $(3, 4)$.

9. True, because there is only an x^2-term and no x-term.

10. False, because if we complete the square for x then the right side will no longer be 4.

12.3 EXERCISES

1. A circle is the set of all points in a plane that lie at a fixed distance from a fixed point.

3. Use $h = 0$, $k = 3$, and $r = 5$ in the standard equation $(x - h)^2 + (y - k)^2 = r^2$ to get the equation $x^2 + (y - 3)^2 = 25$.

5. Use $h = 1$, $k = -2$, and $r = 9$ in the standard equation $(x - h)^2 + (y - k)^2 = r^2$ to get the equation $(x - 1)^2 + (y + 2)^2 = 81$.

7. Use $h = 0$, $k = 0$, and $r = \sqrt{3}$ in the standard equation $(x - h)^2 + (y - k)^2 = r^2$ to get the equation $x^2 + y^2 = 3$.

9. Use $h = -6$, $k = -3$, and $r = 1/2$ in the standard equation $(x - h)^2 + (y - k)^2 = r^2$ to get the equation $(x + 6)^2 + (y + 3)^2 = \frac{1}{4}$.

11. Use $h = 1/2$, $k = 1/3$, and $r = 0.1$ in the standard equation $(x - h)^2 + (y - k)^2 = r^2$ to get the equation
$$\left(x - \tfrac{1}{2}\right)^2 + \left(y - \tfrac{1}{3}\right)^2 = 0.01.$$

13. Compare $(x - 3)^2 + (y - 5)^2 = 2$ with $(x - h)^2 + (y - k)^2 = r^2$ *(where the center is (h, k) and radius is r)*, to get a center at $(3, 5)$ and radius $\sqrt{2}$.

15. Compare $x^2 + \left(y - \tfrac{1}{2}\right)^2 = \tfrac{1}{2}$ with $(x - h)^2 + (y - k)^2 = r^2$ *(where the center is (h, k) and radius is r)*, to get a center at $\left(0, \tfrac{1}{2}\right)$ and radius $\sqrt{\tfrac{1}{2}}$ or $\frac{\sqrt{2}}{2}$.

17. Divide each side by 4 to get $x^2 + y^2 = \frac{9}{4}$, which has center $(0, 0)$ and radius $\sqrt{\frac{9}{4}}$ or $\frac{3}{2}$.

19. Rewrite the equation as $(x-2)^2 + y^2 = 3$, which has center $(2, 0)$ and radius $\sqrt{3}$.

21. The graph of $x^2 + y^2 = 9$ is a circle with radius 3, centered at $(0, 0)$.

23. The graph of $x^2 + (y-3)^2 = 9$ is a circle with radius 3, centered at $(0, 3)$.

25. The graph of $(x+1)^2 + (y-1)^2 = 2$ is a circle with radius $\sqrt{2}$, centered at $(-1, 1)$.

27. The graph of $(x-4)^2 + (y+3)^2 = 16$ is a circle of radius 4, centered at $(4, -3)$.

29. The graph of $\left(x - \frac{1}{2}\right)^2 + \left(y + \frac{1}{2}\right)^2 = \frac{1}{4}$ is a circle of radius $1/2$ and center $(1/2, -1/2)$.

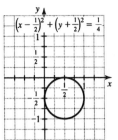

31.
$$x^2 + 4x + y^2 + 6y = 0$$
$$x^2 + 4x + 4 + y^2 + 6y + 9 = 0 + 4 + 9$$
$$(x+2)^2 + (y+3)^2 = 13$$
The circle has radius $\sqrt{13}$ and center $(-2, -3)$.

33.
$$x^2 - 2x + y^2 - 4y - 3 = 0$$
$$x^2 - 2x + 1 + y^2 - 4y + 4 = 3 + 1 + 4$$
$$(x-1)^2 + (y-2)^2 = 8$$
This circle has radius $\sqrt{8} = 2\sqrt{2}$ and center $(1, 2)$.

35.
$$x^2 - 10x + y^2 - 8y = -32$$
$$x^2 - 10x + 25 + y^2 - 8y + 16 = -32 + 25 + 16$$
$$(x-5)^2 + (y-4)^2 = 9$$
This circle has radius 3 and center $(5, 4)$.

37.
$$x^2 - x + y^2 + y = 0$$
$$x^2 - x + \frac{1}{4} + y^2 + y + \frac{1}{4} = \frac{1}{4} + \frac{1}{4}$$
$$\left(x - \frac{1}{2}\right)^2 + \left(y + \frac{1}{2}\right)^2 = \frac{1}{2}$$
Center is $(1/2, -1/2)$ and radius is $\sqrt{\frac{1}{2}}$ or $\frac{\sqrt{2}}{2}$.

39.
$$x^2 - 3x + y^2 - y = 1$$
$$x^2 - 3x + \frac{9}{4} + y^2 - y + \frac{1}{4} = 1 + \frac{9}{4} + \frac{1}{4}$$
$$\left(x - \frac{3}{2}\right)^2 + \left(y - \frac{1}{2}\right)^2 = \frac{7}{2}$$
Center is $(3/2, 1/2)$ and radius $\sqrt{\frac{7}{2}}$ or $\frac{\sqrt{14}}{2}$.

41.
$$x^2 - \frac{2}{3}x + y^2 + \frac{3}{2}y = 0$$
$$x^2 - \frac{2}{3}x + \frac{1}{9} + y^2 + \frac{3}{2}y + \frac{9}{16} = 0 + \frac{1}{9} + \frac{9}{16}$$
$$\left(x - \frac{1}{3}\right)^2 + \left(y + \frac{3}{4}\right)^2 = \frac{97}{144}$$
Center is $(1/3, -3/4)$ and radius is $\sqrt{\frac{97}{144}}$ or $\frac{\sqrt{97}}{12}$.

43. The graph of $x^2 + y^2 = 10$ is a circle centered at $(0, 0)$ with radius $\sqrt{10}$. The graph of $y = 3x$ is a straight line through $(0, 0)$ with slope 3. Use substitution to eliminate y.
$$x^2 + (3x)^2 = 10$$
$$10x^2 = 10$$
$$x^2 = 1$$
$$x = 1 \quad \text{or} \quad x = -1$$
$$y = 3 \quad \text{or} \quad y = -3 \quad \text{Since } y = 3x$$
The solution set is $\{(1, 3), (-1, -3)\}$.

45. The graph of $x^2 + y^2 = 9$ is a circle centered at $(0, 0)$ with radius 3. The graph of $y = x^2 - 3$ is a parabola opening upward.

Use $x^2 = y + 3$ to eliminate x.
$$y + 3 + y^2 = 9$$
$$y^2 + y - 6 = 0$$
$$(y + 3)(y - 2) = 0$$
$$y = -3 \quad \text{or} \quad y = 2$$
If $y = -3$, then $x^2 = -3 + 3 = 0$ or $x = 0$. If $y = 2$, then $x^2 = 2 + 3 = 5$ or $x = \pm\sqrt{5}$. The solution set is
$$\left\{(0, -3), \left(\sqrt{5}, 2\right), \left(-\sqrt{5}, 2\right)\right\}.$$

47. The graph of $(x - 2)^2 + (y + 3)^2 = 4$ is a circle centered at $(2, -3)$ with radius 2. The graph of $y = x - 3$ is a line through $(0, -3)$ with slope 1.

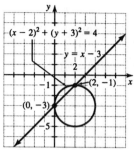

Use substitution to eliminate y.
$$(x - 2)^2 + (x - 3 + 3)^2 = 4$$
$$x^2 - 4x + 4 + x^2 = 4$$
$$2x^2 - 4x = 0$$
$$x^2 - 2x = 0$$
$$x(x - 2) = 0$$
$$x = 0 \quad \text{or} \quad x = 2$$
Since $y = x - 3$
$$y = -3 \qquad y = -1$$
The solution set is $\{(0, -3), (2, -1)\}$.

49. To find the y-intercepts, use $x = 0$ in the equation of the circle.
$$(0 - 1)^2 + (y - 2)^2 = 4$$
$$1 + y^2 - 4y + 4 = 4$$
$$y^2 - 4y + 1 = 0$$
$$y = \frac{4 \pm \sqrt{16 - 4(1)(1)}}{2(1)} = \frac{4 \pm 2\sqrt{3}}{2}$$
$$= 2 \pm \sqrt{3}$$
The y-intercepts are $(0, 2 + \sqrt{3})$ and $(0, 2 - \sqrt{3})$.

51. The radius is the distance from $(2, -5)$ to $(0, 0)$.
$$r = \sqrt{(2 - 0)^2 + (-5 - 0)^2} = \sqrt{4 + 25}$$
$$= \sqrt{29}$$

53. The radius is the distance from $(2, 3)$ to $(-2, -1)$.
$$r = \sqrt{(-2 - 2)^2 + (-1 - 3)^2} = \sqrt{16 + 16}$$
$$= \sqrt{32}$$
The equation of a circle centered at $(2, 3)$ with radius $\sqrt{32}$ is $(x - 2)^2 + (y - 3)^2 = 32$.

55. Substitute $y^2 = 9 - x^2$ into $(x - 5)^2 + y^2 = 9$ to eliminate y.
$$(x - 5)^2 + 9 - x^2 = 9$$
$$x^2 - 10x + 25 + 9 - x^2 = 9$$
$$-10x = -25$$
$$x = 5/2$$
Use $x = 5/2$ in $y^2 = 9 - x^2$ to find y.
$$y^2 = 9 - \left(\frac{5}{2}\right)^2 = \frac{11}{4}$$

$$y = \pm\sqrt{\frac{11}{4}} = \pm\frac{\sqrt{11}}{2}$$

The circles intersect at $\left(\frac{5}{2}, -\frac{\sqrt{11}}{2}\right)$ and $\left(\frac{5}{2}, \frac{\sqrt{11}}{2}\right)$.

57. From the equation of the bore we see that the radius is $\sqrt{83.72}$. The volume is given by $V = \pi r^2 h = \pi \cdot 83.72 \cdot 2874 \approx 755{,}903$ mm^3.

59. Since both x^2 and y^2 are nonnegative for real values of x and y, the only way their sum could be zero is to choose both x and y equal to zero. So the graph consists of $(0, 0)$ only.

61. If we square both sides of $y = \sqrt{1 - x^2}$ we get $x^2 + y^2 = 1$, which is a circle with center $(0, 0)$ and radius 1. In the equation $y = \sqrt{1 - x^2}$, y must be nonnegative. So the graph of $y = \sqrt{1 - x^2}$ is the top half of the graph of $x^2 + y^2 = 1$.

63. B and D can be any real numbers, but A must equal C. So that the radius is positive, we must also have

$$\frac{E}{A} + \left(\frac{B}{2A}\right)^2 + \left(\frac{D}{2A}\right)^2 > 0, \text{ or}$$
$$4AE + B^2 + D^2 > 0.$$

No ordered pairs satisfy $x^2 + y^2 = -9$. So there is no graph.

65. Solve for y to get $y = \pm\sqrt{4 - x^2}$. Then graph $y_1 = \sqrt{4 - x^2}$ and $y_2 = -\sqrt{4 - x^2}$

67. $y = \pm\sqrt{x}$

69.
$$x = y^2 + 2y + 1$$
$$x = (y + 1)^2$$

$$y + 1 = \pm\sqrt{x}$$
$$y = -1 \pm \sqrt{x}$$

12.4 WARM-UPS

1. False, the x-intercepts are $(-6, 0)$ and $(6, 0)$.

2. False, because both x^2 and y^2 must appear in the equation for an ellipse.

3. True, because if the foci coincide then every point on the ellipse will be a fixed distance from one fixed point.

4. True, because if we divide each side of the equation by 2, then it fits the standard equation for an ellipse.

5. True, because if we use $x = 0$ in the equation, we get $y = \pm\sqrt{3}$.

6. False, because both x^2 and y^2 must appear in the equation of a hyperbola.

7. False, because in this hyperbola there are no y-intercepts.

8. True, because it has y-intercepts at $(0, -3)$ and $(0, 3)$.

9. True, because if we divide each side of the equation by 4, then it fits the standard form for the equation of a hyperbola.

10. True, the asymptotes are the extended diagonals of the fundamental rectangle.

12.4 EXERCISES

1. An ellipse is the set of all points in a plane such that the sum of their distances from two fixed points is constant.

3. The center of an ellipse is the point that is midway between the foci.

5. The equation of an ellipse centered at (h, k) is $\frac{(x - h)^2}{a^2} + \frac{(y - k)^2}{b^2} = 1$.

7. The asymptotes of a hyperbola are the extended diagonals of the fundamental rectangle.

9. The graph of $\frac{x^2}{9} + \frac{y^2}{4} = 1$ is an ellipse with x-intercepts $(-3, 0)$ and $(3, 0)$, and y-intercepts $(0, 2)$ and $(0, -2)$.

11. The graph of $\frac{x^2}{9} + y^2 = 1$ is an ellipse with x-intercepts $(-3, 0)$ and $(3, 0)$, and y-intercepts $(0, -1)$ and $(0, 1)$.

13. The graph of $\frac{x^2}{36} + \frac{y^2}{25} = 1$ is an ellipse with x-intercepts $(-6, 0)$ and $(6, 0)$, and y-intercepts $(0, -5)$ and $(0, 5)$.

15. The graph of $\frac{x^2}{24} + \frac{y^2}{5} = 1$ is an ellipse with x-intercepts $(-\sqrt{24}, 0)$ and $(\sqrt{24}, 0)$, and y-intercepts $(0, -\sqrt{5})$ and $(0, \sqrt{5})$.

17. The graph $9x^2 + 16y^2 = 144$ is an ellipse with x-intercepts $(-4, 0)$ and $(4, 0)$, and y-intercepts $(0, -3)$ and $(0, 3)$.

19. The graph of $25x^2 + y^2 = 25$ is an ellipse with x-intercepts $(-1, 0)$ and $(1, 0)$, and y-intercepts $(0, -5)$ and $(0, 5)$.

21. The graph of $4x^2 + 9y^2 = 1$ is an ellipse with x-intercepts $(1/2, 0)$ and $(-1/2, 0)$, and y-intercepts $(0, 1/3)$ and $(0, -1/3)$.

23. The graph of $\frac{(x-3)^2}{4} + \frac{(y-1)^2}{9} = 1$ is an ellipse centered at $(3, 1)$. The 4 in the denominator indicates that the ellipse passes through points that are 2 units to the right and 2 units to the left of the center: $(5, 1)$ and $(1, 1)$. The 9 in the denominator indicates that the ellipse passes through points that are 3 units above and 3 units below the center: $(3, 4)$ and $(3, -2)$.

25. The graph of $\frac{(x+1)^2}{16} + \frac{(y-2)^2}{25} = 1$ is an ellipse centered at $(-1, 2)$. The 16 in the denominator indicates that the ellipse passes through points that are 4 units to the right and 4 units to the left of the center: $(3, 2)$ and $(-5, 2)$. The 25 in the denominator indicates that the ellipse passes through points that are 5 units above and 5 units below the center: $(-1, 7)$ and $(-1, -3)$.

27. The graph of $(x - 2)^2 + \frac{(y + 1)^2}{36} = 1$ is an ellipse centered at $(2, -1)$. The 1 in the denominator indicates that the ellipse passes through points that are 1 unit to the right and 1 unit to the left of the center: $(3, -1)$ and $(1, -1)$. The 36 in the denominator indicates that the ellipse passes through points that are 6 units above and 6 units below the center: $(2, 5)$ and $(2, -7)$.

29. The graph of $\frac{x^2}{4} - \frac{y^2}{9} = 1$ is a hyperbola centered at $(0, 0)$ with x-intercepts at $(-2, 0)$ and $(2, 0)$. There are no y-intercepts. Use 9 to determine the size of the fundamental rectangle. The fundamental rectangle passes through $(0, 3)$ and $(0, -3)$. extend the diagonals of the rectangle to determine the asymptotes. The hyperbola opens to the left and right. The equations of the asymptotes are $y = \pm \frac{3}{2}x$.

31. The graph of $\frac{y^2}{4} - \frac{x^2}{25} = 1$ is a hyperbola with y-intercepts $(0, -2)$ and $(0, 2)$. The fundamental rectangle passes through the y-intercepts, $(-5, 0)$, and $(5, 0)$. The extended diagonals determine the asymptotes, $y = \pm \frac{2}{5}x$.

33. The graph of $\frac{x^2}{25} - y^2 = 1$ is a hyperbola with x-intercepts $(-5, 0)$ and $(5, 0)$. The fundamental rectangle passes through the intercepts, $(0, -1)$, and $(0, 1)$. The extended diagonals determine the asymptotes. The equations of the asymptotes are $y = \pm \frac{1}{5}x$.

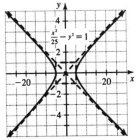

35. The graph of $x^2 - \frac{y^2}{25} = 1$ is a hyperbola with x-intercepts $(-1, 0)$ and $(1, 0)$. Plot $(0, 5)$ and $(0, -5)$ for the fundamental rectangle and extend the diagonals. The equations of the asymptotes are $y = \pm 5x$.

37. Divide each side of $9x^2 - 16y^2 = 144$ by 144 to get $\dfrac{x^2}{16} - \dfrac{y^2}{9} = 1$. The graph is a hyperbola with x-intercepts at $(-4, 0)$ and $(4, 0)$. The fundamental rectangle passes through the y-intercepts, $(0, -3)$, and $(0, 3)$. The extended diagonals determine the asymptotes. The equations of the asymptotes are $y = \pm \dfrac{3}{4}x$.

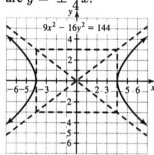

39. The graph of $x^2 - y^2 = 1$ is a hyperbola with x-intercepts $(-1, 0)$ and $(1, 0)$. The fundamental rectangle passes through the y-intercepts, $(0, -1)$, and $(0, 1)$. The extended diagonals determine the asymptotes, $y = \pm x$.

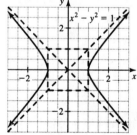

41. The graph of $\dfrac{(x - 2)^2}{4} - (y + 1)^2 = 1$ is a hyperbola centered at $(2, -1)$ and opening left and right. It is a transformation of $\dfrac{x^2}{4} - y^2 = 1$, which has its fundamental rectangle through $(\pm 2, 0)$ and $(0, \pm 1)$. Move the fundamental rectangle 2 to the right and down 1 unit. Then draw the asymptotes and the hyperbola.

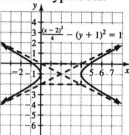

43. The graph of
$$(x + 1)^2/16 - (y - 1)^2/9 = 1$$
is a hyperbola centered at $(-1, 1)$ and opening left and right. It is a transformation of $x^2/16 - y^2/9 = 1$, which has its fundamental rectangle through $(\pm 4, 0)$ and $(0, \pm 3)$. Move the fundamental rectangle 1 to the left and up 1 unit. Then draw the asymptotes and the hyperbola.

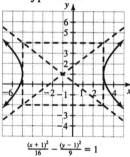

$$\dfrac{(x + 1)^2}{16} - \dfrac{(y - 1)^2}{9} = 1$$

45. The graph of
$$(y - 2)^2/9 - (x - 4)^2/4 = 1$$
is a hyperbola centered at $(4, 2)$ and opening up and down. It is a transformation of $y^2/9 - x^2/4 = 1$, which has its fundamental rectangle through $(\pm 2, 0)$ and $(0, \pm 3)$. Move the fundamental rectangle 4 to the right and up 2 units. Then draw the asymptotes and the hyperbola.

$$\dfrac{(y - 2)^2}{9} - \dfrac{(x - 4)^2}{4} = 1$$

47. First graph the hyperbola and the ellipse.

If we add the equations the variable y will be eliminated and we get the following equation.
$$\frac{x^2}{4} + x^2 = 2$$
Now multiply both sides by 4:
$$x^2 + 4x^2 = 8$$
$$5x^2 = 8$$
$$x^2 = \frac{8}{5}$$
$$x = \pm\sqrt{\frac{8}{5}} = \pm\frac{2\sqrt{10}}{5}$$
Use $x^2 = 8/5$ in the second equation.
$$\frac{8}{5} - \frac{y^2}{9} = 1$$
$$-\frac{y^2}{9} = -\frac{3}{5}$$
$$y^2 = \frac{27}{5}$$
$$y = \pm\sqrt{\frac{27}{5}} = \pm\frac{3\sqrt{15}}{5}$$
The graphs intersect at $\left(\frac{2\sqrt{10}}{5}, \frac{3\sqrt{15}}{5}\right)$, $\left(\frac{2\sqrt{10}}{5}, -\frac{3\sqrt{15}}{5}\right)$, $\left(-\frac{2\sqrt{10}}{5}, \frac{3\sqrt{15}}{5}\right)$, and $\left(-\frac{2\sqrt{10}}{5}, -\frac{3\sqrt{15}}{5}\right)$.

49. The graphs are an ellipse and a circle.

From the graph we can see that there are no points of intersection. If we eliminate y by substituting $y^2 = 1 - x^2$ into the first equation, we get the following equation.
$$\frac{x^2}{4} + \frac{1 - x^2}{16} = 1$$
Now multiply both sides by 16:
$$4x^2 + 1 - x^2 = 16$$
$$3x^2 = 15$$

$$x^2 = 5$$
From $y^2 = 1 - x^2$, we get $y^2 = -4$, which has no real solution. There are no points of intersection.

51. The graphs are a circle and a hyperbola.

It appears that there are 4 points of intersection. To find them, add the equations to get the following equation.
$$2x^2 = 5$$
$$x^2 = \frac{5}{2}$$
$$x = \pm\sqrt{\frac{5}{2}} = \pm\frac{\sqrt{10}}{2}$$
From $y^2 = 4 - x^2$, we get $y^2 = 4 - \frac{5}{2} = \frac{3}{2}$, or
$$y = \pm\sqrt{\frac{3}{2}} = \pm\frac{\sqrt{6}}{2}.$$
The graphs intersect at $\left(\frac{\sqrt{10}}{2}, \frac{\sqrt{6}}{2}\right)$, $\left(\frac{\sqrt{10}}{2}, -\frac{\sqrt{6}}{2}\right)$, $\left(-\frac{\sqrt{10}}{2}, \frac{\sqrt{6}}{2}\right)$, and $\left(-\frac{\sqrt{10}}{2}, -\frac{\sqrt{6}}{2}\right)$.

53. The graphs are an ellipse and a circle.

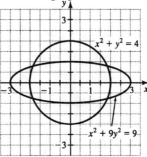

The graphs appear to intersect at 4 points. To find the points, substitute $y^2 = 4 - x^2$ into the first equation.
$$x^2 + 9(4 - x^2) = 9$$
$$-8x^2 + 36 = 9$$
$$-8x^2 = -27$$
$$x^2 = \frac{27}{8}$$
$$x = \pm\sqrt{\frac{27}{8}} = \pm\frac{3\sqrt{6}}{4}$$
Use $x^2 = 27/8$ in $y^2 = 4 - x^2$.

$$y^2 = 4 - \frac{27}{8} = \frac{5}{8}$$
$$y = \pm\sqrt{\frac{5}{8}} = \pm\frac{\sqrt{10}}{4}$$

The graphs intersect at $\left(\frac{3\sqrt{6}}{4}, \frac{\sqrt{10}}{4}\right)$, $\left(\frac{3\sqrt{6}}{4}, -\frac{\sqrt{10}}{4}\right)$, $\left(-\frac{3\sqrt{6}}{4}, \frac{\sqrt{10}}{4}\right)$, and $\left(-\frac{3\sqrt{6}}{4}, -\frac{\sqrt{10}}{4}\right)$.

55. Graph the parabola and the ellipse.

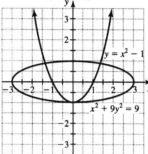

The graphs appear to intersect at 3 points. To find them, substitute $x^2 = y + 1$ into the first equation.
$$y + 1 + 9y^2 = 9$$
$$9y^2 + y - 8 = 0$$
$$(9y - 8)(y + 1) = 0$$
$$y = \frac{8}{9} \quad \text{or} \quad y = -1$$

If $y = -1$, then $x^2 = -1 + 1 = 0$ or $x = 0$. If $y = 8/9$, then
$$x^2 = \frac{8}{9} + 1 = \frac{17}{9} \quad \text{or} \quad x = \pm\frac{\sqrt{17}}{3}.$$

The three points of intersection are $\left(\frac{\sqrt{17}}{3}, \frac{8}{9}\right)$, $\left(-\frac{\sqrt{17}}{3}, \frac{8}{9}\right)$, and $(0, -1)$.

57. Graph the hyperbola and the line.

To find the points of intersection, substitute $x = 2y + 2$ into the first equation.
$$9(2y + 2)^2 - 4y^2 = 36$$

$$9(4y^2 + 8y + 4) - 4y^2 = 36$$
$$36y^2 + 72y + 36 - 4y^2 = 36$$
$$32y^2 + 72y = 0$$
$$4y^2 + 9y = 0$$
$$y(4y + 9) = 0$$
$$y = 0 \quad \text{or} \quad y = -9/4$$

If $y = 0$, then $x = 2(0) + 2 = 2$. If $y = -9/4$, then $x = 2(-9/4) + 2 = -5/2$. The two points of intersection are $(2, 0)$ and $(-5/2, -9/4)$.

59. a) From the graph it appears that the boat is approximately at $(2.5, 1.5)$

b) Substitute $x^2 = 1 + 3y^2$ into $4y^2 - x^2 = 1$.
$$4y^2 - (1 + 3y^2) = 1$$
$$y^2 = 2$$
$$y = \pm\sqrt{2}$$
$$x^2 = 1 + 3(\pm\sqrt{2})^2 = 1 \pm 3(2) = 1 \pm 6$$
$$= 7 \text{ or } -5$$

If $x^2 = 7$, then $x = \pm\sqrt{7}$. If $x^2 = -5$, there is no real solution. Since the boat is in the first quadrant, the boat is at $(\sqrt{7}, \sqrt{2})$.

12.5 WARM-UPS

1. False, because $x^2 + y = 4$ is the equation of a parabola.

2. True, because if we divide each side by 9 we get the standard form for an ellipse.

3. True, because it can be written as $y^2 - x^2 = 1$.

4. True, because $2(0)^2 - 0 < 3$ is correct.

5. False, because $0 > 0^2 - 3(0) + 2$ is incorrect.

6. False, because the origin is on the parabola $x^2 = y$.

7. False, because $(0, 0)$ does not satisfy the inequality.

8. True, because $x^2 + y^2 = 4$ is a circle of radius 2 centered at $(0, 0)$, and the points inside this circle satisfy the inequality.

9. True, because both $0^2 - 4^2 < 1$ and $4 > 0^2 - 2(0) + 3$ are correct.

10. True, because both $0^2 + 0^2 < 1$ and $0 < 0^2 + 1$ are correct.

12.5 EXERCISES

1. Graph the parabola $y = x^2$. Since $(0, 4)$ satisfies $y > x^2$, shade the region containing $(0, 4)$.

3. First graph the parabola $y = x^2 - x$. Since $(5, 0)$ satisfies $y < x^2 - x$, we shade the region **containing $(5, 0)$.**

5. First graph the parabola $y = x^2 - x - 2$. Since $(0, 5)$ satisfies $y > x^2 - x - 2$, shade the region containing $(0, 5)$.

7. First graph the circle $x^2 + y^2 = 9$ using a solid curve. Since $(0, 0)$ satisfies $x^2 + y^2 \leq 9$, shade the inside of the circle.

9. First graph the ellipse $x^2 + 4y^2 = 4$ using a dashed curve. Since $(0, 0)$ does not satisfy the inequality $x^2 + 4y^2 > 4$, we shade the region outside the ellipse.

11. First divide both sides of the inequality by 36.

$$\frac{x^2}{9} - \frac{y^2}{4} < 1$$

Now graph the hyperbola $\frac{x^2}{9} - \frac{y^2}{4} = 1$. Testing a point in each of the three regions, we find that only points in the region containing the origin satisfy $4x^2 - 9y^2 < 36$.

13. First graph the circle centered at $(2, 3)$ with radius 2, using a dashed curve. Since $(2, 3)$ satisfies $(x - 2)^2 + (y - 3)^2 < 4$, shade the region inside the circle.

15. First graph the circle centered at $(0, 0)$ with radius 1. Since $(0, 0)$ does not satisfy $x^2 + y^2 > 1$, shade the region outside the circle.

17. Graph the hyperbola $4x^2 - y^2 = 4$, using a dashed curve. After testing a point in each of the three regions, we see that only points in the region containing $(0, 0)$ fail to satisfy $4x^2 - y^2 > 4$.

19. Graph the hyperbola $y^2 - x^2 = 1$. After testing a point in each of the three regions, we see that only points in the region containing $(0, 0)$ satisfy $y^2 - x^2 \leq 1$.

21. The graph of $y = x$ is a line with slope 1 and y-intercept $(0, 0)$. Since $(5, -5)$ satisfies $x > y$, we shade the region below the line.

23. Graph the circle $x^2 + y^2 = 9$ and the line $y = x$. The graph of $x^2 + y^2 < 9$ is the region inside the circle. the graph of $y > x$ is the region above the line. The intersection of these two regions is shown as follows.

We could have used test points to determine which of the 4 regions determined by the circle and the line satisfies both inequalities.

25. The graph of $x^2 - y^2 = 1$ is a hyperbola. The graph of $x^2 + y^2 = 4$ is a circle of radius 2. Points that satisfy $x^2 + y^2 < 4$ are inside the circle. The hyperbola divides the plane into 3 regions. Points in the 2 regions not containing $(0, 0)$ are the points that satisfy $x^2 - y^2 > 1$. The intersection of these regions is shown in the following graph.

27. The graph of $y = x^2 + x$ is a parabola opening upward, with x-intercepts at $(0, 0)$ and $(-1, 0)$. The graph of $y > x^2 + x$ is the region inside this parabola. The graph of $y = 5$ is a horizontal line through $(0, 5)$. The graph of $y < 5$ is the region below this horizontal line. The points that satisfy both inequalities are the points inside the parabola and below $y = 5$, as shown in the following graph.

29. The graph of $y = x + 2$ is a line with y-intercept $(0, 2)$ and slope 1. The graph of $y \geq x + 2$ is the region above and including the line. The graph of $y = 2 - x$ is a line with y-intercept $(0, 2)$ and slope -1. The graph of $y \leq 2 - x$ is the region below and including this line. The region above the first line and below the second is shown in the following graph. We could have tested a point in each of the 4 regions determined by the two lines. We would find that only points in the region shown satisfy both inequalities.

266

31. The graph of $4x^2 - y^2 = 4$ is a hyperbola opening to the left and right. Points that satisfy $4x^2 - y^2 < 4$ are in the region between the two branches of the hyperbola. The graph of $x^2 + 4y^2 = 4$ is an ellipse passing through $(0, 1)$, $(0, -1)$, $(2, 0)$, and $(-2, 0)$. Points that satisfy $x^2 + 4y^2 > 4$ are outside the ellipse. Points that satisfy both inequalities are outside the ellipse and between the two branches of the hyperbola.

$x^2 + 4y^2 > 4$ and $4x^2 - y^2 < 4$

33. The graph of $x - y = 0$ is the same as the line $y = x$ (through $(0, 0)$ with slope 1). Points that satisfy $x - y < 0$ are above the line $y = x$. The graph of $y + x^2 = 1$ is the same as the parabola $y = -x^2 + 1$, which opens downward. Points that satisfy $y + x^2 < 1$ are below the parabola. The points that satisfy both inequalities, below the parabola and above the line, are shown in the following graph.

35. The graph of $y = 5x - x^2$ is a parabola opening downward. Points that satisfy $y < 5x - x^2$ are the points below the parabola. The graph of $x^2 + y^2 = 9$ is a circle of radius

3. Points that satisfy $x^2 + y^2 < 9$ are inside the circle. Points that satisfy both inequalities are the points inside the circle and below the parabola.

37. Points that satisfy $y \geq 3$ are above or on the horizontal line $y = 3$. Points that satisfy $x \leq 1$ are on or to the left of the vertical line $x = 1$. Points that satisfy both inequalities are shown in the following graph. The points graphed are the points with x-coordinate less than or equal to 1, and y-coordinate greater than or equal to 3.

39. The graph of $4y^2 - 9x^2 = 36$ is a hyperbola opening up and down. Points that satisfy $4y^2 - 9x^2 < 36$ are the points between the two branches of the hyperbola. The graph of $x^2 + y^2 = 16$ is a circle of radius 4. Points that satisfy $x^2 + y^2 < 16$ are inside the circle. The points that satisfy both inequalities are indicated as follows.

$x^2 + y^2 < 16$ and $4y^2 - 9x^2 < 36$

41. The graph of $y = x^2$ is a parabola opening upward. Points that satisfy $y < x^2$ are below this parabola. The graph of $x^2 + y^2 = 1$ is a

267

circle of radius 1. Points that satisfy $x^2 + y^2 < 1$ are inside the circle. The graph of the system of inequalities consists of points that are below the parabola and inside the circle.

43. Let $x =$ the number of paces he walked to the east and $y =$ the number of paces he walked to the north. So $x + y \geq 50$, $x^2 + y^2 \leq 50^2$, and $y > x$. The graph of this system follows.

45. From the graphs it appears that there is no intersection of the two regions and there is no solution to the system.

Enriching Your Mathematical Word Power
1. c   **2.** a   **3.** d   **4.** a   **5.** c
6. d   **7.** b   **8.** d   **9.** c   **10.** a

CHAPTER 12   REVIEW

1. The graph of $y = x^2$ is a parabola and the graph of $y = -2x + 15$ is a straight line.

Use substitution to eliminate y.
$$x^2 = -2x + 15$$
$$x^2 + 2x - 15 = 0$$

$$(x + 5)(x - 3) = 0$$
$$x = -5 \text{ or } x = 3$$
$$y = 25 \quad y = 9 \text{ Since } y = x^2$$
The solution set is $\{(3, 9), (-5, 25)\}$.
3. First graph $y = 3x$ and $y = 1/x$.

Use substitution to eliminate y.
$$3x = \frac{1}{x}$$
$$3x^2 = 1$$
$$x^2 = \frac{1}{3}$$
$$x = \pm \sqrt{\frac{1}{3}} = \pm \frac{\sqrt{3}}{3}$$
If $x = \frac{\sqrt{3}}{3}$, then $y = \sqrt{3}$. If $x = -\frac{\sqrt{3}}{3}$, then $y = -\sqrt{3}$. The solution set is
$$\left\{ \left(\frac{\sqrt{3}}{3}, \sqrt{3} \right), \left(-\frac{\sqrt{3}}{3}, -\sqrt{3} \right) \right\}.$$
5. The second equation can be written as $x^2 = 3y$. Substitute this equation into the first equation.
$$3y + y^2 = 4$$
$$y^2 + 3y - 4 = 0$$
$$(y + 4)(y - 1) = 0$$
$$y = -4 \text{ or } y = 1$$
If $y = -4$, then $x^2 = 3(-4)$ has no solution. If $y = 1$, then $x^2 = 3(1)$ gives us $x = \pm \sqrt{3}$. The solution set is $\left\{ (\sqrt{3}, 1), (-\sqrt{3}, 1) \right\}$.

7. Use substitution to eliminate y.
$$x^2 + (x + 2)^2 = 34$$
$$x^2 + x^2 + 4x + 4 = 34$$
$$2x^2 + 4x - 30 = 0$$
$$x^2 + 2x - 15 = 0$$
$$(x + 5)(x - 3) = 0$$
$$x = -5 \text{ or } x = 3$$
$$y = -3 \quad y = 5 \text{ Since } y = x + 2$$
The solution set is $\{(-5, -3), (3, 5)\}$.
9. Use substitution to eliminate y.
$$\log(x - 3) = 1 - \log(x)$$
$$\log(x - 3) + \log(x) = 1$$
$$\log(x^2 - 3x) = 1$$

$$x^2 - 3x = 10$$
$$x^2 - 3x - 10 = 0$$
$$(x - 5)(x + 2) = 0$$
$$x = 5 \text{ or } x = -2$$

If $x = 5$, then $y = \log(5 - 3) = \log(2)$. If $x = -2$, we get a logarithm of a negative number. So the solution set is $\{(5, \log(2))\}$.

11. Use substitution to eliminate x.
$$y^2 = 2(12 - y)$$
$$y^2 = 24 - 2y$$
$$y^2 + 2y - 24 = 0$$
$$(y + 6)(y - 4) = 0$$
$$y = -6 \text{ or } y = 4$$

If $y = -6$, then $-6 = x^2$ has no solution. If $y = 4$, then $x^2 = 4$ and $x = \pm 2$. The solution set is $\{(2, 4), (-2, 4)\}$.

13. $\sqrt{(3 - 1)^2 + (3 - 1)^2} = \sqrt{4 + 4} = 2\sqrt{2}$

15. $\sqrt{(2 - (-4))^2 + (-8 - 6)^2}$
$= \sqrt{36 + 196} = 2\sqrt{58}$

17. For $y = x^2 + 3x - 18$, use $x = -b/(2a)$ to find the x-coordinate of the vertex. The vertex is $(-\frac{3}{2}, -\frac{81}{4})$ and the axis of symmetry is $x = -3/2$. Since $a = 1$, $p = 1/4$. The focus is $(-\frac{3}{2}, -20)$ and the directrix is $y = -\frac{41}{2}$.

19. For $y = x^2 + 3x + 2$, use $x = -b/(2a)$ to find the x-coordinate of the vertex. The vertex is $(-\frac{3}{2}, -\frac{1}{4})$ and the axis of symmetry is $x = -3/2$. Since $a = 1$, $p = 1/4$. The focus is $(-\frac{3}{2}, 0)$ and the directrix is $y = -\frac{1}{2}$.

21. For $y = -\frac{1}{2}(x - 2)^2 + 3$ the vertex is $(2, 3)$ and the parabola opens down. The axis of symmetry is $x = 2$. Since $a = -1/2$, $p = -1/2$. The focus is $(2, \frac{5}{2})$ and the directrix is $y = \frac{7}{2}$.

23.
$$y = 2x^2 - 8x + 1$$
$$y = 2(x^2 - 4x) + 1$$
$$y = 2(x^2 - 4x + 4 - 4) + 1$$
$$y = 2(x^2 - 4x + 4) - 8 + 1$$
$$y = 2(x - 2)^2 - 7$$
Vertex $(2, -7)$

25.
$$y = -\frac{1}{2}x^2 - x + \frac{1}{2}$$
$$y = -\frac{1}{2}(x^2 + 2x) + \frac{1}{2}$$
$$y = -\frac{1}{2}(x^2 + 2x + 1 - 1) + \frac{1}{2}$$
$$y = -\frac{1}{2}(x + 1)^2 + \frac{1}{2} + \frac{1}{2}$$
$$y = -\frac{1}{2}(x + 1)^2 + 1$$
Vertex $(-1, 1)$

27. The graph of $x^2 + y^2 = 100$ is a circle centered at $(0, 0)$ with radius 10.

29. The graph of $(x - 2)^2 + (y + 3)^2 = 81$ is a circle centered at $(2, -3)$ with radius 9.

31.
$$9y^2 + 9x^2 = 4$$
$$x^2 + y^2 = \frac{4}{9}$$

The graph is a circle with center $(0, 0)$ and radius $2/3$.

33. The equation of a circle with center (h, k) and radius r is $(x - h)^2 + (y - k)^2 = r^2$. The equation of a circle with center $(0, 3)$ and radius 6 is $x^2 + (y - 3)^2 = 36$.

35. The equation of a circle with center $(2, -7)$ and radius 5 is $(x - 2)^2 + (y + 7)^2 = 25$.

37. The graph of $\frac{x^2}{36} + \frac{y^2}{49} = 1$ is an ellipse with x-intercepts at $(-6, 0)$ and $(6, 0)$, y-intercepts at $(0, -7)$ and $(0, 7)$, and centered at the origin.

39. The graph of $25x^2 + 4y^2 = 100$ is an ellipse with x-intercepts $(-2, 0)$ and $(2, 0)$, y-intercepts $(0, -5)$ and $(0, 5)$, and centered at the origin.

41. The graph of $\frac{x^2}{49} - \frac{y^2}{36} = 1$ is a hyperbola with x-intercepts $(-7, 0)$ and $(7, 0)$. The fundamental rectangle passes through the x-intercepts and the points $(0, -6)$ and $(0, 6)$. Extend the diagonals of the fundamental rectangle to get the asymptotes and draw a hyperbola opening to the left and right.

43. Write $4x^2 - 25y^2 = 100$ as $\frac{x^2}{25} - \frac{y^2}{4} = 1$. The graph has x-intercepts at $(-5, 0)$ and $(5, 0)$. The fundamental rectangle passes through the x-intercepts and the points $(0, -2)$ and $(0, 2)$. Extend the diagonals of the rectangle to get the asymptotes and draw a hyperbola opening to the left and right.

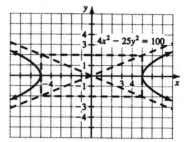

45. First graph the line $4x - 2y = 3$. It goes through $(0, -3/2)$ and $(3/4, 0)$. Since $(0, 0)$ fails to satisfy $4x - 2y > 3$, we shade the region not containing $(0, 0)$.

47. Write $y^2 < x^2 - 1$ as $x^2 - y^2 > 1$. Graph the hyperbola $x^2 - y^2 = 1$ and test a point in each of the three regions: $(-5, 0)$, $(0, 0)$, and $(5, 0)$. Since $(-5, 0)$ and $(5, 0)$ satisfy the inequality, we shade the regions containing those points.

$$y^2 < x^2 - 1$$

49. Write $4x^2 + 9y^2 > 36$ as $\frac{x^2}{9} + \frac{y^2}{4} > 1$.

First graph the ellipse $\frac{x^2}{9} + \frac{y^2}{4} = 1$ through $(-3, 0)$, $(3, 0)$, $(0, -2)$, and $(0, 2)$. Since $(0, 0)$ fails to satisfy the inequality, we shade the region outside of the ellipse.

51. The graph of $y < 3x - x^2$ is the region below the parabola $y = 3x - x^2$. The graph of $x^2 + y^2 < 9$ is the region inside the circle $x^2 + y^2 = 9$. Points that are inside the circle and below the parabola are shown in the following graph.

53. The set of points that satisfy $4x^2 + 9y^2 > 36$ is the set of points outside the ellipse $4x^2 + 9y^2 = 36$. The set of points that satisfy $x^2 + y^2 < 9$ is the set of points inside the circle $x^2 + y^2 = 9$. The solution set to the system consists of points that are outside the ellipse and inside the circle, as shown in the following graph.

55. The equation $x^2 = y^2 + 1$ is the equation of a hyperbola because it could be written as $x^2 - y^2 = 1$.

57. The equation $x^2 = 1 - y^2$ is the equation of a circle because it could be written as $x^2 + y^2 = 1$.

59. The equation $x^2 + x = 1 - y^2$ is the equation of a circle because we could write $x^2 + x + y^2 = 1$, and then complete the square to get the standard equation for a circle.

61. The equation $x^2 + 4x = 6y - y^2$ is the equation of a circle because we could write it as $x^2 + 4x + y^2 - 6y = 0$, and then complete the squares for both x and y to get the standard equation of a circle.

63. The equation is the equation of a hyperbola in standard form.

65. The equation $4y^2 - x^2 = 8$ is the equation of a hyperbola because we could divide by 8 to get the standard equation $\frac{y^2}{2} - \frac{x^2}{8} = 1$.

67. Write $x^2 = 4 - y^2$ as $x^2 + y^2 = 4$ to see that it is the equation of a circle of radius 2 centered at the origin.

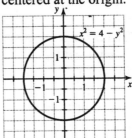

69. Write $x^2 = 4y + 4$ as $y = \frac{1}{4}x^2 - 1$ to see that it is the equation of a parabola opening upward with vertex at $(0, -1)$.

71. Write $x^2 = 4 - 4y^2$ as $\frac{x^2}{4} + y^2 = 1$ to see that it is the equation of an ellipse centered at $(0, 0)$ and passing through $(0, -1)$, $(0, 1)$, $(-2, 0)$, and $(2, 0)$.

271

73. Write $x^2 = 4 - (y - 4)^2$ as $x^2 + (y - 4)^2 = 4$ to see that it is the equation of a circle of radius 2 centered at $(0, 4)$.

75. The radius of the circle is the distance between $(0, 0)$ and $(3, 4)$.
$$r = \sqrt{(3 - 0)^2 + (4 - 0)^2} = \sqrt{9 + 16}$$
$$= \sqrt{25} = 5$$
The equation of the circle centered at $(0, 0)$ with radius 5 is $x^2 + y^2 = 25$.

77. Use the center $(-1, 5)$ and radius 6 in the form $(x - h)^2 + (y - k)^2 = r^2$ to get the equation $(x + 1)^2 + (y - 5)^2 = 36$.

79. The vertex is half way between the focus and directrix at $(1, 3)$. Since the distance from $(1, 3)$ to $(1, 4)$ is 1, we have $p = 1$, and $a = 1/4$. So the equation is
$$y = \tfrac{1}{4}(x - 1)^2 + 3.$$

81. Since the vertex is below the focus, $p = 1/4$ and $a = 1$. The equation is $y = 1(x - 0)^2 + 0$ or $y = x^2$.

83. Since the vertex is $(0, 0)$ the parabola has equation $y = ax^2$. Since $(3, 2)$ is on the parabola, $2 = a \cdot 3^2$, or $a = 2/9$. The equation is $y = \tfrac{2}{9}x^2$.

85. Substitute $y = -x + 1$ into $x^2 + y^2 = 25$ to eliminate y.
$$x^2 + (-x + 1)^2 = 25$$
$$x^2 + x^2 - 2x + 1 = 25$$
$$2x^2 - 2x - 24 = 0$$
$$x^2 - x - 12 = 0$$
$$(x - 4)(x + 3) = 0$$
$$x = 4 \quad \text{or} \quad x = -3$$
$$y = -3 \qquad y = 4 \quad \text{Since } y = -x + 1$$
The solution set is $\{(4, -3), (-3, 4)\}$

87. If we add the two equations to eliminate y, we get the equation $5x^2 = 25$. Solving this equation gives $x^2 = 5$ or $x = \pm\sqrt{5}$. Use $x^2 = 5$ in the first equation.
$$4(5) + y^2 = 4$$
$$y^2 = -16$$

There are no real numbers that satisfy the system of equations. The solution set is $\emptyset$.

89. Let $x =$ the length and $y =$ the width. We can write the following 2 equations.
$$2x + 2y = 16$$
$$xy = 12$$
The first equation can be written as $y = 8 - x$. Substitute $y = 8 - x$ into the second equation.
$$x(8 - x) = 12$$
$$-x^2 + 8x - 12 = 0$$
$$x^2 - 8x + 12 = 0$$
$$(x - 6)(x - 2) = 0$$
$$x = 6 \quad \text{or} \quad x = 2$$
$$y = 2 \qquad y = 6 \quad \text{Since } y = 8 - x$$
So the length is 6 feet and the width is 2 feet.

CHAPTER 12 TEST

1. The graph of $x^2 + y^2 = 25$ is a circle of radius 5 centered at $(0, 0)$.

2. The graph of $\dfrac{x^2}{16} - \dfrac{y^2}{25} = 1$ is a hyperbola centered at the origin. the x-intercepts are $(-4, 0)$ and $(4, 0)$. The fundamental rectangle passes through the x-intercepts, $(0, -5)$, and $(0, 5)$. Extend the diagonals of the fundamental rectangle to obtain the asymptotes. The hyperbola opens to the left and right.

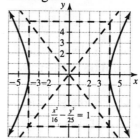

3. The graph of $y^2 + 4x^2 = 4$ is an ellipse with x-intercepts at $(-1, 0)$ and $(1, 0)$. Its y-intercepts are $(0, -2)$ and $(0, 2)$.

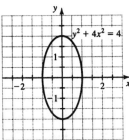

4. The graph of $y = x^2 + 4x + 4$ is a parabola opening upward with y-intercept $(0, 4)$. We can write this equation as $y = (x + 2)^2$. In this form we see that the vertex is $(-2, 0)$.

5. Write $y^2 - 4x^2 = 4$ as $\frac{y^2}{4} - x^2 = 1$. The graph is a hyperbola with y-intercepts at $(0, -2)$ and $(0, 2)$. The fundamental rectangle passes through the y-intercepts, $(-1, 0)$, and $(1, 0)$. Extend the diagonals to get the asymptotes. The hyperbola opens up and down.

6. The graph of $y = -x^2 - 2x + 3$ is a parabola opening downward. The vertex is at $(-1, 4)$. The y-intercept is $(0, 3)$. The x-intercepts are $(-3, 0)$ and $(1, 0)$.

7. First graph the hyperbola $\frac{x^2}{9} - \frac{y^2}{9} = 1$. Since $(0, 0)$ satisfies the inequality, the region containing $(0, 0)$ is shaded.

8. First graph the circle $x^2 + y^2 = 9$, centered at $(0, 0)$ with radius 3. Since $(0, 0)$ fails to satisfy $x^2 + y^2 > 9$, we shade the region outside the circle.

9. The graph of $y = x^2 - 9$ is a parabola opening upward. Its vertex is $(0, -9)$. To find its x-intercepts solve $x^2 - 9 = 0$. The x-intercepts are $(-3, 0)$ and $(3, 0)$. The graph of $y > x^2 - 9$ is the region containing the origin because $(0, 0)$ satisfies the inequality.

10. The graph of $x^2 + y^2 < 9$ is the region inside the circle $x^2 + y^2 = 9$. To find the graph of $x^2 - y^2 > 1$ first graph the hyperbola $x^2 - y^2 = 1$. By testing points, we can see that the two regions not containing the origin satisfy $x^2 - y^2 > 1$. The points in these two regions that are also inside the circle are the points that satisfy both inequalities of the system.

273

$x^2 - y^2 > 1$ and $x^2 + y^2 < 9$

11. The graph of $y = -x^2 + x$ is a parabola opening downward. Points below this parabola satisfy $y < -x^2 + x$. The graph of $y = x - 4$ is a line through $(0, -4)$ with slope 1. Points below this line satisfy $y < x - 4$. The solution set to the system of inequalities consists of points below the parabola and below the line.

$y < -x^2 + x$ and $y < x - 4$

12. Use substitution to eliminate y.
$$x^2 - 2x - 8 = 7 - 4x$$
$$x^2 + 2x - 15 = 0$$
$$(x + 5)(x - 3) = 0$$
$$x = -5 \text{ or } x = 3$$
$$y = 27 \qquad y = -5 \text{ Since } y = 7 - 4x$$
The solution set is $\{(-5, 27), (3, -5)\}$.

13. Substitute $x^2 = y$ into $x^2 + y^2 = 12$.
$$y + y^2 = 12$$
$$y^2 + y - 12 = 0$$
$$(y + 4)(y - 3) = 0$$
$$y = -4 \text{ or } y = 3$$
If $y = -4$, the $x^2 = -4$ has no solution. If $y = 3$, then $x^2 = 3$ or $x = \pm\sqrt{3}$. The solution set is $\{(\sqrt{3}, 3), (-\sqrt{3}, 3)\}$.

14. $\sqrt{(-1-1)^2 + (4-6)^2} = \sqrt{4+4}$
$$= \sqrt{8} = 2\sqrt{2}$$

15. Complete the square to get the equation into the standard form.
$$x^2 + 2x + \quad y^2 + 10y \quad = 10$$
$$x^2 + 2x + 1 + y^2 + 10y + 25 = 10 + 1 + 25$$
$$(x + 1)^2 + (y + 5)^2 = 36$$

The center is $(-1, -5)$ and the radius is 6.

16. For $y = x^2 + x + 3$ the x-coordinate of the vertex is
$$x = \frac{-b}{2a} = \frac{-1}{2(1)} = -\frac{1}{2}.$$
Use $x = -1/2$ in $y = x^2 + x + 3$, to get the vertex $(-\frac{1}{2}, \frac{11}{4})$. Since $a = 1$, we have $p = 1/4$. The focus is $(-\frac{1}{2}, 3)$ and the directrix is $y = \frac{5}{2}$. The axis of symmetry is $x = -\frac{1}{2}$ and the parabola opens up.

17.
$$y = \frac{1}{2}x^2 - 3x - \frac{1}{2}$$
$$y = \frac{1}{2}(x^2 - 6x) - \frac{1}{2}$$
$$y = \frac{1}{2}(x^2 - 6x + 9 - 9) - \frac{1}{2}$$
$$y = \frac{1}{2}(x^2 - 6x + 9) - \frac{9}{2} - \frac{1}{2}$$
$$y = \frac{1}{2}(x - 3)^2 - 5$$

18. The radius of the circle is the distance between $(-1, 3)$ and $(2, 5)$.
$$r = \sqrt{(-1-2)^2 + (3-5)^2} = \sqrt{9+4}$$
$$= \sqrt{13}$$
The equation of the circle with center $(-1, 3)$ and radius $\sqrt{13}$ is $(x + 1)^2 + (y - 3)^2 = 13$.

19. Let x = the length and y = the width. We can write the following two equations.
$$2x + 2y = 42$$
$$xy = 108$$
If we solve $2x + 2y = 42$ for y, we get $y = 21 - x$.
$$x(21 - x) = 108$$
$$-x^2 + 21x - 108 = 0$$
$$x^2 - 21x + 108 = 0$$
$$(x - 9)(x - 12) = 0$$
$$x = 9 \text{ or } x = 12$$
$$y = 12 \qquad y = 9 \text{ Since } y = 21 - x$$
The length is 12 feet and the width is 9 feet.

Making Connections
Chapters 1 - 12

1. The graph of $y = 9x - x^2$ is a parabola that opens downward. Solve $9x - x^2 = 0$ to find the x-intercepts.
$$x(9 - x) = 0$$
$$x = 0 \text{ or } x = 9$$
The x-intercepts are $(0, 0)$ and $(9, 0)$.

The x-coordinate of the vertex is
$$x = \frac{-(9)}{2(-1)} = \frac{9}{2}.$$
The vertex is $(9/2, 81/4)$.

2. The graph of $y = 9x$ is a line through $(0, 0)$ with slope 9.

3. The graph of $y = (x - 9)^2$ is a parabola opening upward, with vertex at $(9, 0)$.

4. Write $y^2 = 9 - x^2$ as $x^2 + y^2 = 9$ to see that it is the equation of a circle of radius 3 centered at the origin.

5. The graph of $y = 9x^2$ is a parabola opening upward with vertex at $(0, 0)$.

6. The graph of $y = |9x|$ is v-shaped. It contains the points $(0, 0)$, $(-1, 9)$, and $(1, 9)$.

7. Write $4x^2 + 9y^2 = 36$ as $\frac{x^2}{9} + \frac{y^2}{4} = 1$ to see that it is the equation of an ellipse through $(-3, 0)$, $(3, 0)$, $(0, 2)$, and $(0, -2)$.

8. Write $4x^2 - 9y^2 = 36$ as $\frac{x^2}{9} - \frac{y^2}{4} = 1$ to see that it is the equation of a hyperbola with x-intercepts $(-3, 0)$ and $(3, 0)$. The fundamental rectangle goes through the x-intercepts, $(0, -2)$, and $(0, 2)$. Extend the diagonals of the rectangle to get the asymptotes.

9. The graph of $y = 9 - x$ is a line through $(0, 9)$ with slope -1.

10. The graph of $y = 9^x$ goes through $(0, 1)$, $(1, 9)$, and $(-1, 1/9)$.

11. $(x + 2y)^2 = x^2 + 2(x)(2y) + (2y)^2$
$$= x^2 + 4xy + 4y^2$$

12. $(x + y)(x^2 + 2xy + y^2)$
$$= x(x^2 + 2xy + y^2) + y(x^2 + 2xy + y^2)$$
$$= x^3 + 2x^2y + xy^2 + x^2y + 2xy^2 + y^3$$
$$= x^3 + 3x^2y + 3xy^2 + y^3$$

13. $(a + b)^3 = (a + b)(a^2 + 2ab + b^2)$
$$= a(a^2 + 2ab + b^2) + b(a^2 + 2ab + b^2)$$
$$= a^3 + 2a^2b = ab^2 + a^2b + 2ab^2 + b^3$$
$$= a^3 + 3a^2b + 3ab^2 + b^3$$

14. $(a - 3b)^2 = a^2 - 2(a)(3b) + (3b)^2$
$$= a^2 - 6ab + 9b^2$$

15. $(2a + 1)(3a - 5) = 6a^2 + 3a - 10a - 5$
$$= 6a^2 - 7a - 5$$

16. $(x - y)(x^2 + xy + y^2)$
$$= x(x^2 + xy + y^2) - y(x^2 + xy + y^2)$$
$$= x^3 + x^2y + xy^2 - x^2y - xy^2 - y^3$$
$$= x^3 - y^3$$

17. Multiply the second equation by -2 and add the result to the first equation.
$$2x - 3y = -4$$
$$\underline{-2x - 4y = -10}$$
$$-7y = -14$$
$$y = 2$$
Use $y = 2$ in $x + 2y = 5$ to find x.
$$x + 2(2) = 5$$
$$x = 1$$
The solution set is $\{(1, 2)\}$.

18. Substitute $y = 7 - x$ into $x^2 + y^2 = 25$.

$$x^2 + (7 - x)^2 = 25$$
$$x^2 + 49 - 14x + x^2 = 25$$
$$2x^2 - 14x + 24 = 0$$
$$x^2 - 7x + 12 = 0$$
$$(x - 3)(x - 4) = 0$$
$$x = 3 \quad \text{or} \quad x = 4$$
$$y = 4 \qquad y = 3 \quad \text{Since } y = 7 - x$$
The solution set is $\{(3, 4), (4, 3)\}$.

19. Adding the first and second equations to eliminate z, we get $3x - 3y = 9$. Adding the second and third equations to eliminate z, we get $2x - y = 4$. Divide $3x - 3y = 9$ by -3 and add the result to $2x - y = 4$.
$$-x + y = -3$$
$$\underline{2x - y = 4}$$
$$x = 1$$
Use $x = 1$ in $-x + y = -3$.
$$-1 + y = -3$$
$$y = -2$$
Use $x = 1$ and $y = -2$ in $x + y + z = 2$.
$$1 + (-2) + z = 2$$
$$z = 3$$
The solution set is $\{(1, -2, 3)\}$.

20. Substitute $y = x^2$ into $y - 2x = 3$.
$$x^2 - 2x = 3$$
$$x^2 - 2x - 3 = 0$$
$$(x - 3)(x + 1) = 0$$
$$x = 3 \quad \text{or} \quad x = -1$$
$$y = 9 \qquad y = 1 \quad \text{Since } y = x^2$$

The solution set is $\{(-1, 1), (3, 9)\}$.

21. $ax + b = 0$
$$ax = -b$$
$$x = -\frac{b}{a}$$

22. Use the quadratic formula with $a = w$, $b = d$, and $c = m$.
$$x = \frac{-d \pm \sqrt{d^2 - 4wm}}{2w}$$

23.
$$A = \tfrac{1}{2}h(B + b)$$
$$2A = h(B + b)$$
$$2A = hB + hb$$
$$2A - bh = hB$$
$$B = \frac{2A - bh}{h}$$

276

24. $\frac{1}{x} + \frac{1}{y} = \frac{1}{2}$

$2xy\left(\frac{1}{x} + \frac{1}{y}\right) = 2xy\left(\frac{1}{2}\right)$

$2y + 2x = xy$

$2y = xy - 2x$

$2y = x(y - 2)$

$x = \frac{2y}{y - 2}$

25. $L = m + mxt$

$L = m(1 + xt)$

$m = \frac{L}{1 + xt}$

26. $y = 3a\sqrt{t}$

$(y)^2 = (3a\sqrt{t})^2$

$y^2 = 9a^2 t$

$\frac{y^2}{9a^2} = t$

$t = \frac{y^2}{9a^2}$

27. First find the slope.

$m = \frac{-3 - 1}{2 - (-4)} = \frac{-4}{6} = -\frac{2}{3}$

Use point-slope form with $(2, -3)$.

$y - (-3) = -\frac{2}{3}(x - 2)$

$y + 3 = -\frac{2}{3}x + \frac{4}{3}$

$y = -\frac{2}{3}x - \frac{5}{3}$

28. Write $2x - 4y = 5$ as $y = \frac{1}{2}x - \frac{5}{4}$. The slope of any line perpendicular to this line is -2. The line through $(0, 0)$ with slope -2 is $y = -2x$.

29. The radius is the distance between $(2, 5)$ and $(-1, -1)$.

$r = \sqrt{(-1 - 2)^2 + (-1 - 5)^2} = \sqrt{9 + 36}$

$= \sqrt{45}$

The equation of circle with center $(2, 5)$ and radius $\sqrt{45}$ is $(x - 2)^2 + (y - 5)^2 = 45$.

30. Use completing the square to get the equation into standard form for a circle.

$x^2 + 3x + y^2 - 6y = 0$

$x^2 + 3x + \frac{9}{4} + y^2 - 6y + 9 = 0 + \frac{9}{4} + 9$

$(x + \frac{3}{2})^2 + (y - 3)^2 = \frac{45}{4}$

The center is $(-\frac{3}{2}, 3)$ and radius is $\sqrt{\frac{45}{4}}$, or $\frac{3\sqrt{5}}{2}$.

31. $2i(3 + 5i) = 6i + 10i^2 = -10 + 6i$

32. $i^6 = i^4 \cdot i^2 = 1(-1) = -1$

33. $(2i - 3) + (6 - 7i) = 3 - 5i$

34. $(3 + i\sqrt{2})^2 = 9 + 6i\sqrt{2} + 2i^2$

$= 7 + 6i\sqrt{2}$

35. $(2 - 3i)(5 - 6i) = 10 - 15i - 12i + 18i^2$

$= -8 - 27i$

36. $(3 - i) + (-6 + 4i) = -3 + 3i$

37. $(5 - 2i)(5 + 2i) = 25 - 4i^2 = 29$

38. $\frac{2 - 3i}{2i} = \frac{(2 - 3i)(-i)}{2i(-i)} = \frac{-2i + 3i^2}{2(1)}$

$= -\frac{3}{2} - i$

39. $\frac{(4 + 5i)(1 + i)}{(1 - i)(1 + i)} = \frac{4 + 9i + 5i^2}{2}$

$= \frac{-1 + 9i}{2} = -\frac{1}{2} + \frac{9}{2}i$

40. $\frac{4 - \sqrt{-8}}{2} = \frac{4 - 2i\sqrt{2}}{2} = 2 - i\sqrt{2}$

41. a) $m = \frac{250 - 200}{0.30 - 0.40} = \frac{50}{-0.1} = -500$

$q - 250 = -500(x - 0.30)$

$q - 250 = -500x + 150$

$q = -500x + 400$

b) $R = qx = (-500x + 400)x$

$R = -500x^2 + 400x$

c) The graph is a parabola through $(0, 0)$, $(0.80, 0), (0.40, 80)$

d) The maximum revenue occurs at the vertex of the parabola.

$x = \frac{-400}{2(-500)} = 0.40$

The maximum revenue occurs when bananas are $0.40 per pound.

e) To find the maximum revenue use the formula for revenue with $x = 0.40$.

$R = -500(0.40)^2 + 400(0.40) = 80$

The maximum revenue is $80 when the bananas are $0.40 per pound.

13.1 WARM-UPS

1. True, because the formula $a_n = 2n$ will produce even numbers when n is a positive integer.

2. True, because the formula $a_n = 2n - 1$ will produce odd numbers when n is a positive integer.

3. True, by the definition of sequence.

4. False, because the domain of a finite sequence is the set of positive integers less than or equal to some fixed positive integer.

5. False, because if $n = 1$, $a_1 = (-1)^2 \cdot 1^2 = 1$.

6. False, because the independent variable is n.

7. True.

8. False, because the 6th term is $a_6 = (-1)^7 2^6 = -64$.

9. True.

10. True, because the sequence is $a_n = 2^n$ and $a_{10} = 2^{10} = 1024$.

13.1 EXERCISES

1. A sequence is a list of numbers.

3. A finite sequence is a function whose domain is the set of positive integers less than or equal to some fixed positive integer.

5. The terms of the sequence a_n are found by squaring the integers from 1 through 8, because of the formula $a_n = n^2$. So the terms are 1, 4, 9, 16, 25, 36, 49, and 64.

7. $b_1 = \dfrac{(-1)^1}{1} = -1$, $b_2 = \dfrac{(-1)^2}{2} = \dfrac{1}{2}$,

$b_3 = \dfrac{(-1)^3}{3} = -\dfrac{1}{3}$, $b_4 = \dfrac{(-1)^4}{4} = \dfrac{1}{4}$,

$b_5 = \dfrac{(-1)^5}{5} = -\dfrac{1}{5}$, etc.

The 10 terms of the sequence are
$-1, \dfrac{1}{2}, -\dfrac{1}{3}, \dfrac{1}{4}, -\dfrac{1}{5}, \dfrac{1}{6}, -\dfrac{1}{7}, \dfrac{1}{8}, -\dfrac{1}{9}$, and $\dfrac{1}{10}$.

9. $c_1 = (-2)^{1-1} = 1$, $c_2 = (-2)^{2-1} = -2$, $c_3 = (-2)^{3-1} = 4$, $c_4 = (-2)^{4-1} = -8$, $c_5 = (-2)^{5-1} = 16$
The five terms are 1, -2, 4, -8, and 16.

11. $a_1 = 2^{-1} = \dfrac{1}{2}$, $a_2 = 2^{-2} = \dfrac{1}{4}$,
$a_3 = 2^{-3} = \dfrac{1}{8}$, $a_4 = 2^{-4} = \dfrac{1}{16}$, etc.
The six terms are $\dfrac{1}{2}, \dfrac{1}{4}, \dfrac{1}{8}, \dfrac{1}{16}, \dfrac{1}{32}$, and $\dfrac{1}{64}$.

13. $b_1 = 2(1) - 3 = -1$, $b_2 = 2(2) - 3 = 1$,
$b_3 = 2(3) - 3 = 3$, $b_4 = 2(4) - 3 = 5$,
$b_5 = 2(5) - 3 = 7$, $b_6 = 2(6) - 3 = 9$,
$b_7 = 2(7) - 3 = 11$
The seven terms are $-1, 1, 3, 5, 7, 9$, and 11.

15. $c_1 = 1^{-1/2} = 1$, $c_2 = 2^{-1/2} = \dfrac{1}{\sqrt{2}}$,
$c_3 = 3^{-1/2} = \dfrac{1}{\sqrt{3}} = \dfrac{\sqrt{3}}{3}$, $c_4 = 4^{-1/2} = \dfrac{1}{2}$,
$c_5 = 5^{-1/2} = \dfrac{1}{\sqrt{5}} = \dfrac{\sqrt{5}}{5}$

The five terms are $1, \dfrac{\sqrt{2}}{2}, \dfrac{\sqrt{3}}{3}, \dfrac{1}{2}$, and $\dfrac{\sqrt{5}}{5}$.

17. $a_1 = \dfrac{1}{1^2 + 1} = \dfrac{1}{2}$, $a_2 = \dfrac{1}{2^2 + 2} = \dfrac{1}{6}$,
$a_3 = \dfrac{1}{3^2 + 3} = \dfrac{1}{12}$, $a_4 = \dfrac{1}{4^2 + 4} = \dfrac{1}{20}$

The first four terms are $\dfrac{1}{2}, \dfrac{1}{6}, \dfrac{1}{12}$, and $\dfrac{1}{20}$.

19. $b_1 = \dfrac{1}{2(1) - 5} = -\dfrac{1}{3}$,
$b_2 = \dfrac{1}{2(2) - 5} = -1$,
$b_3 = \dfrac{1}{2(3) - 5} = 1$, $b_4 = \dfrac{1}{2(4) - 5} = \dfrac{1}{3}$

The first four terms are $-\dfrac{1}{3}, -1, 1$, and $\dfrac{1}{3}$.

21. $c_1 = (-1)^1(1 - 2)^2 = -1$,
$c_2 = (-1)^2(2 - 2)^2 = 0$,
$c_3 = (-1)^3(3 - 2)^2 = -1$,
$c_4 = (-1)^4(4 - 2)^2 = 4$

The first four terms are $-1, 0, -1$, and 4.

23. $a_1 = \dfrac{(-1)^{2(1)}}{1^2} = 1$, $a_2 = \dfrac{(-1)^{2(2)}}{2^2} = \dfrac{1}{4}$,
$a_3 = \dfrac{(-1)^{2(3)}}{3^2} = \dfrac{1}{9}$, $a_4 = \dfrac{(-1)^{2(4)}}{4^2} = \dfrac{1}{16}$

The first four terms are $1, \dfrac{1}{4}, \dfrac{1}{9}$, and $\dfrac{1}{16}$.

25. This sequence is a sequence of odd integers starting at 1. Even integers are represented by $2n$ and odd integers are all one less than an even integer. Try the formula $a_n = 2n - 1$. To see that it is correct, find a_1 through a_5.

27. A sequence of alternating ones and negative ones is obtained by using a power of -1. Since we want a_1 to be positive when $n = 1$, we use the $n + 1$ power on -1. A formula for the general term is $a_n = (-1)^{n+1}$.

29. This sequence is a sequence of even integers starting at 0. Even integers can be generated by $2n$, but to make the first one 0 (when $n = 1$), we use the formula $a_n = 2n - 2$.

31. This sequence is a sequence consisting of the positive integral multiples of 3. Try the formula $a_n = 3n$ to see that it generates the appropriate numbers.

33. Each term in this sequence is one larger than the corresponding term in the sequence of Exercise 31. Try the formula $a_n = 3n + 1$ to see that it generates the given sequence.

35. To get the alternating signs on the terms of the sequence, we use a power of -1. Next we observe that the numbers 1, 2, 4, 8, 16 are powers of 2. The formula $a_n = (-1)^n 2^{n-1}$ will produce the given sequence. Remember that n always starts at 1, so in this case we use $n - 1$ as the power of 2 to get $2^0 = 1$.

37. Notice that the numbers 0, 1, 4, 9, 16, ... are the squares of the nonnegative integers: 0^2, $1^2, 2^2, 3^2, 4^2, ...$. We would use n^2 to generate squares, but since n starts at 1, we use $a_n = (n - 1)^2$ to get the given sequence.

39. After the first penalty the ball is on the 4-yard line. After the second penalty the ball is on the 2-yard line. After the third penalty the ball is on the 1-yard line, and so on. The sequence of five terms is 4, 2, 1, $\frac{1}{2}$, $\frac{1}{4}$.

41. To find the amount of increase, we take 5% of $34,940 to get $1747 increase. We could just multiply $34,940 by 1.05 to obtain the new price. In either case the price the next year is $36,687. To find the price for the next year we multiply the last year's price by 1.05:

1.05($36,687) = $38,521. Repeating this process gives us the prices $36,687, $38,521, $40,447, $42,470, $44,593, and $46,823 as the price of the truck through 2008.

43. Of the $1 million, 80% is respent in the community. So $800,000 is respent. Of the $800,000 we have 80% respent in the community. So $640,000 is respent in the community, and so on. The first four terms of the sequence are $1,000,000, $800,000, $640,000, $512,000.

45. Possible vertical repeats are 27 in., 13.5 in., 9 in., 6.75 in., and 5.4 in.

47. If we put the word great in front of the word grandparents 35 times, then we have 2^{37} of this type of relative. Use a calculator to find that $2^{37} \approx 137,438,953,500$. This is certainly larger than the present population of the earth.

51. **a)** Use a calculator to find
$$a_{100} = (0.999)^{100} = 0.9048,$$
$$a_{1000} = (0.999)^{1000} = 0.3677,$$
and $\quad a_{10000} = (0.999)^{10000} = 0.00004517$.

b) a_n approaches zero as n gets larger and larger.

13.2 WARM-UPS

1. True, because of the definition of series.

2. False, the sum of a series can be any real number.

3. False, there are 9 terms in 2^3 through 10^3.

4. False, because the terms in the first series are opposites of the terms in the second series.

5. False, the ninth term is
$$\frac{(-1)^9}{(9+1)(9+2)} = -\frac{1}{110}.$$

6. True, because the terms are -2 and 4 and the sum is 2.

7. True, because of the distributive property.

8. True, because the notation indicates to add the number 4 five times.

9. True, because $2i + 7i = 9i$.

10. False, because in the series on the left side the 1 is added in three times and in the series on the right side the 1 is only added in once.

13.2 EXERCISES

1. Summation notation provides a way of writing a sum without writing out all of the terms.

3. A series is the indicated sum of the terms of a sequence.

5. $\displaystyle\sum_{i=1}^{4} i^2 = 1^2 + 2^2 + 3^2 + 4^2$

$= 1 + 4 + 9 + 16 = 30$

7. $\displaystyle\sum_{j=0}^{5}(2j - 1) = (2 \cdot 0 - 1) + (2 \cdot 1 - 1)$

$+ (2 \cdot 2 - 1) + (2 \cdot 3 - 1)$
$+ (2 \cdot 4 - 1) + (2 \cdot 5 - 1)$
$= -1 + 1 + 3 + 5 + 7 + 9 = 24$

9. $\displaystyle\sum_{i=1}^{5} 2^{-i} = 2^{-1} + 2^{-2} + 2^{-3} + 2^{-4} + 2^{-5}$

$= \frac{1}{2} + \frac{1}{4} + \frac{1}{8} + \frac{1}{16} + \frac{1}{32} = \frac{31}{32}$

11. $\displaystyle\sum_{i=1}^{10} 5i^0 = 5(1)^0 + 5(2)^0 + 5(3)^0 + 5(4)^0$

$+ 5(5)^0 + 5(6)^0 + 5(7)^0$
$+ 5(8)^0 + 5(9)^0 + 5(10)^0$
$= 5 + 5 + 5 + 5 + 5 + 5 + 5 + 5 + 5 + 5$
$= 50$

13. $\displaystyle\sum_{i=1}^{3}(i - 3)(i + 1) = (1 - 3)(1 + 1)$

$+ (2 - 3)(2 + 1) + (3 - 3)(3 + 1)$
$= -4 + (-3) + 0 = -7$

15. $\displaystyle\sum_{j=1}^{10}(-1)^j$

$= (-1)^1 + (-1)^2 + ... + (-1)^{10}$
$= -1 + 1 - 1 + 1 - 1 + 1 - 1$
$+ 1 - 1 + 1$
$= 0$

17. The sum of the first six positive integers is written as $\displaystyle\sum_{i=1}^{6} i$. There are other ways to indicate this series in summation notation but this is the simplest.

19. To get the signs of the terms to alternate from positive to negative we use a power of -1. To get odd integers we use the formula $2i - 1$. So this series is written

$\displaystyle\sum_{i=1}^{6}(-1)^i(2i - 1)$.

21. This series consists of the squares of the first six positive integers. It is written in summation notation as $\displaystyle\sum_{i=1}^{6} i^2$.

23. This series consists of the reciprocals of the positive integers. If the index i goes from 1 to 4, we must use $2 + i$ to get the numbers 3 through 6. So the series is written $\displaystyle\sum_{i=1}^{4}\frac{1}{2 + i}$.

25. The terms of this series are logarithms of positive integers. If the index i goes from 1 to 3, we must use $i + 1$ to get the numbers 2 through 4. The series is written $\displaystyle\sum_{i=1}^{3}\ln(i + 1)$.

27. Since the subscripts range from 1 through 4, we let i range from 1 through 4: $\displaystyle\sum_{i=1}^{4} a_i$.

29. The subscripts on x range from 3 through 50, so i ranges from 1 through 48: $\displaystyle\sum_{i=1}^{48} x_{i+2}$.

31. The subscripts on w range from 1 through n, so i ranges from 1 through n: $\displaystyle\sum_{i=1}^{n} w_i$.

33. If $j = 0$ when $i = 1$, then $i = j + 1$ and j ranges from 0 through 4: $\displaystyle\sum_{j=0}^{4}(j + 1)^2$

35. If $j = 1$ when $i = 0$, then $i = j - 1$ and j ranges from 1 through 13. If we substitute $i = j - 1$ into $2i - 1$ we get $2i - 1 = 2(j - 1) - 1 = 2j - 3$. So the series is written $\displaystyle\sum_{j=1}^{13}(2j - 3)$.

37. If $j = 1$ when $i = 4$, then $i = j + 3$ and j ranges from 1 through 5. Substitute $j + 3$ for i to get $\displaystyle\sum_{j=1}^{5}\frac{1}{j + 3}$.

39. If $j = 0$ when $i = 1$, then $i = j + 1$ and j ranges from 0 through 3. The exponent $2i + 3$ becomes $2(j + 1) + 3 = 2j + 5$. The series is written as $\sum_{j=0}^{3} x^{2j+5}$.

41. If $j = 0$ when $i = 1$, then $i = j + 1$ and j ranges from 0 through $n - 1$. Replacing i by $j + 1$ gives us the series $\sum_{j=0}^{n-1} x^{j+1}$.

43. $\sum_{i=1}^{6} x^i = x + x^2 + x^3 + x^4 + x^5 + x^6$

45. $\sum_{j=0}^{3} (-1)^j x_j$
$= (-1)^0 x_0 + (-1)^1 x_1 + (-1)^2 x_2 + (-1)^3 x_3$
$= x_0 - x_1 + x_2 - x_3$

47. $\sum_{i=1}^{3} i x^i = 1x^1 + 2x^2 + 3x^3$
$\qquad\qquad = x + 2x^2 + 3x^3$

49. On the first jump he has moved $\frac{1}{2}$ yard. On the second jump he moves $\frac{1}{4}$ yard. On the third jump he moves $\frac{1}{8}$ yard, and so on. To express the reciprocals of the powers of 2, we use 2^{-i}. His total movement after nine jumps is expressed as the series $\sum_{i=1}^{9} 2^{-i}$.

51. $\sum_{i=1}^{4} 1,000,000(0.8)^{i-1}$.

53. A sequence is basically a list of numbers. A series is the indicated sum of the terms of a sequence.

13.3 WARM-UPS

1. False, because the common difference is -2 $(1 - 3 = -2)$.
2. False, because the difference between consecutive terms is not constant.
3. False, because the difference between the consecutive terms is sometimes 2 and sometimes -2.
4. False, because the nth term is given by $a_n = a_1 + (n - 1)d$.
5. False, because the second must be 7.5, making $d = 2.5$ and the fourth term 12.5.

6. False, because if the first is 6 and the third is 2, the second must be 4 in order to have a common difference.
7. True, because that is the definition of arithmetic series.
8. True, because the series is $5 + 7 + 9 + 11 + 13$ and the common difference is 2.
9. True, because of the formula for the sum of an arithmetic series.
10. False, because there are 11 even integers from 8 through 28 inclusive and the sum is $\frac{11}{2}(8 + 28)$.

13.3 EXERCISES

1. An arithmetic sequence is one in which each term after the first is obtained by adding a fixed amount to the previous term.
3. An arithmetic series is an indicated sum of an arithmetic sequence.
5. The common difference is $d = 6$ and the first term is $a_1 = 0$. Use the formula $a_n = a_1 + (n - 1)d$ to find that the nth term is $a_n = 0 + (n - 1)6 = 6n - 6$.
7. The common difference is $d = 5$ and the first term is 7. The nth term is $a_n = 7 + (n - 1)5 = 5n + 2$.
9. The common difference is $d = 2$ and the first term is $a_1 = -4$. Use the formula $a_n = a_1 + (n - 1)d$ to find that the nth term is $a_n = -4 + (n - 1)2 = 2n - 6$.
11. The common difference is $d = 1 - 5$ $= -4$ and the first term is $a_1 = 5$. The nth term is $a_n = 5 + (n - 1)(-4) = -4n + 9$.
13. The common difference is $d = -9 - (-2)$ $= -7$ and the first term is -2. The nth term is $a_n = -2 + (n - 1)(-7) = -7n + 5$.

15. The common difference is $d = -2.5 - (-3) = 0.5$ and the first term is -3. The nth term is $a_n = -3 + (n - 1)(0.5) = 0.5n - 3.5$.

17. The common difference is $d = -6.5 - (-6) = -0.5$ and the first term is -6. The nth term is $a_n = -6 + (n - 1)(-0.5)$ $= -0.5n - 5.5$.

19. $a_1 = 9 + (1 - 1)4 = 9$,
$a_2 = 9 + (2 - 1)4 = 13$,
$a_3 = 9 + (3 - 1)4 = 17$,
$a_4 = 9 + (4 - 1)4 = 21$,
$a_5 = 9 + (5 - 1)4 = 25$
The first five terms of the arithmetic sequence are 9, 13, 17, 21, and 25.

21. $a_1 = 7 + (1 - 1)(-2) = 7$,
$a_2 = 7 + (2 - 1)(-2) = 5$,
$a_3 = 7 + (3 - 1)(-2) = 3$,
$a_4 = 7 + (4 - 1)(-2) = 1$,
$a_5 = 7 + (5 - 1)(-2) = -1$
The first five terms of the arithmetic sequence are 7, 5, 3, 1, and -1.

23. $a_1 = -4 + (1 - 1)3 = -4$,
$a_2 = -4 + (2 - 1)3 = -1$,
$a_3 = -4 + (3 - 1)3 = 2$,
$a_4 = -4 + (4 - 1)3 = 5$,
$a_5 = -4 + (5 - 1)3 = 8$
The first five terms of the arithmetic sequence are $-4, -1, 2, 5$, and 8.

25. $a_1 = -2 + (1 - 1)(-3) = -2$,
$a_2 = -2 + (2 - 1)(-3) = -5$,
$a_3 = -2 + (3 - 1)(-3) = -8$,
$a_4 = -2 + (4 - 1)(-3) = -11$,
$a_5 = -2 + (5 - 1)(-3) = -14$
The first five terms of the arithmetic sequence are $-2, -5, -8, -11$, and -14.

27. $a_1 = -4(1) - 3 = -7$,
$a_2 = -4(2) - 3 = -11$,
$a_3 = -4(3) - 3 = -15$,
$a_4 = -4(4) - 3 = -19$,
$a_5 = -4(5) - 3 = -23$
The first five terms of the arithmetic sequence are $-7, -11, -15, -19$, and -23.

29. $a_1 = 0.5(1) + 4 = 4.5$,
$a_2 = 0.5(2) + 4 = 5$, $a_3 = 0.5(3) + 4 = 5.5$,
$a_4 = 0.5(4) + 4 = 6$, $a_5 = 0.5(5) + 4 = 6.5$
The first five terms of the arithmetic sequence are 4.5, 5, 5.5, 6, and 6.5.

31. $a_1 = 20(1) + 1000 = 1020$,
$a_2 = 20(2) + 1000 = 1040$,
$a_3 = 20(3) + 1000 = 1060$,
$a_4 = 20(4) + 1000 = 1080$,
$a_5 = 20(5) + 1000 = 1100$
The first five terms of the arithmetic sequence are 1020, 1040, 1060, 1080, and 1100.

33. Use $a_1 = 9$, $n = 8$, and $d = 6$ in the formula $a_n = a_1 + (n - 1)d$.
$$a_8 = 9 + (8 - 1)6 = 51$$

35. Use $a_1 = 6$, $a_{20} = 82$, and $n = 20$ in the formula $a_n = a_1 + (n - 1)d$.
$82 = 6 + (20 - 1)d$
$82 = 6 + 19d$
$76 = 19d$
$4 = d$

37. Use $a_7 = 14$, $d = -2$, and $n = 7$ in the formula $a_n = a_1 + (n - 1)d$.
$14 = a_1 + (7 - 1)(-2)$
$14 = a_1 - 12$
$26 = a_1$

39. From the fact that the fifth term is 13 and the first term is -3, we can find the common difference.
$13 = -3 + (5 - 1)d$
$13 = -3 + 4d$
$16 = 4d$
$4 = d$
Use $a_1 = -3$, $n = 6$, and $d = 4$ in the formula $a_n = a_1 + (n - 1)d$.
$$a_6 = -3 + (6 - 1)4 = 17$$

41. Use $a_1 = 1$, $a_{48} = 48$, and $n = 48$ in the formula $S_n = \frac{n}{2}(a_1 + a_n)$.
$$S_{48} = \frac{48}{2}(1 + 48) = 1176$$

43. To find n, use $a_1 = 8$, $d = 2$, and $a_n = 36$ in the formula $a_n = a_1 + (n - 1)d$.
$36 = 8 + (n - 1)2$
$36 = 8 + 2n - 2$
$30 = 2n$
$15 = n$
Use $a_1 = 8$, $a_{15} = 36$, and $n = 15$ in the formula $S_n = \frac{n}{2}(a_1 + a_n)$.
$$S_{15} = \frac{15}{2}(8 + 36) = 330$$

45. To find n, use $a_1 = -1$, $d = -6$, and $a_n = -73$ in the formula $a_n = a_1 + (n - 1)d$.
$-73 = -1 + (n - 1)(-6)$
$-78 = -1 - 6n + 6$
$-78 = -6n$
$13 = n$
Use $a_1 = -1$, $a_{13} = -73$, and $n = 13$ in the formula $S_n = \frac{n}{2}(a_1 + a_n)$.
$$S_{13} = \frac{13}{2}(-1 + (-73)) = -481$$

47. To find n, use $a_1 = -6$, $d = 5$, and $a_n = 64$ in the formula $a_n = a_1 + (n-1)d$.

$$64 = -6 + (n-1)5$$
$$64 = -6 + 5n - 5$$
$$75 = 5n$$
$$15 = n$$

Use $a_1 = -6$, $a_{15} = 64$, and $n = 15$ in the formula $S_n = \frac{n}{2}(a_1 + a_n)$.

$$S_{15} = \frac{15}{2}(-6 + 64) = 435$$

49. To find n, use $a_1 = 20$, $d = -8$, and $a_n = -92$ in the formula $a_n = a_1 + (n-1)d$.

$$-92 = 20 + (n-1)(-8)$$
$$-92 = 20 - 8n + 8$$
$$-120 = -8n$$
$$15 = n$$

Use $a_1 = 20$, $a_{15} = -92$, and $n = 15$ in the formula $S_n = \frac{n}{2}(a_1 + a_n)$.

$$S_{15} = \frac{15}{2}(20 + (-92)) = -540$$

51. $\displaystyle\sum_{i=1}^{12}(3i - 7) = -4 + (-1) + \ldots + 29$

Use $a_1 = -4$, $a_{12} = 29$, and $n = 12$ in the formula $S_n = \frac{n}{2}(a_1 + a_n)$.

$$S_{12} = \frac{12}{2}(-4 + 29) = 150$$

53. $\displaystyle\sum_{i=1}^{11}(-5i + 2) = -3 + (-8) + \ldots + (-53)$

Use $a_1 = -3$, $a_{11} = -53$, and $n = 11$ in the formula $S_n = \frac{n}{2}(a_1 + a_n)$.

$$S_{11} = \frac{11}{2}(-3 + (-53)) = -308$$

55. Use $a_1 = \$22{,}000$, $n = 7$, and $d = \$500$ in the formula $a_n = a_1 + (n-1)d$.

$$a_7 = 22{,}000 + (7-1)500 = \$25{,}000$$

57. The students read 5 pages the first day, 7 pages the second day, 9 pages the third day, and so on. To find the number they read on the 31st day, let $n = 31$, $d = 2$, and $a = 5$ in the formula $a_n = a_1 + (n-1)d$.

$$a_{31} = 5 + (31-1)2 = 65$$

To find the sum $5 + 7 + 9 + \ldots + 65$, use $n = 31$, $a_1 = 5$, and $a_{31} = 65$ in the formula $S_n = \frac{n}{2}(a_1 + a_n)$.

$$S_{31} = \frac{31}{2}(5 + 65) = 1085$$

59. The only sequence that does not have a common difference is (b) and so it is not arithmetic.

13.4 WARM-UPS

1. False, because the ratio of two consecutive terms is not constant.
2. False, there is a common ratio of 2 between adjacent terms.
3. True, because the general form for a geometric sequence is $a_n = a_1 r^{n-1}$.
4. True, because if $n = 1$, then $3(2)^{-1+3} = 12$.
5. True, because $a_1 = 12$ and $a_2 = 6$ gives $r = 1/2$.
6. True, because of the definition of geometric series.
7. False, because we have a formula for the sum of a finite geometric series.
8. False, because $a_1 = 6$.
9. True, because this is the correct formula for the sum of all of the terms of an infinite geometric series with first term 10 and ratio 1/2.
10. False, because there is no sum for an infinite geometric series with a ratio of 2.

13.4 EXERCISES

1. A geometric sequence is one in which each term after the first is obtained by multiplying the preceding term by a constant.
3. A geometric series is an indicated sum of a geometric sequence.
5. The approximate value of r^n when n is large and $|r| < 1$ is 0.
7. Since the first term is 1/3 and the common ratio is 3, the nth term is $a_n = \frac{1}{3}(3)^{n-1}$.
9. Since the first term is 64 and the common ratio is 1/8, the nth term is $a_n = 64\left(\frac{1}{8}\right)^{n-1}$.
11. Since the first term is 8 and the common ratio is $-1/2$, the nth term is
$$a_n = 8\left(-\frac{1}{2}\right)^{n-1}.$$
13. Since the first term is 2 and the common ratio is $-4/2 = -2$, the nth term is
$$a_n = 2(-2)^{n-1}.$$

15. Since the first term is $-1/3$ and the common ratio is $(-1/4)/(-1/3) = 3/4$, the nth term is $a_n = -\frac{1}{3}\left(\frac{3}{4}\right)^{n-1}$.

17. $a_1 = 2(1/3)^{1-1} = 2,$
$a_2 = 2(1/3)^{2-1} = 2/3,$
$a_3 = 2(1/3)^{3-1} = 2/9,$
$a_4 = 2(1/3)^{4-1} = 2/27,$
$a_5 = 2(1/3)^{5-1} = 2/81$
The first 5 terms are 2, $\frac{2}{3}$, $\frac{2}{9}$, $\frac{2}{27}$, and $\frac{2}{81}$.

19. $a_1 = (-2)^{1-1} = 1,$
$a_2 = (-2)^{2-1} = -2,$ $\quad a_3 = (-2)^{3-1} = 4,$
$a_4 = (-2)^{4-1} = -8,$ $\quad a_5 = (-2)^{5-1} = 16$
The first 5 terms are $1, -2, 4, -8,$ and 16.

21. $a_1 = 2^{-1} = 1/2,$
$a_2 = 2^{-2} = 1/4,$ $\quad a_3 = 2^{-3} = 1/8,$
$a_4 = 2^{-4} = 1/16,$ $\quad a_5 = 2^{-5} = 1/32$
The first 5 terms are $\frac{1}{2}$, $\frac{1}{4}$, $\frac{1}{8}$, $\frac{1}{16}$, and $\frac{1}{32}$.

23. $a_1 = (0.78)^1 = 0.78,$
$a_2 = (0.78)^2 = 0.6084,$
$a_3 = (0.78)^3 = 0.4746,$
$a_4 = (0.78)^4 = 0.3702,$ $a_5 = (0.78)^5 = 0.2887$
The first 5 terms are $0.78, 0.6084, 0.4746,$
$0.3702,$ and 0.2887.

25. Use $a_4 = 40$, $n = 4$, and $r = 2$ in the
formula $a_n = a_1 r^{n-1}$.
$$40 = a_1(2)^{4-1}$$
$$40 = 8a_1$$
$$5 = a_1$$

27. Use $a_4 = 2/9$, $n = 4$, and $a_1 = 6$ in the
formula $a_n = a_1 r^{n-1}$.
$$\tfrac{2}{9} = 6r^{4-1}$$
$$\tfrac{1}{27} = r^3$$
$$\tfrac{1}{3} = r$$

29. Use $r = 1/3$, $n = 4$, and $a_1 = -3$ in the
formula $a_n = a_1 r^{n-1}$.
$$a_4 = -3\left(\tfrac{1}{3}\right)^{4-1} = -3\left(\tfrac{1}{27}\right) = -\tfrac{1}{9}$$

31. Use $r = 1/2$, $a_1 = 1/2$, and $a_n = 1/512$
in the formula $a_n = a_1 r^{n-1}$ to find n.
$$\tfrac{1}{512} = \tfrac{1}{2}\left(\tfrac{1}{2}\right)^{n-1}$$
$$\tfrac{1}{2^9} = \left(\tfrac{1}{2}\right)^n$$
$$n = 9$$
Use $n = 9$, $a_1 = 1/2$, and $r = 1/2$ in the
formula $S_n = \dfrac{a_1(1 - r^n)}{1 - r}$.
$$S_9 = \frac{\tfrac{1}{2}\left(1 - \left(\tfrac{1}{2}\right)^9\right)}{1 - \tfrac{1}{2}} = 1 - \frac{1}{512} = \frac{511}{512}$$

33. Use $n = 5$, $a_1 = 1/2$, and $r = -1/2$ in
the formula $S_n = \dfrac{a_1(1 - r^n)}{1 - r}$.
$$S_5 = \frac{\tfrac{1}{2}\left(1 - \left(-\tfrac{1}{2}\right)^5\right)}{1 - \left(-\tfrac{1}{2}\right)} = \frac{\tfrac{1}{2}(\tfrac{33}{32})}{\tfrac{3}{2}} = \frac{11}{32}$$

35. First determine the number of terms.
Since $r = 2/3$, the nth term is $30\left(\tfrac{2}{3}\right)^{n-1}$.
Solve
$$30\left(\tfrac{2}{3}\right)^{n-1} = \frac{1280}{729}$$
$$\left(\tfrac{2}{3}\right)^{n-1} = \frac{128}{2187} = \left(\tfrac{2}{3}\right)^7$$
$$n - 1 = 7$$
$$n = 8$$
Use $n = 8$, $a_1 = 30$, and $r = 2/3$ in the
formula $S_n = \dfrac{a_1(1 - r^n)}{1 - r}$.
$$S_8 = \frac{30\left(1 - \left(\tfrac{2}{3}\right)^8\right)}{1 - \left(\tfrac{2}{3}\right)} = \frac{30(\tfrac{6305}{6561})}{\tfrac{1}{3}}$$
$$= \frac{63050}{729} \approx 86.4883$$

37. $\displaystyle\sum_{i=1}^{10} 5(2)^{i-1} = S_{10} = \frac{5\left(1 - (2)^{10}\right)}{1 - (2)}$
$$= \frac{5(-1023)}{-1} = 5115$$

39. $\displaystyle\sum_{i=1}^{6} (0.1)^i = S_6 = \frac{0.1\left(1 - (0.1)^6\right)}{1 - (0.1)}$
$$= \frac{0.1(0.999999)}{0.9} = 0.111111$$

41. $\displaystyle\sum_{i=1}^{6} 100(0.3)^i = S_6$
$$= \frac{100(0.3)\left(1 - (0.3)^6\right)}{1 - (0.3)}$$
$$= \frac{100(0.3)(1 - (0.3)^6)}{0.7} = 42.8259$$

43. Use $a_1 = 1/8$ and $r = 1/2$ in the formula
for the sum of an infinite geometric series
$S = \dfrac{a_1}{1 - r}$.
$$S = \frac{\tfrac{1}{8}}{1 - \tfrac{1}{2}} = \frac{\tfrac{1}{8}}{\tfrac{1}{2}} = \frac{1}{4}$$

45. Use $a_1 = 3$ and $r = 2/3$ in $S = \dfrac{a_1}{1 - r}$.
$$S = \frac{3}{1 - \tfrac{2}{3}} = \frac{3}{\tfrac{1}{3}} = 9$$

47. Use $a_1 = 4$ and $r = -1/2$ in $S = \frac{a_1}{1-r}$.
$$S = \frac{4}{1-(-\frac{1}{2})} = \frac{4}{\frac{3}{2}} = \frac{8}{3}$$

49. Use $a_1 = 0.3$ and $r = 0.3$ in $S = \frac{a_1}{1-r}$.
$$S = \frac{0.3}{1-0.3} = \frac{0.3}{0.7} = \frac{3}{7}$$

51. Use $a_1 = 3$ and $r = 0.5$ in $S = \frac{a_1}{1-r}$.
$$S = \frac{3}{1-0.5} = \frac{3}{0.5} = 6$$

53. Use $a_1 = 0.3$ and $r = 0.1$ in $S = \frac{a_1}{1-r}$.
$$S = \frac{0.3}{1-0.1} = \frac{0.3}{0.9} = \frac{1}{3}$$

55. Use $a_1 = 0.12$ and $r = 0.01$ in
$S = \frac{a_1}{1-r}$.
$$S = \frac{0.12}{1-0.01} = \frac{0.12}{0.99} = \frac{12}{99} = \frac{4}{33}$$

57. We want the sum of the geometric series $2000(1.12)^{45} + 2000(1.12)^{44} + ...$
$$+ 2000(1.12).$$
Note that the last deposit is made at the beginning of the 45th year and earns interest for only one year. Rewrite the series as $2000(1.12) + 2000(1.12)^2 + ... + 2000(1.12)^{45}$ where the $a_1 = 2000(1.12)$, $n = 45$, and $r = 1.12$.
$$S_{45} = \frac{2000(1.12)\left(1 - (1.12)^{45}\right)}{1 - 1.12}$$
$$= \$3,042,435.27$$

59. We want the sum of the finite geometric series $1 + 2 + 4 + 8 + 16 + ... + 2^{30}$, which has 31 terms and a ratio of 2.
$$S_{31} = \frac{1(1 - 2^{31})}{1 - 2} = 2^{31} - 1$$
$$= 2,147,483,647 \text{ cents}$$
$$= \$21,474,836.47$$

61. Use $r = 0.80$, $a_1 = 1,000,000$ in $S = \frac{a_1}{1-r}$.
$$S = \frac{1,000,000}{1 - 0.80} = \$5,000,000$$

63. Only sequence (d) is not geometric because it is the only one that does not have a constant ratio.

65. Use $a_1 = 24/100 = 0.24$ and $r = 1/100 = 0.01$ in the formula for $S = \frac{a_1}{1-r}$.
$$S = \frac{0.24}{1 - 0.01} = \frac{0.24}{0.99} = \frac{8}{33}$$

13.5 WARM-UPS

1. False, because there are 13 terms in a binomial to the 12th power.

2. False, because the 7th term has variable part $a^6 b^6$.

3. False, because if $x = 1$ the equation is incorrect.

4. True, because the signs alternate in any expansion of a difference.

5. True, because we can obtain it from the 7th line.

6. True, because $1 + 4 + 6 + 4 + 1 = 2^4$.

7. True, because of the binomial theorem.

8. True, because $2^n = (1 + 1)^n$
$$= \sum_{i=0}^{n} \frac{n!}{(n-i)!i!} 1^{n-i} 1^i = \sum_{i=0}^{n} \frac{n!}{(n-i)!i!},$$
and the last sum is the sum of the coefficients in the nth row.

9. True, by definition of 0! and 1!.

10. True, because $\frac{7 \cdot 6 \cdot 5 \cdot 4 \cdot 3 \cdot 2 \cdot 1}{5 \cdot 4 \cdot 3 \cdot 2 \cdot 1 \cdot 2 \cdot 1} = 21$

13.5 EXERCISES

1. The sum obtained for a power of a binomial is called a binomial expansion.

3. The expression $n!$ is the product of the positive integers from 1 through n.

5. $\frac{5!}{2!3!} = \frac{5 \cdot 4}{2} = 10$

7. $\frac{8!}{5!3!} = \frac{8 \cdot 7 \cdot 6}{3 \cdot 2 \cdot 1} = 56$

9. The coefficients in the 5th row of Pascal's triangle are 1, 5, 10, 10, 5, 1. Use these coefficients with the pattern for the exponents.
$$(r + t)^5$$
$$= r^5 + 5r^4t + 10r^3t^2 + 10r^2t^3 + 5rt^4 + t^5$$

11. The coefficients in the 3rd row are 1, 3, 3, 1. Use these coefficients with the pattern for the exponents, and alternate the signs.
$$(m - n)^3 = m^3 - 3m^2n + 3mn^2 - n^3$$

13. Use the coefficients 1, 3, 3, 1 and let $y = 2a$ in the binomial theorem.
$$(x + 2a)^3$$
$$= 1x^3(2a)^0 + 3x^2(2a)^1 + 3x(2a)^2 + (2a)^3$$
$$= x^3 + 6ax^2 + 12a^2x + 8a^3$$

15. Use the coefficients 1, 4, 6, 4, 1 in the binomial theorem.

$$(x^2 - 2)^4 = (x^2)^4 - 4(x^2)^3 2 + 6(x^2)^2 2^2$$
$$- 4x^2 2^3 + 1(x^2)^0 2^4$$
$$= x^8 - 8x^6 + 24x^4 - 32x^2 + 16$$

17. Use the coefficients from the 7th line 1, 7, 21, 35, 35, 21, 7, 1 and alternate the signs of the terms.

$$(x - 1)^7 = x^7 - 7x^6 + 21x^5 - 35x^4 + 35x^3$$
$$- 21x^2 + 7x - 1$$

19. Use the binomial theorem to write the first 4 terms of $(a - 3b)^{12}$.

$$\frac{12!}{12!0!} a^{12} b^0 - \frac{12!}{11!1!} a^{11} b^1 + \frac{12!}{10!2!} a^{10} b^2$$
$$- \frac{12!}{9!3!} a^9 b^3$$
$$= a^{12} - 36a^{11} b + 594a^{10} b^2 - 5940a^9 b^3$$

21. Use the binomial theorem to write the first 4 terms of $(x^2 + 5)^9$.

$$\frac{9!}{9!0!} (x^2)^9 5^0 + \frac{9!}{8!1!} (x^2)^8 5^1 + \frac{9!}{7!2!} (x^2)^7 5^2$$
$$+ \frac{9!}{6!3!} (x^2)^6 5^3$$
$$= x^{18} + 45x^{16} + 900x^{14} + 10500x^{12}$$

23. Use the binomial theorem to write the first 4 terms of $(x - 1)^{22}$.

$$\frac{22!}{22!0!} x^{22} 1^0 - \frac{22!}{21!1!} x^{21} 1^1 + \frac{22!}{20!2!} x^{20} 1^2$$
$$- \frac{22!}{19!3!} x^{19} 1^3$$

$$= x^{22} - 22x^{21} + 231x^{20} - 1540x^{19}$$

25. Use the binomial theorem to write the first 4 terms of $\left(\frac{x}{2} + \frac{y}{3}\right)^{10}$.

$$\frac{10!}{10!0!} \left(\frac{x}{2}\right)^{10} \left(\frac{y}{3}\right)^0 + \frac{10!}{9!1!} \left(\frac{x}{2}\right)^9 \left(\frac{y}{3}\right)^1$$
$$+ \frac{10!}{8!2!} \left(\frac{x}{2}\right)^8 \left(\frac{y}{3}\right)^2 + \frac{10!}{7!3!} \left(\frac{x}{2}\right)^7 \left(\frac{y}{3}\right)^3$$
$$= \frac{x^{10}}{1024} + \frac{5x^9 y}{768} + \frac{5x^8 y^2}{256} + \frac{5x^7 y^3}{144}$$

27. Use the formula for the kth term of $(x + y)^n$ with $k = 6$ and $n = 13$.

$$\frac{13!}{(13 - 6 + 1)!(6 - 1)!} a^{13-6+1} w^{6-1}$$
$$= \frac{13!}{8!5!} a^8 w^5 = 1287a^8 w^5$$

29. Use the formula for the kth term with $k = 8$ and $n = 16$.

$$\frac{16!}{(16 - 8 + 1)!(8 - 1)!} m^{16-8+1} (-n)^{8-1}$$
$$= \frac{16!}{9!7!} m^9 (-n)^7 = -11440m^9 n^7$$

31. Use the formula for the kth term with $k = 4$ and $n = 8$.

$$\frac{8!}{(8 - 4 + 1)!(4 - 1)!} x^{8-4+1} (2y)^{4-1}$$
$$= \frac{8!}{5!3!} x^5 (2y)^3 = 56x^5 8y^3 = 448x^5 y^3$$

33. Use the formula for the kth term with $k = 7$ and $n = 20$.

$$\frac{20!}{(20 - 7 + 1)!(7 - 1)!} (2a^2)^{20-7+1} b^{7-1}$$
$$= \frac{20!}{14!6!} (2a^2)^{14} b^6$$
$$= 635{,}043{,}840a^{28} b^6$$

35. Use $n = 8$, $x = a$, and $y = b$ in the binomial theorem with summation notation.

$$(a + m)^8 = \sum_{i=0}^{8} \frac{8!}{(8 - i)!\, i!} a^{8-i} m^i$$

37. Use $n = 5$, $x = a$, and $y = -2x$ in the binomial theorem with summation notation.

$$(a + (-2x))^5 = \sum_{i=0}^{5} \frac{5!}{(5 - i)!\, i!} a^{5-i} (-2x)^i$$
$$= \sum_{i=0}^{5} \frac{5!\,(-2)^i}{(5 - i)!\, i!} a^{5-i} x^i$$

39. $(a + (b + c))^3$
$$= a^3 + 3a^2(b + c) + 3a(b + c)^2 + (b + c)^3$$
$$= a^3 + 3a^2 b + 3a^2 c + 3ab^2 + 6abc + 3ac^2$$
$$+ b^3 + 3b^2 c + 3bc^2 + c^3$$
$$= a^3 + b^3 + c^3 + 3a^2 b + 3a^2 c + 3ab^2 + 3ac^2$$
$$+ 3b^2 c + 3bc^2 + 6abc$$

Enriching Your Mathematical Word Power
1. a 2. d 3. c 4. b 5. a
6. c 7. d 8. b 9. d 10. a

CHAPTER 13 REVIEW

1. $a_1 = 1^3$, $a_2 = 2^3$, $a_3 = 3^3$, $a_4 = 4^3$, $a_5 = 5^3$

The terms of the sequence are 1, 8, 27, 64, 125.

3. $c_1 = (-1)^1(2 \cdot 1 - 3) = 1$,
$c_2 = (-1)^2(2 \cdot 2 - 3) = 1$,
$c_3 = (-1)^3(2 \cdot 3 - 3) = -3$,
$c_4 = (-1)^4(2 \cdot 4 - 3) = 5$,
$c_5 = (-1)^5(2 \cdot 5 - 3) = -7$
$c_6 = (-1)^6(2 \cdot 6 - 3) = 9$
The terms are 1, 1, -3, 5, -7, 9.

5. $a_1 = -\frac{1}{1} = -1$, $a_2 = -\frac{1}{2}$, $a_3 = -\frac{1}{3}$
The first three terms are $-1, -\frac{1}{2}, -\frac{1}{3}$.

7. $b_1 = \frac{(-1)^{2 \cdot 1}}{2 \cdot 1 + 1} = \frac{1}{3}$, $b_2 = \frac{(-1)^{2 \cdot 2}}{2 \cdot 2 + 1} = \frac{1}{5}$
$b_3 = \frac{(-1)^{2 \cdot 3}}{2 \cdot 3 + 1} = \frac{1}{7}$
The first three terms are $\frac{1}{3}, \frac{1}{5}$, and $\frac{1}{7}$.

9. $c_1 = \log_2(2^{1+3}) = \log_2(2^4) = 4$
$c_2 = \log_2(2^{2+3}) = \log_2(2^5) = 5$
$c_3 = \log_2(2^{3+3}) = \log_2(2^6) = 6$
The first three terms are 4, 5, and 6.

11. $\sum_{i=1}^{3} i^3 = 1^3 + 2^3 + 3^3 = 36$

13. $\sum_{n=1}^{5} n(n-1)$
$= 1(1-1) + 2(2-1) + 3(3-1)$
$\quad + 4(4-1) + 5(5-1)$
$= 0 + 2 + 6 + 12 + 20 = 40$

15. The terms in the series are reciprocals of even integers. Even integers are usually represent as $2i$, but to get 4 in the denominator when $i = 1$, we use $2(i+1)$.
$$\sum_{i=1}^{\infty} \frac{1}{2(i+1)}$$

17. The terms in this series are the squares of integers. Squares are usually represented as i^2, but to get the first term 0 when $i = 1$, we use $(i-1)^2$.
$$\sum_{i=1}^{\infty} (i-1)^2$$

19. To get alternating signs for the terms, we use a power of -1. If we use $(-1)^i$, then $i = 1$ makes the first term negative. So we use $(-1)^{i+1}$.
$$\sum_{i=1}^{\infty} (-1)^{i+1} x_i$$

21. $a_1 = 6 + (1-1)5 = 6$
Since the common difference is 5, the first four terms are 6, 11, 16, and 21.

23. $a_1 = -20 + (1-1)(-2) = -20$
Since the common difference is -2, the first four terms are $-20, -22, -24$, and -26.

25. $a_1 = 1000(1) + 2000 = 3000$
Since the common difference is 1000, the first four terms are 3000, 4000, 5000, and 6000.

27. Use $a_1 = 1/3$, $d = 1/3$, and the formula $a_n = a_1 + (n-1)d$.
$$a_n = \tfrac{1}{3} + (n-1)\tfrac{1}{3} = \frac{n}{3}$$

29. Use $a_1 = 2$, $d = 2$, and the formula $a_n = a_1 + (n-1)d$.
$$a_n = 2 + (n-1)(2) = 2n$$

31. Use $a_1 = 1$, $a_{24} = 24$, $n = 24$, and the formula $S_n = \frac{n}{2}(a_1 + a_n)$.
$$S_{24} = \frac{24}{2}(1 + 24) = 300$$

33. Use $a_1 = 1/6$, $d = 1/3$, $a_n = 11/2$, and the formula $a_n = a_1 + (n-1)d$ to find n.
$$\tfrac{11}{2} = \tfrac{1}{6} + (n-1)\tfrac{1}{3}$$
$$33 = 1 + (n-1)2$$
$$32 = 2n - 2$$
$$34 = 2n$$
$$17 = n$$
Now use $n = 17$, $a_1 = 1/6$, $a_{17} = 11/2$, and the formula $S_n = \frac{n}{2}(a_1 + a_n)$ to find the sum.
$$S_{17} = \frac{17}{2}\left(\tfrac{1}{6} + \tfrac{11}{2}\right) = \frac{17}{2}\left(\frac{34}{6}\right) = \frac{289}{6}$$

35. Use $a_1 = -1$, $a_7 = 11$, $n = 7$, and the formula $S_n = \frac{n}{2}(a_1 + a_n)$ to find the sum.
$$S_7 = \frac{7}{2}(-1 + 11) = 35$$

37. $a_1 = 3\left(\frac{1}{2}\right)^{1-1} = 3$, $a_2 = 3\left(\frac{1}{2}\right)^{2-1} = \frac{3}{2}$,
$a_3 = 3\left(\frac{1}{2}\right)^{3-1} = \frac{3}{4}$, $a_4 = 3\left(\frac{1}{2}\right)^{4-1} = \frac{3}{8}$
The first four terms are 3, $\frac{3}{2}$, $\frac{3}{4}$, and $\frac{3}{8}$.

39. $a_1 = 2^{1-1} = 1$, $a_2 = 2^{1-2} = \frac{1}{2}$
$a_3 = 2^{1-3} = \frac{1}{4}$, $a_4 = 2^{1-4} = \frac{1}{8}$
The first four terms are 1, $\frac{1}{2}$, $\frac{1}{4}$, and $\frac{1}{8}$.

41. $a_1 = 23(10)^{-2(1)} = 0.23,$
$a_2 = 23(10)^{-2(2)} = 0.0023,$
$a_3 = 23(10)^{-2(3)} = 0.000023,$
$a_4 = 23(10)^{-2(4)} = 0.00000023$

The first four terms of the geometric sequence are 0.23, 0.0023, 0.000023, and 0.00000023.

43. Use $a_1 = 1/2$, $r = 6$, and the formula $a_n = a_1 r^{n-1}$.
$$a_n = \tfrac{1}{2}(6)^{n-1}$$

45. Use $a_1 = 7/10$, $r = 1/10$, and the formula $a_n = a_1 r^{n-1}$.
$$a_n = 0.7(0.1)^{n-1}$$

47. Use $a_1 = 1/3$, $r = 1/3$, $n = 4$, and the formula $S_n = \dfrac{a_1(1 - r^n)}{1 - r}$.
$$S_4 = \frac{\frac{1}{3}\left(1 - \left(\frac{1}{3}\right)^4\right)}{1 - \frac{1}{3}} = \frac{\frac{1}{3}\left(\frac{80}{81}\right)}{\frac{2}{3}} = \frac{40}{81}$$

49. Use $a_1 = 0.3$, $r = 0.1$, $n = 10$, and the formula $S_n = \dfrac{a_1(1 - r^n)}{1 - r}$.
$$S_{10} = \frac{0.3\left(1 - (0.1)^{10}\right)}{1 - 0.1} = \frac{0.3(0.9999999999)}{0.9}$$
$$= 0.3333333333$$

Your calculator may not give ten 3's after the decimal point, but doing this computation without a calculator does give ten 3's and this is the exact answer.

51. Use $a_1 = 1/4$, $r = 1/3$, and the formula for the sum of an infinite geometric series $S = \dfrac{a_1}{1 - r}$.
$$S = \frac{\frac{1}{4}}{1 - \frac{1}{3}} = \frac{\frac{1}{4}}{\frac{2}{3}} = \frac{3}{8}$$

53. Use $a_1 = 18$, $r = 2/3$, and the formula for the sum of an infinite geometric series $S = \dfrac{a_1}{1 - r}$.
$$S = \frac{18}{1 - \frac{2}{3}} = \frac{18}{\frac{1}{3}} = 54$$

55. The coefficients for the fifth power of a binomial are 1, 5, 10, 10, 5, and 1.
$$(m + n)^5 = m^5 + 5m^4 n + 10m^3 n^2$$
$$+ 10m^2 n^3 + 5mn^4 + n^5$$

57. The coefficients for the third power of a binomial are 1, 3, 3, and 1. Alternate the signs because it is a difference to a power.
$$(a^2 - 3b)^3 = 1(a^2)^3(3b)^0 - 3(a^2)^2(3b)^1$$
$$+ 3(a^2)^1(3b)^2 - 1(a^2)^0(3b)^3$$
$$= a^6 - 9a^4 b + 27a^2 b^2 - 27b^3$$

59. Use $n = 12$ and $k = 5$ in the formula for the kth term.
$$\frac{12!}{(12 - 5 + 1)!(5 - 1)!} x^{12-5+1} y^{5-1} = \frac{12!}{8!4!} x^8 y^4$$
$$= 495x^8 y^4$$

61. Use $n = 14$ and $k = 3$ in the formula for the kth term.
$$\frac{14!}{(14 - 3 + 1)!(3 - 1)!} (2a)^{14-3+1}(-b)^{3-1}$$
$$= \frac{14!}{12!2!} (2a)^{12}(-b)^2 = 372{,}736a^{12}b^2$$

63. Use the binomial theorem expressed in summation notation, with $n = 7$.
$$(a + w)^7 = \sum_{i=0}^{7} \frac{7!}{(7 - i)!\, i!}\, a^{7-i} w^i$$

65. The sequence has neither a constant difference nor a constant ratio. So it is neither arithmetic nor geometric.

67. There is a constant difference of 3. So the sequence is an arithmetic sequence.

69. There is a constant difference of 2. So the sequence is an arithmetic sequence.

71. Use $a_1 = 6$, $n = 4$, $a_4 = 1/30$, and the formula $a_n = a_1 r^{n-1}$.
$$\frac{1}{30} = 6r^{4-1}$$
$$\frac{1}{180} = r^3$$
$$r = \sqrt[3]{\frac{1}{180}} = \frac{1}{\sqrt[3]{180}}$$

73. $\displaystyle\sum_{i=1}^{5} \frac{(-1)^i}{i!} = \frac{(-1)^1}{1!} + \frac{(-1)^2}{2!} + \frac{(-1)^3}{3!}$
$$+ \frac{(-1)^4}{4!} + \frac{(-1)^5}{5!}$$
$$= -1 + \frac{1}{2} - \frac{1}{6} + \frac{1}{24} - \frac{1}{120}$$

75. This is the summation notation for the binomial expansion of $(a + b)^5$.
$$\frac{5!}{5!0!} a^5 b^0 + \frac{5!}{4!1!} a^4 b^1 + \frac{5!}{3!2!} a^3 b^2 + \frac{5!}{2!3!} a^2 b^3$$
$$+ \frac{5!}{1!4!} a^1 b^4 + \frac{5!}{0!5!} a^0 b^5$$
$$= a^5 + 5a^4 b + 10a^3 b^2 + 10a^2 b^3 + 5ab^4 + b^5$$

77. There are 26 terms because in the expansion of $(x + y)^n$ there are $n + 1$ terms.

79. The first \$3000 earns interest for 16 years. The second \$3000 earns interest for 15 years, and so on. The last \$3000 earns interest for 1 year. The total in the account at the end of 16 years is the sum of the following series.
$3000(1.1) + 3000(1.1)^2 + \ldots + 3000(1.1)^{16}$
This is a geometric series with $n = 16$, $a_1 = 3000(1.1)$ and $r = 1.1$.

$$S_{16} = \frac{3000(1.1)\left(1 - (1.1)^{16}\right)}{1 - 1.1} = \$118,634.11$$

81. We compute a new balance 16 times by multiplying by $1 + 0.10$ each time.
$3000(1.10)^{16} = \$13,784.92$

CHAPTER 13   TEST

1. $a_1 = -10 + (1 - 1)6 = -10$
$a_2 = -10 + (2 - 1)6 = -4$
$a_3 = -10 + (3 - 1)6 = 2$
$a_4 = -10 + (4 - 1)6 = 8$
The first four terms are $-10, -4, 2$, and 8.

2. $a_1 = 5(0.1)^{1-1} = 5$
$a_2 = 5(0.1)^{2-1} = 0.5$
$a_3 = 5(0.1)^{3-1} = 0.05$
$a_4 = 5(0.1)^{4-1} = 0.005$
The first four terms are $5, 0.5, 0.05$, and 0.005.

3. $a_1 = \frac{(-1)^1}{1!} = -1$, $a_2 = \frac{(-1)^2}{2!} = \frac{1}{2}$,
$a_3 = \frac{(-1)^3}{3!} = -\frac{1}{6}$, $a_4 = \frac{(-1)^4}{4!} = \frac{1}{24}$
The first four terms are $-1, \frac{1}{2}, -\frac{1}{6}$, and $\frac{1}{24}$.

4. $a_1 = \frac{2(1) - 1}{(1)^2} = 1$, $a_2 = \frac{2(2) - 1}{(2)^2} = \frac{3}{4}$,
$a_3 = \frac{2(3) - 1}{(3)^2} = \frac{5}{9}$, $a_4 = \frac{2(4) - 1}{(4)^2} = \frac{7}{16}$
The first four terms are $1, \frac{3}{4}, \frac{5}{9}$, and $\frac{7}{16}$.

5. The sequence is an arithmetic sequence with $a_1 = 7$ and $d = -3$. So the general term is $a_n = 7 + (n - 1)(-3) = 7 - 3n + 3$
$= 10 - 3n$.

6. This sequence is a geometric sequence with $a_1 = -25$, and $r = -1/5$. So the general term is $a_n = -25\left(-\frac{1}{5}\right)^{n-1}$.

7. This sequence is a sequence of even integers, which we can represent as $2n$. To get the signs to alternate, we use a power of -1. So the general term is $a_n = (-1)^{n-1}2n$.

8. This sequence is a sequence of squares of the positive integers. So the general term is $a_n = n^2$.

9. $\sum\limits_{i=1}^{5}(2i + 3) = 2(1) + 3 + 2(2) + 3$
$+ 2(3) + 3 + 2(4) + 3 + 2(5) + 3$
$= 5 + 7 + 9 + 11 + 13$

10. $\sum\limits_{i=1}^{6}5(2)^{i-1} = 5(2)^{1-1} + 5(2)^{2-1}$
$+ 5(2)^{3-1} + 5(2)^{4-1} + 5(2)^{5-1} + 5(2)^{6-1}$
$= 5 + 10 + 20 + 40 + 80 + 160$

11. $\sum\limits_{i=0}^{4}\frac{4!}{(4-i)!i!}m^{4-i}q^i = \frac{4!}{4!0!}m^4q^0$
$+ \frac{4!}{3!1!}m^3q^1 + \frac{4!}{2!2!}m^2q^2$
$+ \frac{4!}{1!3!}m^1q^3 + \frac{4!}{0!4!}m^0q^4$
$= m^4 + 4m^3q + 6m^2q^2 + 4mq^3 + q^4$

12. Use $a_1 = 9$, $a_{20} = 66$, and $n = 20$ in the formula for the sum of an arithmetic series.
$S_{20} = \frac{20}{2}(9 + 66) = 10(75) = 750$

13. Use $a_1 = 10$, $n = 5$, and $r = 1/2$ in the formula for the sum of a finite geometric series.

$$S_5 = \frac{10\left(1 - \left(\frac{1}{2}\right)^5\right)}{1 - \frac{1}{2}} = \frac{10\left(\frac{31}{32}\right)}{\frac{1}{2}} = \frac{155}{8}$$

14. Use $a_1 = 0.35$ and $r = 0.93$ in the formula for the sum of an infinite geometric series.
$S = \frac{0.35}{1 - 0.93} = \frac{0.35}{0.07} = 5$

15. Use $a_1 = 2$, $a_{100} = 200$, and $n = 100$ in the formula for the sum of a finite arithmetic series.
$S_{100} = \frac{100}{2}(2 + 200) = 50(202) = 10,100$

16. Use $a_1 = 1/4$ and $r = 1/2$ in the formula for the sum of an infinite geometric series.
$S = \frac{1/4}{1 - 1/2} = \frac{1/4}{1/2} = \frac{1}{2}$

17. Use $a_1 = 2$, $r = 1/2$ and $a_n = 1/128$ to find n:

$$2\left(\frac{1}{2}\right)^{n-1} = \frac{1}{128}$$
$$\left(\frac{1}{2}\right)^{n-1} = \frac{1}{256} = \frac{1}{2^8}$$
$$n - 1 = 8$$
$$n = 9$$

Use $a_1 = 2$, $n = 9$, and $r = 1/2$ to find the sum of the 9 terms.

$$S_9 = \frac{2(1 - (1/2)^9)}{1 - 1/2} = \frac{511}{128} \approx 3.9922$$

18. Use $a_1 = 3$, $a_5 = 48$, $n = 5$ in the formula for the general term of a geometric sequence.

$$48 = 3r^{5-1}$$
$$16 = r^4$$
$$\pm 2 = r$$

19. Use $a_1 = 1$, $a_{12} = 122$, $n = 12$, and the formula for the general term of an arithmetic sequence.

$$122 = 1 + (12 - 1)d$$
$$121 = 11d$$
$$11 = d$$

20. Use $n = 15$ and $k = 5$ in the formula for the kth term of a binomial expansion.

$$\frac{15!}{(15 - 5 + 1)!(5 - 1)!}r^{15-5+1}(-t)^{5-1}$$
$$= \frac{15!}{11!4!}r^{11}t^4 = 1365r^{11}t^4$$

21. Use $n = 8$ and $k = 4$ in the formula for the kth term of a binomial expansion.

$$\frac{8!}{(8 - 4 + 1)!(4 - 1)!}(a^2)^{8-4+1}(-2b)^{4-1}$$
$$= \frac{8!}{5!3!}a^{10}(-2)^3b^3 = -448a^{10}b^3$$

22. $800(1.10)^1 + 800(1.10)^2 + \ldots$
$$+ 800(1.10)^{25}$$
$$= \frac{800(1.10)(1 - 1.10^{25})}{1 - 1.10} = \$86,545.41$$

Making Connections

Chapters 1 - 13

1. $f(3) = 3^2 - 3 = 9 - 3 = 6$

2. $f(n) = n^2 - 3$

3. $f(x + h) = (x + h)^2 - 3$
$$= x^2 + 2xh + h^2 - 3$$

4. $f(x) - g(x) = x^2 - 3 - (2x - 1)$
$$= x^2 - 2x - 2$$

5. $g(f(3)) = g(6) = 2(6) - 1 = 11$

6. $(f \circ g)(2) = f(g(2)) = f(3) = 3^2 - 3 = 6$

7. $m(16) = \log_2(16) = 4$

8. $(h \circ m)(32) =$
$h(m(32)) = h(5) = 2^5 = 32$

9. $h(-1) = 2^{-1} = 1/2$

10. $h^{-1}(8) = \log_2(8) = 3$

11. $m^{-1}(0) = 2^0 = 1$

12. $(m \circ h)(x) = m(h(x)) = m(2^x)$
$$= \log_2(2^x) = x$$
So $(m \circ h)(x) = x$.

13. If y varies directly as x, then $y = kx$. Since $y = -6$ when $x = 4$, we have $-6 = 4k$, or $k = -3/2$. When $x = 9$, we can use the original formula with $k = -3/2$.

$$y = -\frac{3}{2}(9) = -\frac{27}{2}$$

14. If a varies inversely as b, then $a = k/b$. If $a = 2$ when $b = -4$, we can find k.

$$2 = \frac{k}{-4}$$
$$-8 = k$$

To find a, use $k = -8$ and $b = 3$.

$$a = \frac{-8}{3} = -\frac{8}{3}$$

15. If y varies directly as w and inversely as t, then $y = (kw)/t$. Use $y = 16$, $w = 3$, and $t = -4$ to find k.

$$16 = \frac{k(3)}{-4}$$
$$-64 = 3k$$
$$\frac{-64}{3} = k$$

To find y, use $k = -64/3$, $w = 2$, and $t = 3$.

$$y = -\frac{64(2)}{3(3)} = -\frac{128}{9}$$

16. If y varies jointly as h and the square of r, then $y = khr^2$. Use $y = 12$, $h = 2$, and $r = 3$ to find k.

$$12 = k(2)(3)^2$$
$$12 = 18k$$
$$\frac{2}{3} = k$$

To find y, use $k = 2/3$, $h = 6$, and $r = 2$.

$$y = \frac{2}{3}(6)(2)^2 = 16$$

17. The graph of $x > 3$ is the region to the right of the vertical line $x = 3$. The graph of $x + y < 0$ is the region below the line $y = -x$. The region to the right of $x = 3$ and below $y = -x$ is shown in the following graph.

18. The inequality $|x - y| \geq 2$ is equivalent to the compound inequality $x - y \geq 2$ or $x - y \leq -2$. The graph of $x - y \geq 2$ is the region on or below the line $y = x - 2$. The graph of $x - y \leq -2$ is the region on or above the line $y = x + 2$. Since the word or is used, the graph of the compound inequality is the union of these two regions as shown in the following diagram.

19. The graph of $y < -2x + 3$ is the region below the line $y = -2x + 3$. The graph of $y > 2^x$ is the region above the curve $y = 2^x$. Since the word and is used, the graph of the compound inequality is the intersection of these two regions, the points that lie above $y = 2^x$ and below $y = -2x + 3$. We could have used a test point in each of the four regions to see which region satisfies both inequalities.

20. The inequality $|y + 2x| < 1$ is equivalent to $-1 < y + 2x < 1$. This inequality is also written as
$$y + 2x > -1 \quad \text{and} \quad y + 2x < 1$$
$$y > -2x - 1 \quad \text{and} \quad y < -2x + 1$$
The graph of $y > -2x - 1$ is the region above the line $y = -2x - 1$. The graph of $y < -2x + 1$ is the region below the line $y = -2x + 1$. The points that satisfy both inequalities are the points that lie between these two parallel lines.

21. The graph of $x^2 + y^2 = 4$ is a circle of radius 2 centered at the origin. Since $(0, 0)$ satisfies the inequality $x^2 + y^2 < 4$, we shade the region inside the circle.

22. The graph of $x^2 - y^2 = 1$ is a hyperbola with x-intercepts $(-1, 0)$ and $(1, 0)$. The fundamental rectangle passes through the x-intercepts and $(0, 1)$ and $(0, -1)$. Extend the diagonals for the asymptotes. The hyperbola opens to the left and right. Test a point in each region to see that only points in the region containing the origin satisfy the inequality.

23. Graph the curve $y = \log_2(x)$ and shade the region below the curve to show the graph of $y < \log_2(x)$.

24. Write $x^2 + 2y < 4$ as $y < -\frac{1}{2}x^2 + 2$, to see that the boundary is a parabola opening downward with vertex at $(0, 2)$.

25. The graph of $\frac{x^2}{4} + \frac{y^2}{9} < 1$ is the region inside the ellipse $\frac{x^2}{4} + \frac{y^2}{9} = 1$. The graph of $y > x^2$ is the region above the parabola $y = x^2$. The points that satisfy the compound inequality are inside the ellipse and above the parabola as shown in the diagram.

26. $\frac{a}{b} + \frac{b}{a} = \frac{a(a)}{b(a)} + \frac{b(b)}{a(b)} = \frac{a^2 + b^2}{ab}$

27. $1 - \frac{3}{y} = \frac{y}{y} - \frac{3}{y} = \frac{y-3}{y}$

28. $\frac{x-2}{x^2-9} - \frac{x-4}{x^2-2x-3}$

$= \frac{(x-2)(x+1)}{(x-3)(x+3)(x+1)}$

$\quad - \frac{(x-4)(x+3)}{(x-3)(x+1)(x+3)}$

$= \frac{x^2 - x - 2}{(x-3)(x+3)(x+1)}$

$\quad - \frac{x^2 - x - 12}{(x-3)(x+3)(x+1)}$

$= \frac{10}{(x-3)(x+3)(x+1)}$

29. $\frac{x^2 - 16}{2x+8} \cdot \frac{4x^2 + 16x + 64}{x^3 - 16}$

$\frac{(x-4)(x+4)}{2(x+4)} \cdot \frac{4(x^2 + 4x + 16)}{x^3 - 16}$

$= \frac{2(x^3 - 64)}{x^3 - 16}$

30. $\frac{(a^2 b)^3}{(ab^2)^4} \cdot \frac{ab^3}{a^{-4}b^2} = \frac{a^6 b^3}{a^4 b^8} \cdot \frac{ab^3}{a^{-4}b^2}$

$= \frac{a^7 b^6}{b^{10}} = \frac{a^7}{b^4}$

31. $\frac{x^2 y}{(xy)^3} \div \frac{xy^2}{x^2 y^4} = \frac{x^2 y}{x^3 y^3} \cdot \frac{x^2 y^4}{xy^2}$

$= \frac{x^4 y^5}{x^4 y^5} = 1$

32. $8^{2/3} = (\sqrt[3]{8})^2 = 2^2 = 4$

33. $16^{-5/4} = \frac{1}{(\sqrt[4]{16})^5} = \frac{1}{(2)^5} = \frac{1}{32}$

34. $-4^{1/2} = -\sqrt{4} = -2$

35. $27^{-2/3} = \frac{1}{(\sqrt[3]{27})^2} = \frac{1}{3^2} = \frac{1}{9}$

36. $-2^{-3} = -\frac{1}{2^3} = -\frac{1}{8}$

37. $2^{-3/5} \cdot 2^{-7/5} = 2^{-10/5} = 2^{-2} = \frac{1}{2^2} = \frac{1}{4}$

38. $5^{-2/3} \div 5^{1/3} = 5^{-\frac{2}{3} - \frac{1}{3}} = 5^{-1} = \frac{1}{5}$

39. $(9^{1/2} + 4^{1/2})^2 = (3 + 2)^2 = 5^2 = 25$

40. a) Age 4 years 3 months is 4.25 years.

$h(4.25)$

$= 79.041 + 6.39(4.25) - e^{3.261 - 0.993(4.25)}$

$= 105.8$ cm

105.8 cm $\cdot \frac{1 \text{ in}}{2.54 \text{ cm}} = 41.7$ in.

c) Using a graphing calculator we get that a child who has a height of 80 cm has an age of 1.3 years

14.1 WARM-UPS

1. False, because there are $26 \cdot 10 = 260$ codes available.

2. False, because there are $(24)^3$ different sorority names possible.

3. True, because of the fundamental counting principle ($4 \cdot 5 \cdot 4 = 80$).

4. True, because the number of permutations of 4 objects is $4! = 4 \cdot 3 \cdot 2 \cdot 1 = 24$.

5. True, because $\frac{100!}{98!} = 100 \cdot 99 = 9900$.

6. False, because $P(100, 98) = \frac{100!}{2!}$, which is not equal to 9900.

7. True, because
$$P(5, 2) = \frac{5!}{3!} = \frac{5 \cdot 4 \cdot 3 \cdot 2 \cdot 1}{3 \cdot 2 \cdot 1} = 20.$$

8. True, because $P(18, 0) = \frac{18!}{18!} = 1$.

9. False, because if there are 5 possibilities on each of 10 questions, then the number of ways to answer is 5^{10}.

10. True, because there are 2 choices on each of the 10 questions.

14.1 EXERCISES

1. A tree diagram shows the number of different ways that a sequence of events can occur.

3. A permutation is an ordering or arrangement of distinct objects in a linear manner.

5. By the fundamental counting principle, the number of routes is $3 \cdot 4 = 12$.

7. Since there are 4 aces, 4 kings, 4 queens, 4 jacks, and 4 tens to choose from, there are $4^5 = 1024$ different hands consisting of one of each of these cards.

9. For each optional item there are 2 choices, whether to get it or not. Since $2^8 = 256$, there are 8 optional items.

11. There are 5 subjects that can be arranged in any order. The number of permutations of 5 things is $5! = 120$. So there are 120 different orders possible for doing his homework.

13. The number of ways to pick a first, second, and third school from 12 is $P(12, 3) = 12 \cdot 11 \cdot 10 = 1320$.

15. From the 34 available shows, 3 are to be selected and shown in some order. Since there will be a first, second, and third show, there are $P(34, 3) = 34 \cdot 33 \cdot 32 = 35{,}904$ different schedules possible.

17. Since there are 4 possible choices for each of 20 questions, the number of ways to mark the answers is $4^{20} = 1.0995 \times 10^{12}$.

19. In this problem we are counting the number of ways to seat the 37 people selected, and not just who will get on the flight. The number of ways to arrange 37 out of 39 people in seats is $P(39, 37) = \frac{39!}{2!} = 1.01989 \times 10^{46}$.

21. Since there are 3 choices for the exterior, 3 choices for the bedrooms, 2 choices for the fireplace (to have it or not), 2 choices for the number of baths, and 5 choices for the kitchen, there are $3 \cdot 3 \cdot 2 \cdot 2 \cdot 5 = 180$ different houses available.

23. If the first digit cannot be zero, then there are only 9 choices for the first digit. Since there are 10 choices on each of the remaining 6 positions, there are $9 \cdot 10^6 = 9{,}000{,}000$ different phone numbers available.

25. The number of ways to arrange 7 students in a row is $P(7, 7) = 7! = 5040$.

27. Since there are 10 choices for each of the 6 digits and 26 choices for the single letter, there are $26 \cdot 10^6 = 26{,}000{,}000$ different license plates possible. If the single letter can occur any where but last, there are 6 positions the single letter can occur in. So there are $6 \cdot 26 \cdot 10^6 = 156{,}000{,}000$ different plates possible.

29. For each of the 3 letters we must choose whether or not it should be in the subset. So we have 2 choices on each item to get $2^3 = 8$ possible subsets. They are $\{a, b, c\}$, (a, b), $\{a, c\}$, $\{b, c\}$, $\{a\}$, $\{b\}$, $\{c\}$, and $\emptyset$.

31. On each of the 10 true-false questions there are 2 choices and on each of the 10 multiple choice questions there are 5 choices. So the number of different ways to mark the answers is
$$2^{10} \cdot 5^{10} = (2 \cdot 5)^{10} = 10^{10} = 1 \times 10^{10}.$$

33. The number of ways to rearrange 4 letters is $P(4, 4) = 4! = 24$. The permutations follow.

MILK, MIKL, MKIL, MKLI, MLIK, MLKI
IMLK, IMKL, IKLM, IKML, ILMK, ILKM
LMIK, LMKI, LKIM, LKMI, LIMK, LIKM
KLIM, KLMI, KILM, KIML, KMIL, KMLI

35. $P(8, 3) = \dfrac{8!}{(8-3)!} = \dfrac{8!}{5!} = 8 \cdot 7 \cdot 6$
$$= 336$$

37. $P(52, 0) = \dfrac{52!}{(52-0)!} = \dfrac{52!}{52!} = 1$

39. $\dfrac{P(10, 4)}{4!} = \dfrac{10!}{4!(10-4)!} = \dfrac{10!}{4!6!}$
$$= \dfrac{10 \cdot 9 \cdot 8 \cdot 7}{4 \cdot 3 \cdot 2 \cdot 1} = 210$$

41. $\dfrac{P(12, 3)}{3!} = \dfrac{12!}{3!(12-3)!} = \dfrac{12!}{3!9!}$
$$= \dfrac{12 \cdot 11 \cdot 10}{3 \cdot 2 \cdot 1} = 220$$

43. $\dfrac{14!}{3!11!} = \dfrac{14 \cdot 13 \cdot 12}{3 \cdot 2 \cdot 1} = 364$

45. $\dfrac{98!}{95!3!} = \dfrac{98 \cdot 97 \cdot 96}{3 \cdot 2 \cdot 1} = 152{,}096$

14.2 WARM-UPS

1. True, because we are not concerned with the order in which they are answered, but just which two are answered.

2. True, because we are looking for the number of subsets of size one from a set of size 3.

3. False, because $C(4, 12)$ is meaningless.

4. False, because $C(10, 6) = \dfrac{P(10, 6)}{6!}$. So there are more permutations of 10 things taken 3 at a time than combinations of 10 things taken 3 at a time.

5. False, because if they all get called superior there are $C(10, 3)$ ways to make the choice.

6. True, because the number of ways to select 4 students out of 20 to stay after is the same as the number of ways to select 16 out of 20 to let go.

7. True, because $P(9, 4) = \dfrac{9!}{5!}$ and $C(9, 4) = \dfrac{9!}{5!4!}$.

8. False, because $P(8, 3) = \dfrac{8!}{5!}$ and $\binom{8}{5} = C(8, 5) = \dfrac{8!}{5!3!}$.

9. True, because $C(100, 1) = \dfrac{100!}{99!1!} = 100$.

10. True, because $P(8, 3) = \dfrac{8!}{5!} = \dfrac{8!}{1!1!1!5!}$.

14.2 EXERCISES

1. A combination of n things taken r at a time is a subset of size r taken from a set of n elements.

3. In a labeling problem we want the number of ways of labeling n object with n labels, where some of the labels are identical.

5. The number of 5-card poker hands is the same as the number of subsets of size 5 from a set of size 52, $C(52, 5) = \dfrac{52!}{5!47!} = 2{,}598{,}960$.

7. The number of ways to select 3 candidates from 5 is $C(5, 3) = \dfrac{5!}{2!3!} = 10$

9. The number of ways to select 5 out of 20 and give them all the same treatment is $C(20, 5) = \dfrac{20!}{5!15!} = 15{,}504$.

11. The number of ways to select 6 objects from a set of 49 is $C(49, 6) = \dfrac{49!}{6!43!} = 13{,}983{,}816$.

13. The coefficient of $w^3 y^4$ is $C(7, 3) = \dfrac{7!}{3!4!} = 35$.

15. The number of ways to choose 3 objects from 6 and arrange them in order is $P(6, 3) = \dfrac{6!}{3!} = 6 \cdot 5 \cdot 4 = 120$.

17. The number of ways to choose 3 questions out of 8 and give them all the same treatment (answer them) is $C(8, 3) = \dfrac{8!}{3!5!} = 56$.

19. The number of ways to order 12 examination papers is $12! = 479{,}001{,}600$.

21. The number of ways to select 5 seats and give them all the same treatment is
$$C(150, 5) = \dfrac{150!}{5!145!}$$
$$= \dfrac{150 \cdot 149 \cdot 148 \cdot 147 \cdot 146}{5 \cdot 4 \cdot 3 \cdot 2 \cdot 1}$$
$$= 591{,}600{,}030.$$

23. Each selection of 2 points determines a unique line segment. For 3 points on a circle, the number of distinct line segments is $C(3, 2) = \dfrac{3!}{2!1!} = 3$. For 4 points on a circle, the number of line segments is $C(4, 2) = 6$. For 5 points on a circle, the number of distinct line segments is $C(5, 2) = 10$. For n points on a circle, the number of distinct line segments is $C(n, 2)$.

25. Since there are 6 outcomes on each die, there are $6 \cdot 6 = 36$ different ordered pairs that can occur as the result of tossing a pair of dice.

27. Since there are 2 outcomes on each of the 3 tosses, there are $2^3 = 8$ different ordered triples that can occur as the result of tossing a coin 3 times.

29. Since $C(n, r) = \dfrac{n!}{(n-r)!r!}$ and

$$C(n, n-r) = \dfrac{n!}{(n-r)!(n-(n-r))!}$$
$$= \dfrac{n!}{(n-r)!r!},$$

we have $C(n, r) = C(n, n-r)$.

31. The number of ways to choose 2 classes from 4 is $C(4, 2) = 6$. The number of ways to choose 2 classes from 5 is $C(5, 2) = 10$. So the number of ways to choose the math classes and the history classes is
$C(4, 2) \cdot C(5, 2) = 60$.

33. The number of ways to select 4 students to get A's is $C(20, 4)$. The number of ways to select 5 students from the remaining 16 to get B's is $C(16, 5)$. The number of ways to select 11 of the remaining 11 to get C's is $C(11, 11)$. So the number of ways to make all 3 selections is $C(20, 4) \cdot C(16, 5) \cdot C(11, 11)$
$= \dfrac{20!}{4!16!} \cdot \dfrac{16!}{5!11!} \cdot \dfrac{11!}{11!0!} = 21,162,960$.

35. The number of ways to march in the order BFBFBFBF is $4 \cdot 4 \cdot 3 \cdot 3 \cdot 2 \cdot 2 \cdot 1 \cdot 1 = 576$.

37. Since we are labeling the papers with the labels A, B, and C, there are $\dfrac{16!}{6!3!7!} = 960,960$ ways to assign the grades.

39. We label the 52 cards with 13 Bob, 13 Carol, 13 Ted, and 13 Alice labels. The number of ways to do this is
$\dfrac{52!}{13!13!13!13!} = 5.36447 \times 10^{28}$. The number of seconds in a trillion years is
$60 \cdot 60 \cdot 24 \cdot 365 \cdot 10^{12} = 3.15360 \times 10^{19}$. So the number of ways to deal out the 52 cards is larger than the number of seconds in a trillion years.

41. The letters in TOYOTA can be arranged in $\dfrac{6!}{2!2!1!1!} = 180$ different ways.
The letters in ALGEBRA can be arranged in $\dfrac{7!}{2!1!1!1!1!1!1!} = 2520$ different ways.

The letters in STATISTICS can be arranged in $\dfrac{10!}{3!3!2!1!1!} = 50400$ different ways.

43. The number of ways to label the 20 sales persons with the 20 car labels is
$\dfrac{20!}{6!5!8!1!} = 698,377,680$.

45. The number of ways to label the 9 sofas with the discount labels is $\dfrac{9!}{4!4!1!} = 630$.

47. The coefficient of $a^3 b^2 c^5$ is
$\dfrac{10!}{3!2!5!} = 2520$.

49. The coefficient of $a^3 b^5$ is $\dfrac{8!}{0!3!5!} = 56$.

14.3 WARM-UPS

1. False, because $P(\mathrm{E}) = \dfrac{n(\mathrm{E})}{n(\mathrm{S})}$.

2. False, because there are $2^3 = 8$ equally likely outcomes for tossing 3 coins.

3. True, because there are 4 equally likely outcomes and on 3 of those we can say that at least one tail has occurred.

4. True, because at least one 4 occurs on 11 out of the 36 equally likely outcomes.

5. False, because 33/32 is larger than 1 and no probability is larger than 1.

6. False, because the complement to the event of getting exactly 3 heads is getting at least one tail.

7. True, because the complement of getting exactly 3 tails is getting at least one head.

8. False, because the odds are the ratio of getting snow to not getting snow. In this case the odds are 80 to 20 or 4 to 1.

9. False, because if the odds in favor are 2 to 3 then the probability of the event is 2/5 and the probability of the complement is 3/5.

10. False, because the ratio of 1/2 to 1/3 is equivalent to the ratio 6/2 to 6/3 or 3 to 2.

14.3 EXERCISES

1. An experiment is process for which the outcomes are uncertain.

3. An event is a subset of a sample space.

5. The addition rule says that if A and B are events in a sample space, then
$P(A \cup B) = P(A) + P(B) - P(A \cap B)$.

7. If a single die is tossed then the sample space consists of 6 equally likely outcomes $\{1, 2, 3, 4, 5, 6\}$.

a) $P(\text{a number larger than 3}) = \frac{3}{6} = \frac{1}{2}$

b) $P(\text{a number less than or equal to 5}) = \frac{5}{6}$

c) $P(\text{a number other than 6}) = \frac{5}{6}$

d) $P(\text{a number larger than 7}) = \frac{0}{6} = 0$

e) $P(\text{a number smaller than 9}) = \frac{6}{6} = 1$

9. If a pair of coins is tossed then the sample space is $\{HH, HT, TH, TT\}$.

a) $P(\text{exactly 2 heads}) = \frac{1}{4}$

b) $P(\text{at least one tail}) = \frac{3}{4}$

c) $P(\text{exactly 2 tails}) = \frac{1}{4}$

d) $P(\text{at most 1 tail}) = \frac{3}{4}$

11. If a pair of dice is tossed, there are 36 equally likely outcomes:

1,1 1,2 1,3 1,4 1,5 1,6
2,1 2,2 2,3 2,4 2,5 2,6
3,1 3,2 3,3 3,4 3,5 3,6
4,1 4,2 4,3 4,4 4,5 4,6
5,1 5,2 5,3 5,4 5,5 5,6
6,1 6,2 6,3 6,4 6,5 6,6

a) $P(\text{a pair of twos}) = \frac{1}{36}$

b) $P(\text{at least one 2}) = \frac{11}{36}$

c) $P(\text{a sum of 7}) = \frac{6}{36} = \frac{1}{6}$

d) $P(\text{a sum greater than 1}) = \frac{36}{36} = 1$

e) $P(\text{a sum less than 2}) = \frac{0}{36} = 0$

13. If a single ball is selected from the jar of 12 balls, the sample space is
$\{R_1, R_2, R_3, B_1, B_2, B_3, B_4, G_1, G_2, G_3, G_4, G_5\}$

a) $P(\text{the ball is red}) = \frac{3}{12} = \frac{1}{4}$

b) $P(\text{the ball is not blue}) = \frac{8}{12} = \frac{2}{3}$

c) $P(\text{the ball is either red or green}) = \frac{8}{12} = \frac{2}{3}$

d) $P(\text{the ball is neither red nor green})$
$$= \frac{4}{12} = \frac{1}{3}$$

15. Since two balls are to be selected without replacement, the number of possible outcomes is $10 \cdot 9 = 90$. There are 90 ordered pairs of numbers in the sample space.

a) The event $(1, 2)$ is one of the 90 outcomes to the experiment. So $P(1, 2) = \frac{1}{90}$.

b) The sum of the numbers is 3 if either $(1, 2)$ or $(2, 1)$ occurs. So $P(\text{sum is 3}) = \frac{2}{90} = \frac{1}{45}$.

c) The sum is 6 occurs for $(1,5)$, $(2,4)$, $(4,2)$, and $(5,1)$. So $P(\text{sum is 6}) = \frac{4}{90} = \frac{2}{45}$.

17. There are $C(52, 5) = 2,598,960$ possible poker hands.

a) The ace, king, queen, jack, and ten of spades is one of the 2,598,960 possible hands. So the probability of drawing that particular hand is $\frac{1}{2,598,960}$.

b) Since there are 4 twos, 4 threes, 4 fours, 4 fives, and 4 sixes, there are $4^5 = 1024$ hands that consist of one of each of these cards. So the probability of drawing a hand containing 1 two, 1 three, 1 four, 1 five, and 1 six is
$$\frac{1024}{2,598,960}.$$

19. $P(\text{H or W})$
$$= P(\text{H}) + P(\text{W}) - P(\text{H and W})$$
$$= 0.38 + 0.65 - 0.24 = 0.79$$

21. For 3 children there are 8 possible outcomes. $\{BBB,BBG,BGB,BGG,GBB,GBG, GGB, GGG\}$
There are no outcomes that consist of 3 boys and 3 girls.
$P(\text{3B or 3G}) = P(\text{3B}) + P(\text{3G})$
$$= \frac{1}{8} + \frac{1}{8} = \frac{1}{4}$$

23. Since there are 4 aces, $P(\text{A}) = \frac{4}{52}$. Since there are 13 hearts, $P(\text{H}) = \frac{13}{52}$. Since there is one card that is both an ace and a heart, $P(\text{A and H}) = \frac{1}{52}$.
$P(\text{A or H}) = P(\text{A}) + P(\text{H}) - P(\text{A and H})$
$$= \frac{4}{52} + \frac{13}{52} - \frac{1}{52} = \frac{16}{52} = \frac{4}{13}$$

25. Since not surviving is the complement of surviving, $P(\text{not surviving})$
$$= 1 - P(\text{surviving}) = 1 - 0.005 = 0.995.$$

27. For a pair of dice there are 36 equally likely outcomes.

a) The probability of getting a pair of fours is $\frac{1}{36}$.

b) The probability of not getting a pair of fours is $1 - \frac{1}{36} = \frac{35}{36}$.

c) Getting at least one number that is not a four is the same event as not getting a pair of fours. So the probability is $\frac{35}{36}$.

29. Let E be the event that Edna comes ashore within 30 miles of Charleston. Since $P(E) = 0.60$, $P(\overline{E}) = 0.40$. The odds in favor of E are 0.60 to 0.40 or 3 to 2.

31. Let U be the event that the stock market goes up. Since $P(U) = \frac{3}{5}$, $P(\overline{U}) = \frac{2}{5}$.

a) The odds in favor of the stock market going up are $\frac{3}{5}$ to $\frac{2}{5}$ or 3 to 2

b) The odds against the stock market going up are 2 to 3.

33. a) The odds against the Black Hawks winning their next game are 1 to 3.

b) The probability that the Black Hawks win their next game is $\frac{3}{3+1} = \frac{3}{4}$.

35. The probability of getting exactly 2 heads in 3 tosses of a coin is $\frac{3}{8}$. The probability of not getting exactly 2 heads in 3 tosses of a coin is $\frac{5}{8}$. So the odds in favor of getting exactly 2 heads are $\frac{3}{8}$ to $\frac{5}{8}$ or 3 to 5.

37. The probability that the sum is 8 is $\frac{5}{36}$. The probability that the sum is not 8 is $\frac{31}{36}$. So the odds in favor of 8 are $\frac{5}{36}$ to $\frac{31}{36}$ or 5 to 31.

39. The probability of winning with a single ticket is $\frac{1}{1,000,000}$ and the probability of not winning with a single ticket is $\frac{999,999}{1,000,000}$. So the odds in favor of winning are 1 to 999,999.

41. If the odds in favor of an event are 1 to 31, then the probability of the event is
$$\frac{1}{1+31} = \frac{1}{32}.$$

43. a) $\frac{1}{C(44,6)} = \frac{1}{7,059,052}$

b) $\frac{7,059,051}{7,059,052}$

c) 1 to 7,059,051

Enriching Your Mathematical Word Power
1. c **2.** a **3.** b **4.** d **5.** b
6. a **7.** d **8.** b **9.** d **10.** a

CHAPTER 14 REVIEW

1. Since there are 15 questions with 3 choices each, there are 3^{15} or 14,348,907 ways to answer the questions.

3. The number of possible 3-letter words is $P(6,3) = \frac{6!}{3!} = 120$.

5. Since there are 3 choices for the newspaper ad, 12 choices for the radio ad, and 4 choices for the television ad, there are $3 \cdot 12 \cdot 4 = 144$ ways to place the 3 ads.

7. The number of ways to choose 4 cities from 8 is $C(8,4) = \frac{8!}{4!4!} = 70$.

9. a) The number of ways to choose 4 people out of 9 is $C(9,4) = \frac{9!}{5!4!} = 126$.

b) The number of ways to choose 4 people out of 5 Democrats is $C(5,4) = \frac{5!}{4!1!} = 5$.

c) The number of ways to choose 2 Democrats and 2 Republicans is
$C(5,2) \cdot C(4,2) = \frac{5!}{3!2!} \cdot \frac{4!}{2!2!} = 60$.

11. The coefficient of $x^7 y^3$ is
$C(10,7) = \frac{10!}{7!3!} = 120$.

13. The coefficient of $w^2 x^3 y^6$ is
$\frac{11!}{2!3!6!} = 4620$.

15. The number of different arrangements for the word ARKANSAS is $\frac{8!}{3!2!1!1!1!} = 3360$. The number of different arrangements for the word MARCH is $5! = 120$.

17. Since there are 2 possibilities for each child, there are $2^6 = 64$ possible families.

19. The number of ways to mark the answers to a 12 question true-false test is $2^{12} = 4096$. Since only one of these ways is all correct, the probability of getting all correct is $\frac{1}{4096}$. Since only one of the ways to mark the answers is all incorrect, the probability that all are wrong is $\frac{1}{4096}$.

21. a) Since there are 5 green beans among the 14 in the jar, the probability of drawing a green bean is $\frac{5}{14}$.

b) Since there are 9 beans that are either yellow or red, the probability of drawing a yellow or red bean is $\frac{9}{14}$.

c) Since there are no beans that are blue, the probability of drawing a blue bean is 0.
d) Since all of the beans are not purple, the probability of drawing a bean that is not purple is 1.

23. There are 8 possible outcomes for the sexes of 3 children. The probability of getting 3 boys is $\frac{1}{8}$ and the probability of not getting 3 boys is $\frac{7}{8}$. So the odds in favor of getting 3 boys are $\frac{1}{8}$ to $\frac{7}{8}$ or 1 to 7.

25. The probability of drawing a Democrat is 0.60 and the probability of not getting a Democrat is 0.40. So the odds in favor of getting a Democrat are 0.60 to 0.40 or 3 to 2.

27. $P(\text{M or E})$
$$= P(\text{M}) + P(\text{E}) - P(\text{M and E})$$
$$= 0.70 + 0.60 - 0.40 = 0.9$$

29. $8! = 8 \cdot 7 \cdot 6 \cdot 5 \cdot 4 \cdot 3 \cdot 2 \cdot 1 = 40{,}320$

31. $\frac{5!}{3!} = \frac{5 \cdot 4 \cdot 3 \cdot 2 \cdot 1}{3 \cdot 2 \cdot 1} = 5 \cdot 4 = 20$

33. $\frac{9!}{3!3!3!} = \frac{9 \cdot 8 \cdot 7 \cdot 6 \cdot 5 \cdot 4 \cdot 3 \cdot 2 \cdot 1}{3 \cdot 2 \cdot 1 \cdot 3 \cdot 2 \cdot 1 \cdot 3 \cdot 2 \cdot 1}$
$$= 1680$$

35. $C(8, 6) = \frac{8!}{6!2!} = \frac{8 \cdot 7}{2} = 28$

37. $P(8, 4) = \frac{8!}{4!} = 8 \cdot 7 \cdot 6 \cdot 5 = 1680$

39. $C(8, 1) = \frac{8!}{7!1!} = 8$

CHAPTER 14 TEST

1. The number of different configurations is $4 \cdot 6 \cdot 3 = 72$.

2. The number of permutations of 20 students taken 3 at a time is
$P(20, 3) = 20 \cdot 19 \cdot 18 = 6840$.

3. The number of ways to choose 3 attractions from 5 is $C(5, 3) = \frac{5!}{3!2!} = 10$.

4. The number of ways to choose 2 boys from 12 is $C(12, 2) = 66$. The number of ways to choose 2 girls from 8 is $C(8, 2) = 28$. So the number of teams possible is $66 \cdot 28 = 1848$.

5. The number of ways to select 4 workers out of 12 to all get the same disease is
$C(12, 4) = \frac{12!}{4!8!} = 495$.

6. The coefficient of $a^3 b^8$ is
$$C(11, 3) = \frac{11!}{8!3!} = 165.$$

7. The coefficient of $w^4 x^3 y^2 z^5$ is

$$\frac{14!}{4!3!2!5!} = 2{,}522{,}520.$$

8. The number of different permutations of the letters in ALASKA are $\frac{6!}{3!1!1!1!} = 120$.

9. There are 4 equally likely outcomes when a pair of coins is tossed: {HH, HT, TH, TT}. The probability that at least on head occurs is $\frac{3}{4}$.

10. At least one head or at least one tail occurs on all 4 of the 4 equally likely outcomes. So the probability is 1.

11. The probability of drawing a spade in one draw is $\frac{13}{52}$ and the probability of not drawing a spade in one draw is $\frac{39}{52}$. So the odds in favor of drawing a spade are 13 to 39 or 1 to 3.

12. The probability of getting 3 boys is $\frac{1}{8}$ and the probability of not getting 3 boys is $\frac{7}{8}$. So the odds against 3 boys are $\frac{7}{8}$ to $\frac{1}{8}$ or 7 to 1.

13. a) Since the odds are 1 to 20 in favor of the Saints going to the Super Bowl, the odds are 20 to 1 against the Saints going to the Super Bowl.

b) The probability is $\frac{1}{1 + 20}$ or $\frac{1}{21}$ that the Saints go to the Super Bowl.

c) The probability is $\frac{20}{21}$ that the Saints do not go to the Super Bowl.

14. There are 36 ordered pairs for the equally likely outcome for tossing a pair of dice.

a) At least one die shows a 3 on 11 of the outcomes. So $P(\text{A}) = \frac{11}{36}$.

b) The sum is 7 on 6 of the outcomes. So $P(\text{B}) = \frac{6}{36} = \frac{1}{6}$.

c) The sum is 7 and at least one die shows a 3 occurs for the outcomes (3, 4) and (4, 3). So $P(\text{A} \cap \text{B}) = \frac{2}{36} = \frac{1}{18}$.

d) $P(\text{A} \cup \text{B}) = P(\text{A}) + P(\text{B}) - P(\text{A} \cap \text{B})$
$$= \frac{11}{36} + \frac{6}{36} - \frac{2}{36} = \frac{15}{36} = \frac{5}{12}$$

15. $C(100, 2) = \frac{100!}{98!2!} = \frac{100 \cdot 99}{2} = 4950$

16. $P(200, 3) = \frac{200!}{197!} = 200 \cdot 199 \cdot 198$
$$= 7{,}880{,}400$$

17. $\frac{10!}{3!4!3!} = \frac{10 \cdot 9 \cdot 8 \cdot 7 \cdot 6 \cdot 5}{3 \cdot 2 \cdot 1 \cdot 3 \cdot 2 \cdot 1} = 4200$

299

303

304

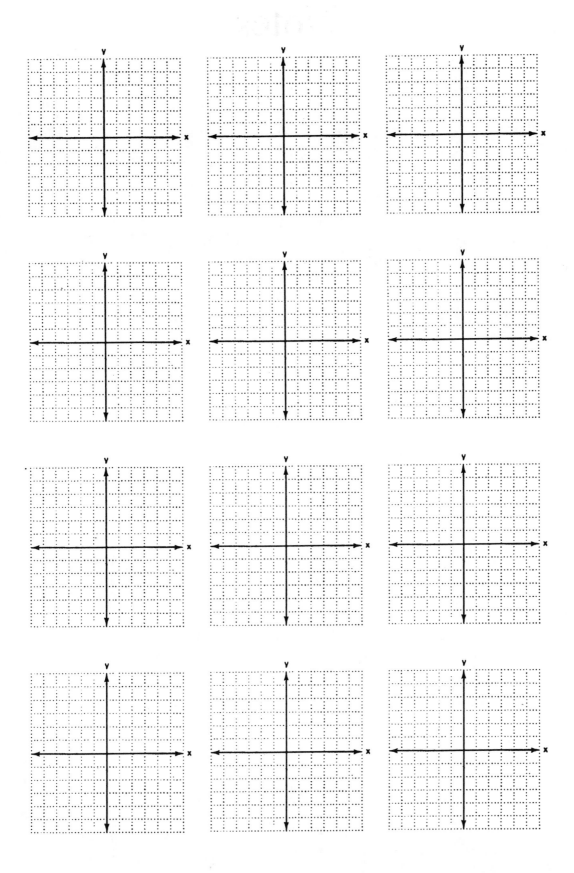

309

Notes

Notes

Notes

Notes

Notes

Notes